WITHDRAWN

# Wide Energy

# Bandgap

# Electronic
Devices

# Fan Ren
### University of Florida, USA

# John C. Zolper
### DARPA, USA

# Wide Energy

# Bandgap

# Electronic
### Devices

**World Scientific**
*New Jersey • London • Singapore • Hong Kong*

*Published by*

World Scientific Publishing Co. Pte. Ltd.

5 Toh Tuck Link, Singapore 596224

*USA office:* Suite 202, 1060 Main Street, River Edge, NJ 07661

*UK office:* 57 Shelton Street, Covent Garden, London WC2H 9HE

**Library of Congress Cataloging-in-Publication Data**
Wide energy bandgap electronic devices / edited by Fan Ren, John C. Zolper.
 p. cm.
 Includes bibliographical references and index.
 ISBN 981-238-246-1
 1. Wide gap semiconductors. 2. Power semiconductors. 3. Gallium arsenide
semiconductors. 4. Silicon carbide. I. Ren, Fan. II. Zolper, J. C.

 TK7871.85.W533 2003
 621.3815'2--dc21

                                          2003041123

**British Library Cataloguing-in-Publication Data**
A catalogue record for this book is available from the British Library.

Printed in Singapore by World Scientific Printers (S) Pte Ltd

## PREFACE

Two of the most successful technologies in existence today have created the Si integrated circuits (ICs) industry and the data storage industry. Both continue to advance at a rapid pace. In the case of ICs, the number of transistors on a chip doubles about every 18 months according to Moore's Law. For magnetic hard disk drive technology, a typical desktop computer drive today has a 40 Gbyte/disk capacity, whereas in 1995 this capacity was $\sim$ 1 Gbyte/disk. Since 1991, the overall bit density on a magnetic head has increased at an annual rate of 60% to 100% and is currently $\sim$ 10.7 Gbits/in$^2$. The integrated circuits operate by controlling the flow of carriers through the semiconductor by applied electric fields. The key parameter therefore is the charge on the electrons or holes. For the case of magnetic data storage, the key parameter is the spin of the electron, as spin can be thought of as the fundamental origin of magnetic moment. The characteristics of ICs include high speed signal processing and excellent reliability, but the memory elements are volatile (the stored information is lost when the power is switched off, as data is stored as charge in capacitors, i.e. DRAMs). A key advantage of magnetic memory technologies is that they are non-volatile since they employ ferromagnetic materials which by nature have remanence.

The recent emergence of SiC and GaN-based devices promises a further revolution in areas as diverse as high speed terrestrial and satellite-based communication systems, advanced radar, integrated sensors, solar-blind ultraviolet detectors, high temperature electronics, power flow control in hybrid electric vehicles and utility power switching, and non-volatile spintronic memory or signal processing devices. The materials properties of both semiconductor systems make them ideally suited to operation at elevated temperatures and at current or voltage levels inaccessible to Si. For example, there is a strong interest in developing wide bandgap power devices for use in the electric power utility industry. With the onset of deregulation in the industry, there will be increasing numbers of trans-actions on the power grid in the United States, with different companies buying and selling power. The main applications are in the primary distri-bution system (100 ~ 2000 kVA) and in subsidiary transmission systems (1 ~ 50 MVA). A major problem in the current grid is momentary voltage sags, which affect motor drives, computers, and digital controls. Therefore, a system for eliminating power sags and switching transients would dramati-cally improve power quality. For example, it is estimated that a two-second outage at a large computer center can cost US$600,000 or more, and an outage of less than one cycle, or a voltage sag of 25% for two cycles, can cause a microprocessor to malfunction. In particular, computerized tech-nologies have led to strong consumer demands for less expensive electricity, premium quality power, and uninterruptible power. There is a strong inter-est in wide bandgap semiconductor gas sensors for applications including fuel leak detection in spacecraft. In addition, these detectors would have dual use in automobiles and aircraft, fire detectors, exhaust diagnosis, and emissions from industrial processes. GaN and SiC are capable of operat-ing at much higher temperatures than more conventional semiconductors such as Si because of their large bandgap (3.4 eV for GaN, 3.26 eV for the 4H–SiC poly-type vs. 1.1 eV for Si). Simple Schottky diode or field-effect transistor structure fabricated in GaN and SiC are sensitive to a number of gases, including hydrogen and hydrocarbons. The sensing mechanism is thought to be the creation of a polarized layer on the semiconductor sur-face by hydrogen atoms diffusing through the metal contact. One additional attractive attribute of GaN and SiC is the fact that gas sensors based on this material could be integrated with high-temperature electronic devices on the same chip. High electron mobility transistors (HEMTs) fabricated in the AlGaN/GaN material system show great potential for both space-borne and terrestrial applications where high power and elevated temperature operation are needed. For satellite-based communication systems, weather forecasting, remote sensing, or nuclear industry applications, the radiation resistance of electronics devices must be established. One would expect the AlGaN/GaN system to be even more radiation-hard than the more conven-tional AlGaAs/GaAs heterostructure due to the higher displacement ener-gies in the nitrides. Reports have shown significant (> 50%) decreases in transconductance of AlGaN/GaN HEMTs irradiated with 1.8 MeV protons to doses of $10^{14}$ cm$^{-2}$, while smaller changes (< 30%) in transconductance in similar HEMTs exposed to higher energy (40 MeV) protons at lower doses ($5 \times 10^{10}$ cm$^{-2}$) have been observed.

In this book, we bring together numerous experts in the field to review the progress in SiC and GaN electronic devices and novel detectors. Professor Morkoc reviews the growth and characterization of nitrides, followed by chapters from Professor Shur, Professor Karmalkar, and Professor Gaska on high electron mobility transistors, Professor Pearton and co-workers on ultra-high breakdown voltage GaN-based rectifiers, and the group led by Professor Abernathy on emerging MOS devices in the nitride system. Dr. Baca from Sandia National Laboratories and Dr. Chang from Agilent review the use of mixed Group V nitrides as the base layer in novel heterojunction bipolar transistors. There are three chapters on SiC, including Professor Skowronski on growth and characterization, Professor Chow on power Schottky and $p$-$i$-$n$ rectifiers, and Professor Cooper on power MOSFETs. Professor Dupuis and Professor Campbell give an overview of short wavelength, nitride-based detectors. Finally, Jihyun Kim and co-workers describe recent progress in wide bandgap semiconductor spintronics where one can obtain room temperature ferromagnetism and exploit the spin of the electron in addition to its charge. These chapters provide a summary of the current state-of-the-art in SiC and GaN and identify future areas of development. The remarkable improvements in material quality and device performance in the last few years show the promise of these technologies for areas that Si cannot operate because of its smaller bandgap. We feel that this collection of chapters provide an excellent introduction to the field and is an outstanding reference for those performing research on wide bandgap semiconductors.

F. Ren
University of Florida
Gainesville, FL, USA


J. C. Zolper
Defense Advanced Research Projects Agency
Washington D.C., USA

# CONTENTS

# CHAPTER 1

# GROWTH OF III-NITRIDE SEMICONDUCTORS AND THEIR CHARACTERIZATION

Hadis Morkoç

*Department of Electrical Engineering and Department of Physics*
*Virginia Commonwealth University*
*601 Main St. P.O. Box 23284-3072, Richmond, VA 23284*

## Contents

## 1. Introduction

Semiconductor nitrides (AlN, GaN, and InN) are promising materials for optoelectronic devices and high power/temperature electronics.[1-8] They cover an energy bandgap range of 1.9 to 6.2 eV, from red to ultraviolet (UV) wavelengths. Specifically, nitrides are suitable for such applications as surface acoustic wave devices,[9] UV detectors,[10,11] UV and visible light emitting diodes (LEDs),[12-14] and laser diodes (LDs)[15] for digital data read-write applications. Nitride semiconductor-based LEDs have found applications in displays, lighting, indicator lights, advertisements, and traffic signs/signals. Nitride based short wavelength lasers are crucial for high-density optical read and write technologies. When used as UV sensors in jet engines, automobiles, and furnaces (boilers), nitride detectors allow for optimal fuel efficiency and control over the effluents produced for a cleaner environment. Moreover, UV sensors that operate in the solar blind region (260 to 290 nm) would have high sensitivity as the ozone layer absorbs solar radiation at these wavelengths.[11,16] Already, AlGaN based detectors with Al mole fractions approaching the solar blind region of the spectrum have been fabricated into arrays for imaging. Detector arrays, of 32 × 32 pixels, have been fabricated and tested already.[17]

GaN's large bandgap, high dielectric breakdown field, fortuitously good electron transport properties[18-20] (an electron mobility possibly in excess of 2000 $cm^2V^{-1}s^{-1}$ and a peak velocity approaching $3 \times 10^7$ $cms^{-1}$ at room temperature), and favorable thermal conductivity make these materials attractive for high power/temperature electronic devices.[21] GaN/AlGaN MODFETs prepared by MBE on SiC substrates exhibited a total power level of 6.3 W at 10 GHz from a 1-mm wide device.[22] What is note worthy is that the power level is not really thermally limited as the power density extrapolated from a 0.1 mm device is 6.5 W. When four of these devices were power combined in a single stage amplifier, an output power of 22.9 W with a power added efficiency of 37% was obtained at 9 GHz.[23] Equally impressive is the noise figure of 0.85 dB at 10 GHz with an associated gain of 11 dB. The drain breakdown voltages in these quarter-micron-gate devices are about 60 V, which are in part responsible for such a record performance.[24]

Nitride semiconductors have been deposited by vapor phase epitaxy (i.e. both hydride VPE[25] [HVPE] which has been developed for thick GaN layers and organometallic VPE[26] [OMVPE] which has been developed for heterostructures), and, in a vacuum, by a slew of variants of molecular beam epitaxy (MBE).[16] All the high performance light emitters, which require high quality InGaN, have been produced by OMVPE. On the other hand, MBE has been very successful in producing structures which do not require InGaN. Some examples are FETs and detectors. With its innate refined control over growth parameters, its *in situ* monitoring capability, and the uniformity of the films, MBE is well suited for depositing heterostructures and gaining insights to the deposition/incorporation mechanisms. MBE's control over the growth parameters is such that any structure can be grown in any sequence. The structures based on conventional compound semiconductors, such as IR lasers for CD players, surface emitting vertical cavity lasers, and high-performance pseudomorphic MODFETs, have all been produced very successfully, most of them commercially, by MBE. Nitride growth, however, requires much higher temperatures than those used in producing conventional Groups III–V semiconductors for which MBE systems were initially designed. With design improvements, such as handling relatively high substrate temperatures, having sources to function without self destruction in a nitrogen environment, particularly ammonia, and RF nitrogen sources which are nearly ion free and clean, MBE began to produce

very high mobility bulk, and two dimensional electron gas structures. However, they are grown on MOCVD and HVPE buffer layers because the
growth mode of MBE, which is coveted in all other applications, has not
lead to layers on foreign substrates with very low defect concentrations.

The room-temperature electron mobility values in bulk GaN grown with
HVPE to a thickness of 60 $\mu$m were reported to be 950 cm$^2$/Vs.[27] That
reported for MOCVD grown layers was also in excess of 900 cm$^2$/Vs,[28]
although the temperature dependence of the mobility in this particular
sample was rather unique. Early MBE layers exhibited mobilities as
high as 580 cm$^2$/Vs on SiC substrates, which at that time were not as
commonly used as compared to recent times.[29] Typically, however, the
MBE-grown films produce much lower mobility values in the range of 100–
300 cm$^2$/Vs.[30] The lower mobilities have been attributed to both high
dislocation densities[30–32] and elevated levels of point defects[33,34] in the
GaN films.

Dislocations are an important scattering mechanism in films having
dislocation densities above $1 \times 10^8$ cm$^{-2}$.[30,31] Depending on the particulars
of the growth and the substrate preparation, GaN films grown by MBE
typically have dislocation densities in the range of $5 \times 10^9$–$5 \times 10^{10}$ cm$^{-2}$.[30]
With refined procedures, however, dislocation densities in the range of
$8 \times 10^8$–$2 \times 10^9$ cm$^{-2}$ can be obtained when grown directly on sapphire with
AlN or GaN buffer layers without any lateral epitaxial growth. Dislocation
reduction is really the key to achieving high mobility GaN which goes to
the heart of the need for a buffer layer and/or early stages of growth. Based
on the premise that the [002] X-ray diffraction is affected by screw dislocations and the [104] peak by edge dislocations and the fact that RF-nitrogen
grown MBE layers produce excellent [002] peaks (with full width at half
maximum values in the 40–120 arcseconds range) while the [104] peaks are
wider and weaker (in the 180–300 arcseconds range), one could conclude
that the majority of the dislocations in MBE layers are the propagating
edge type of dislocations.

The strength of MBE, that is producing two-dimensional (2D) growth,
does not bode well in dislocation reduction as the edge dislocations which
propagate along the *c*-axis go right through the sample, though the detailed
picture is somewhat dependent on the particulars of the growth such as
pitted or pit-free growth which depends on the Group V/III ratio. Some sort
of 3D growth at the early stages of the growth, as in the case of growth from
vapor, followed by a smoothing layer, would help reduce the dislocation
density. The other option is to use HVPE or MOCVD buffer layers for MBE

growth. This approach has led to record or near record bulk (1150 cm$^2$/Vs at RT)[35] and 2DEG (53,500 cm$^2$/Vs at 4.2 K) mobilities,[36] with more recent mobilities being about 73,000 cm$^2$/Vs at LHe$_2$ temperature. It is clear that buffer layers grown by the vapor phase epitaxy method help eliminate the main problem associated with MBE, which is the poor quality of the buffer layer. The other long-standing obstacle for MBE, difficulties associated with sapphire and SiC substrate preparation, has been eliminated. In the case of sapphire, a high-temperature anneal in O$_2$ environment produces atomically smooth and damage-free surface.[37] In the case of SiC, some form of H$_2$ etching at elevated temperatures removes the surface damage caused by polishing as in the case of sapphire.[38] Controlling the Ga/N ratio and the substrate temperature causes the dislocation density across the homoepitaxial interface to remain constant.[39,40] While the above two investigations are related to RF MBE, ammonia MBE has produced electron mobilities as high as 73,000 cm$^2$/Vs (at 4.2 K) when grown on bulk GaN wafer grown under high pressure and temperature conditions.[41]

The evolution of GaN development followed an uneven path in that the focused activity took place in early nineteen seventies and since about the middle of nineteen nineties. There have been systematic efforts to grow InN, GaN and AlN, and their ternaries (to a lesser extent their quaternaries) by vapor phase epitaxial deposition and recently by MBE using reactive nitrogen from a plasma source or from ammonia. In the following, we will review the vapor phase deposition method and its use in the growth of nitride semiconductors. This will be followed by MBE growth of the same.

## 2. Substrates, Growth Methods, and Growth Experiments

Due to lack of native GaN substrates, GaN based nitride semiconductors have been grown epitaxially over a number of substrates including 3-inch diameter sapphire, thus monocrystalline layers can be obtained over large areas. This is in spite of the fact that efforts have gone on to produce GaN[42] and AlN[43] bulk materials for investigating their properties as well as potential use as substrates for epitaxy. Despite this limitation, remarkable progress in the growth of high quality epitaxial III-nitride films by a variety of methods such as Hydride Vapor Phase Epitaxy,[25,44] Organometallic Vapor Phase Epitaxy (VPE) (inorganic VPE),[45] Reactive Molecular Beam Epitaxy (MBE)[2,13,46] and bulk crystal growth from Ga solution have recently been achieved. By far the most frequently used methods are the VPE methods. The most successful among the VPE methods is the metal-organic chemical vapor deposition (MOCVD). With the exception of FETs,

MOCVD is the primary method employed in the investigation and production of optoelectronic devices, such as LEDs and lasers, albeit the quality of MBE films comes close to that grown by MOCVD. The inorganic VPE was the first method used to grow epitaxial III-N semiconductors, but it was nearly abandoned later.[47] The technique, however, got revived recently by growing very high quality and thick buffer layers and templates with the growth of device structures to follow by MBE and MOCVD.[25,48,49] In spite of the rapid development of III-N technology, many problems remain to be overcome. The main technological issue has been and remains to be the lack of native substrates and resultant lack of high crystalline quality of films. In this chapter, substrates used for growth and related issues, growth by various techniques employed, and characterization will be discussed.

## 2.1. *Primer on Substrates*

Nitride films grown by MBE have been deposited on a variety of substrates such as Si, sapphire, SiC, and ZnO. Many other substrates such as $MgAl_2O_4$, $LiGaO_2$, $LiAlO_2$, etc. have also been used.[1,5] Sapphire is the usual substrate on which high quality GaN was epitaxially grown for the first time. The surface of the substrates used have to be prepared for epitaxial growth, a process which includes degreasing followed by chemical etch when possible. Surfaces of as-received sapphire and SiC substrates contain mechanical polishing damage which must be removed. Chemical etches are not yet available for this purpose. Consequently, a high temperature treatment under a controlled environment is employed, as will be discussed below.

The procedure for degreasing is the same for all the four substrates discussed above, namely Si, sapphire, SiC, and ZnO. The substrate is first dipped in a solution of (TriChlorEthane) TCE, kept at 300°C, for 5 minutes. It is then rinsed for 3 minutes each in acetone and methanol. This is followed by a 3 minute rinse in deionized (DI) water. The above process is repeated three times to complete the degreasing process. The substrates are then etched, the procedure for which is substrate dependent.

### 2.1.1. *Si*

For wurtzitic growth, (111) plane Si is used, (001) surface is also used for cubic GaN growth, albeit in a few cases. As-received Si surface is already excellent and removal of only a very thin surface layer, using the RCA etch, followed by hydrogenation of surface dangling bonds is carried out.

This is accomplished by immersing Si for 10 minutes in a 1 : 1 : 5 solution of HCl : $H_2O_2$ : $H_2O$ kept at 60°C, which grows a porous oxide, followed by a rinse in deionized water. The resulting oxide layer is then removed by dipping the substrate in a 10 : 1 solution of $H_2O$ : HF for 20 seconds. The hydrogenation process takes place through a short exposure of the wafer to an HF solution.

### 2.1.2. *SiC*

For SiC substrates, roughly 3 microns of the epilayer surface are removed using a hot KOH solution (300–350°C) as the etchant. This is followed by a DI rinse for 3 minutes and the wafer is blow-dried by $N_2$. This approach is not a desirable one in that the resultant surface is very rough. Defective regions on the surface get attacked by this etch rather rapidly leading to very rough surfaces. The SiC substrate then undergoes an oxidation and passivation procedure. The substrate is immersed for 5 minutes in a 5 : 3 : 3 solution of HCl : $H_2O_2$ : $H_2O$ at 60°C, followed by a $\sim$ 30 second DI rinse. The resulting oxide layer is then removed by dipping the substrate, for 20 seconds, in a 10 : 1 solution of $H_2O$ : HF. This procedure is repeated several times after which the substrate should not be exposed to the atmosphere for no longer than 30 minutes before another oxidation-passivation procedure would be required.

Skipping the chemical etching step leaves the surface with damage left from mechanical polishing. Techniques have been developed to remove the surface damage by plasma or vapor etching. One is the mechanical chemical polish, which has made substantial progress recently,[50] and the other is etching in H and Cl environments at very high temperatures such as 1500°C. The surface morphology of an as-received SiC substrate that underwent a standard mechanical chemical polish (MCP), and a special MCP procedure (see Ref. 50), is shown in Fig. 1. The group of the author, in collaboration with the group of E. Janzen at Linköping University, was able to H etch Leyl SiC followed by MODFET growth. These devices did not exhibit the negative differential resistance characteristic of the sapphire substrates. However, SiC substrates prepared by the sublimation method did not appear to survive this high temperature H etching process. Those investigating various issues dealing with SiC have recently developed and exploited the *in situ* H etching[38] and HCl etching processes with great success.[51] An AFM image of a SiC substrate polished by a high temperature H treatment (1700°C) by the Feenstra group[52] is shown in Fig. 2, which clearly

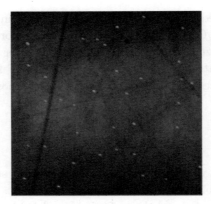

**Fig. 1.** (a) AFM image of as-received SiC surface following a standard mechanical chemical polish. Image size is $10 \times 10 \; \mu m^2$, and vertical scale is 50 nm. Note presence of scratches.

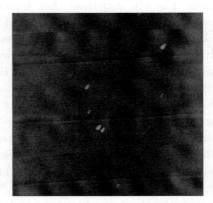

**Fig. 1.** (b) AFM image of SiC surface after a mechanical chemical polish performed at Eagle Picher. Image $10 \times 10 \; \mu m$, vertical 5 nm. Note that scratches are no longer present.

shows the well ordered and unbroken atomic terraces indicative of superb surface quality.

### 2.1.3. *Sapphire*

Among the substrates described above, sapphire remains the most frequently used substrate for GaN epitaxial growth so far mostly due to its low price, the availability of large area crystals of good quality, its transparent nature, stability at high temperatures, and a fairly mature growth

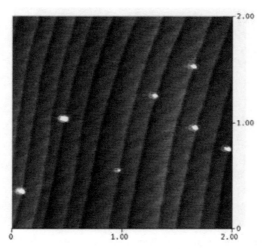

**Fig. 2.** AFM image of SiC surface after an hydrogen etching step. Image $2 \times 2$ $\mu$m, vertical 1.5 nm. Note that scratches are no longer present. Hydrogen polishing was performed in the laboratory of Prof. R. Feenstra of CMU.

technology. The orientation order of the GaN films grown on the main sapphire plans {basal, $c$-plane (0001), $a$-plane ($11\bar{2}0$), and $r$-plane ($1\bar{1}02$)} by ECR-MBE has been studied in great detail.[53,54]

The calculated lattice mismatch between the basal GaN and the basal sapphire plane is larger than 30%. However, the actual mismatch is smaller ($\sim$ 15%), because the small cell of Al atoms on the basal sapphire plane is rotated by 30° with respect to the larger sapphire unit cell. It is on this plane that the best films were grown with relatively small in- and out-of plane misorientation. In general, films on this plane show either none or nearly none of the cubic GaN phase.

For $Al_2O_3$ substrates, 3 : 1 solution of $H_2SO_4$ : $H_3PO_4$ is used as the etchant. The substrate is dipped in this solution, kept at 300°C, for 20 minutes. This is followed by a rinse in DI water for 3 minutes. Though the hot etch removes some material, the resultant surface still bears the scratches caused by mechanical polish. Shown in Fig. 3 are AFM images of two $c$-plane sapphire surfaces, before (Fig. 3(a)) and after (Fig. 3(b)) the chemical etching. It is very clear that the etching did not remove the scratches from the surface that were introduced during the polishing process. Obviously, smoothness at the atomic level is not seen even though the RMS roughness was reduced from 0.323 nm (Fig. 3(a)) to 0.211 nm (Fig. 3(b)). To overcome this problem, a high temperature annealing step

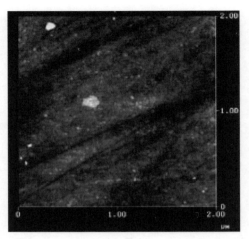

**Fig. 3.** (a) AFM image of an as-received sapphire substrate. Note scratches caused by mechanical chemical polish.

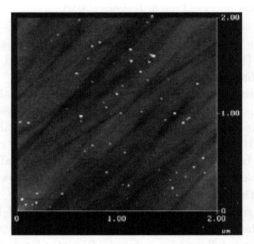

**Fig. 3.** (b) AFM image of a sapphire substrate after a 180°C etch in sulfuric/phosphoric acid. Some improvements are apparent, but the scratches remain and are accented to some extent.

has been employed which gives rise to atomically smooth surfaces. A very high temperature annealing investigation of sapphire substrates was recently undertaken. Annealing experiments in air at 1000, 1100, 1200, 1300, and 1380°C (the ceiling of the furnace employed) for 30 and 60 minutes periods were conducted to determine the best conditions with the aid of

AFM images of the finished surface. This was followed by observation of RHEED patterns once in an MBE system.

A small, but progressive, improvement was observed on reduction of scratches up to 1300°C. However, the anneals at 1380°C for 1 hour lead to scratch free and smooth surfaces to the point where the only noticeable feature in AFM images are the atomic steps which are about 0.15 nm in

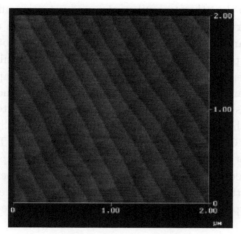

**Fig. 4.** AFM image of sapphire following a 1380°C −1 h anneal in atmosphere. Atomically flat surface is clearly visible. Atomic step heights are about 0.15 nm, which represent the only roughness in the image. The diagonal lines, from left to right, are the artifacts of AFM.

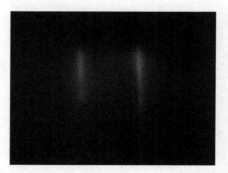

**Fig. 5.** A RHEED image at about 800°C of an annealed sapphire at 1380°C for one hour. Clear streaky RHEED pattern observed at temperatures as low as 600°C clearly indicates that the high temperature annealing step produces clean epi-ready surfaces. Without the annealing procedure, the RHEED images are not as clear, elongated and reproducible.

height. AFM images indicated that the 1380°C anneal for one hour leads to atomically smooth surfaces as shown in Fig. 4.[37] RHEED images typically show extended and bright rods associated with sapphire at temperatures as low as 600°C during the ramp up as shown in Fig. 5.

### 2.1.4. *ZnO*

ZnO is similar to the other two substrates and yet very different. Similar in that it shares very similar problems, chiefly among them are the scratches caused by mechanical polishing. Chemical etches have not been developed to deal with this problem. In fact, it may even not be possible to accomplish this task with chemical etches. However, as in the case of sapphire, annealing in oxygen appears to lead to smooth surfaces. An anneal at 900°C for three hours leads to a roughness in the atomic bilayer step scale which is about the best that can be achieved, as shown in Fig. 6. Annealing in air at 850 and 950°C also led to very smooth surfaces.[55] This implies that the process is not very temperature sensitive and that reproducible results should be obtained without extreme control over the temperature employed. The RHEED patterns also, once placed in an MBE system, bear this out by showing sharp $1 \times 1$ rods during the ramp up of the temperature, as shown in Fig. 7.

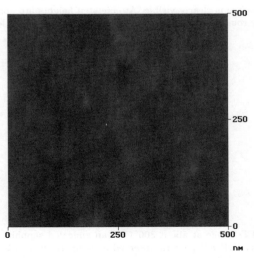

**Fig. 6.** AFM image of a ZnO surface after a three-hour annealing procedure in air at $T_a = 900$°C. The root-mean-square roughness (RMS) is about 0.1 nm. The image is 500 nm × 500 nm.

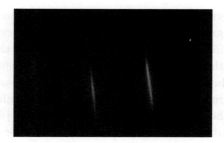

**Fig. 7.** RHEED image of ZnO taken at 780°C.

### 2.1.5. *AlN and GaN*

The most satisfactory methods of growing high purity single-crystal Group III nitrides would be those in which these nitrides are themselves used as the starting materials. Though this has been a goal for many researchers for quite sometime, the progress has been painstakingly slow. However, recent advances in hydride VPE grown templates look very promising. Slack and McNelly[56] suggested a technique in which high purity AlN powder from Al metal using $AlF_3$ as an intermediate product can be produced. The AlN powder can be converted to single-crystals by sublimation in a closed tungsten crucible or in an open tube with a gas flow. The main problem with these growth techniques may be the surface oxidation of the powder. If this oxidation is carefully avoided, then AlN can be produced only with 100 ppm of oxygen, and lower amounts of other impurities. The purest AlN prepared by Slack and McNelly[56] using tungsten crucible had 350 ppm oxygen. However, when W or Re was used for the crucible, very little contamination of the AlN with metal impurities was found. Further the crystals grown in W or Re crucibles were generally of uniform amber color indicating that indeed both oxygen and carbon contamination were scrupulously minimized by using W and Re crucibles. Growth on AlN for substrates has picked up momentum of late with notable progress in size. Efforts are also underway to grow epitaxial layers on AlN substrates.

On the GaN side, which is harder due to the much higher nitrogen overpressure, Leszczynski *et al.*[57] and Grzegory *et al.*[58] performed nitride crystal growth from the solution under high $N_2$ pressure. The experiments were carried out in a gas pressure chambers of internal diameter 30 mm with a furnace of 14 mm (1500°C) or 10 mm (1800°C) and with a boron nitride (BN) crucible containing Al, Ga or In. The temperatures were stabilized with a precision better than 1 degree. Attempts were made to optimize the pressure range for growth. With this optimization, the crystals were grown only at a pressure for which the nitride was stable over the whole temperature range along the crucible. In the case of GaN, single crystals were grown from the solution in liquid Ga under an $N_2$ pressure of 8–17 kbar at temperatures ranging between 1300–1600°C. The quasilinear temperature gradient in 5 to 24 hour processes was 30–100°C/cm. Typical synthesis solid diffusion crystallization was the observed mechanism for crystal formation. The nucleation and growth of single-crystal GaN took place through the process of thin polycrystalline GaN film on the Ga surface, and through its dissociation and transport into the cooler part of the crucible. The same group also carried out cursory experiments with growth of AlN and InN. At high $N_2$ pressure, the synthesis rate of AlN was high, and the synthesis rate of InN was extremely low. The rate of AlN was so high that, at pressure lower than 6.5 kbar, thermal explosion took place during heating of bulk Al sample. Due to low stability, the crystallization rate of AlN at 1600–1800°C was marginally low. On the other hand, due to (low temperature) kinetic and (low stability) thermodynamical barriers, crystal growth experiments for InN resulted in very small crystallites (5–50 microns) when grown particularly by slow cooling of the system from the temperatures exceeding the stability for InN.

An examination of the crystal morphology indicated that it (e.g. morphology) depends on the pressure, temperature range, and supersaturation during growth. For pressures and temperatures lying deeply in GaN stability field (e.g. higher pressure and lower temperature), the crystals are hexagonal prisms elongated in $c$-direction. At the conditions close to the equilibrium curve, the dominating shape of the crystals is hexagonal plate. The crystals grown slowly (slower than 0.1 mm/h), e.g. at smaller temperature gradients, exhibited higher crystal quality. These were transparent, slightly yellowish, and had flat mirror-like surfaces. Typical widths (FWHM) of X-ray rocking curves for $(004)$CuK$\alpha$ reflection are 23–32 arcseconds. Note that these curves are significantly narrower than the corresponding curves of heteroepitaxial GaN films grown by MBE or MOCVD techniques on non-nitride substrates.

It was observed that, probably due to nonuniform distribution of nitrogen in the solution across growing crystal face, the quality of the GaN crystals deteriorates with increasing growth rate (high supersaturation) and with dimensions of the crystals. It was apparent especially when the size of the face becomes comparable to the size of the crucible. The deterioration of quality of the 5–10 mm crystals grown with a rate of 0.5–1 mm/h was demonstrated by the broadening of the rocking curve.

### 2.1.6. *LiGaO$_2$ and LiAlO$_2$ substrates*

LiGaO$_2$ is orthorhombic with space group Pna21. The lattice parameters are $a = 5.402$ Å, $b = 6.372$ Å and $c = 5.007$ Å with a density of 4.175 and hardness of 7.5. The thermal expansion coefficients are $\Delta a = 6 \times 10^{-6}$ C$^{-1}$, well matched to GaN, $\Delta b = 9 \times 10^{-6}$ C$^{-1}$ and $\Delta c = 7 \times 10^{-6}$ C$^{-1}$. Optically, it is biaxial with transparency from 0.3 to 6 $\mu$m and refractive indices of $n_a = 1.7617$, $n_b = 1.7311$ and $n_c = 1.7589$ at 620 nm. A proper orientation of this crystal matches GaN, but it remains to be seen if that orientation leads to successful growth. The crystal has no natural cleavage planes.

The structure of LiGaO$_2$ is similar to the wurtzitic structure, but because Li and Ga have different ionic radii, the crystal has orthorhombic structure. The atomic arrangement in the (001) face is hexagonal, which promotes the epitaxial growth of (0001) GaN, so that the epitaxial relationship (0001) GaN/(001) LiGaO$_2$ is expected. The distance between the nearest cations in LiGaO$_2$ is in the range of 3.133–3.189 Å, while the distance between nearest anions is in the range of 3.021–3.251 Å. The lattice mismatch between (0001) GaN on (001) LiGaO$_2$ is then only 1–2%.

LiAlO$_2$ is tetragonal with space group P41212. The lattice constants are $a = 5.1687$ Å and $c = 6.2679$ Å with a density of 2.615 and hardness of 8. The thermal expansion coefficients are $\Delta a = 7.1 \times 10^{-6}$ C$^{-1}$, which will cause compressive strain in GaN, and $\Delta c = 15 \times 10^{-6}$ C$^{-1}$. Optically, it is uniaxial with transparency from 0.2 to 4 $\mu$m and refractive indices of $n_e = 1.6014$ and $n_o = 1.6197$ at 633 nm. The crystal has no natural cleavage planes requiring, as in the case of LGO below, the facet development with chemically assisted ion etching or chemical mechanical polishing. Of course the former is better in terms of performance and cost reduction. The structure of interest is the high temperature form (or $g$ form) of the compound. It melts congruently around 1700°C and is stable at room temperature. Single crystal can be grown by the conventional Czochralsky melt pulling method.

The epitaxy relationships between GaN and LiAlO$_2$ are expected to be $(01\bar{1}0)/(100)$ LiAlO$_2$ with $[2\bar{1}\bar{1}0]//[001]$ LiAlO$_2$. Unlike Al$_2$O$_3$ and 6H–SiC substrates with very smooth surfaces, excepting scratches caused by mechanical chemical polishing, the LiAlO$_2$ substrate exhibited a wave-like

surface with equidistant grooves about 10 nm deep, which could have originated from the mechanical surface polishing.

## 3. Nitride Growth Techniques

### 3.1. *MBE*

MBE is an extremely versatile technique for preparing thin semiconductor
heterostructures owing to the control over the growth parameters which it
offers, and the inherent *in situ* monitoring capablity. Application of MBE
to GaN and its related alloys has been treated in detail earlier.[59] Nearly
identical material can also be found in another review.[6] In this process, thin
films are formed on a heated substrate through various reactions between
thermal molecular beams (atomic beams in the case of RF activated
nitrogen) of the constituent elements and the surface species on the substrate, but intrinsic to the substrate. The composition of the epilayer and
its doping level depend on the arrival rates of the constituent elements
and dopants. The typical growth rate of 1 $\mu$m per hour, or slightly more
than one monolayer per second ($mls^{-1}$), is sufficiently low to allow for
surface migration of the impinging species on the surface. In the case of
growth along the $\langle 111 \rangle$ for cubic and $c$-directions for wurtzitic systems,
one monolayer constitutes a bi-layer. MBE growth is carried out under
conditions which are governed primarily by the kinetics. This allows the
preparation of many different structures that are otherwise not possible to
attain. This is in contrast to MOCVD of conventional compound semiconductors, such as GaAs. However, in the case of GaN, even the MOCVD
has elements of kinetics.[60] In nitride growth by MBE, the metal species
is provided by Ga, In, and Al metal sources, the dopants are provided by
pure Si for $n$-type and Mg for $p$-type using conventional Knudsen effusion
cells that are heated to sufficient temperatures for the desired growth rate,
composition and doping levels. On the other hand, nitrogen is an inert
gas and is one of the least reactive gases because of its large molecular
cohesive energy (946.04 kJmol = 9.8 eV per $N_2$ molecule). Because of the
triple bond between the two nitrogen atoms, dissociation of one molecule
into two reactive nitrogen atoms requires a large amount of energy which
cannot be provided by thermal means.

In a plasma environment, however, and at reduced pressures, a
significant dissociation of the nitrogen molecules takes place. Atomic
nitrogen is reactive even at room temperature and bonds with many metals. Consequently, Group III nitrides can be grown by plasma-assisted

molecular beam epitaxy, where the plasma-induced fragmentation of nitrogen molecules is combined with the evaporation of metal atoms from effusion cells. In this vein, MBE growth of GaN has been reported by electron cyclotron resonance microwave-plasma-assisted molecular beam epitaxy (ECR-MBE).[5] Several laboratories in the past have attempted reactive molecular beam epitaxy (RMBE) growth in which $N_2$ or $NH_3$ was decomposed on the substrate surface.[61-65]

The MBE deposition system for nitrides consists essentially of a conventional MBE chamber, but with added equipment such as a compact ECR source or a compact RF source and ammonia, as shown in Fig. 8. These compact nitrogen sources are small enough to be mounted on one of the effusion cell port flanges. Some systems are equipped with both RF and ammonia sources such as the ones available in author's laboratory for maximum versatility. RF sources operate at several frequencies such as 13.56 MHz and 2.45 GHz. The latter has gained dominance because of its compact nature and production of mostly the N radicals with very little if any ion content. As alluded to earlier, reactive nitrogen can also be supplied to the surface from an ammonia source, $NH_3$.[66,67] Ammonia initially was thought of not being compatible with MBE system as it reacts with metals such as copper used in the flange and valve seals, which is a simple problem and has been dealt with effectively. In addition, hydrogen released from ammonia makes hot metal components fragile. This too can

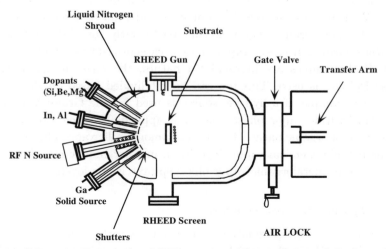

**Fig. 8.** Schematic diagram of an MBE system used for nitride growth in the author's laboratory.

be and has been dealt with effectively. Another apprehension about $NH_3$ is its reactivity with oxygen. This too does not seem to be a problem of concern and the vacuum systems are of UHV type and oxygen background is negligible.

Plasma and ammonia sources are operated with flow rates, which produce a system pressure of anywhere between $2 \times 10^{-6}$ and $1 \times 10^{-4}$ Torr, the actual value being dependent on the particulars of the system. The RF plasma sources can be operated at power levels between 90 and 500 W and pressures of 10–100 Torr in the plasma chamber itself. Well-designed RF sources can have cracking efficiencies as high as 10%, though in most operating conditions the efficiency is about 1%. In both ammonia and RF activated growth, typical pressures are in the low $10^{-5}$ Torr regime, which can nearly be construed as collision free growth mode. In recently designed RF sources, the emission wavelength of the plasma concurs with the assertion that only the non-ionic species with low energies, about 2 eV, are generated in large quantities. This energy is well below the 24 eV damage threshold energy predicted for GaN.[68] The same, however, does not hold for ECR sources in which the kinetic energy of the nitrogen ions[69] can be high and could result in surface atom displacement.

## 3.2. *Vapor Phase Epitaxy*

VPE has long been employed for the growth of many semiconductor structures. With ongoing source developments and improved reactor designs, this technique became very powerful, particularly for GaN and related materials. Growth from the vapor phase is categorized based on the sources used. If the sources are inorganic in nature, the term inorganic Vapor Phase Epitaxy is used. This too can be subdivided based on the sources used. For example, if a hydride source is used for the Group V element, the term hydride vapor phase epitaxy is applied. If at least some of the sources are of organic nature, the term organometallic vapor phase eptixay, organometallic chemical vapor deposition, metalorganonic chemical vapor deposition, or metalorganic vapor phase epitaxy is employed.

### 3.2.1. *Hydride vapor phase epitaxy*

The most successful epitaxial growth technique in the early investigations of the Group III nitrides, which also led to the first epitaxial form of GaN, was a vapor transport method by Maruska and Tietjen.[49] Their approach involved flowing HCl vapor over metallic Ga, which forms gaseous GaCl, which is then reacted downstream at the substrate with $NH_3$.

Using mass spectroscopy, Ban[70] showed that the relevant chemical reactions on the source and deposition sides controlling the growth are

$$2Ga(l) + 2HCl(g) \rightarrow 2GaCl(g) + H_2(g)$$

$$GaCl(g) + NH_3(g) \rightarrow GaN(s) + HCl(g) + H_2(g).$$

The supply of GaCl is controlled by the Ga boat temperature and the flow rates of the HCl gas and the carrier gas. The reactor in this method is heated by a two-zone resistance furnace with the region containing the Ga boat being kept at a different temperature than the region holding the substrate for reaction, as shown in Fig. 9. The Ga zone temperature impacts the growth rate a great deal. The Ga source temperature is held at a constant temperature between 850 and 900°C. The reaction efficiency of HCl with Ga is near unity. Though depends on the reactor itself, the typical flow rates are about tens of sccm for HCl, 1 l/min for $NH_3$ and 2 l/min for the carrier gas. GaN films can be grown with rates up to 1 mmh$^{-1}$ on

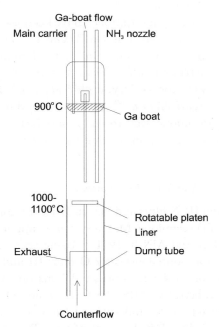

**Fig. 9.** Schematic diagram of a two-zone HVPE reactor, which utilizes Ga and ammonia sources, used for nitride growth. Courtesy of R. J. Molnar.

$Al_2O_3$ substrates at atmospheric pressure. However, those high growth rates deplete the Ga source material rather quickly and lead to very rough surfaces with columnar growth. The substrate zone temperature varies between 1050 and 1200°C. At lower substrate zone temperatures, the growth rate decreases exponentially due to the decreasing pyrolysis efficiency of the GaCl and $NH_3$. On the other hand, at higher substrate zone temperatures thermally induced decomposition of reactants reduces the growth rate. The hydrogen ambient also aids this reduction in that competing processes such as $GaH_x$ would take place.

Using this method, the first single crystal GaN thin films were realized. The growth rate was quite high (0.5 $\mu$m/min.), which allowed the growth of extremely thick films, whose properties were less influenced by the thermal and lattice mismatches with the substrate. However, GaN grown by this technique had very high background $n$-type carrier concentrations, typically $\sim 10^{19}$ cm$^{-3}$. Maruska *et al.*[71] later showed that Zn and Mg dopant incorporation could be achieved by the simultaneous evaporation of the dopant source in the HCl stream. Others have since used the technique of Maruska and Tietjen for GaN growth[72-74] and several others[75] have extended it to grow high quality AlN.

Many other variations of this approach have been used. These variations include[76-80] gaseous sources such as GaCl, $GaCl_2$, $GaCl_3$, $Ga(C_2H_5)_2Cl$, $GaCl_2NH_3$, $AlC_3$, $AlBr_3$, $InCl_3$, and $GaBr_3$ for Group III element(s) as reactants for $NH_3$. Pastrnak *et al.*[81] elected to react $N_2$ with $GaCl_3$, $AlCl_3$ and $InCl_3$ in their CVD process. Dryburgh[82] grew AlN from AlSe and $N_2$. By introducing $PH_3$, Igarashi *et al.*[83] achieved several percent P incorporation in GaN. The technique was recently popularized by R. Molnar of Lincoln Laboratory who began producing high quality GaN.[84]

### 3.2.2. *Metal organic chemical vapor epitaxy*

High quality epitaxial III-N films and heterostructures for devices have been succeeded by MOCVD technique. Manasevit *et al.* applied this technique to the deposition of GaN and AlN in 1971.[85] Using trimethylgallium (TMG) and ammonia ($NH_3$) as source gases for Groups III and V species, respectively, the authors obtained $c$-axis oriented films on sapphire (0001) and on 6H–SiC (0001) substrates. MIS-like LEDs followed, albeit they relied on deep states induced by Zn and suffered from very low efficiencies due to poor crystalline quality. Development of Low Temperature (LT) buffer layers addressed the quality issue some.[86] The technique improved over the years to the point that undoped GaN films with a low background carrier concentration of $5 \times 10^{16}$ cm$^{-3}$ and with an X-ray symmetric peak FWHM of $\sim 30$ arcseconds have been grown.[87] The X-ray data should be treated with caution, as the symmetric peak is not as sensitive to the edge dislocation as much. MOCVD has been used for the development of LEDs,[88] lasers,[89] transistors[90] and detectors.[91]

The best MOCVD reactors for Group III nitride film growth incorporate laminar flow at high operating pressures and separate inlets for the nitride precursors, and ammonia to minimize predeposition reactions. A successful, two-flow MOCVD reactor is shown in Fig. 10.[92] The main flow composed of reactant gases with a high velocity is directed through the nozzle parallel to a rotating substrate. The sub-flow gas composed of nitrogen and hydrogen is directed perpendicular to the substrate. The purpose of the flow normal to the substrate surface is to bring the reactant gases in contact with the substrate, and to suppress thermal convection effects. Hydrogen is the carrier gas of choice. A rotating susceptor was used to enhance uniformity of the deposited films. If one goes with the premise that smallest rocking curve half width implies an all around good quality, GaN films can claim this quality. These films with one of the narrowest rocking curve with full width at half maximum, FWHM, values of 37 s$^{-1}$ were grown with a modified EMCORE GS 3200 UTM reactor. This reactor generally incorporates separate inlets for ammonia and the nitride precursor, all are normal to the substrate surface which rotates at speeds over 1000 rpm, and a laminar flow cell to

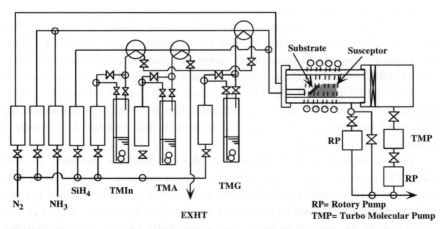

**Fig. 10.** Schematic of a horizontal MOCVD system, similar to that used in Prof. I. Akasaki Laboratory.

assure uniform growth[117] (informal reports indicate this value is down to under 30 arseconds).

MOCVD reactors incorporating new concepts have been designed to grow layers at lower temperatures. These technologies utilize an activated form of nitrogen in order to lower deposition temperatures of Group III nitrides. That these technologies are interesting is apparent, for example, from deposition of polycrystalline and amorphous GaN films at temperatures lower than 350°C by plasma enhanced CVD. Epitaxial GaN and AlN have been grown by variants of methods activating nitrogen, such as laser-assisted CVD, remote plasma-enhanced CVD, atomic layer epitaxy with $NH_3$ cracked by a hot filament, with ammonia catalytic decomposed, photo-assisted CVD, and ECR plasma-assisted CVD. However, none of these approaches have been able to produce material comparable in quality to standard MOCVD systems, and consequently, they did not really become a player in the field.

As for the mechanism involved, growth of nitride semiconductors by MOCVD relies on the transport of organometallic precursor gases, and hydrides for the nitrogen source, and reacting them on or near the surface of a heated substrate. The deposition is through pyrolysis. The underlying chemical mechanisms are complex and involve a set of gas phase and surface reactions. Although MOCVD has long been assumed to be a thermodynamically equilibrium process, nitride MOCVD process may involve kinetics as well. The fundamental understanding of the processes involved is still evolving and as such reaction mechanism and the related kinetic rate parameters are poorly understood.

The deposition of epitaxial nitride layers by MOCVD involves the reaction of metal-containing In, Ga or Al gasses with ammonia, $NH_3$. Commonly, the metal-containing gases are trimethygallium ($Me_3Ga$), trimethylindium ($Me_3In$), or trimethyaluminum ($Me_3Al$). Radicals, reactive by most definitions, react in the gas phase with donors containing acidic hydrogen, such as $NH_3$, and form adducts. The key here is to eliminate the unwanted radicals by forming stable molecules followed by their removal from the reaction region.

The aforementioned treatment of the mechanisms involved in the MOCVD process clearly indicates that any precursor must balance the requirement of volatility and stability, which often counter one another, to be transported to the surface and decomposed for deposition. To put it another way, these precursors must have appropriate reactivity to decompose thermally into the desired solid and to generate readily removable gaseous side products. Ideally the precursors should be non-pyrophoric, water and oxygen insensitive, noncorrosive and nontoxic. The trialkyls, trimethylgallium (TMG)[93] and triethylgallium (TEG), trimethylaluminum (TMA),[94] trimethylindium (TMI),[95] and others are usually used as III-metal precursors. Ammonia ($NH_3$),[96] hydrazine ($N_2H_4$),[97-99] monomethylhydrazine $(CH_3)N_2H_2$[100,101] dimethylhydrazine $(CH_3)_2N_2H_2$ have all been used to varying degrees of success as nitrogen precursors. Although trialkyl compounds (TMA, TMG, TMI, etc.) are pyrophoric and extremely water and oxygen sensitive, and ammonia is highly corrosive, much of the best material grown today is produced by conventional MOCVD by reacting these compounds with $NH_3$ at substrate temperatures in the neighborhood of 1000°C.[102-109]

Investigators have reacted trimethylgallium (TMG),[110] triethylgallium (TEG), and GaCl[111,112] with $NH_3$ plasma. Sheng *et al.*[113] reacted trimethylaluminum (TMA) and NH3 in the presence of hydrogen plasma. Wakahara *et al.*[114] grew InN by reacting trimethylindium (TMI) with microwave activated $N_2$. Eremin *et al.*[115] used nitrogen to transport metallic Ga to the reaction zone where it was reacted with active nitrogen. These commonly employed precursors at least satisfy the criteria of sufficient volatility and appropriate reactivity. During growth of nitrides employing trialkyl precursors, adduct formation with ammonia with TMA and TMG is well documented. Usually mixing at room temperature, adduct formation between TMG or TEG and ammonia is complete in less than 0.2 seconds. The resulting adduct $Ga(CH_3)_3-NH_3$ has a vapor pressure of 0.92 Torr at room temperature, while the vapor pressure of $Ga(C_2H_5)_3 : NH_3$ is much lower.

To counter early beliefs that stability of ammonia, and the requisite relatively high temperature growth can be alleviated with the use of other more volatile nitrogen sources were explored. For example, Fujieda *et al.*[97] replaced $NH_3$ with $N_2H_4$, and observed that significantly smaller amount of $N_2H_4$ was required to maintain the same growth rate. However, they also noted that the CVD growth rate was limited by the decomposition of TMG, so that the benefits of $N_2H_4$ are limited. Matloubian and Gershenzon[116] used TMA and $NH_3$, and a substrate temperature range of 673–1473 K to grown nitrides. Single-crystal AlN films were obtained only at 1473 K. Dupuie and Gulari[117] reported that the presence of a hot filament near the substrate increased the growth rate of AlN grown with TMA and $NH_3$ by two orders of magnitude. However, the use of a hot filament immediately raises concerns about residual contamination, most prominently oxygen, which was not addressed by the author.

A case in point for the elimination of the unwanted radicals by forming stable molecules followed by their removal from the reaction region is that of AlN growth from mixtures of methyl alkyls which may proceed by the formation of an intermediate gas phase adduct (MeAlNH$_3$), followed by the elimination of CH$_4$. The exact path may be that co-adsorption of Me$_3$Al and NH$_3$ at room temperature generates surface adduct species such as (Me$_2$AlNH$_3$) and adsorbed NH$_3$.[118] As the substrate temperature is raised above 320°C, the appearance of vibrational bands corresponding to AlN indicates the formation of extended (Al–N) networks on the surface. These Al(NH$_2$)$_2$Al species finally eliminate H$_2$ at the surface to form AlN.[119] The possible chemical reactions in the process are

$$2\text{Me}_2\text{AlNH}_3(\text{adducts}) \rightarrow \text{MeAl}(\text{NH}_2)_2\text{AlMe}(\text{adducts})$$

$$+ 2\text{CH}_4(g) \quad for \quad T = 270\text{–}330°C$$

$$(\text{MeAl.NH}_2)_2\text{AlMe}(\text{adducts}) \rightarrow \text{Al.NH}_2\text{Al.}(\text{adducts})$$

$$+ 2\text{CH}_4(g) \quad for \quad T = 330\text{–}780°C$$

and

$$\text{Al.NH}_2\text{Al.}(\text{adducts}) \rightarrow 2\text{AlN}(\text{adduct}) + 2\,\text{H}_2(g)\,.$$

As for GaN, investigations are relatively limited, but it would be fair to assume that processes similar to that with AlN growth are most likely in place. Adducts of Ga compounds are weaker electron acceptors than the corresponding Al adduct compounds, and therefore these adducts may not be abundant due to re-dissociation in the hot zone. It may be due to this that successful GaN growth by MOCVD requires very large V/III ratios which favor adducts formation. Thermal stability of NH$_3$, although low compared to that of N$_2$, could be partially responsible for the use of high substrate temperatures, typically above 550°C for InN and above 1000°C for GaN and AlN. The high growth temperature necessitated by the process itself associated with high nitrogen vapor pressure over GaN lead to the inevitable loss of nitrogen from the nitride film. This may also be the path to carbon contamination from the decomposition of the organic radical during metalorganic pyrolysis. The loss of nitrogen can be alleviated by the use of high V/III gas ratios during the deposition, particularly for InGaN (e.g. > 2000 : 1).

Assuming that high substrate temperatures represent a problem in relation to ammonia, which seems reasonable — particularly in early days, various alternative approaches can be and have been taken. One approach is to use alternative nitrogen precursors which are thermally less stable than $NH_3$. Hydrazine ($N_2H_4$), which is a larger and less stable molecule, has been used in combination with TMAl to deposit AlN at temperatures as low as 220°C. However, hydrazine is toxic, unstable, and not as pure as $NH_3$. Consequently, a compromise between quality and substrate temperature must be made. Researchers took the quality/purity as the primary parameter and stayed with $NH_3$. More recently, other nitrogen sources such as tert-butylamine (t-BuNH$_2$),[120] isopropylamine (i-PrNH$_2$) and trimethylsilylazide (TMeSiN$_3$) have been used with TMAl or t-Bu$_3$Al to deposit AlN films at lower substrate temperatures (400–600°C) and reduced V/III gas ratios (5 : 1 − 70 : 1).[121] However, the deposited films were invariably contaminated with high levels of residual carbon (up to 11 at. %).

Deposition of InN is nearly an intractable problem even for low temperature deposition processes such as MBE let alone MOCVD, which is a much higher temperature process. Needless to say that InN is a very challenging problem for MOCVD because the decomposition temperature is close to the minimum temperature for efficient thermal activation of ammonia, about 550°C (N–H bond strength, 3.9 eV) and pyrolysis of the organicmetallic source.[98,122] There are two reports of alternative single-source precursors for InN. The solid state pyrolysis of polymeric [In(NH$_2$)$_3$],[123] and air-stable, nonpyrophoric and volatile single-source precursor (N$_3$)Ga[(CH$_2$)$_3$NMe$_2$]$_2$.[124] Growth rates between 0.3 and 5 $\mu$m/h were obtained in the substrate temperature range of 300 to 450°C and a reactor pressure of $2 \times 10^{-4}$ Torr.

Hydrogen and, to a lesser extent, nitrogen are predominantly used as the transport gas. They can influence the chemical reaction mechanism of Et$_3$M or Me$_3$M in the gas phase by changing the reaction temperature of the metalorganic compounds or the concentration of reaction products. Hydrogen at the surface of the growing film can influence the growth rate and the structural properties.[125,126] To obtain a basic understanding of the role of hydrogen in GaN growth, the possible sources of hydrogen and the influence of hydrogen on the chemical reaction mechanism in the gas phase and at the surface are unique and important issues in the context of MOCVD.

Pyrolysis of highly concentrated $NH_3$ in the presence of $H_2$ as the carrier gas results in a high concentration of molecular and atomic hydrogen near the substrate surface. Since the growth temperatures above 900°C are employed which are higher than the decomposition temperature of the metalorganic compounds and their hydrocarbon ligands, the conditions for the desired bond breaking between the metal atom and the methyl or ethyl groups of the precursors are in place. However, the same can also lead to pyrolysis of the hydrocarbons with incorporation of hydrogen and carbon into the films. In this vein, decomposition of $Me_3Ga$ (TMG) and $Et_3Ga$ in hydrogen and nitrogen atmospheres using a quadrupole mass analyser has been investigated.[127] The decomposition reaction of the metalorganic precursors was found to be strongly affected by the presence of molecular hydrogen. Decomposition of $Me_3Ga$ occurs at 400°C and 500°C, in $H_2$ and $N_2$, respectively. Similarly, decomposition $Et_3M$ occurs at 260°C and 300°C in $H_2$ and $N_2$, respectively. Clearly, molecular hydrogen reduces the reaction temperature. The reaction mechanisms are through hydrolysis for $Me_3Ga$ in $H_2$, homolytic fission for $Me_3Ga$ in $N_2$, and $\beta$-elimination for $Et_3M$ in both $H_2$ and $N_2$. By changing the reaction temperature and the reaction mechanism of the metalorganic precursor, the partial pressure of hydrogen affects the deposition rate of GaN and therefore the structural properties of the resulting film, especially at low growth temperatures.[6]

## 4. Growth of Nitrides by Various Techniques

### 4.1. *Growth by MBE*

Growth by MBE differs greatly based on whether layers are grown on a foreign substrate such as Si, sapphire, SiC, and ZnO, or on GaN either in the form of a template or a bulk platelet. Growth on a foreign substrate generally requires high temperature growth as low temperature growths generally lead to good surface morphologies and large concentrations of edge dislocations propagating along the *c*-direction. Though substrate temperature measurements are not reliable, what is implied here is that 800°C and above would be high, while 650–750°C would be medium temperature. Note that in growth on sapphire and other foreign substrates the growth may commence with high or low temperature nitridation and low temperature buffer layers.[128] For growth on already existing GaN templates, the issues are very different in that the dislocation reduction measures and growth nucleation on a foreign substrate are not the main concern. What becomes important is then the parameters giving rise to high mobility or some such performance parameter.[35] In this case, it is universally agreed that medium temperature growth under fairly Ga-rich conditions lead to best surface morphologies and mobilities.[129]

If GaN is grown directly on sapphire the quality of the crystals is not as good as that is grown on small GaN templates due to large lattice and thermal expansion coefficient mismatch and, mostly due to non-ideal nucleation. The degradation of the quality is evident by a wide X-ray rocking curve (10 arcminutes typically), rough surface (containing hillocks), high electron concentration (up to $10^{19}$ cm$^{-3}$), and considerable yellow emission in the luminescence spectra. To improve the GaN layer quality, low temperature AlN or GaN buffer layers are usually grown.[130,131] Although, the low temperature AlN and GaN buffer layers were initially used solely for MOCVD, their use in the MBE process was shown to be highly beneficial also. The use of buffer layers became a standard technique for obtaining good quality GaN and AlGaN films.

A short period of nitridation of the sapphire substrate precedes the buffer layer growth, generally between 10 and 30 minutes, or until

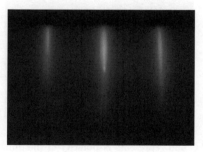

**Fig. 11.** (a) RHEED image taken along the [11$\bar{2}$0] direction of AlN, rotated about 30° with respect to that of sapphire, during nitridation of sapphire.

**Fig. 11.** (b) RHEED image taken along the [11$\bar{2}$0] direction of AlN during AlN growth on sapphire.

the RHEED image indicates a transition from sapphire to AlN with the accompanying 30° rotation. During the nitridation process a thin $AlO_{1-x}N_x$ film may form on the substrate surface,[132] transforming $Al_2O_3$, in a natural way, through $AlO_{1-x}N_x$ into AlN and the epitaxial films grow on AlN which has nearly the same lattice constant as GaN (mismatch 2–4% in $c$-plane). This is why the best films until now have been obtained on sapphire with AlN buffer layer. It is obvious that nitridation parameters can greatly influence the properties of the epitaxial layer.[133,134] A set of RHEED patterns showing the nitridation, AlN and GaN layers, all in sequence, stages is shown in Figs. 11 and 12. Recently, GaN buffer layers have also been used. In general, AlN and GaN buffer layers grown at relatively high temperatures lead to Ga and N-polarity films, respectively.[135] Those grown at lower temperatures can lead to either polarity, depending on the particulars, as will be discussed in some detail later in this manuscript. Shown in

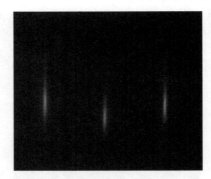

**Fig. 12.** (a) RHEED image of GaN deposited on a nitridated sapphire, taken along $[11\bar{2}0]$ direction at 800°C and represents N-polarity of the film.

**Fig. 12.** (b) RHEED image of AlN grown on GaN of Fig. 10(a), taken at 800°C along the $[11\bar{2}0]$ azimuth and represents a N-polarity film. The bright spots represent the specular reflection associated with each diffraction.

Fig. 12, is a series of RHEED images of GaN buffer layers deposited directly on sapphire and AlN grown on that GaN buffer layer. These images are for an N-polarity film and upon cooling do not show the characteristic $2 \times 2$ pattern observed in Ga-polarity films. Instead it showed a $1 \times 1$ pattern although, a $3 \times 3$ diffraction should have been observed. A $1 \times 1$ pattern can also be observed for a Ga-polarity film for a range of Group III/V ratio, but under the conditions used, Ga-polarity films lead to strong $2 \times 2$ patterns.

Despite the large lattice mismatch between the sapphire substrate and GaN, and stacking mismatch, reasonably high quality layers grown on sapphire substrates by MBE permitted the fabrication of a variety of devices, such as Schottky diodes[3] and modulation doped FETs (MODFET),[3,136–139] multiple InGaN/AlGaN quantum wells[140,141] and observation of photopumped stimulated emission at 425 K.[142] Although the layer quality on sapphire is almost unmatched, sapphire (0001) substrates are not most suitable for GaN-based lasers, because the cleavage planes in sapphire substrate are not perpendicular to the surface. Therefore, the use of sapphire (0001) substrate requires the fabrication of laser mirror facets by reactive ion etching or other more complicated cleavage techniques.[143] The cleavage problem can be avoided using an *a*-plane (2110) sapphire substrate. However, the growth on this plane has been much less studied.

Much work has been done regarding growth of nitrides by MBE on sapphire substrates which can be found in other reviews.[144,145] Only a succinct review of growth on various substrates, namely the polarity issues, use of novel techniques to reduce defect density such as quantum dot filtering of dislocations, and high quality GaN on GaN templates will be discussed here.

### 4.1.1. *Polarity of MBE grown layers*

MBE with its inherent control over the growth parameters can be used to interrogate certain structural and electrical processes in the crystal. One such topic is the polarity of the films since the *c*-plane of sapphire is a polar surface and GaN does not share the stacking order with sapphire. Consequently, GaN grown on sapphire could either be terminated with Ga ([0001] direction) or N ([000$\bar{1}$] direction).[146] Being non-centro-symmetric due to its wurtzitic and ionic structure, nitrides exhibit large piezoelectric effects when under stress along the *c*-direction. Moreover, spontaneous polarization charges also appear at the heterointerfaces due to the different degrees of ionicity of the various binary and ternary nitrides.[147] The signs of spontaneous polarization and piezoelectric polarization depend on the polarity of the film. This charge and its sign must be known and controlled in electronic devices, particularly in modulation doped FETs. For example, electric field caused by polarization effects can increase or decrease interfacial free carrier concentration causing the gate potential needed to vary drastically.[148] The polarity of the film can also have an impact on the effective band discontinuities.[149]

GaN layers grown by MBE can be either N or Ga-polarity, and each can be grown under Ga and N-rich conditions. In the case of growth with ammonia as the nitrogen source, in conjunction with high temperature growth > 800°C, very nitrogen rich conditions lead to better quality films in all respects i.e. PL, X-ray diffraction and electron mobility. This is in part due to simultaneous removal of material which leads to the elimination of regions that are not of as high quality. In addition, the tips of the clusters are partially etched away, which leads to smoother surfaces.

In the early stages of development, multiple polarities have been confirmed in epitaxial GaN layers on sapphire substrates by convergent beam electron diffraction.[150] Hemispherically scanned X-ray photoelectron diffraction,[151] collision ion scattering spectroscopy,[152] convergent beam electron diffraction, and etching techniques have been used to determine the polarity of GaN. With more insight, and control over the MBE growth process, reports emerged as to which growth parameters lead to what polarity. It has been implied that GaN buffer layers grown at 700°C lead to N-polarity with poor layer quality, and that the N-polarity samples are characterized with much higher background concentration and overall inferior quality.[153] Others reported N-polarity when grown directly on sapphire, and Ga-polarity when grown with an AlN buffer layer inserted prior to GaN growth.[154,155] Moreover, reflection high energy electron diffraction (RHEED) capability allows the MBE grower to determine the polarity of some of the films by inspecting the surface reconstruction during cool down.[156] Surface charge sensitive Electric Force Microscopy for determining the polarity of the sample and its distribution on the surface has also been reported.[157] The polarity impacts the dopant incorporation also in that the N-face is more amenable to incorporation.[158] Ga-polar GaN has also been reported to be more resistant to chemical etches such as molten NaOH+KOH (200°C).

For polarity investigations, unintentionally doped GaN layers were grown by molecular beam epitaxy (MBE) on *c*-plane of sapphire substrates using RF activated N.[159] It should be mentioned that a low temperature nitridation, in combination with low temperature buffer growth, has been reported to lead to interfaces void of cubic GaN and improved quality.[160] Four groups of samples were grown and investigated. The first and second sets utilized GaN buffer layers grown near 500°C, and near 800°C. The third and fourth groups utilized AlN buffer layers grown near 500°C, and 890–920°C. Following the buffer layers, typically 1 $\mu$m thick GaN layers were grown at a substrate temperature between 720 and 850°C with growth rates in the range of 300 to 1000 nm per hour under N-limited (Ga-rich) conditions. In addition to the *in situ* RHEED images, AFM, X-ray diffraction, and hot $H_3PO_4$ at 160°C were employed to confirm the polarity assignment.

Layers with high temperature GaN buffer layers (around 770°C or higher) invariably turned out to be N-polarity regardless whether a static or graded substrate temperature was employed during the buffer growth. Upon cooling, the RHEED pattern indicated only the bulk $1 \times 1$ structure, though others have reported higher order reconstruction. Conversely, layers with AlN buffers grown in the temperature range of 880–960°C with thicknesses in the range of 8–35 nm and growth rates of 40–60 nm/h led to Ga-polarity. Consequently, a $(2 \times 2)$ pattern RHEED pattern was observed upon cool-down at temperatures ranging between 280–650°C depending on the V/III ratio employed for the mail layer.

The low substrate temperature buffer growth, $T_s \leq 650°C$ but primarily around 500°C, resulted in layers with either polarity with either GaN or AlN buffer layers. The GaN structures with 100–150 nm thick GaN buffer layers at a growth rate of about 600 nm/h led to Ga-polarity with the characteristic $2 \times 2$ pattern upon cool-down after the entire structure was completed. However, when the thickness of the buffer layer was reduced to 30–60 nm, keeping the growth rate the same, the layers turned out to be of mixed polarity with a faint $(2 \times 2)$ RHEED reconstruction observed upon cool-down. When about 110–220 nm thick buffer layers grown at 500°C with 220 nm per hour growth rate were used, the resultant layers were of N-polarity with only the $1 \times 1$ reconstruction observed during cool-down.

The AlN buffer layers grown at $T_s \leq 650°C$, but primarily around 500°C, exhibited Ga or N-polarity depending on the growth conditions. When 10–15 nm thick buffer layers grown at a growth rate of 60 nm/h were employed, Ga-polarity resulted. However, when 2.5–22 nm thick buffer layers were employed with a growth rate of 15–25 nm/h, N-polarity resulted.

The typical surface morphologies of as-grown Ga-polar films with different buffer layers are presented in Fig. 13. In this case, a high temperature AlN buffer layer tends to result in a smooth, but pitted

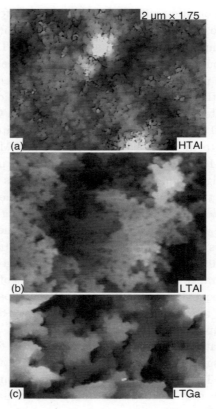

**Fig. 13.** Ga-polarity samples obtained under three different buffer layer and growth conditions: high temperature AlN buffer layer (a), low temperature AlN buffer layer (b), and low temperature GaN buffer layer (c).

layer (Fig. 13(a)), consistent with the Group III/V ratio employed. Higher Group III/V ratios generally lead to disappearance of the pits. A low temperature AlN buffer layer grown at a high rate leads to a Ga-polar surface morphology with irregular stepped terraces, often with pits and/or a rough surface (Fig. 13(b)). When a low temperature GaN buffer layer grown at high rate was used, we found a similar morphology to that shown in Fig. 1(b) with a more drastic variation in terrace height and shape (Fig. 13(c)).

The surface morphologies of as-grown N-polar films with different buffer layers are presented in Fig. 14. With a high temperature GaN buffer layer, the film morphology is that of non-coalesced columns (Fig. 14(a)).

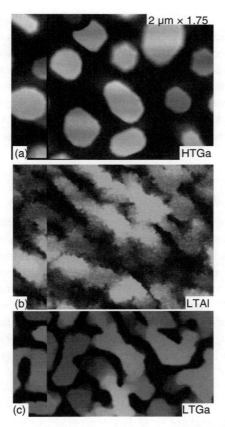

**Fig. 14.** N-polarity samples obtained under three different buffer layer and growth conditions: high temperature GaN buffer layer (a), low temperature AlN buffer layer (b), and low temperature GaN buffer layer (c). The rough surface morphology can be improved by growing the top GaN layer at lower temperature such as 720°C instead of 800°C, which was the case here.

In general, smoother morphologies with stepped terraces were found when a low temperature AlN buffer layer grown at a low rate was used (Fig. 14(b)). Using low temperature GaN buffer layers grown at a low rate, the morphologies vary from extremely rough surfaces with non-coalesced columns to a surface shown in Fig. 14(c), where very tall columns and terraces are separated by deep troughs.

The simple model that explains the N-polarity with GaN buffer layer calls for Ga to form [0001] or the long bond to the O-surface of sapphire. In the same vein, the Ga-polarity results with AlN buffer layers when O leaves the surface and N forms the [0001], or the long bond with Al of sapphire. This simple model, while consistent with high temperature buffer layers, does not explain our results with low temperature AlN and GaN buffer layers which led to either polarity depending on the growth conditions. Detailed investigations are necessary to gain an insight as to the mechanisms involved. High growth rates mainly leading to Ga-polarity and low growth rates to N-polarity would indicate that there must be some atomic exchange or interaction that may be suppressed or promoted by large growth rates, depending on the case.

### 4.1.2. *Properties of nitrogen polarity layers*

GaN grown with N-polarity on sapphire substrates appears to be considered of lower quality in many respects including structural, optical and electrical properties. It has been implied that GaN buffer layers grown at 700°C lead to N-polarity with poor layer quality, and that the N-polarity samples are characterized with much higher background concentration and overall inferior quality.[161] Recent data appear to indicate that the electrical, structural, and optical properties of nitrogen polarity films have improved considerably, though not yet comparable to Ga-polarity samples in general. Needless to say what is worst is a mixture of polarity across the wafer, which would lead to increased carrier scattering due to flipping in the polarization field.

An N-polarity can be obtained by first growing a GaN buffer layer, either at high temperature or at low temperatures with low growth rates, as described in the previous section. Several types of buffer layers were considered. In one group, the buffer layers were grown at 500°C and in the other they were grown in a ramped fashion from 600 to 800°C. In the third, they were grown commencing at about 900°C and ending at a temperature anywhere between 800 to 900°C, this figure being determined by the growth temperature of the final GaN layer. In some layers $NH_3$ as well as RF activated N were used.

Shown in Fig. 15 are the low temperature PL spectrum of N-polarity layer on a buffer layer grown at 500°C and 800°C with both RF and $RF+NH_3$ as the nitrogen source. The spectra were taken with an unfocused laser beam to relatively enhance the contribution by defects such as the notorious yellow luminescence. The salient features of all three spectra are that the excitonic transitions are strong and the defect peaks are relatively weak. This is quite customary of N-polarity samples. However, one

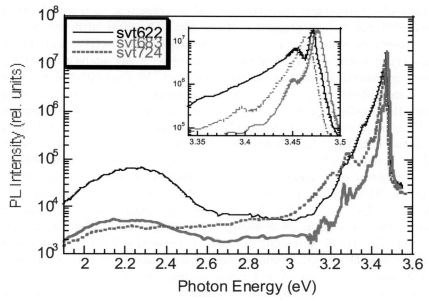

**Fig. 15.** Low temperature PL of GaN layers grown with GaN buffer layers taken at 30 K with unfocused He-Cd laser beam to enhance the defect-related transitions. Max at 3.471 eV (sample # 622 with low temperature GaN buffer layer); 3.476 eV (sample # 683 with medium temperature GaN buffer layer (800°C)); 3.464 eV (sample # 724 with medium temperature GaN buffer layer (800°C) and grown with both NH3 and RF nitrogen impinging simultaneously on the surface). The Full Width at Half Maximum (FWHM) values = 6.6 meV (# 622); 10 meV (# 683); 13.5 meV (# 724).

should mention that the N-polarity samples, while having comparable or slightly lower dislocation density as compared to Ga-polarity samples, tend to exhibit higher concentration of inversion domains.

### 4.1.3. *X-ray characterization of GaN*

The structural film quality is commonly assessed through the rocking curve full width at half maximum (FWHM) for the GaN (002) peak. The pitfall is that this diffraction is sensitive to planarity of the surface and is sensitive to defects that affect adversely that planarity. Often times, half width of this diffraction is used to draw conclusions about the overall quality of GaN. Edge dislocations do not hurt that planarity, but are detrimental otherwise. In fact, the best films in terms of optical and electrical performance, and device performance are those that exhibit (002) peak halfwidths of about 5 arcminutes. A recent high-resolution X-ray investigation[162] indicates that the in-plane structural properties of GaN is consistent with the observed electrical and optical properties of the film. The films with very sharp, 45 arcseconds, symmetric rocking curves showed poor electrical and optical properties which is consistent with the in-plane X-ray analysis indicating poor asymmetric diffraction. On the other hand, the films with 5 arcminutes or so showed excellent optical and electrical properties, which are consistent with the excellent in-plane X-ray structure indicating narrow asymmetric diffraction. From the out-of-plane XRD data and the cross sectional TEM images of the same films it appears that the sharp X-ray rocking curve is likely a result of planar defects[163] and not due to any reduction in defects.

The effect of threading dislocation structure on the X-ray diffraction peak width in epitaxial GaN films has been studied.[162,164] Quantitative defect analysis shows that the threading dislocations are predominantly pure edge dislocations lying along the $c$-axis, as will be detailed in this review. The specific threading dislocation geometry will lead to distortions of only specific crystallographic planes. Edge dislocations will distort only ($hkl$) planes with either $h$ or $k$ non-zero. Rocking curves on off-axis ($hkl$) planes will be broadened, while symmetric (001) rocking curves will be insensitive to pure edge threading dislocations content in the film. Screw threading dislocations, with [001] directions, have a pure shear strain field that distort all ($hkl$) planes with $l$ non-zero. The dramatic broadening of the asymmetric (102) rocking curve evidences a high density of pure edge threading dislocations. Therefore the rocking curves widths for off-axis reflections are a more reliable indicator of the structural quality of GaN films. The structural properties of GaN film grown on sapphire differ for the in-plane and out-of-plane structural features. The measured in-plane coherence lengths are smaller and the rocking curve widths are larger than those in the direction normal to the plane. It implies that optical and electrical properties are anisotropic along the film plane and plane normal directions. The in-plane structure is more closely related to electron mobility and optical properties than those in the plane-normal direction, suggesting that asymmetric diffraction, as opposed to out-of-plane diffraction, should be weighed more heavily.

### 4.1.4. *Defect filtering with Quantum Dots*

Lateral Epitaxial Overgrowth in MBE is not successful, as growth rates in the lateral direction are not much different from vertical ones. The same is true for selective growth. Consequently, other methods must be developed for reducing extended defects. One method that has shown great potential is the imbedding of a quantum dot region within the buffer layer for terminating dislocations.[165] Utilizing defect delineating chemical etching (which has been verified to be reliable on control samples) on the GaN films grown with and without GaN/AlN QDs, the etch pit density in GaN layers grown on top of Qdots has been shown to be reduced from $\sim 10^{10}$ cm$^{-2}$ to $\sim 10^7$ cm$^{-2}$.[166] Preliminary cross-sectional TEM images show that dislocations are blocked by the Qdot region. The mechanisms responsible for this remain to be investigated. The details of QD preparation and resultant optical properties have been reported elsewhere and will not be repeated here.[167,168]

The density of dislocations in the epilayers with and without Qdot insertion in their buffer layers was examined by wet chemical etch and atomic force microscopy (AFM). Hot (160°C) phosphoric acid ($H_3PO_4$) was used as the chemical etchant. It has been demonstrated that this etchant only attacks the defect sites in Ga-polar (0001) GaN surface and that etch pit density determined as such is comparable to the dislocation density determined by TEM, Photo Electric Chemical (PEC) etching, and KOH etching in control samples.[169,170] The pits, mostly in hexagonal shape, appear after the surface is etched. To reiterate, the pit density measured by AFM has been correlated to the dislocation density in the film and the validity of the etching process for defect delineation has been established previously for HVPE samples by comparing to TEM investigations.[170]

A typical AFM image of the as grown and the etched surface for the sample without QDots is shown in Figs. 16(a) and 16(b), respectively. The vertical scale is 30 nm for image Fig. 16(a) and 20 nm for Fig. 16(b). The size of both images is 1 $\mu$m × 1 $\mu$m. The etching time for the image Fig. 16(b) is 15 seconds. The result shows the high density of defects. A very rough estimate of the dislocation density by accounting the number of the etched pits is on the order of $10^{10}$ cm$^{-2}$. This is a typical value for the GaN films grown by MBE on sapphire substrates using AlN as the buffer layer, which are inherently thin.

An AFM image of the as grown and the etched surface for the sample with Qdot insertion in its buffer layer is shown in Figs. 17(a) and 17(b), respectively. The vertical scale is 30 nm for Fig. 17(a) and 50 nm for Fig. 15(b). The size is also 1 $\mu$m × 1 $\mu$m for Fig. 17(a) but 10 $\mu$m × 10 $\mu$m for Fig. 17(b). The etching time for the Fig. 17(b) is 12 minutes. The change in the etching time in this case only affects the pit size but not the density. A pit density, $\sim 4 \times 10^7$ cm$^{-2}$ can be observed. As compared to the sample

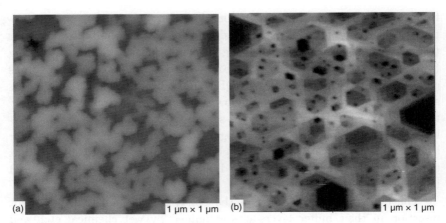

**Fig. 16.** AFM images of sample A (sample # 743). (a) As-grown surface. Vertical scale is 30 nm. (b) Etched at 160°C H₃PO₄ for 15 seconds. Vertical scale is 20 nm.

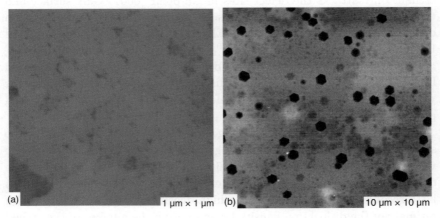

**Fig. 17.** AFM images of sample C (sample # 744). (a) As-grown surface, vertical scale is 30 nm. (b) Etched at 160°C H₃PO₄ for 12 minutes. Vertical scale is 50 nm. Note a pit density of $\sim 4 \times 10^7$ cm$^{-2}$.

without Qdot insertion, the sample without Qdot insertion, a reduction in the dislocation density by nearly three orders of magnitudes in the sample with Qdots is evident.

Preliminary cross-sectional images indicate the effectiveness of the quantum dots as shown in Fig. 18 where the dislocations emanating from the substrate-epi interface are blocked by the quantum dot region. Since the core energy of edge dislocations is minimum along the $c$-direction,

**Fig. 18.** Cross-sectional TEM images of a quantum dot composite grown on GaN which was in turn grown on an AlN/GaN/AlN/Sapphire template. Astonishingly, the edge dislocations that are present in the GaN terminate in the quantum dot region. The images were provided by Prof. D. J. Smith of Ariziona State University.

approaches that have been successful in conventional compound semi-conductors have met with failure in GaN. It is believed that the quantum dot region is unique and the mechanisms of defect reductions must be explored and the method itself holds great promise once optimized. The strain fields, nanolateral overgrowth, generation of local defects (bridging and thus annihilating dislocations) are among the likely causes of dislocation termination. In the case of lateral growth on the $SiO_2$ pat-terned GaN surface, it was found that the dislocations from the GaN were blocked by the $SiO_2$ and the lateral growth on the $SiO_2$ shows a very low dislocation density. In the QDot case, the dislocations extending from the film/substrate interface appear to be interrupted by GaN QDs. For example, the dislocation lines that have an in-plane component may loop around the QDs and no longer extend into the sample surface. Dislocations may also terminate at the surface of the dots. The partially lateral growth of the GaN QDs and the AlN spacer layer may also have contribution to the observed low dislocation density. The detailed mechanism of the dis-location interaction with GaN/AlN QD layers is not yet well understood at this time and is a subject of further investigation.

### 4.1.5. *Growth by MBE on Si substrates*

Si substrates are very attractive not only because of their high quality, availability and low cost, but also for the possibility of integration of Si-based electronics with wide-band gap semiconductor devices. Nitridation as carried out on sapphire is not performed because of the possibility of either stoichiometric or non-stoichiometric SiN formation ($Si_3N_4$ is the stoichiometric form of it). Several growth initiation techniques can be employed. One approach utilizes the growth of a thin AlN layer directly on Si followed by GaN deposition. GaN deposition directly on Si is not generally performed in MBE as Ga evaporates rather readily from the Si surface at typical MBE deposition temperatures. The other method involves the deposition of a sub to one monolayer of Al deposition at about 650°C

**Fig. 19.** RHEED image obtained on GaN grown on an AlN/Si composite during cooldown along the $[11\bar{2}0]$ azimuth. Note the $2 \times 2$ pattern indicative of Ga polarity and long N bond to the Si surface.

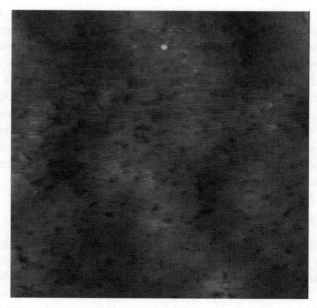

**Fig. 20.** AFM image of GaN grown on Si substrate at about 620°C. The image size is $2 \times 2$ $\mu$m$^2$, and the vertical scale is 10 nm.

followed by nitridation to convert it to AlN which is then followed by regular deposition of AlN. Shown in Fig. 19 is the RHEED pattern obtained on GaN grown on Si followed by an AlN buffer layer. The $2 \times 2$ pattern is indicative of a Ga-polarity sample which implies that N forms the long bonds to Si. The AFM image of the same is shown in Fig. 20. The XRD measurements indicated (002) symmetric and (102) asymmetric peark FWHM values of 15.9 and 10.9 arcminutes, respectively. PL measurements indicated half width values of about 19 meV at 30 K.

Cubic GaN is generally grown on Si (100) substrates,[171] while Si (111) are employed as the substrates for the wurtzitic GaN growth.[172,173] Both phases, wurtzitic and zincblende, are frequently detected accompanied with a large number of extended defects such as dislocations, stacking faults and twins. GaN grown on Si (001) is predominantly cubic. A 30 nm GaN low temperature buffer layer can accommodate the 17% lattice mismatch between the film and substrate by a combination of misoriented domains and misfit dislocations.[174] Beyond the buffer layer, highly oriented domains separated by inversion domain boundaries are found by transmission electron microscopy (TEM). Stacking faults, microtwins and localized regions of wurtzitic structure are major defects in the film. The films grown on Si (111) have a predominantly wurtzitic structure with a presence of twinned cubic phase.

Marked improvement of the wurtzitic GaN film quality grown by ECR MBE on Si (111) was observed when a surface preparation procedure that makes atomically flat terraces was used.[175] Wide atomically flat terraces were created by etching using solution of $7 : 1$ NH$_4$F : HF, greatly increasing the distance between surface steps and thus decreasing their density. The latter reduced the number of stacking mismatch boundaries of the epitaxial GaN films, improving their crystalline quality, as it was attested by XRD and PL measurements.

The AlN buffer layer was likewise grown on Si (111) substrate prior to the GaN epitaxy. The good quality of the epilayers was attested by presence of free exciton in low temperature PL. The PL decay times of excitonic emission were in the range of those reported for high quality GaN grown on sapphire.[176] Quality of GaN grown on Si is somewhat but not that much inferior to that grown on sapphire. In fact, GaN quantum dots with very high efficiency have been grown on Si substrates.[177,178] Additionally, with the advent of lateral epitaxial overgrowth utilizing SiC/Si composites for tempelate, GaN with quality comparable to those on sapphire substrates has been prepared by organometallic vapor phase epitaxy.

### 4.1.6. *Growth by MBE on SiC substrates*

The development of growth technology on SiC is less developed than that on sapphire, because large surface area SiC substrates became available commercially only a short time ago, and because of their prohibitive cost. As a substrate, 6H–SiC presents an advantage in that the lattice constant and thermal expansion coefficient are closer to that of GaN as compared to sapphire. The relatively small lattice mismatch for the basal plane is 3.5% for GaN and nearly a perfect match for AlN. The problems with SiC are the lack of stacking match and, an appropriate chemical etch for its surface. Lin *et al.*[179] proposed a two stage cleaning of the SiC surface. In the first step, the surface is hydrogen passivated using a HF dip before introduction into the vacuum system. Secondly, the substrate is treated with hydrogen plasma, which reduces the CO level (oxygen-carbon bonding) to below the X-ray photoemission detection limit. Upon heating in the MBE chamber the SiC substrates were observed to have a sharp $(1 \times 1)$ surface reconstruction. GaN epilayers deposited on AlN buffer layers showed sharp X-ray diffraction and PL peaks. The early attempts in the authors' laboratory to grow GaN layers on SiC met with difficulty due to the surface damage roughness, though occasionally very high mobility samples could be obtained.[29,180]

GaN or AlN buffer layers are usually grown on SiC substrates.[181] The minimal defect density, the highest mobility, and the best PL characteristics of GaN/SiC epitaxial layers were obtained on the optimized GaN buffer layers. The stress related phenomena in GaN films grown on SiC are much smaller than those of films grown on sapphire.[182] The defect structure of GaN films grown with and without AlN buffer layer on 6H–SiC by ECR MBE have been studied by Smith *et al.*[183] Threading defects are identified as double position boundaries originating at the substrate-film interfaces. The density of these defects was related to the smoothness of the surface.

With the development of high resistivity SiC substrates and the high thermal conductivity of SiC, this substrate material became a popular one for the growth of high power FET structures. The heat dissipation is a major problem, however, in GaN MODFETs on sapphire substrates as the thermal conductivity of this substrate is about 0.3 W/cmK; may even be somewhat lower. To make the matters worse, the thermal conductivity decreases rapidly as the temperature increases. Consequently, devices show a decreasing drain current (negative differential output conductance) as the drain bias is increased, and needless to say, the power performance is degraded. To overcome this, one must either remove the sapphire substrate followed by mounting the structure on a substrate with better thermal conductivity, employ flip-chip mounting, or grow the structure on a substrate with better thermal conductivity. Among the substrates with good thermal conductivity are Si and in particular SiC. Layers on Si, unless ELO with SiC interlayers is performed, are, however, not of as high quality as one would like. This leaves SiC substrates which are expensive and suffer from inferior surface polishing damage due to the hardness of SiC. The early attempts in authors' laboratory to grow GaN layers on SiC met with some difficulty due to this surface damage, although very high mobilities could be obtained occasionally.[183]

Two approaches can be employed to remove the surface damage. One is the mechanical chemical polish, which is very slow in coming, and the other is etching in H and Cl environment at very high temperatures such as 1500°C. The H cleaning process has been adopted as a standard procedure for MBE growth of GaN on SiC[52] with cleaning temperature of about 1700°C. The author's in collaboration with the group of E. Janzen at Linkping University were able to H etch Leyl SiC followed by MODFET growth. These devices did not exhibit the negative differential resistance characteristic of the sapphire substrates. However, SiC substrates prepared by the sublimation method did not appear to survive this high temperature H etching process. Researchers have exploited the *in situ* H etching process[184] and HCl etching process.[185]

A comparison of GaN epilayers grown on sapphire and SiC substrates with AlN buffer layers by ECR MBE revealed the superiority of SiC substrates.[29] Structural quality as measured by the XRD rocking curve, the photoluminescence and the mobility were superior for the films grown on SiC. The films grown on SiC had a higher mobility (580 cm$^2$/Vs) than those grown on sapphire (230 cm$^2$/Vs), using AlN buffer layers for both cases. The electron mobility of 580 cm$^2$/Vs of film grown on SiC is still one of the highest values ever reported in all MBE grown GaN. It should however be pointed out that unless the surface damage is removed from SiC surface, growth on SiC by MBE is not competitive. As mentioned earlier, H polishing of SiC at elevated temperatures, as high as 1700°C, lead to atomically smooth and clean surface with well-defined terraces that are bi-atomic step in height.

Growth of GaN by MBE is much simpler on SiC than it is on sapphire once proper surface cleaning procedures have been worked out. GaN on SiC has recently been grown in author's laboratory using both RF and ammonia for N source. Beginning with RF for N source, the SiC substrate shows well-defined $1 \times 1$ pattern indicative of the efficacy of H polishing. Once the thermal desorption process is completed, the substrate temperature is reduced to the growth temperature which is about $650°C$. At that point, both Ga and N source were directed to the substrate with fluxes clearly placing the surface under very Ga-rich conditions. During the first few seconds of the growth, the RHEED pattern is of a very well defined $3 \times 3$ pattern. Soon afterwards, the pattern evolves to a $1 \times 1$ pattern which is typical of GaN. Upon cooling after growth, the pattern gives way to a $2 \times 2$ pattern. Layers grown with ammonia show different RHEED images initially. At first, the pattern is spotty, and in a short time evolves into a typical $1 \times 1$ pattern.

Post growth evaluation included AFM, X-ray and PL investigations. Shown in Fig. 21 is an AFM image of GaN grown on H polished SiC with RF N source. Generally the surface is characterized as smooth overall but with broken steps indicative of structural defects. GaN layers were also grown in H polished SiC using ammonia for the nitrogen source. The XRD measurements for ammonia grown GaN indicated (002) symmetric and (102) asymmetric peak FWHM values of 3.81 and 10.5 arcminutes, respectively, despite the 0.23 $\mu$m thickness of the layer. The PL measurements indicated half width values of about 10 meV and an efficiency of 0.3% at 30 K which is very good.

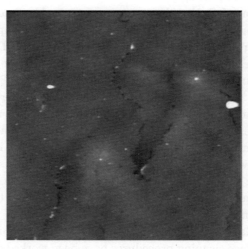

**Fig. 21.** AFM image of GaN grown on H polished SiC with RF N source at about $650°$C. The image size is $5 \times 5$ $\mu$m$^2$, rms = 0.904 nm and the vertical scale is 10 nm.

With recent advent of porous SiC (PSiC) substrates which have been reported to lead to SiC epitaxial layers with better quality, as compared to standard SiC substrates, attempts have been made to utilize this method for GaN growth as well. The state of the p-SiC is such that as-received samples had to to be converted to porous SiC as opposed to the damage-free H polished SiC substrate surfaces which are defect free in terms of extended defects. Nitride growth experiments were also carried out on porous SiC, as this may lead to better interfacial transition from SiC to GaN. It is thought that the nanosized pores would lead to lateral epitaxial growth, which will be discussed in detail in the MOCVD section, with coalescence in the nanoscale. This is thought to lead to a defect free coalescence of islands very early on during the growth as they grow out. This activity has its genesis in SiC epitaxial growth,[186] and has recently been extended into the arena of GaN growth on PSiC as well.[187] In SiC, screw dislocations prevent high blocking voltages from being obtained and appear to be the main suspect that is preventing large-area devices from being realized. The preliminary results obtained in SiC layers grown on porous SiC are quite compelling, the enthusiasm for which is spilling over into GaN.

Growth on PSiC substrates has been undertaken in author's laboratory in an effort to first find out whether films with better quality can be grown and second what the pathways are. The experiments which are preliminary in nature followed a growth sequence as in the case of growth on SiC with ammonia. Simply, following the thermal outgassing process, the substate temperature is lowered to about 650°C and growth with ammonia under Ga-rich conditions commenced. Ammonia is thermally cracked with aid of catalysis by Ga and as such the amount of N available for growth is very temperature dependent. In this particular experiment, the substrate temperature is low for an efficient catalysis and thus the growth rate is low and temperature dependent. A series of experiments with varying ammonia flow rate have been conducted. The layers have been investigated with AFM, X-ray and PL techniques.

The AFM images from porous SiC substrates were examined first. Figure 22 presents an image from one of the substrate (M1029-D3) before GaN growth. The image size is $2 \times 2$ $\mu m^2$ and the vertical scale is 15 nm. Pores are clearly seen on the surface just like pits. The density of pores is about $\sim 3 \times 10^{10}$ cm$^{-2}$. The pores have diameters of $\sim 30$ nm and are very shallow (depth of $\sim 3$ nm). Except for the pores, the overall surface is flat with a roughness (root-mean-square or rms) value of 1.7 nm. Shown in Fig. 23 is the RHEED image of a GaN grown on H polished SiC with ammonia as the source at about 650°C during cooldown at 440°C showing a 2 × 2 reconstruction indicative of a Ga-polar film. Figure 24 shows the AFM image of the same GaN layer on PSiC substrate. Despite the 0.15 $\mu$m thickness of the film, the X-ray data indicated 4.40 and 13.7 arcminutes FWHM for the [002] and [102] diffraction peaks, respectively. The PL FWHM value for this 0.15 $\mu$m thick film was about 14 meV and the PL efficiency was 0.13%, both at 30 K. The results are very preliminary and do not yet indicate any trends.

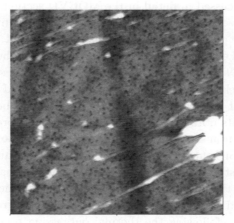

**Fig. 22.** AFM image measured from a porous SiC substrate (M1029D3). The image size is 2 × 2 $\mu$m$^2$. The vertical scale is 15 nm. The roughness (rms) value is 1.7 nm.

**Fig. 23.** RHEED image taken along the [11$\bar{2}$0] azimuth of GaN grown on H polished SiC with ammonia as the source at about 650°C during cooldown at 440°C showing a 2 × 2 reconstruction indicative of a Ga-polar film.

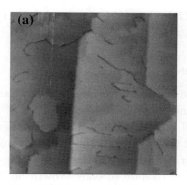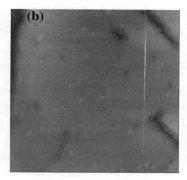

**Fig. 24.** AFM image of GaN grown on PSiC with ammonia as the nitrogen source. Note the atomic, albeit broken, steps indicative of the generally smooth surfaces.

Steady improvement in power performance of GaN/AlGaN MODFETs has led to very recent results at HRL laboratories with record breaking performance.[189] Typical DC characteristics include 600 mA/mm current performance and > 60 V drain breakdown voltage. The current gain cut-off and maximum power gain cut-off frequencies measured were about 48 and 100 GHz, respectively, for −5.5 V and 12.5 V gate and drain bias voltages, respectively. A minimum noise figure of 0.85 dB at 10 GHz with an associated gain of 11 dB is simply remarkable.

At HRL laboratories, a 6.3 W of CW output power was obtained at 10 GHz from a 1 mm wide transistor device.[188] More importantly, the power density remained nearly constant as the device size was scaled upward from a 0.1 mm width, where the device exhibits 6.5 W/mm, to 1.0 mm. These record-setting transistors were epitaxially grown AlGaN/GaN heterostructures on semi-insulating SiC substrates by MBE. HRL has developed a growth process using MBE which has virtually eliminated material defects common to other reported GaN devices, thereby enabling the scaling. MBE growth also produces device characteristics with less than 5% standard deviation over the 2-inch diameter SiC substrate, a six-fold improvement over previously reported results.

Following the same tradition, the researchers at HRL laboratories have expanded their work to amplifiers with several cells and showed very good power scalability up to 2 mm of total gate periphery.[188] Using 250 nm gate devices, a CW output power of 22.9 W with an associated power-added efficiency of 37% was measured for an amplifier at 9 GHz with four 1 mm gate periphery devices. Furthermore, the same authors[188] also showed a

CW power density of 4 W/mm at 20 GHz that is the state of the art for any three terminal solid state device at this frequency.

### 4.1.7. *Growth by MBE on ZnO substrates*

ZnO is considered a promising substrate for the III-N semiconductors since it has a close match for GaN $c$ and $a$-planes and a close stacking order.[189,190] ZnO was used as a buffer layer for GaN growth but the films were inferior to the quality of the films with AlN and GaN buffer layers.[191,192] Matsuoka *et al.* have used ZnO substrates to grow lattice matched $In_{0.22}Ga_{0.78}N$ alloys.[193] The most comprehensive study to date of GaN growth on ZnO was accomplished by MBE.[194] The high quality of the resultant films has been confirmed through photoluminescence and reflectance measurements. The width of excitonic transitions ($\sim$ 8 meV) was in the same range as the best results obtained on sapphire by MBE at the time. The yellow luminescence was totally absent from the PL spectra. X-ray measurements as well as polarized optical data confirm that the GaN epilayers are better oriented with respect to the substrate axis than those on sapphire and SiC. More work is needed to delineate intrinsic properties from extraneous issues, such as defective substrate surface morphology, which in time may be overcome.

### 4.1.8. *Growth by MBE on LiAlO₂ and LiGaO₂ substrates*

As mentioned earlier, the (111) surface of lithium gallate lattice closely matches GaN. Even though, there are difficulties with surface preparation and stability of the material in the presence of H and high temperatures, not to mention the poor thermal conductivity, efforts have been expended to take advantage of this material for substrates. Promising results from material growths on lithium gallate including aluminum, gallium and indium-nitride alloys by MBE have recently been reported.[195] Second order reconstructions of the GaN surface on LGO, indicating smooth, well-ordered surfaces during growth, accompanied the report. From a structural point of view, the X-ray reciprocal space mapping indicated higher structural quality than GaN on SiC and sapphire while TEM data indicate $6 \times 10^8$ cm$^{-2}$ threading dislocation density.[195] By improving the surface finish of LGO via polishing, the authors noted linear, as opposed to spiral, step flow growth of AlGaN/GaN during MBE growth. In this mode of growth, mono-layer terraces are observed via AFM with 30 nm of $Al_{0.25}Ga_{0.75}N$ on a 2.4 $\mu$m thick GaN layer, in contrary to the spiral step flow growth observed previously for MBE grown GaN samples on MOCVD buffer layers.[196] Additionally, the dislocation density derived from the small pits in the AFM image result in a threading dislocation density of $4$–$5 \times 10^8$ cm$^{-3}$, which is in close agreement with the TEM data presented above. The FWHM of omega-2theta X-ray diffraction rocking curves as low as 85 arcseconds for a 1 $\mu$m thick film were measured with a $\langle 0004 \rangle$ reflection. However, one must be weary of this reflection as it is sensitive mostly to screw dislocations which are miniscule in density compared to edge dislocations. Asymmetric reflections, such as [102] reflection, would be a better gauge of the quality of the film.

### 4.1.9. *Growth by MBE on GaN templates*

In the absence of large area GaN substrates in the conventional sense, the basic research of the homoepitaxial GaN growth has been on small templates GaN platelets grown from the liquid phase under high hydrostatic pressure and at high temperatures. Growth was performed both by MOCVD[197–199] and MBE[200,201] on the aforementioned GaN platelets. In addition, GaN epitaxial layers have been grown by MBE on MOCVD[35] and HVPE[36] prepared templates, and by MOCVD on HVPE.[202] More on this topic will be given shortly. Such studies serve to establish benchmark values for the optoelectronic properties of thin GaN films on native substrates. *p*- and *n*-type doping were achieved and a *p-n* junction was realized by MOCVD on GaN crystals grown from Ga solution. The deposition temperature was 1000–1050°C. The width of rocking curves for epitaxial layers was practically the same as that of the substrate. The photoluminescent spectra were dominant by exciton emission at low temperatures, pointing to high quality of the epitaxial layers.

There are, in general, two possible approaches for suitable substrates for homoepitaxial growth: The desirable one is to grow a large bulk crystal, cut it and polish slices. This method is widely used for conventional semiconductors. It cannot be applied easily to GaN bulk growth, since only small pieces of GaN can be grown only at high temperatures and very high pressures (tens of kbar), taking into account that GaN begins to decompose at 800°C. The second possibility that has been explored is the growth of thick GaN films on foreign substrates (sapphire, SiC or Si). It is well known that the quality of the heteroepitaxial film becomes better as the thickness gets larger. Very popular epitaxial deposition methods, such as MOCVD and MBE, have very slow growth rates, a few of microns per hour at best, and cannot be used for the growth of thick films. As discussed earlier, the inorganic CVD method has high growth rates (up to 100 $\mu$m/h or larger). Thus, it is argued that the best way to grow homoepitaxial GaN film is to use a two-stage growth, a thick GaN substrate grown by inorganic CVD in one-two hours followed by the device layer grown by MOCVD or by MBE.

Early efforts of homoepitaxy relied on growth by MBE on buffer layers grown by MOCVD or HVPE. Due to smooth surfaces obtained by these two vapor techniques and avoidance of complications brought about by heteroepitaxy, atomically smooth layer could be grown by MBE. This permitted delineation of regimes leading to smooth and rough surfaces, namely Ga-rich and N-rich growth conditions. We should mention that prior to GaN, growth by MBE was always conducted under slightly group V-rich conditions.

The schematic diagram shown in Fig. 25 delineates the Ga and N-rich growth regions for GaN samples on GaN templates grown by MBE on MOCVD buffer layers, as a function of substrate growth temperature.[129] The temperatures involved are higher than the re-evaporation of Ga from the surface and higher temperature end of the data require extremely high Ga fluxes to maintain smooth surfaces. The figure makes the point that there are two regions, N-rich and Ga-rich (Ga-droplet) separated by a transition region. The N-region is characterized by rough surfaces and Ga-rich

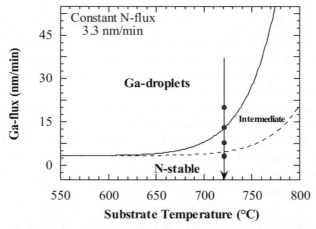

**Fig. 25.** Growth diagram defining the Ga-flux conditions and substrate temperatures for the Ga-droplet, intermediate, and N-stable growth regimes at a constant N flux of 2.8 nm/minutes. The circles define the conditions used for the samples grown in this study. After Ref. 130. Courtesy of J. Speck.

region represents smooth surfaces. Within the Ga-droplet regime, excess Ga accumulates on the surface during growth in the form of large droplets (5–20 $\mu$m in diameter). In the intermediate regime, excess Ga forms an adlayer of Ga on the surface without the formation of Ga-droplets. Films grown in the N-stable regime are Ga deficient and the growth rate is limited by the available Ga flux. The boundary line between the Ga-droplet and intermediate regimes was measured previously and found to have an Arrhenius dependence with activation energy of 2.8 eV corresponding to the activation energy for Ga desorption from liquid Ga.[129,203] The lines shown are only estimation for illustrative purposes.

The homoepitaxial films grown by the MBE method showed a superior quality of PL and structural characteristics, in terms of dislocation densities, with respect to heteroepitaxial layer grown on sapphire.[41] The dislocation density was estimated to be $10^7$–$10^8$ cm$^{-2}$ for homoepitaxial film as compared to $10^{10}$–$10^{11}$ cm$^{-2}$ for the films grown on sapphire. We should however mention that SMD like defects originating at the GaN substrate interface can in general propagate to the surface. When epitaxial GaN is attempted on this surface, these defects would propagate again through the layer.[204,205]

As mentioned previously, homoepitaxy also includes growth on GaN templates or on epitaxial layers grown by HVPE of MOCVD. In those cases, we should mention that, in addition to misfit dislocations, other threading defects originating at the GaN substrate interface, such as edge dislocations, screw dislocations and mixed dislocations, propagate to the surface unless they run into one another and loop, and the burgers vectors cancel, leading to dislocation looping.[206]

GaN and related heterostructures have been grown by MBE on GaN/sapphire templates prepared by MOCVD.[129] In this particular work, the samples grown under nitrogen rich conditions are not conductive, whereas the others are, and had room temperature mobilities approaching 1200 cm$^2$V$^{-1}$s$^{-1}$. The mobility increases as the Ga flux increases and turns over above a certain Group III/V ratio where the Ga-droplets begin to form. The regions where the Ga-droplets form are of lower quality and degrade the overall mobility measured. The optimum properties are reached when this Ga-adlayer coverage is maximized without the formation of Ga-droplets (i.e. at the highest Ga flux within the intermediate regime).

While on this topic, let us discuss the growth of heterojunctions grown on templates in turn prepared by HVPE. The specific example is the much-studied AlGaN/GaN two-dimensional electron gases (2DEGs) system. Samples over a broad range of electron densities, ranging from $n_s = 6.9 \times 10^{11}$ to $1.1 \times 10^{13}$ cm$^{-2}$, were grown with the best mobility of 53,300 cm$^2$/Vs at a density of $2.8 \times 10^{12}$ cm$^{-2}$ and temperature of $T = 4.2$ K.[36] Magnetotransport studies on these samples display exceptionally clean signatures of the Quantum Hall effect. The investigation of the dependence of the 2DEG mobility on the carrier concentration suggests that the low-temperature mobility in these AlGaN/GaN heterostructures is currently limited by the interplay between the charged dislocation scattering and the interface roughness. The typical MBE overgrowth consisted of a 0.5 $\mu$m undoped GaN buffer layer capped by approximately 30 nm of Al$_{0.09}$Ga$_{0.91}$N. Changing the thickness of the Al$_{0.09}$Ga$_{0.91}$N caused a variation of the sheet carrier concentration (formed by electrons released from defect centers on or near the surface of Al$_{0.09}$Ga$_{0.91}$N), which is the screening charge in response to polarization charges present at the interface. Recently, these mobility figures were extended to about 73,000 cm$^2$V$^{-1}$s$^{-1}$ and beyond by MBE growth on HVPE GaN layers using a RF N-source, and also on GaN bulk platelets using ammonia as the reactive nitrogen source.

Nominally undoped 1.5 $\mu$m layer was recently grown by MBE on Ga face of freestanding GaN template.[207] The template in turn was grown by HVPE on a $c$-plane sapphire substrate and separated from the substrate by laser lift-off. Before the overgrowth, the GaN template was mechanically polished, dry-etched and finally etched in molten KOH. A 30 nm thick undoped Al$_x$Ga$_{1-x}$N cap layer with $x = 12\%$ has been deposited on top of the 1.5 $\mu$m GaN layer. Part of the $10 \times 10$ mm surface of the sample was mechanically shuttered during the MBE process, so that both GaN substrate and MBE-overgrown layer could be studied under the same experimental conditions.

Steady-state PL was excited with a He-Cd laser (325 nm), dispersed by a 1200 rules/mm grating in a 0.5 m monochromator and detected by a photo multiplier tube. The best resolution of the PL set-up was about 0.2 meV, the photon energy was calibrated with a mercury lamp accounting for the refraction index of air (1.0003). The temperature was varied from 15 to 300 K in the closed cycle cryostat. The excitation density in the range of $10^{-4}$ to 100 W/cm$^2$ was obtained by using unfocused (2 mm diameter) and focused (0.1 mm diameter) laser beam attenuated with neutral density filters.

Structural properties of the GaN template and overgrown MBE epilayer were characterized by high-resolution X-ray diffraction and atomic force microscopy (AFM). The full widths at half maximum (FWHM) of the omega scan (rocking curve) for the [002], [102] and [104] directions are 53, 145 and 54 arcseconds, respectively, for the MBE-grown area and 52, 137 and 42 arcseconds, respectively, for the GaN substrate. The similar characteristics of the substrate alone and the substrate with MBE overgrowth indicate similar crystal quality of GaN both in the substrate and overgrown layer. A comprehensive characterization of the quality of the GaN templates has been reported elsewhere[208] and will not be repeated here. Suffice it to say that the template quality is unmatched in terms of transport and optical properties, and extended and point defects. The surfaces of both substrate and MBE-overgrown layer are very smooth, with the root-mean-square roughness of about 0.2 nm in the $2 \times 2$ $\mu$m AFM images.

Excitonic PL spectra of the substrate and MBE-overgrown area are shown in Fig. 26. The main features of the PL spectra are similar, yet a few distinctions can be noticed. A peak at 3.4673 eV is attributed to the $A$-exciton bound to unidentified shallow acceptor $(A^0, X_A^{n=1})$.[209] The most intense peaks at 3.4720 and 3.4728 eV are attributed to the $A$-exciton bound to two neutral shallow donors, $D_1$ and $D_2$ : $(D_1^0, X_A^{n=1})$ and $(D_2^0, X_A^{n=1})$. The FWHM of these peaks is 0.7 meV at 15 K and their positions are the same with an accuracy of 0.2 meV for the substrate and different points of the overgrown layer. Ratio between the intensities of the $(D_1^0, X_A^{n=1})$

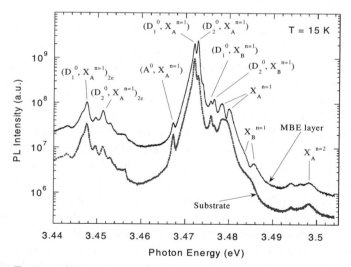

**Fig. 26.** Excitonic PL spectrum for GaN in the area of substrate and in the MBE-overgrown part. Excitation density is 100 W/cm$^2$.

and $(D_2^0, X_A^{n=1})$ peaks is different in the substrate and overgrown layer (Fig. 26). This feature helped us to identify four other peaks related to the excitons bound to the same shallow donors. Namely, the peaks at 3.4758 and 3.4766 eV, having the same intensity ratio and energy separation as the $(D_1^0, X_A^{n=1})$ and $(D_2^0, X_A^{n=1})$ peaks are attributed to the $B$ exciton bound to the $D_1$ and $D_2$ donors: $(D_1^0, X_B^{n=1})$ and $(D_2^0, X_B^{n=1})$, in agreement with Refs. 209 and 210. Note that in the previous works only a single peak was observed in this region $(3.476 \pm 0.002$ eV) of PL spectrum.[211-213] Our data rule out the alternative assignment relating this peak to the $A$-exciton bound to an ionized donor $(D^+, X_A^{n=1})$.[211,213] Indeed, the energy separation between the $(D_1^0, X_B^{n=1})$ and $(D_2^0, X_B^{n=1})$ peaks observed in our study is only 0.8 meV, whereas the peak separation of the $(D_1^+, X_A^{n=1})$ and $(D_2^+, X_A^{n=1})$ excitons should be equal to the difference between the ionization energies of the donors which is about 4 meV as will be estimated below.

Two other peaks, also related to donor bound excitons (DBE), are observed at 3.4475 and 3.4512 eV and attributed to the so-called two-electron transitions: $(D_1^0, X_A^{n=1})_{2e}$ and $(D_2^0, X_A^{n=1})_{2e}$. This type of the DBE recombination, first observed in GaP,[214] involves radiative recombination of one electron with a hole leaving the neutral donor with second electron in an excited state. One of these peaks (3.451 eV) has previously been reported[209,215] and is attributed to Si.[215] In the effective-mass approximation, the donor excitation energy from the ground to the $n = 2$ state equals $3/4$ of the donor binding energy $E_D$. Consequently, from the energy separation between the principal DBE line and the associated two-electron satellite we can find the binding energies of two shallow donors: $E_{D1} = 4/3 \times 24.5 = 32.6$ meV, $E_{D2} = 4/3 \times 21.6 = 28.8$ meV. It appears reasonable that the peak at 3.443 eV is the two-electron transition related to the second excited state $(n = 3)$ of the $D_1$ donor since its separation from the $(D_1^0, X_A^{n=1})$ peak is $8/9 \times 32.6 = 29.0$ meV.

Free excitons (FE) related to the $A$ and $B$ valence bands ($X_A^{n=1}$ and $X_B^{n=1}$) are identified at about 3.479 and 3.484 eV, respectively, in the PL spectrum of the GaN substrate (Fig. 26). The $X_A^{n=1}$ peak has an asymmetrical shape, and the $X_B^{n=1}$ peak is seen as a shoulder. In the MBE-overgrown epilayer these peaks are split: a doublet with maxima at 3.4782 and 3.4799 eV and a doublet with the high energy maximum at 3.4856 eV (the low-energy maximum is seen as a shoulder in Fig. 26). With decreasing excitation intensity the double peaks merge together and the shape of the $X_A^{n=1}$ peak becomes similar to that in the GaN substrate. Note that in the substrate, the shape and position of the $X_A^{n=1}$ peak are identical in the excitation density range from 1 to 100 W/cm$^2$. Similiar splitting has been observed earlier and attributed to a manifestation of the polariton branches.

From the separation between the FE and DBE peaks we find the binding energies of the DBEs related to the $D_1$ and $D_2$ donors as 7.0 and 6.2 meV, respectively. According to empirical Haynes rule, the binding energy of the DBE is proportional to the binding energy of the corresponding donor. The proportionality constant ($\alpha$) for the $D_1$ and $D_2$ donors is found to be 0.215 that is close to the result of Meyer ($\alpha = 0.2 \pm 0.01$).[211] In addition to the dominant DBE doublet, we observed a weak peak at 3.474 eV (Fig. 26). Assuming that this peak is the DBE ($D_3^0, X_A^{n=1}$) with the localization energy of 5.0 meV, we can estimate the binding energy of the corresponding donor, $D_3$, to be $E_{D3} = 5/\alpha = 23.3$ meV. Furthermore, a weak peak at about 3.456 eV can be attributed to the two-electron satellite of the ($D_3^0, X_A^{n=1}$) peak. The origin of the weak peaks at 3.4496 and 3.4529 eV remains unclear. These peaks may be two-electron satellites of some unresolved DBE lines or excitons bound to deeper acceptors.[217]

With increasing temperature, all peaks related to the bound excitons quench, while the quenching of the free excitons is negligible up to 50 K (Fig. 27). This result supports the above assignments of the peaks. Excitons

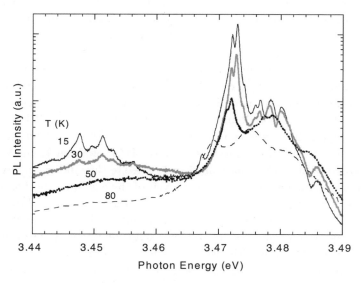

**Fig. 27.**   PL spectrum of the MBE-overgrown GaN layer at different temperatures.

bound to the $D_1$ donor quench faster than the excitons bound to the $D_2$ donor, which is consistent with their binding energies.

In the range from 3.49 to 3.50 eV, we resolved three peaks (Fig. 26), which are often observed in this region, and sometimes denoted as $X_1$, $X_2$ and $X_3$.[209,213,217,218] It is believed that they all are related to the free $A$-exciton and the highest energy one ($X_3$) is clearly established to be the $n = 2$ excited state of the $A$-exciton ($X_A^{n=2}$).[213,217,218] From the energy positions of the $X_A^{n=1}$ and $X_A^{n=2}$ peaks (3.479 and 3.4983 eV, respectively), we can find the $A$-exciton binding energy in the hydrogen model as 25.7 meV and the band gap $E_g = 3.5047$ eV. Note that the observed positions of the PL peaks from the substrate and the overgrown epilayer are the same with accuracy of 0.2 meV. As compared to the PL spectra from the strain-free GaN,[209] the PL peaks observed in our sample are shifted to higher energy by about 1 meV, indicating small residual strain in the GaN template.

The PL spectra of the substrate part and MBE-overgrown area are very similar in wide range of the photon energies. Besides multiple sharp peaks related to the above-discussed excitonic transitions and their less resolved LO phonon replicas at energy separations that are multiples of 92 meV, the spectra contain much weaker shallow donor-acceptor pair (DAP) emission with the main peak at about 3.26 eV, which is related to transitions from a shallow donor to unidentified shallow acceptor. With increasing excitation power, maximum of the shallow DAP band shifted from 3.255 to 3.260 eV, which can be related to saturation of emission from distant pairs having longer lifetime. With increasing temperature from 15 to 60 K, the shallow DAP band quenches giving the way to transitions from the conduction band to the shallow acceptor ($e$-$A$ transitions) with the main maximum at about 3.282 eV (Fig. 28). From the separation between the DAP and $e$-$A$ peaks in the low-excitation limit one can estimate the ionization energy $E_D$ of the shallow donor predominantly participating in the DAP transitions. Accounting for the average kinetic energy ($kT/2$) of free electrons (1.5 meV at 35 K) and the Coulomb interaction in the DAP of about 7.5 meV,[219] we obtained $E_D = 32$ meV. This value is very close to the binding energy of the $D_1$ donor obtained above from the analysis of the excitonic spectrum. Note that position of the DAP band is the same in the substrate and overgrown layer within an accuracy of about 0.5 meV. This result may be explained by the fact that at low temperature and low excitation intensity the occupation of the deeper $D_1$ donor with electrons is more favorable.

It is commonly believed that Si and O are the dominant shallow donors in undoped GaN grown by different techniques.[210,215] The binding energy

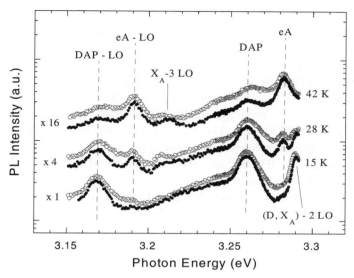

**Fig. 28.** PL spectrum of the shallow DAP band at different temperatures in the area of substrate (open circles) and in the MBE-overgrown part (filled circles). Excitation density is 0.1 W/cm².

of the Si donor of about 29 meV has been extrapolated from the magnetic field dependence of the infrared absorption in the Si-doped GaN.[220] Mireles and Ulloa calculated the value of 30.4 meV for Si in GaN using an effective mass theory approach that accounts for the central cell corrections.[221] These values are close to the binding energy of our $D_2$ donor (28.8 meV) and thus we assigned it to Si. Note that a shallower binding energy of Si donors (22 meV) has been obtained from PL study of Si-doped GaN.[223] Such low value seems questionable for the substitutional Si donor and the defect observed in Ref. 222 may be related to some complex or extended defect.[221] Furthermore, we propose that our $D_3$ donor is the same defect observed in Si-doped GaN by Jayapalan *et al.*[222] Since oxygen is expected to be somewhat deeper donor than silicon,[215] we attribute the $D_1$ donor (32.6 meV) to oxygen. This result is consistent with the calculated binding energy of oxygen donor (31.4 meV).[221]

## 4.2. *Growth by HVPE*

Molnar *et al.*[25,223] were able to demonstrate that HVPE is capable of growing thick GaN films on sapphire substrates with greatly improved structural and electronic properties, as compared to heteroepitaxial GaN films grown by MOCVD or MBE on sapphire. To achieve such properties, the sapphire substrates were treated *in situ* with GaCl or ZnO.[224] GaN films up to sufficient thicknesses to be self-supporting templates when removed from sapphire. Another group headed by Dr. Y.-J. Park at Samsung Advanced Institute of Technology has been producing free-standing templates with this technique as well. These layers and templates exhibit the lowest impurity concentration with the highest electron mobilities of any GaN so far. The deposition temperature is generally about 1050°C with growth rates of 15 $\mu$m per hour.

The variants of HVPE have also been explored and used for growth of GaN and its ternaries with AlN. Growth by a modified VPE process dubbed the sublimation sandwich method (SSM) was reported by Wetzel *et al.*[225] and Fischer *et al.*[226] (Fig. 29). Initially, the GaN films were grown from metallic Ga and ammonia on (0001) 6H–SiC, using a modification of the sandwich method described previously by Vodakov *et al.*[227] In this approach, the quartz reactor contains a Ga cell for each substrate for multi-wafer processing with one ammonia stream only. The gap between the substrate and the Ga source is typically about 5 mm. The ammonia flow rate through the gap is very high 25 to 50 l/m at atmospheric pressure. Under these conditions, there is an effective mass transport of Ga vapor and nitrogen to the surface of the substrates. At growth temperatures between 1170 and 1270°C, GaN layers were obtained at growth rates up to 0.3 mm per hour.

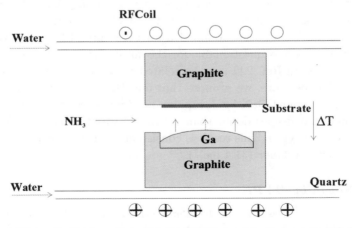

**Fig. 29.** Schematic diagram of a proximity HVPE vessel for the growth of GaN at very high growth rates, approaching 0.3 mm per hour.

### 4.2.1. *TEM investigation*

Cross-sectional and plan-view TEM methods have been employed to characterize GaN layers grown by HVPE.[25,228] The HVPE layers investigated were grown on sapphire coated with ZnO in a chloride-transport HVPE vertical reactor.[229] The density of misfit dislocations at the layer-substrate interface was investigated by High Resolution Electron Microscopy (HREM). This exercise led to a misfit dislocation density at the layer/substrate interface of about $2 \times 10^{13}$ cm$^{-2}$.

Conventional TEM studies indicate that the dominant defects present to be the threading dislocations. Bright field TEM images, recorded under multi-beam conditions in order to image all dislocations with different Burgers vectors, were used to estimate the density of these dislocations with resultant figures to be of the order of about $10^{10}$ cm$^{-2}$ near the layer/substrate interface. The density of threading dislocations gradually decreased away from the interface and for the 55 $\mu$m thick layer it reached a value of about $10^{8}$ cm$^{-2}$ at the surface. The density of dislocations determined form plan-view samples was in very good agreement with those extrapolated from cross-sectional samples. The threading dislocation densities are plotted as a function of distance from the interface in Fig. 30. A gradual

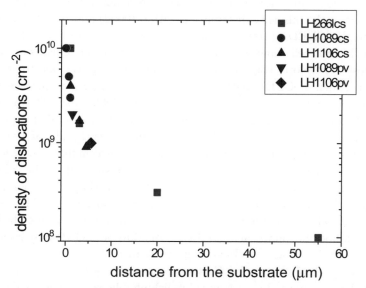

**Fig. 30.** Density of threading dislocations vs. distance from the interface in HVPE films. Courtesy of J. Jasinski and Z. Liliental-Weber.

decrease of density of these dislocations with increase of the distance from the substrate shown on this plot indicates a gradual improvement of layer quality with thickness.

The relative numbers of different types of threading dislocations (edge, screw and mixed) were investigated and two types of dark field images — with (0002) and (−2110)-type reflections were recorded. In the first type of images, only the screw and mixed dislocations are visible whereas on the second one, only the edge and mixed dislocations are observed. The results led to the conclusion that no specific type of threading dislocation dominates and that all three types of threading dislocations (edge, screw and mixed) are present in HVPE GaN layers in comparable densities.

A freestanding GaN template was also investigated by TEM. The sample studied[230] was obtained from Samsung and was grown by HVPE on sapphire to a thickness of 300 $\mu$m and separated from the sapphire substrate by laser-induced lift off.[231] The GaN layer was then mechanically polished and dry etched on the Ga-face to obtain a smooth nearly epi-ready surface, whereas the N-face was only mechanically polished. Three specimens were prepared for TEM studies: one cross-sectional and two plane-view (from both template sides) samples. The polarity on the two sides of the GaN template was determined by the well-established method of convergent beam electron diffraction (CBED). Since GaN is non-centro-symmetric, the difference in the intensity distribution within (0002) and (000$\bar{2}$) diffraction discs in the CBED pattern can be attributed to Ga and N distributions within the unit cell. However, this intensity distribution depends on sample thickness. Therefore, correct use of this method requires one to compare the experimental CBED patterns with patterns simulated for the thickness indicated by the pattern in the central, (0000) disc. To apply this method to the investigation of the sample indicated, several (for different thicknesses) [1$\bar{1}$00] zone axis CBED patterns on each side of the cross-sectional specimen were recorded and compared with simulated patterns.[230] The CBED patterns obtained on the side previously next to the substrate indicate that it is of [000$\bar{1}$], N-polarity which means that a long bond along the $c$-axis is from N to Ga. The polarity determination by CBED is consistent with chemical etching experiments in which the N-face etched very rapidly in hot phosphoric acid ($H_3PO_4$). In addition, Schottky barriers fabricated on this surface exhibited a much reduced Schottky barrier height (0.75 eV vs. 1.27 eV on the Ga-face), only after some 30–40 $\mu$m of the material was removed by mechanical polishing followed by chemical etching to remove the damage caused by the first mechanical polish.[232]

The density of threading dislocations determined from the plan-view sample was estimated to be about $4 \pm 1 \times 10^7$ cm$^{-2}$. These threading dislocations were observed also in cross-section. Few of them are clearly visible in bright field images as shown in Fig. 31. The density of these dislocations determined from cross-section was found to be about $3 \pm 1 \times 10^7$ cm$^{-2}$, which is in good agreement, within experimental error, with that obtained from the plan-view sample. For comparison, a density of about $1 \times 10^7$ cm$^{-2}$ was obtained by etching the N-face in $H_3PO_4$ for 15 seconds at 160°C followed by counting the etch pits on several Atomic Force Microscopy (AFM) images, details of which are discussed in the section entitled "Structural analysis by etching". Most of these threading dislocations are of mixed Burger's vector because they are visible in bright field images with $g$-vector parallel and perpendicular to the $c$-axis (Fig. 31). However, one needs to be careful with such a conclusion because of the very low statistics (very few dislocations observed within the electron transparent area).

TEM studies of a plan-view specimen prepared for the Ga-face side revealed a nearly defect-free surface. Very few dislocations were found on

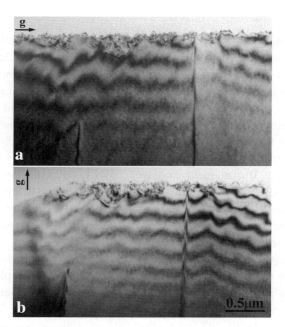

**Fig. 31.** Bright field TEM micrographs of a cross-section sample near the N-face side for the $g$-vectors perpendicular (a) and parallel (b) to the $c$-axis. Note that both dislocations are visible in both images. Courtesy of J. Jasinski and Z. Liliental-Weber.

this surface. Based on the plan-view study, the density of these dislocations was estimated to be much less than $1 \times 10^7$ cm$^{-2}$, however due to the very low statistics there is a relatively large uncertainty for this estimation. In cross-sectional study one could not find any threading dislocation within the electron transparent area and based on this information one can estimate that density of these dislocations is less than about $0.5 \times 10^6$ cm$^{-2}$. Hot $H_3PO_4$ acid defect revealing process followed by several AFM images up to 50 $\mu$m $\times$ 50 $\mu$m indicated a dislocation count of about $5 \times 10^5$ cm$^{-2}$.

The significantly lower dislocation density of the G-face side with respect to that near the N-face was most likely due to dislocation interaction within the layer. This observation is remarkable in the sense that dislocation reduction through dislocation interaction has not been very successful in GaN. A very low density of threading dislocations in the template compares very favorably to values measured in standard HVPE GaN layers[229,233] which are several orders of magnitude larger, and indicate a very high structural quality of the free-standing GaN template under investigation.

### 4.2.2. *X-ray diffraction*

High-resolution X-ray rocking curves were measured by a Philips X'Pert MRD system equipped with a four-crystal Ge (220) monochromator and Cu K$_{\alpha 1}$ line of X-ray source. The instrumental resolution was better than 10 arcseconds in the diffraction geometry in effect. The X-ray diffraction rocking curves from both the Ga and N-faces were examined for different X-ray beam slit widths, from 2 mm down to 0.02 mm, which correspond proportionally to the spot size on the sample due to the highly collimated nature of the beam.

Among some 10 wafers investigated, HVPE epitaxial layers exhibited (0002) peaks with FWHM values ranging from 5.8 to 9.3 arcminutes. The (10–14) asymmetric peak FWHM values were narrower and ranged from 3.9 to 5.2. Sharper asymmetric peaks are indicative of reduced dislocations such as the pure edge varieties.

A crystallographic analysis by high-resolution X-ray diffraction rocking curves (omega scans) with different slit widths of a GaN template[234] shows a very narrow FWHM on the Ga-face of the freestanding template, down to 69 arcseconds for (0002) reflection (at a slit width of 20 $\mu$m), and 103 arcseconds for (10–14) (at a slit width of 100 $\mu$m as the signal gets to be very small for smaller slit widths, and much longer data collection times are required). The FWHM for the N-face is 160 arcseconds for

(0002) direction (at a slit width of 20 $\mu$m) and 140 arcseconds for (10–14) direction (at a slit width of 100 $\mu$m). The superior quality of the Ga-face over that of the N-face agrees well with the large deviation of extended defect density as determined from the etching experiments. As mentioned above, lattice distortions from dislocations with screw component would contribute to the broadening of the (0002) peak, and the lattice distortion from the edge dislocations contributes to the width of the asymmetric peak.[235,236] Comparing with the reported X-ray data,[237,238] of HVPE GaN, the FWHMs of which are typically in the range of 5–10 arcminutes for the (0002) peak and 4–5 arcminutes for the (10–14) peak, the density of both type of dislocations are dramatically reduced in the free-standing GaN template.

With the increase of the slit width, not only the FWHM is increased, but also a non-gaussian multi-peak feature emerges. Specifically, when the slit size is increased from 0.02 mm to 2 mm, the FWHM of the (0002) peak on the Ga-face increases from 69 arcseconds to 20.6 arcminutes; when the slit size is increased from 0.1 mm to 2 mm, the FWHM of the (10–14) peak increased from 103 arcseconds to 24 arcminutes. A more recent Samsung GaN template exhibited much better X-ray diffraction characteristics. The FWHM of [002], [102], [104] peaks were 53, 145, and 54 arcseconds. A source beam slit width of 20 $\mu$m used for the [002] diffraction and 50 $\mu$m for the [104] and [102] diffractions, as the signal intensity for these asymmetric peaks is smaller.

This sort of broadening has been attributed to the tilt and twist[236] that result in sub-millimeter scale mosaic spread and cause the observed dependence on the slit size. This seem at first not plausible as the X-ray diffraction system uses a highly collimated Cu $K_{\alpha 1}$ point source. Instead, the low defect concentration observed by TEM and defect delineating etches in these templates point to another source, namely bowing.[239] This is despite the fact that GaN was removed from sapphire and a casual observer could indeed make the assumption that it is relaxed with no cause for bowing. Though, the exact genesis is not known at this time, bowing could also be caused by the surface mechanical polish, which affects both the Ga- and N-faces. By assuming the FWHM at beam width of 0.02 mm to be the intrinsic broadening of GaN, an upper limit of the bowing radius of 1.20 m or perhaps slightly larger was estimated. It has been reported[240] that a bowing radius of about 0.8 m was found in the HVPE grown GaN thick films ($\sim$ 275 $\mu$m) before separation from sapphire substrate, and a bowing radius of about 4 m after separation. As mentioned above, the sample underwent a mechanical polishing process, which could be the reason for the relatively large bowing.

### 4.2.3. *Impurities and SIMS*

Impurities may be transported on the surface of the substrate, with their genesis in the chemicals used, and/or introduced during the deposition process. The latter may be due to the environment in which the deposition takes place and impurities present in the source material used. GaN is no exception and what exacerbates the situation is that there are many structural and point defects present in addition to impurities. The lack of extremely high crystalline quality takes away the intrinsic resistance against incorporation of impurities. Among the chemical impurities elements such as O, Si and C are among the ones that have received a good deal of attention. The former two in GaN form shallow donors while the latter is a deep level with acceptor tendencies.

While oxygen and silicon have received a great deal of attention as the main source of unintentional impurities, the role of carbon is not well understood. This is in spite of the fact that there is a massive amount of carbon present in MOCVD. Computations predict carbon to be a deep acceptor — but its behavior could be significantly affected by complexing with other impurities or defects. This could come to light when MOCVD buffer layers are employed and is relevant in cases when this method is used for the production of electronic devices. In addition to being substitutional and well behaved impurities, these species can also form complexes with native defects, complicating the analyses. Since the conductivity is of paramount importance for devices, it is imperative that impurity incorporation in GaN is examined.

Unwanted impurities impede efforts to attain high performance electronic and to some extent optoelectronic device operation. Secondary-ion-mass-spectroscopy (SIMS) measurements demonstrate that it is difficult to reduce common impurities such as C, Si, and O, much below the $10^{17}$ cm$^{-3}$ level in thick GaN layers, $10^{16}$ cm$^{-3}$ in free-standing templates, and somewhat worse in thin epitaxial layers grown directly on lattice mismatched substrates. The picture is somewhat better in films grown by MBE on MOCVD and HVPE templates. The SIMS data can shed light on whether the measured donor and acceptor concentrations, by Hall, are due to impurities or native defects or both.

As mentioned above, chemical impurities can best be probed with SIMS. The technique is refined and calibrated to the point where chemical impurity concentrations down to about $10^{15}$ cm$^{-3}$ can be quantitatively determined. Since there are issues surrounding GaN being doped with residual oxygen in substitutional or interstitials sites next to a vacancy for example, SIMS is a really important player. In Figs. 32 and 33, SIMS profiles in an HVPE layer on sapphire substrate and a freestanding GaN template are shown. For the latter, profiles from both the Ga- and N-faces are shown. Specific impurities probed were nominally O, C, and Si. In the HVPE layer, both O and Si concentrations drop rapidly away from the surface — mainly due in part to the artifact of the technique and in part due to condensates on the surface, down to about $10^{17}$ cm$^{-3}$ for Si and high $10^{16}$ cm$^{-3}$ for O. The Ga-face profile in the Samsung template indicates levels below mid $10^{16}$ cm$^{-3}$ for all three of the impurities. The picture is different for the N-side, however, as this side was juxtaposed to the substrate during growth and was mechanically polished after laser separation. The impurity concentration is some 1–3 orders of magnitude higher than the case for the Ga-face. Particularly, the concentration of O and C is high,

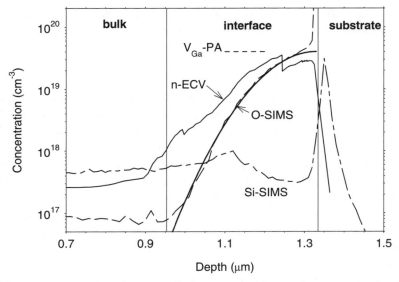

**Fig. 32.** SIMS profiles in a HVPE GaN layer on sapphire substrate showing the depth distributions of O and Si concentrations. The average $n$ and true $n$ (from Hall measurements) in HVPE GaN on Al$_2$O$_3$ are also shown. Courtesy of D. C. Look.

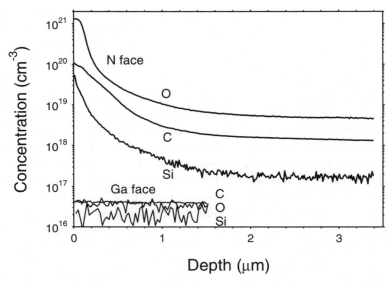

**Fig. 33.** SIMS profiles in a HVPE GaN template showing the depth distributions of O, C, and Si concentrations from both N- and Ga-faces with Ga-face figures being below mid $10^{16}$ cm$^{-3}$. Courtesy of D. C. Look.

albeit some drop occurs deeper in the film. Data suggest that impurity incorporation is dependent on the structural quality of the film.

### 4.2.4. *Ionized impurities*

Ionized charge concentration can best be determined by Hall measurements. To this end, both the epitaxial layers and a template grown by HVPE, unintentionally doped, were characterized by variable temperature Hall measurements, in some cases in the form of differential Hall measurements. These measurements were coupled with calculations of the temperature dependent mobility and SIMS measurements in an attempt to determine the acceptor and donor concentrations, and likely causes for those impurities. A highly conductive layer at the original sapphire-GaN interface is commonly observed in HVPE layers,[241] perhaps even other forms of growth. This manifests itself as a degenerate layer with relatively high carrier concentration (about $3 \times 10^{17}$ cm$^{-2}$) being maintained to temperatures as low as 15 K. In the absence of this layer, the net electron concentration practically would approach nil due to relatively large donor binding energies in GaN. This layer can be dealt with by either assuming a two-layer conduction in the model calculations or by removing the suspect region by etching. More accurate results can be obtained by extending the two-layer approach to successive Hall measurement at each step of certain amount of the layer removal, as will be discussed below. Since the Samsung template was a freestanding one, some 30 $\mu$m material was removed from the N-face, which represents the original epi-substrate interface. The additional benefit of such etching is that the damage caused during mechanical polishing is removed.

In HVPE GaN grown on $Al_2O_3$, after a few microns of growth, the quality improves; however, the interface region is known to be highly defective and conductive, and will thus have a strong influence on the measured electrical properties.[241] In the last few years, several schemes have been proposed to study the bulk electrical properties independently of the interface properties. The first model[241] treated the interface region as a well-defined temperature-independent degenerate gas. A bi-layer Hall analysis can then be used to determine the mobility and electron concentration of the structure away from the interface indicating room temperature mobilities of about 1000 cm$^2$/Vs, and electron concentrations of $7 \times 10^{17}$ cm$^{-3}$. The room temperature electron mobility in a more recent HVPE layer may be as high as 1400 cm$^2$/Vs near the surface. To gain a better insight, the electro-chemical capacitance-voltage method was employed to study the interfacial region.[242] In this method the charge depth profile near the surface is successively performed while the epilayer is progressively etched away. When used in conjunction with SIMS analysis, it becomes evident that O, not Si, may be the dominant donor in the interfacial region.

For the case of the Samsung template, after removing some 30 $\mu$m material from the back N-face, high electron mobilities both at room temperature ($\mu = 1425$ cm$^2$/Vs at $T = 273$ K) and below were measured ($\mu = 7385$ cm$^2$/Vs at 48.2 K). The temperature dependencies of both the Hall mobility and electron concentration pointed to the absence of such a second conducting layer after etching. Therefore, the analysis was based on a single layer model with a single donor level.[234,243]

The temperature dependence of the calculated electron mobility, by Boltzmann Transport Equation (BTE) based theory,[244] was fitted to the experiments in concert with the temperature dependence of the electron concentration and charge neutrality. For accuracy, Raman modes were measured and the recent unscreened acoustic deformation potential used. The latter was obtained from samples with very high low temperature 2DEG mobilities where the acoustic phonon scattering is prominent.

Excellent fitting of the mobility in the Samsung template in the entire temperature range measured was obtained which resulted in ionized donor and total acceptor concentrations of $1.48 \times 10^{16}$ cm$^{-3}$ and $2.4 \times 10^{15}$ cm$^{-3}$, respectively. As mentioned previously, a singly ionized acceptor was assumed.

An accurate calculation of the mobility requires e.g. the numerical iterative solution of the Boltzmann transport equation (BTE).[244] For simplicity, one can use an alternative approach by considering the limiting effect of each scattering mechanism on mobility as if they are independent of each other. Assuming that scattering events are independent of each other, the total mobility can be approximated by the Matthiesen's rule in the form of $1/\mu = \sum(1/\mu_i)$. The mobility limited by ionized impurity scattering takes into account the screening by free carriers. The donor impurity concentration ($N_D$) and the acceptor concentration ($N_A$) are important fitting parameters.[245,246] Scattering by polar optical phonons is inelastic in nature so a quasi-analytical treatment has been derived[247] for reducing the complexity of calculations. The mobilities limited by acoustic-mode deformation potential scattering and by piezoelectric scattering can be expressed[248] with the acoustic deformation potential[249] and piezoelectric constant in wurtzitic GaN.

The mobility limited by polar optical phonon scattering, acoustic phonon scattering and piezoelectric effects are independent of impurity levels, so their temperature dependence is universal for GaN. Needless to say, the impurity scattering is very sensitive to ionized impurity concentration, especially in the low temperature range. This is because acceptors are completely ionized, while scattering related to lattice vibrations are frozen out at low temperatures. The mobility limited by neutral impurities ($N_D$–$N_A$–$n$) is also counted in the total mobility by using the Erginsoy's expression.[250] The details of this treatment for the Samsung material can be found in a paper by Yun *et al.*[234] The present treatment will report on the exact calculation by BTE with more refined and accurately measured Hall data.

An acoustic deformation potential of $E_{ds} = 8.54$ eV was used without screening. If screening is included, a value of 12 is better but is not applicable here. This is a controversial in that the low temperature mobility of the 2DEG GaN/AlGaN system is considered affected by screening. To fit the experimental and the calculated mobility, screened mobility is used which causes the deformation potential to go up to 12 eV. In the instant case, one is dealing with a three-dimensional system with low electron concentration. Therefore, the deformation potential without screening, which is about 8.54 eV, has been used in the simulations as a fixed parameter. The phonon energies used were measured in this sample as $A1(LO) = 737.0$ cm$^{-1}$; $A1(TO) = 532.5$ cm$^{-1}$; $E1(LO) = 745.0$ cm$^{-1}$; $E1(TO) = 558.5$ cm$^{-1}$. The high frequency dielectric constant was calculated using the Lyddane-Sachs-Teller relation expression $\varepsilon_0 = \varepsilon_\infty(\omega_{LO}/\omega_{TO})$. The Debye temperature was also calculated. The other parameters were kept the same as those reported previously. Experimental and calculated, through Rode's iterative BTE method, mobilities as a function of temperature are shown in Fig. 34. The calculations are shown for three different values of the acceptor concentrations for elucidating sensitivity of the mobility to the acceptor concentration. The best fit is $2.4 \times 10^{15}$ cm$^{-3}$ giving the best fit with an accuracy of $0.5 \times 10^{15}$ cm$^{-3}$. The acceptor concentration agrees well with the Ga vacancy concentration measured by Prof. Sarineen in the very same sample. This implies that acceptors in the GaN template are not of impurity origin, but rather of point defect origin.

The temperature dependent electron concentration was also fit with one donor model, using an acceptor concentration of $2.4 \times 10^{15}$ cm$^{-3}$, as

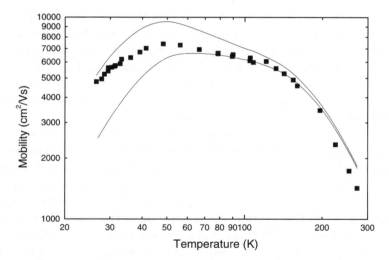

**Fig. 34.** Experimental mobility vs. temperature data (open circles) of the free-standing bulk GaN sample. Solid lines indicate the calculated values, using Rode's[244] iterative method of Boltzmann Transport Equation, for acceptor concentrations of $3.4 \times 10^{16}$ cm$^{-3}$, $2.4 \times 10^{16}$ cm$^{-3}$, and $1.4 \times 10^{16}$ cm$^{-3}$. The best fit is obtained for an acceptor concentration of $2.4 \times 10^{16}$ cm$^{-3}$. No adjustable fitting parameter was used.

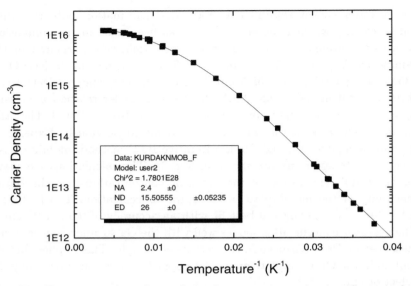

**Fig. 35.** Temperature dependence of the electron concentration in a Samsung free-standing GaN template. One donor model fits the data well with a donor activation energy of 26 meV.

shown in Fig. 35. The quality of the fit over the entire temperature range indicates that the one donor model is quite satisfactory. It is important that the measurements are made at very low temperatures to obtain a good and reliable fit. Previous efforts were frustrated because of highly conductive interfacial layer, which is nearly temperature independent. Since the N-face was etched sufficiently to remove the conductive layer, the measurements represented the actual picture in the bulk.

The acceptors in the $n$-type semiconductor under study are all ionized and negatively charged. The charge balance requires that the total positive charge which is just the ionized donor concentration be equal to the total negative charge which is the sum of the total acceptor concentration, since all are ionized, and the electron concentration. The charge neutrality must be satisfied at all temperatures. Consideration of the above led to a donor concentration of $1.55 \times 10^{16}$ cm$^{-3}$ and a donor ionization (binding) energy of 25.5 meV. The value of the donor activation energy is the screened value. The donor binding energy determined from the above fitting corresponds to the screened binding energy which is related to the binding energy in a dilute semiconductor through $E_D = E_D^0 - \alpha N_D^{1/3}$. Here $E_D^0$ is the binding energy in the dilute limit and $\alpha$ is an empirical screening parameter, which is dependent on the semiconductor. Assuming the value[1,243] of $\alpha$ to be $\alpha = 2.1 \times 10^{-8}$ meV cm, the unscreened donor binding energy for intrinsic GaN should be about 30.7 meV. The value of 25.5 meV is consistent with the above expression and represents the highest value for the measured donor binding energy in any GaN reported thus far.

### 4.2.5. *Growth of AlGaN*

Aluminum nitride films can be grown through the reaction of $AlCl_3$ with $NH_3$ or ammonolysis of $AlCl_3NH_3$.[251] In this pyrolytic process, the reactant gasses are supplied to the reaction zone either separately or in the form of the compound $AlCl_3NH_3$ in appropriate composition. In early varieties, vertical reactors and substrate temperatures in the range of 700 to 1300°C were used for pyrolytic decomposition of reactant species which resulted in deposition of AlN films. Evaporation temperatures for $AlCl_3$ and $AlCl_3$ $NH_3$ are 80 to 140°C and 180 to 400°C, respectively, which are conveniently accessible. Deposition rates in the range of 0.6 to 6 $\mu$m per hour have been attained.

Following the binary growth, $Al_xGa_{1-x}N$ ($x = 0.45$) layers were grown on sapphire substrates at 1050°C by Baranov *et al.*[252] A mixture of GaCl, AlCl and ammonia was introduced to obtain the ternary material. The growth was initiated with GaN which was then followed by the deposition of AlGaN. Simply raising the temperature of the Al source from 600 to 750°C increased the Al content. All undoped AlGaN samples showed $n$-type conductivity with temperature-independent electron concentrations and mobilities between 77 and 300 K. With increasing Al content up to values of $x = 0.4$ or greater, the electron concentration decreased from $5 \times 10^{19}$ cm$^{-3}$ to $1 \times 10^{19}$ cm$^{-3}$ and the mobility from 100 to 10 cm$^2$/Vs.

The results obtained thus far indicate that high quality GaN films and templates can be grown on sapphire and other substrates by HVPE. These layers and templates served the basis for further growth on them by MBE and MOCVD. As such the technique is very useful. The extent to which the method can be useful could be expanded if high resistivity films can also be obtained. Due to the high temperatures employed, it has not yet been possible to grow InGaN ternary with this method. Similarly, $p$-type doping has so far been lacking which means that HVPE technique alone is not in a position to produce device structures requiring $pn$ junctions. Though, Si and O incorporation from the walls has been an issue, the high quality of the recent layers indicates that this issue is not as serious as it used to be.

## 4.3. *Growth by MOCVD*

At the time of this writing, MOCVD is the workhorse of GaN and related materials growth even though the growth mechanisms are not well understood. As mentioned in the MOCVD method section, the substrate temperatures well in excess of 900°C are required to obtain single crystalline high-quality GaN films, and the GaN films with the best electrical and optical properties are grown at 1050°C or even higher. Such high temperatures in $H_2$ ambient provide very formidable challenges. Metalalkyls need to be kept below pyrolysis temperature until just before the reaction zone, ammonia and metalalkyls would have to be kept separate until just before the reaction zone as well.

High growth temperatures associated with the MOCVD process require that the substrates used do not decompose at the deposition temperatures. Sapphire, Si and SiC are among the substrates that meet this criterion. The high N/Ga flux ratio in used in order to minimize the nitrogen loss is another issue. The high temperatures have a contradictory effect on the crystal quality. On one hand the surface mobility of atoms and thus quality of grown film is higher, for higher substrate temperatures. On the other hand, high vapor pressure of N over Ga and particularly In brings the growth process very close to the dissociation temperature. Not as serious, but post growth cooling introduces more strain and thus more structural defects may be introduced during cooling.

At substrate temperatures exceeding 1100°C the dissociation of GaN results in voids in the growth layer. Use of dimetilhydrosin (DMHy)[253] as a nitrogen source permits to lower deposition temperature, since DMHy decomposes at lower temperatures ($\sim 500°C$) than ammonia ($\sim 800°C$). The film quality is not high enough, however. Rocking curve FWHM of 150–250 arcseconds and electron concentration and mobility of 5 × $10^{19}$ cm$^{-3}$ and 48 cm$^2$/Vs, respectively, have been obtained.

Nitrogen vacancies are widely considered to be the primary cause of large background electron concentration in GaN. So, an improvement in crystal growth must involve more complete incorporation of nitrogen into the crystal lattice. In an effort to achieve this complete incorporation of nitrogen into the crystal lattice, several investigators[254,255] substituted the more reactive hydrazine in favor of ammonia. Andrews and Littlejohn[256] tried $Ga(C_2H_5)_3NH_3$, which already has a Ga–N bond, as a source material. Others used plasma excitation of the nitrogen in CVD growth environment, which proved to be the most popular method of increasing the reactivity of the nitrogen. In the CVD growth approach, several variations have been attempted in the presence of plasma. Suitable precursors in the metalorganic CVD (MOCVD) growth must exhibit sufficient volatility and stability to be transported to the surface. These precursors should also have appropriate reactivity to decompose thermally into the desired solid and to generate readily removable gaseous side products. Ideally the precursors should be nonpyrophoric, water and oxygen insensitive, noncorrosive and nontoxic.

In contrast to the MBE growth, plasma excitation of the nitrogen species has not proven necessary in CVD growth. Although trialkyl compounds (TMA, TMG, TMI, etc.) are pyrophoric and extremely water and oxygen sensitive, and ammonia is highly corrosive, much of the best material grown today is produced by conventional MOCVD by reacting these trialkyl compounds with $NH_3$ at substrate temperatures in the neighborhood of $1000°C$.[257-264] Temperatures in excess of $800°C$ are required to obtain single crystalline high-quality GaN films, and the GaN films with the best electrical and optical properties are grown at $1050°C$. At substrate temperatures exceeding $1100°C$ the dissociation of GaN results in voids in the growth layer. A similar situation is observed also for AlN film growth. To overcome this difficulty, some enhancements have been suggested.

### 4.3.1. *Low temperature buffer layers*

Though low temperature buffer layers for enhanced nucleation were explored much earlier, a milestone was achieved in 1986 when Amano *et al.*[265] reported the use of low temperature buffer layers to enhance the surface morphology, and electrical and optical properties of GaN. This was accomplished by depositing about 50 nm thick AlN film on *c*-sapphire as a nucleation layer. Until then, hexagonal GaN grown by MOCVD was deposited directly on sapphire substrates and suffered from low quality which manifested itself with high electron concentrations approaching $10^{20}$ cm$^{-3}$. In the low temperature buffer layer approach as reported by Amano *et al.*, an AlN layer is first deposited at $600°C$, or thereabouts, followed by a high temperature annealing process in $H_2$ atmosphere and finally, the GaN layer grown at $1000°C$. The surface morphology of the GaN film in this manner was improved to an RMS roughness of about 1 nm which was also free of cracks.

GaN films with low temperature buffer layers displayed much improved quality as measured by X-ray diffraction and photoluminescence. The

FWHM of the double-crystal X-ray rocking curve (0002), obtained for optimum conditions, was 110 arcseconds.[266] The 4 K photoluminescence spectrum measured exhibited a sharp donor-bound exciton transition with a FWHM value of about 1 meV. This was accompanied with the observation of free exciton related transitions in some of the samples. On the transport properties side, the electron concentration was reduced to about 1017 cm$^{-3}$ and with an associated mobility of about 600 cm$^2$/Vs at room temperature for the first time. Such relatively low electron concentration moved the GaN layers from the degenerate state to the non-degenerate state and temperature dependency of the carrier concentration and electron mobility emerged. Consequently, the interpretation of residual donors and scattering mechanisms began.

Attempts were made to determine the evolution of the low temperature AlN buffer layer and the subsequent GaN overlayer. TEM analyses showed that, as expected, GaN is defective particularly at the AlN/GaN interface, and that GaN grew in a columnar fashion.[267] The crystalline structure of the low temperature buffer layer is still shrouded with controversy. The initial stage of growth is very important for the obtaining of heteroepitaxy and the quality of the resulting film. In general three cases of epitaxy are two-dimensional (2D) layer-by-layer mode, three-dimensional (3D) island mode, or mixed (M) mode, and layer-by-layer growth followed by island formation. The first mode results in the smooth surface, while the last two lead to rough surfaces and to low quality epitaxial layers unless one deals with quantum dot-like structures. The mode of growth is determined by many parameters, such as interfacial energy of solid and vapor phase and interfacial energy of vapor phase and substrate, which in turn depend on growth temperature, the bond strength and bond lengths of the substrate and overgrowth material, the rate of impingement of species, surface migration rates of reactants, super saturation of the gas phase, the size of critical nuclei and others. Since the roughness of the nucleation layer is larger at higher growth temperatures, the epitaxial growth is as a rule divided into two steps: Smooth thin ($\sim$ 20 nm) buffer layer grown at low temperatures and main layer grown at higher temperatures.

The AlN buffer layer has amorphous-like structure at the deposition temperature of 600°C but during the ramping process, it is crystallized and exhibits the well known columnar structure[267] as shown in the first panel of Fig. 36. Cross-sectional transmission electron microscopy images of the highly defective initial GaN layer display features similar to those of the AlN buffer layer which suggests columnar fine crystals as shown

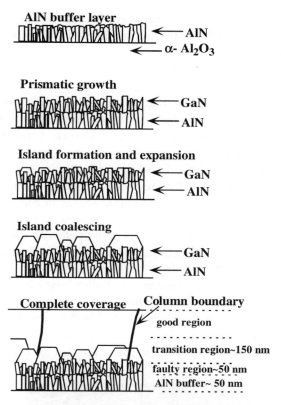

**Fig. 36.** Artistic rendition of the salient structural features as the low temperature AlN buffer layer followed by ramping to high temperature (about 1000–1100°C) and GaN growth evolve. The schematic is based on the analysis of the cross sectional TEM images. Patterned after Ref. 267.

in the artistic rendition of Fig. 36, second panel. Each GaN column is probably grown from a GaN nucleus formed on top of each columnar AlN region. Consequently, a high-density nucleation occurs owing to the high density of the AlN columns. The columns have disordered orientations and as the base of the columns gets larger, the number of columns emanating at the growth surface gradually decreases. Crystalline structure supports the prismatic growth which leads to a general alignment along the $c$ direction with some twist and tilt. The relative "tilt" and "twist", between these columns decreases as the layer thickness increases, showing that the film continues to evolve towards a more ordered structure as growth proceeds.[268] The sub-grain boundaries are more likely the primary defected regions of the epilayer. The small value of strain broadening observed for the (0002), and the (112) reflections indicate that within each GaN column the crystal quality is high.

During the stage where the trapezoidal crystals are formed, as the front area of the column increases by the geometric selection, the growth front follows the $c$-face, as shown in Fig. 36 (3rd from top). Figure 37 shows the changes in surface morphology during the early growth stage of GaN deposition.[267] The stage depicted in Fig. 37(a) corresponds to the generation of the trapezoidal islands where the preferred orientation begins to be established. Subsequently, lateral growth and coalescing of the islands occur in stages as shown in Figs. 37(b) to 37(c). The trapezoidal crystals grow at a higher rate in a transverse direction, as shown in Fig. 36 schematic of the growth front (panel 5), and the islands coalescence. Finally, since the crystallographic directions of all islands agree well with each other with a few degrees of disorder, a smooth GaN layer with a small number of defects results, as seen in Fig. 36 (5th from top) and Fig. 37(d).

The concept of AlN low temperature buffer layers has been extended to include GaN by Nakamura.[269] One may argue that the large lattice mismatch between the low temperature AlN buffer layer and the following GaN layer would cause new defect generation and that the use of low temperature GaN would remove the cause for this generation. As in the

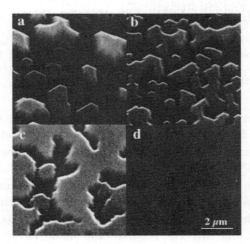

**Fig. 37.** Surface morphology of GaN layers, as observed by scanning electron microscopy, SEM, grown on annealed low temperature AlN buffer as the growth evolves, panels $a$, $b$, $c$, and $d$ correspond to 3, 5, 10 and 60 minutes, respectively. Patterned after Ref. 267.

case of AlN, the crystalline quality of the GaN nucleation layer is initially poor but improves while heating to the growth temperature of the epitaxial GaN layer, commonly above 1000°C.[270]

### 4.3.2. *Lateral epitaxial growth*

To reiterate, heteroepitaxial deposition of GaN on low temperature GaN or AlN buffer layers on $Al_2O_3$ and SiC substrates results in films containing high dislocation densities ($10^8$–$10^{10}$ cm$^{-2}$). As discussed in detail so far in this manuscript and regardless of the deposition method employed, this high concentration of dislocations result from the lattice mismatch between the buffer layer and the film and/or the buffer layer and the substrate. Without any doubt, the high defect concentration limits device performance.

The genesis of epitaxial lateral overgrowth (ELO), lateral epitaxial overgrowth (LEO), selective area growth (SAG), epitaxial lateral over growth (ELOG), all referring to the same concept, can be traced to Si[271] and other conventional semiconductors prior to its use in GaN which broke the bottleneck in relation to GaN based laser longevities. The method relies on growth of GaN on windows opened in a dielectric material such as $SiO_2$ followed by lateral extension and coalescence. The growth conditions are optimized to support the lateral growth until coalescence. The LEO method has been demonstrated in material systems such as Si[271–273] GaAs on Si,[274,275] InGaAs on GaAs,[276] and InP on Si.[277]

Lateral epitaxial overgrowth (LEO) of GaN layers using patterned $SiO_2$ has been reported.[278,279] This was followed by a transmission electron microscopic study which revealed that these overgrown regions of pyramids contained much lower density of dislocations.[280] Using the LEO technique, Nakamura reported laser diode lifetimes of about 10,000 hours at room temperature.[281] This is substantial in the longevity of laser diodes hovered around 300 hours prior to LEO.

In one particular investigation,[278] substrates for the lateral epitaxy studies were prepared by depositing 1.5–2.0 $\mu$m thick GaN layers at 1000°C on high temperature (1100°C) AlN buffer layers, which in turn were grown on 6H–SiC substrates. A $SiO_2$ mask layer (thickness = 1000 Å) was subsequently deposited on each GaN/AlN/6H–SiC(0001) sample and patterned by standard photolithography techniques and etching in a buffered HF solution. The pattern consisted of 3 $\mu$m wide, parallel stripe openings with a 7 $\mu$m pitch, which were oriented along the $\langle 11\bar{2}0 \rangle$ and $\langle 1\bar{1}00 \rangle$ directions in each GaN film. Prior to lateral overgrowth, the patterned samples were dipped in a 50% buffered HCl solution to clean the underlying GaN layer. The lateral overgrowth of GaN was achieved at 1000–1100°C and 45 Torr. Triethylgallium (13–39 $\mu$mol/min) and NH$_3$ (1500 sccm) precursors were used in combination with a 3000 sccm of H$_2$ diluent. The second lateral epitaxial overgrowth was conducted on the first laterally grown layer via the repetition of SiO$_2$ deposition, lithography and lateral epitaxy.

Figure 38 shows the representative cross-sectional SEM images of two GaN stripes along $\langle 11\bar{2}0 \rangle$ and $\langle 1\bar{1}00 \rangle$ directions selectively grown for 60 minutes. Truncated triangular stripes which have $(1\bar{1}00)$ slanted facets and a narrow $(0001)$ top facet were observed for window openings along the $\langle 11\bar{2}0 \rangle$ direction. Rectangular stripes with a $(0001)$ top facet, $(11\bar{2}0)$ vertical side facet, and a $(1\bar{1}00)$ slanted facet appear to have been developed in samples grown on stripes along the $\langle 1\bar{1}00 \rangle$ direction. SEM observations of GaN stripes for different growth durations of up to 3 minute revealed similar morphologies regardless of stripe orientation. The amount of lateral growth exhibited a strong dependence on stripe orientation. Results obtained under various growth conditions showed that the lateral growth rate perpendicular to the $\langle 1\bar{1}00 \rangle$ direction was much faster than those perpendicular to the along $\langle 11\bar{2}0 \rangle$ direction. Dependence of morphological development on the window orientation has been attributed to the stability of the crystallographic planes in the GaN structure. Stripes oriented along the $\langle 11\bar{2}0 \rangle$ direction always had wide $(1\bar{1}00)$ slanted facets and either a very narrow or no $(0001)$ top facet depending on the growth conditions. The stability of the $(1\bar{1}01)$ plane of the wurtzitic GaN, and the associated low growth rate of this plane are thought to be reason for the observations. As shown in Fig. 38(b), the $\{1\bar{1}01\}$ planes of the $\langle 1\bar{1}00 \rangle$ oriented stripes were wavy, which points

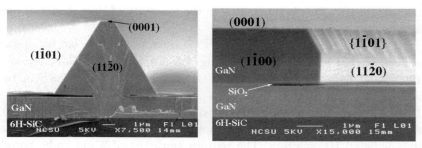

**Fig. 38.** Scanning electron micrographs showing the morphologies of GaN layers grown on stripe openings oriented along (a) $\langle 11\bar{2}0 \rangle$ and (b) $\langle 1\bar{1}00 \rangle$ directions. Courtesy of R. Davis.

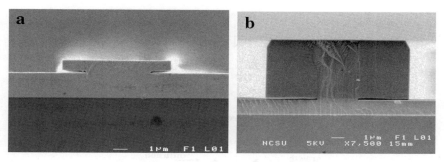

**Fig. 39.** Scanning electron micrographs of ⟨1Ī00⟩ oriented GaN stripes grown at TEG flow rates of (a) 13 $\mu$mol/min and (b) 39 $\mu$mol/min for 60 minutes. Courtesy of R. Davis.

to the co-existence of several Miller indices. It is believed that competitive growth of selected {1Ī01} planes occurs during the deposition which causes these planes to become unstable and their growth rate to increase relative to that of the (1Ī01) plane of stripes oriented along ⟨11Ī20⟩.

Beyond the orientation of the stripes, the flow rate of the metaly alkyl which in this case is TEG, influences the morphological evolution of the GaN stripes as shown in Fig. 39. An increase in the flow increased the growth rate of the stripes in both the lateral and the vertical directions. However, the lateral/vertical growth rate ratio decreased from 1.7 at a TEG flow rate of 13 $\mu$mol/min to 0.86 at 39 $\mu$mol/min. The considerable increase in the concentration of the Ga species on the surface may sufficiently impede their diffusion to the {1Ī01} planes such that chemisorption and GaN growth occur more readily on the (0001) plane. The morphologies of the GaN layers were also a strong function of the growth temperatures. Stripes grown at 1000°C possessed a truncated triangular shape. This morphology gradually changed to the rectangular cross-section as the growth temperature was increased. One can summarize the evolution of LEO growth in Fig. 40.

Single LEO GaN layers with thickness of 2 $\mu$m were obtained using 3 $\mu$m wide stripe openings with 10 $\mu$m pitches and oriented along the ⟨1Ī00⟩ direction (Fig. 41(a)). The growth parameters were 1100°C and a TEG flow rate of 26 $\mu$mol/min. Atomic force microscopy showed that the surfaces of the LEO GaN showed a terraced structure having an average step height of 0.32 nm.

An obvious extension of single LEO is the double LEO. This of course requires, an additional lithographical step following a second growth interruption/exposure to environmental conditions of the GaN template.

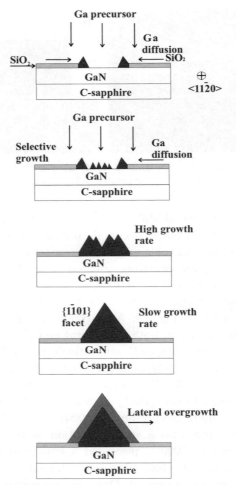

**Fig. 40.** Schematic representation of the evolution of the LEO growth process. At first the growth initiates at the boundary between the mask and GaN followed by eventual expansion of the grown region to cover the entire opening in the mask. This is followed by trapezoidal growth and which in turn followed by lateral growth.

Each black spot in the overgrown double LEO GaN layers shown in Figs. 41(a) and 41(c) is a subsurface void which forms when two growth fronts coalesce. These voids were most often observed using the lateral growth conditions wherein rectangular stripes having vertical {11$\bar{2}$0} side facets developed. The morphologies of the finished surfaces of single and double LEO layers imaged by SEM are shown in Figs. 41(b) and 41(d).

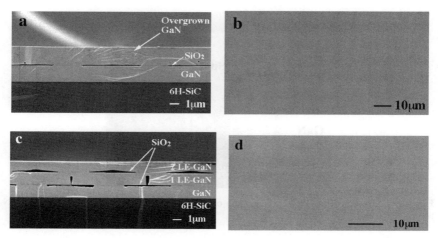

**Fig. 41.** Cross-section and surface SEM micrographs of the first, (a) and (b), and second, (c) and (d), coalesced GaN layers, respectively, grown on 3 μm wide and 7 μm spaced stripe openings oriented along ⟨1$\bar{1}$00⟩. Courtesy of R. Davis.

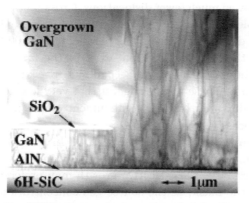

**Fig. 42.** Cross-section TEM micrograph of a section of a laterally overgrown GaN layer epitaxial over-on an SiO₂ mask region. Courtesy of R. Davis.

Surface morphology of the second overgrown layer was comparable to the first layer. Cracks were occasionally observed along the coalesced interface under selected growth conditions, probably due to the thermal mismatch between the GaN layers and the SiO₂ mask.

Cross-sectional TEM image of Fig. 42 shows a typical laterally overgrown GaN. Threading dislocations, originating from the GaN/AlN buffer layer interface, propagate to the top surface of the regrown GaN layer

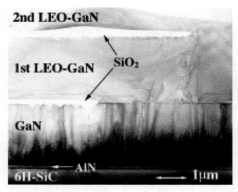

**Fig. 43.** Cross-section TEM micrograph of a section of the second lateral grown GaN layers. Courtesy of R. Davis.

within the window regions of the mask. The dislocation density within these regions, calculated from the plan view TEM micrograph is approximately $10^9$ cm$^{-2}$. By contrast, however, additional microstructural studies of the LEO regions showed much fewer dislocations. Cross-sectional TEM observation of the double LEO sample in the micrograph presented in Fig. 43 shows that a very low density of dislocations parallel to the (0001) plane, formed via bending of threading dislocation, exist in the first and second LEO-GaN layers on the SiO$_2$ masks. The second SiO$_2$ mask is slightly misaligned relative to the first. These results suggest that very low defect density GaN layers can be fabricated by precise alignment of the mask in the second lithographic process.

Let us examine what goes on in LEO. In the particular example discussed above, GaN initially grows selectively on the stripes not covered by the dielectric mask. This selective growth may preferentially seed itself at the edges of the mask, as the source gas concentration is high in these regions because of the lateral vapor and surface diffusion of Ga from masked regions to the window regions. The process could also be enhanced as the edges of the mask could serve as nucleation sites. Following the growth of stripes, lateral growth followed by coalescence of the stripes occurs. Obviously the surface before coalescence is rough in nature.

The growth rate is high in the [0001] direction but the [1$\bar{1}$01] growth rate on the side facets is very low as that surface is stable. The GaN at this stage has a simple morphology composed of two [1$\bar{1}$01] facets on a length scale comparable to the widths of the striped windows, in other words pyramidal form. As the growth continues, the facets begin to laterally grow over the

masked surface. Zheleva *et al.*[280] analyzed the structural defects in these
GaN pyramids. In the region over the windows, the dislocations have a
density comparable to that of the GaN seed layer as mentioned previously.
Most of the extended dislocations propagate in the growth direction through
the GaN. The dislocation density is drastically reduced within a pyramidal
volume which ends at approximately one third of the pyramid height. The
part of the pyramidal region which is above the masked area, contains much
reduced dislocation density.

Others investigating LEO reported dislocation densities in the range of
$10^4$–$10^5$ cm$^{-2}$.[282] We should point out that figures that are this low are not
correctly measured by TEM. The etching methods reported in the HVPE
section of this chapter are much more reliable. In this particular report the
work was done with stripes oriented in the [$1\bar{1}00$] direction in order to yield
a large lateral growth rate/vertical growth rate ratio. The stripe spacing
was varied to give ratios of open width to patterned period of 0.1 to 0.5.
The LEO GaN was bound by the (0001) facet on top and by vertical [$11\bar{2}0$]
sidewalls on the edges, which showed a lateral growth rate of up to 6 $\mu$m/h.
Patterns with 10 $\mu$m stripes and a ratio of open width to patterned period
of 0.5 enabled full coalescence of the overgrown GaN film after 90 minutes of
growth. TEM and AFM (counting the pit density where the bilayer steps
are terminated/broken) observations indicated that the density of mixed
character dislocations reaching the surface of the LEO GaN is in the $10^4$–
$10^5$ cm$^{-2}$ range. To draw the contrast between the overgrown region and the
region directly over the template GaN, a cross-sectional TEM image and a
plan view TEM image around the vicinity of a stripe are shown in Figs. 42
and 44, respectively. Clearly, the LEO process is effective in reducing the
extended defect density.

A cursory interpretation of the LEO results may appear to indicate
as though the defect problem in GaN has largely been solved. However,
the detailed investigations indicate that the wing regions (the regions over
the masked areas) are not defect free when they coalesce. This is in part
caused by sagging of the wings which is much more than that which can
be attributed to SiO$_2$ becoming soft during the growth. Strain has been
suggested as the likely cause. In fact, the wings sag so much so that extra
peaks in X-ray diffraction appear from the surface facets. An additional
issue is that of the Si outdiffusion from the SiO$_2$ mask material which
auto dopes GaN *n*-type. While this may be acceptable for lasers and LEDs
and devices of that kind, it is an impediment for field effect transistors.
Compensation techniques with Ga vacancies which are of *p*-type can be

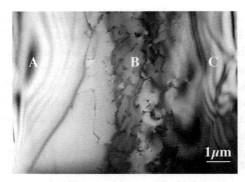

**Fig. 44.** Planview TEM micrograph of a section in the vicinity of the lateral grown GaN and that grown on the GaN template below. Courtesy of R. Davis.

used, but not ideal. It should also be pointed out that the benefit of the LEO technology is limited to regions on the final GaN device layer that are located on the overgrown regions. Moreover, degradation of the vertical thermal conductivity of the composite, due to the dismal thermal conductivity of $SiO_2$, is a limiting factor which must be dealt with in high power electronic devices, and lasers. Nevertheless, LEO has increased the level of device performance considerably even though it is looked on as a temporary measure.

### 4.3.3. *Pendeo epitaxy*

A variant of LEO is termed Pendeo Epitaxy. This method relies full selective growth of GaN on GaN stripes on SiC or Si substrates. The method has the advantage of eliminating the $SiO_2$ on which conventional LEO relies. The group of Prof. Davis has pioneered this approach that is termed, pendeo- (from the Latin: to *hang* or be *suspended*) epitaxy (PE), as a promising new process.[283] In this approach, the growth does not initiate through open windows on the (0001) surface of the GaN seed layer; instead, it is forced to selectively begin on the sidewalls of a tailored microstructure comprised of forms previously etched into this seed layer. Continuation of the pendeo-epitaxial growth of GaN layer until coalescence over and between these forms results in a layer of lower defect-density GaN.

The group of Prof. Davis explored pendeo-epitaxy growth on on-axis 6H–SiC (0001) and on-axis Si (111) substrates. In the former, each seed layer consisted of a 1 $\mu$m thick GaN film grown on a 100 nm thick AlN buffer layer previously deposited on a 6H–SiC(0001) substrate. In the growth on the Si substrates, a 1 $\mu$m 3C–SiC(111) film was initially grown on a very thin 3C–SiC (111) layer produced by conversion of the Si (111) surface at 1360°C for 90 seconds via reaction with $C_3H_8$, with $H_2$ as the carrier gas. The film was subsequently achieved by simultaneously decreasing the flow rate of the $C_3H_8/H_2$ mixture and introducing a $SiH_4/H_2$ mixture. Both the conversion step and the SiC film deposition were achieved using a cold-wall atmospheric pressure chemical vapor deposition (APCVD) reactor. A 100 nm thick AlN buffer layer and a 1 $\mu$m GaN seed layer were subsequently deposited in the manner described above for the 6H–SiC substrates.

Following the growth of the base GaN layer either on 6H–SiC of 3C–SiC coated Si, a 100 nm $Si_3N_4$ mask layer and a Ni (as an mask resistant to nitride etches) were deposited on the seed layers via plasma enhanced CVD, and e-beam evaporation, respectively. The Ni layer was patterned by lithography and sputtering, and the underlying $Si_3N_4$ and nitride layers were patterned by inductively coupled plasma (ICP) etching in windows. The nickel mask layer was removed in a wet etch leaving the $Si_3N_4$ mask in place on top of the stripes, as shown in Fig. 45. It is critical that the etching is complete all the way down to SiC. The stripes used were of rectangular shape oriented along the [1$\bar{1}$00] direction, thus providing a sequence of parallel GaN sidewalls (nominally (11$\bar{2}$0) faces). The windows were 2 and 3 $\mu$ms in width and window-to-window separations were 3 and 7 $\mu$ms. Immediately prior to pendeo-epitaxial growth, the patterned samples were dipped in an acid solution to remove surface contaminants from the walls of the underlying GaN seed structures.

Three primary stages associated with the pendeo-epitaxy are note worthy: (i) initiation of lateral homoepitaxy from the sidewalls of the GaN

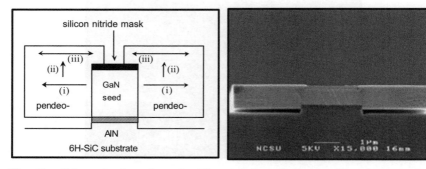

**Fig. 45.** Schematic of pendeo-epitaxial growth from GaN sidewalls and over a silicon nitride mask. Courtesy of R. Davis.

seed, (ii) vertical growth and (iii) lateral growth over the $Si_3N_4$ mask covering the raised stripes. Pendeo-epitaxial growth of GaN was achieved within the temperature range of 1050–1100°C using the same pressure and V/III ratio used for the deposition of the GaN seed layer.

### 4.3.3.1. *Pendeo epitaxy on SiC substrates*

The pendeo-epitaxial phenomenon is made possible by the initiation of growth from a GaN face other than the (0001), and the use of the substrate (in this case SiC) as a pseudo-mask. By capping the seed-forms with a growth mask, the GaN is limited to grow initially and selectively only on the GaN sidewalls. As in the case of conventional LEO, no growth occurs on the $Si_3N_4$ mask. What is unique is that no deposition occurs on the exposed SiC surface areas if higher growth temperatures are employed to enhance lateral growth. In this scenario the Ga- and N-containing species are more likely to either diffuse along the surface or evaporate from both the $Si_3N_4$ mask and the silicon carbide substrate, as confirmed by the cross-sectional SEM image of Fig. 46 where the newly deposited GaN has grown truly suspended (pendeo) from the sidewalls of the GaN seed. Following enhanced lateral growth, vertical growth occurs from the advancing (0001) face of the laterally growing GaN. Once the vertical growth becomes extended to a height greater than the silicon nitride mask, the epitaxial growth coalesces as in the case of conventional LEO.

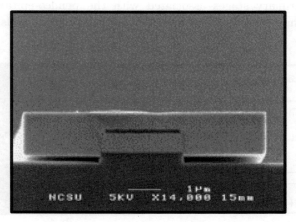

**Fig. 46.** Cross-sectional SEM of a $GaN/Al_{10}Ga_{90}N$ pendeo epitaxial growth structure showing coalescence over the seed mask. Courtesy of R. Davis.

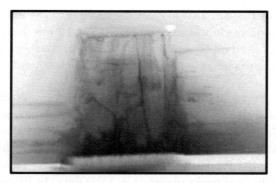

**Fig. 47.** Cross-sectional TEM of a GaN pendeo-epitaxial structure showing confinement of threading dislocations under the seed mask, and a reduction of defects in the regrowth. Courtesy of R. Davis.

A cross-sectional TEM micrograph showing a typical pendeo-epitaxial growth structure is shown in Fig. 47. Threading dislocations extending into the GaN seed structure, originating from the GaN/AlN and AlN/SiC interfaces are clearly visible. The $Si_3N_4$ mask acts as a barrier to direct vertical growth of GaN and thus propagation of these defects into the laterally overgrown pendeo-epitaxial film. Preliminary analyses of the GaN seed/GaN PE and the AlN/GaN PE interfaces revealed evidence of the lateral propagation of the defects; however, there is yet no evidence that the defects reach the (0001) surface where device layers will be grown. As in the case of LEO, there is a significant reduction in the defect density in the regrown areas.

### 4.3.3.2. *Pendeo epitaxy on silicon substrates*

The approach of PE on Si suffers from the three-dimensional nucleation and growth of GaN islands caused by the combination of significant mismatches in lattice parameters, the higher surface energy of GaN and the chemical reactivity of Si with the reactants in the growth environment. To address the above concerns, Davis *et al.*[284] have developed a process similar to those used for growth of GaN on 6H–SiC (0001), but replaced the 6H–SiC substrate with 3C–SiC a (111) transition layer grown on a Si (111) substrate. The atomic arrangement of the (111) plane of 3C–SiC is equivalent to the (0001) plane of 6H–SiC; this facilitates the sequential deposition of a high temperature 2H–AlN(0001) buffer layer of sufficient quality for the GaN seed layer. The rest of the growth and processing steps are similar to that of PE of GaN on 6H–SiC that we just discussed. For the initial demonstrations of PE growth of GaN films on silicon, 0.5 and 2 $\mu$m thick 3C–SiC (111) layers were deposited on 50 mm diameter, 250 $\mu$m thick converted Si (111) substrates. All subsequent research described below used the $\sim$ 2.0 $\mu$m barrier layer.

Figure 48 shows a cross-sectional SEM micrograph of a PE GaN layer grown laterally and vertically from raised GaN stripes etched in a GaN/AlN/3C–SiC/Si(111) substrate and over the silicon nitride mask atop each stripe. As in the case of LEO, tilting or sagging of the wing region as they coalesce is common to this method as well.[285] This tilt manifests itself by the additional peaks observed in XRD data. In PE GaN on Si with 3C–SiC interlayers, the XRD spectrum taken along the [1$\bar{1}$00] direction, parallel to the stripes, consisted of one peak. However, the spectrum along the [11$\bar{2}$0] direction, perpendicular to the stripes, exhibited two superimposed peaks, separated by a tilt of 0.2°. In contrast to LEO, the PE GaN regions that coalesce contain two sets of coalesced growth fronts, namely, over the trenches and over the masks. A more sensitive method of evaluating the tilt in PE grown GaN, such as TEM diffraction, is needed to determine if tilting is present in both of these areas. Analyses of such a study from small areas of coalesced regions in a trench and near the coalesced region over the silicon nitride mask, revealed no evidence of tilt in the laterally grown material over the trenches; however, significant tilt was observed over the masked regions of the GaN seed.

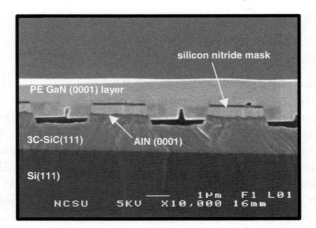

**Fig. 48.** Cross-sectional SEM micrograph of a coalesced PE GaN epilayer deposited on a 3C–SiC/Si(111) substrate. Courtesy of R. Davis.

### 4.3.4. *Selective growth using W masks*

Selective area growth (SAG) and epitaxial lateral overgrowth (ELO) of GaN with tungsten (W) masks[286] and $WN_x$ masks[287] using metalorganic vapor phase epitaxy (MOVPE) and hydride vapor phase epitaxy (HVPE) have also been investigated. The $WN_x$ mask was employed to prevent dissolution of the underlying GaN layer, due to the W catalytic effect. The $WN_x$ mask was produced by nitrogenation of the W film using $NH_3$, which is already present in the reactor, at temperatures higher than 600°C. Thermal stability of $WN_x$ is good and the $WN_x/n$-GaN contact forms a Schottky type.

The selectivity of the GaN growth on the W mask as well as the control $SiO_2$ mask was found to be excellent for both MOVPE and HVPE. The ELO–GaN layers were successfully obtained by HVPE on the stripe patterns along the $\langle 1\bar{1}00 \rangle$ crystal axis with the W mask as well as the $SiO_2$ mask. No voids between the $SiO_2$ mask and the overgrown GaN layer were observed. In contrast, there were triangular voids between the W mask and the overgrown layer. The surface of the ELO–GaN layer was quite uniform for both mask materials. In the case of MOVPE, the ELO layers on the W mask and $SiO_2$ masks are similar for stripes oriented along the $\langle 11\bar{2}0 \rangle$ and $\langle 1\bar{1}00 \rangle$ directions. In other words, no voids were observed between the W or $SiO_2$ mask and the overgrown GaN layer by using MOVPE. As in the case of $SiO_2$ masks, W and $WN_x$ masks led to growth patterns where triangular growth results for stripes along the $\langle 11\bar{2}0 \rangle$ direction, and trancuted triangular growth results for stripes along the $\langle 1\bar{1}00 \rangle$, as shown schematically in Fig. 49.

The details of the experimental methods employed are as follows:[286] The growth experiments were performed using atmospheric pressure HVPE and MOVPE systems on 3.0–4.5 $\mu$m thick (0001) GaN layers grown on the $c$-plane (0001) of sapphire substrates. As usual a low temperature (LT) buffer layer grown by MOVPE was first grown on the substrate. A 120 nm thick W film was deposited on the GaN surface by RF sputtering at room temperature. Stripe windows of 10 $\mu$m wide with a periodicity of 20 $\mu$m were formed on the W film with conventional photolithography and wet chemical etching in $H_2O_2$ at room temperature. In the case of HVPE, GaCl and $NH_3$ were used as the source gases, and $N_2$ was used as the carrier gas. The flow rates of HCl and $NH_3$ were 10 sccm and 0.5 $l$/min, respectively. The growth temperature was 1090°C. In the case of MOVPE, TMG and $NH_3$ were used as the source gases, and $H_2$ was used as the carrier gas. The

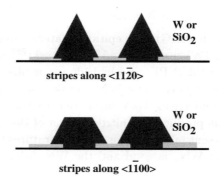

**Fig. 49.** Schematic representation of GaN grown in windows opened in W or $SiO_2$ masks. The triangular growth results for stripes along the $\langle 11\bar{2}0 \rangle$ direction, and trancuted triangular growth results for stripes along the $\langle 1\bar{1}00 \rangle$ direction. Patterned after a figure provided by K. Hiramatsu (Ref. 286).

flow rates of TMG and $NH_3$ were 18.7 $\mu$mol/min and 2.5 $l$/min, respectively. The growth temperature was 1060°C. The nitrogenation of W was performed at 600–950°C in $NH_3$ with $H_2$ or $N_2$ ambient.[286] In order to remove any O from the surface of the W film, the substrate was annealed in a $H_2$ ambient at 400°C for 10 minutes. Then, the substrate was heated to the nitrogenation temperature in $H_2$ and $NH_3$ ambients. Following the nitrogenation of W, the ELO of GaN was attempted by a two-step growth at 950°C and 1050°C at a low pressure (LP) of 300 Torr. The two-step process was to prevent dissociation of GaN in contact with W by completing the first stage at a lower temperature which was 950°C for 30 minutes in this case. In the second step, to bury the W mask easily, a high temperature (1050°C) growth was performed for 90 minutes.

### 4.3.5. *Low temperature buffer insertion*

In parallel to LEO, other techniques such as those that do not require exposure to atmosphere part way through the process and photolithography have been explored. Among them is low temperature interlayer.[288] The LT buffer layers inserted periodically a few times reduce the dislocation density in the top layer. The LT buffer layers are grown using growth conditions very similar if not identical to the standard LT buffer layers. After each LT buffer layers, the structure is automatically annealed in $H_2$ prior to the growth of the high temperature layers. In the process propagating defects collide with the new interfacial defects generated in LT insertion buffer layer and annihilate one another. Defect concentrations have been lowered to about the same figure available by LEO using this method. Recently, this method has been extended to grow AlGaN on GaN layers without the notorious cracking effect observed in tensile strained AlGaN on GaN.[288] The problem is a very serious one for nitride lasers as the cladding layer thickness is limited by cracking. The reduced thickness employed to avoid cracking causes leakage of the optical field to the GaN buffer layer with deleterious effects.

Shown in Figs. 50(a) and 50(b) are images of two GaN layers grown by MOCVD after they have been subjected to a defect staining etch. Corresponding schematic diagrams of the grown layers are also shown. Figure 50(a) is for a sample grown only on a LT buffer layer which in turn is grown on sapphire. While that in Fig. 50(b) has an additional LT buffer layer inserted in the main GaN layer. As the images indicate, the etch pit density dropped from about $10^8$ cm$^{-2}$ to about $5 \times 106$ cm$^{-2}$ when an additional LT buffer layer is inserted.[289]

LT AlN insertion buffer layers (IL) have been examined with a particular attention paid to electrical transport and optical properties of the resulting

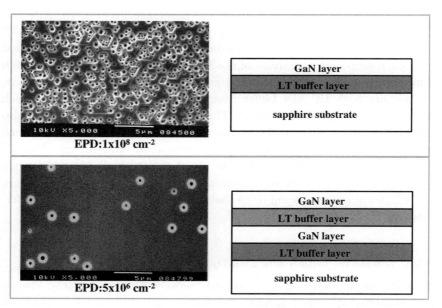

**Fig. 50.** Images of two GaN layers grown by MOCVD after they have been subjected to a defect staining etch, (a) is for a sample is grown only on a LT buffer layer which in turn is grown on sapphire and (b) has an additional LT buffer layer inserted in the main GaN layer. Courtesy of M. Koike.

GaN layer on top.[290] It was shown that as the number of AlN IL/HT GaN layers increased, the electron mobility increased in the top Si-doped GaN layer, nearly as much as by a factor of two from 440 to 725 cm$^2$/Vs. The dependence of the electron mobility on temperature had the characteristic peak at about 100 K before falling off toward lower temperatures. This implies that the mobility was not dominated by any 2DEG which could form at each of the interfaces between the AlN layer and the underlying GaN layer, and that the measured values do indeed show an improvement on the electrical quality of GaN with LT AlN insertion. Cross-sectional transmission electron microscopy images were remarkable in that a significant reduction in the screw dislocation density for GaN films grown on the AlN IL/HT GaN layers was obtained. This is consistent with the assertion that in GaN edge dislocation core energy is minimum along the *c*-direction and it is unlikely that the edge dislocations would be affected by these AlN insertion layers. Of course, the situation is different with screw dislocations. As for the XRD investigation, the symmetric and off-axis linewidths increased as the number of AlN IL/HT GaN layers increased, indicating a greater relative misalignment of the adjacent HT GaN layers.

An interesting observation about the difference between LT AlN and GaN insertion layers is the noteably different residual strain in GaN.[288] It was shown that the in-plane biaxial stress thickness product increased with each successive repetition of the LT GaN insertion. Further, the high temperature (HT) GaN layers were under tensile strain. On the contrary, the picture with LT AlN buffer insertion remained the same with each insertion of LT AlN buffer layers. Threading dislocations (TD) were also measured using plan view TEM analysis. A reduction of TDs with each insertion of LT buffer layer is observed, from about mid $10^9$ cm$^{-2}$ in HT GaN with no LT insertion buffers to about mid to high $10^7$ cm$^{-2}$ when some six LT buffer layers were inserted. While there were some variations, the TD density did not seem to be that dependent on the type of LT buffer, meaning GaN or AlN, used.[288]

### 4.3.6. *Growth on LiAlO$_2$ and LiGaO$_2$ substrates*

To overcome the large lattice mismatch between GaN and Al$_2$O$_3$, Fischer and co-workers[291] used LiAlO$_2$ substrates with 1.5% lattice mismatch and 21% mismatch in the thermal expansion coefficients respectively. As mentioned previously, in conjunction with substrate issues, the epitaxy relationships between GaN and LiAlO$_2$ are expected to be $(01\bar{1}0)/(100)$

LiAlO$_2$ with $(2\bar{1}\bar{1}0)/(001)$ LiAlO$_2$. Unlike Al$_2$O$_3$ and 6H–SiC substrates with very smooth surfaces, the LiAlO$_2$ substrate exhibited a wave-like surface with equidistant grooves about 10 nm deep, which could have originated from the mechanical surface polishing. Because of the high surface roughness, the interface between GaN and LiAlO$_2$ was highly defective and caused three-dimensional growth of the GaN MOCVD of monocrystalline GaN thin films on $\beta$-LiGaO$_2$ substrates was realized by Kung *et al.*[293] As alluded to earlier in the section dealing with substrates, the structure of LiGaO$_2$ is similar to the wurtzitic structure, but because Li and Ga have different ionic radii, the crystal has orthorhombic structure. The atomic arrangement in the (001) face is hexagonal, which promotes the epitaxial growth of (0001) GaN, so that the epitaxial relationship (0001)GaN/(001) LiGaO$_2$ is expected. The distance between the nearest cations in LiGaO$_2$ is in the range of 3.133–3.189 Å, while the distance between nearest anions is in the range of 3.021–3.251 Å. The lattice mismatch between (0001) GaN on (001) LiGaO$_2$ is then only 1–2%. The GaN layers were grown by MOCVD at temperatures between 600 and 1000°C, at deposition rates of about 0.7 $\mu$m h$^{-1}$. The samples had smooth surfaces and the X-ray rocking curve was as narrow as 300 arcseconds, for substrate temperatures of 900°C. All GaN layers were $n$-type ($n$ about 10$^{20}$ cm$^{-3}$) as determined by room-temperature Hall measurements. The electron mobility was about 10 cm$^2$ V$^{-1}$ s$^{-1}$. The high electron concentrations may be due to the incorporation of oxygen, resulting from the decomposition of LiGaO$_2$ at elevated substrate temperatures. This effect would probably be reduced by the use of low-temperature buffer layers.

### 4.3.7. *GaN on GaN templates*

The homoepitaxial growth of GaN has recently been achieved by using GaN (0001) bulk substrate.[293,294] No buffer layer was used for the growth. The GaN bulk substrate, which was used for the growth, was cleaned first by boiling in *aqua regia* for 10 minutes, and then in organic liquid with an ultrasonic cleaner prior to the growth. A horizontal MOCVD reactor was employed at atmospheric pressure to grow the GaN layer. TMG and ammonia were the source gases, and H$_2$ was the carrier gas. For the growth of Mg-doped GaN, bis-cyclopentadienyl magnesium was utilized. To protect the GaN substrate from the escape of nitrogen at high temperatures ($> 800°C$), ammonia was fed into the reactor before the beginning of the substrate heating. The *as grown* $n$-GaN films thus obtained were found to be high quality single-crystal films with good surface morphology. The *as grown* GaN : Mg films showed strong blue cathodoluminescence and photoluminescence spectra peaking at 445 nm even without LEEBI treatment or thermal annealing.

GaN homoepitaxial layers grown by MOCVD on the highly conductive GaN bulk crystals grown at high hydrostatic pressure exhibit smaller free-electron concentrations in the layers in contrast to the substrates which had about $2.5 \times 10^{19}$ cm$^{-3}$ of free electrons.[295] In X-ray diffraction, [0002] peaks for the substrate and epitaxial layers had rocking curves with half widths of about 20 arcseconds. The photoluminescence spectrum exhibited by the epitaxial layer was less than 1 meV wide. GaN layers grown by MOCVD on freestanding GaN templates grown by HVPE also showed excellent properties. X-ray reflections, with FWHM as low as 20 arcseconds, were obtained. The dislocation density was determined to be $2 \times 10^7$ cm$^{-2}$. The lattice mismatch between the GaN substrate and the homoepitaxial layer was below $3 \times 10^{-5}$, and the PL linewidth was about 0.5 meV.[296]

Much more refined experiments followed these early attempts which resulted in GaN with excellent properties as determined by photoluminescence.[297-299] One could realize unstrained GaN layers with dislocation densities comparable to that in the template on which the layers are grown. In the case of platelets prepared by the high-pressure technique, this dislocation density is several orders of magnitude lower than the best conventional heteroepitaxy. Through the use of dry etching techniques for surface preparation, a pathway was blazed to achieve high crystal quality in the overgrown epitaxial film.[297] The layers so grown reveal an exceptional optical quality as determined by a reduction of the low-temperature photoluminescence (PL) linewidth from 5 meV to 0.1 meV and a reduced symmetric XRD diffraction rocking curve width from 400 to 20 arcseconds. The latter is not surprising as the epitaxial layer replicates the structural properties of the template unless the surface preparation is not optimum. Narrow PL linewidths paved the way for observing fine structure of the donor-bound exciton line at 3.471 eV with five fine features inclusive of the excited states of free excitons. Additionally, all three free excitons as well as their excited states were visible in the photoluminescence spectrum at 2 K. A PL scan taken at 2 K is shown in Fig. 51. Moreover, InGaN/GaN MQW structures as well as GaN $p$-$n$ and InGaN/GaN double heterostructure LEDs on GaN bulk single crystal substrates have also been prepared.[297] These particular LEDs were reported to be twice as bright as their counterparts grown directly on sapphire. In addition they exhibited

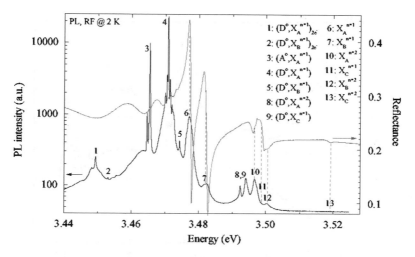

PL: linewidth down to 90 μeV $(A°, X_A^{n=1})$

RF: $X_A$, $X_B$, $X_C$ and their excited states visible

**Fig. 51.** Photoluminescence scan of a GaN layer grown on GaN platelet grown by the high-pressure technique. Courtesy of. M. Kamp.

improved high power characteristics, which are attributed to enhanced crystal quality and increased $p$-type doping.

Though very preliminary, high quality GaN layers have also been grown on freestanding templates which are in turn prepared by HVPE. The PL spectrum of an undoped GaN epilayer grown by MOCVD on such a template is shown in Fig. 52 for a wide range of photon energies. Besides exciton-related peaks, which will be considered later, the spectrum contains two broad bands: A yellow luminescence (YL) band with a maximum at about 2.3 eV and a blue luminescence (BL) band with a maximum at about 3 eV. The YL band is the omnipresent feature in PL spectra of $n$-type GaN grown by different techniques and it is most commonly attributed to a structural defect, namely a complex of the gallium vacancy with oxygen or silicon atom.[300] The BL band is most probably related to the surface states of GaN.[301] Unlike the BL often observed in HVPE and MOCVD grown GaN,[302] the BL in the sample strongly bleached with the laser exposure time, very similar to the behavior observed earlier.[301] The bleaching is attributed to photo assisted desorption of oxygen atoms from the GaN surface. Note that with increasing excitation intensity the relative contribution of the YL and BL bands decreases due to saturation of the

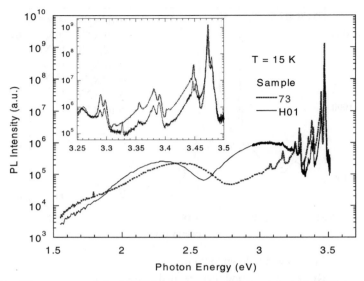

**Fig. 52.** PL spectrum of the MOCVD-overgrown GaN layer (sample H01) in comparison with the spectrum of the freestanding GaN template (sample 73). Excitation density is 0.1 W/cm$^2$. The inset enlarges excitonic part of the spectrum.

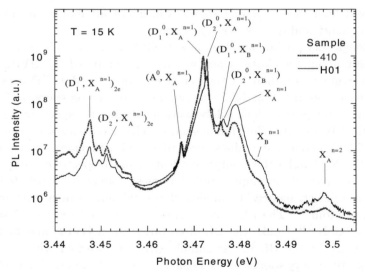

**Fig. 53.** Excitonic PL spectrum of the MOCVD-overgrown GaN layer (sample H01) in comparison with the spectrum of the freestanding GaN template (sample 410). Excitation density is 100 W/cm$^2$.

corresponding defects with holes, and at highest excitation density these bands can be barely detected in the spectrum (intensity is 5 orders of magnitude weaker than the peak intensity of the main exciton peak).

Excitonic spectrum of the undoped GaN epilayer is shown in comparison with the spectrum of the freestanding GaN template in Fig. 53. Positions and shapes of exciton peaks are identical in freestanding substrate, MBE- and MOCVD-overgrown layers and only relative intensities of some peaks slightly differ (Compare Figs. 26 and 53). All discussions and attributions made in Sec. 4.1 for the MBE-overgrown GaN are valid for MOCVD-overgrown GaN.

### 4.3.8. *Growth of ternary alloys*

One important advantage of Group III nitrides as in the case of the arsenide, phosphide and antiminide based compound semiconductors, is the possibility of varying the bandgap, in the limit, between 1.9 and 6.2 eV by the deposition of InGaN and AlGaN alloys. Ternary alloys of GaN, AlN and InN are used in heterostructures ubiquitously. As such their properties and in particular their dependence on the chemical composition is very important. Among the most important parameter of these ternaries is the dependence of their bandgap on the composition. At the current stage of device development, any growth method intended for electronic and optical devices must be able to deal with alloy growth. Below, the properties of nitride ternaries pertinent to growth issues are discussed.

It is clear to those who grow GaN that it is really a challenge to grow high quality material given the lack of native substrates and disparity in the properties of Ga and N atoms. The situation is exacerbated when one adds In to produce InGaN and Al to produce AlGaN. In fact if the hetero-structure contains both ternaries and GaN binary, the growth conditions must be modified throughout the run to suit each layer. The alloy growth becomes rather difficult when about more than 20% of InN or AlN is attempted to be added to the GaN lattice. The quality of AlGaN degrades with increasing AlN mole fraction. Fortuitously, the LT buffer insertion has been some help in that the quality of AlGaN has been improved. As for InGaN, while high quality requires growth temperatures in excess of approximately 800°C, the amount of indium incorporated is limited to low values because of the high volatility of nitrogen and the decomposition of InN and InGaN.

### 4.3.8.1. *AlGaN*

Ternary alloys of wurtzite and zincblende polytypes of GaN with AlN form a continuous alloy system with a wide range of bandgap with a small change in the lattice constant. An accurate knowledge of the compositional dependence of the barrier as well as material is a requisite in attempts to analyze heterostructures in general and quantum wells and super-lattices in particular. The barriers can be formed of AlGaN or GaN, and while dependent on the barrier material, the wells can be formed of GaN or InGaN layers. The compositional dependence of the lattice constant, the direct energy gap, electrical and CL properties of the AlGaN alloys was measured by Yoshida *et al.*[65] The compositional dependence of the principle bandgap of AlGaN can be calculated from the following empirical expression providing that the bowing parameter, $b$, is known accurately:

$$E_g(x) = x\,E_g(\text{AlN}) + (1 - x)E_g(\text{GaN}) - bx(1 - x)$$

where $E_g(\text{GaN}) = 3.4$ eV, $E_g(\text{AlN}) = 6.20$ eV, $x$ is the AlN molar fraction, and $b$ is the bowing parameter and has controversial values as discussed below. A consensus is not attained about the compositional dependence of the AlGaN bandgap, as would be expected since the field is still developing. Independent investigators came up with widely varying conclusions. For example, Yoshida *et al.*[65] concluded that as the Al mole fraction increases the energy bandgap of $\text{Al}_x\text{Ga}_{1-x}\text{N}$ deviates upwards implying a negative value for the bowing parameter, $b$. This contrasts the data of Wickenden *et al.*[303] which support a vanished bowing parameter, $b$. Moreover, Koide *et al.*[304] observed that the bowing parameter is positive and that the bandgap of the alloy deviates downward indicating a positive value for the bowing parameter. In the compositional dependence calculation of Amano *et al.*[305] the contribution of joint density of states and three dimensional exciton of $E_0$ specific point were taken into consideration. With the ability to observe the contribution of both joint density of state and three dimensional exciton close to the bandgap energy, Amano *et al.*[305] claimed to be able to determine the bandgap of AlGaN within an error of less than 10 meV. Since tensile strain in AlGaN layer is less than 0.5%, a strain-induced change in transition energy is expected to be small.

Continuing on with efforts to get at the compositional dependence of AlGaN bandgap, Ochalski *et al.*[306] used photo reflectance (PR) to measure the AlN composition as opposed to photoluminescence since it gives an insight into the joint density of states, which reduces the influence of impurities or band-tailing effects. The optical observed transitions in AlGaN may not be intrinsic in nature, which could cause errors. Using films in which the AlN mole fractions were determined by growth calibrations in conjunction with X-ray measurements, Ochalski *et al.* suggested a quasi-zero band-gap bowing in wurtzite $Al_x Ga_{1-x}N$ alloys. These measurements were based on a series of samples grown by various both molecular beam epitaxy and organometallic vapor deposition methods. Plotted in Fig. 54 are our experimental data, together with results of previous investigations. Obviously, in contrast with GaInN,[307] one finds no bowing of the $x$-dependence of the band-gap in $Al_x Ga_{1-x}N$. This result is consistent with the most recent theoretical investigation.[308] Nevertheless, the theoretical modeling of bowing parameters is not straightforward; as discussed by Wright and Nelson,[308] and the result of such a calculation may be related to intimate details of the models. It should be noted that the literature values of the bowing parameter vary between $-0.4$ to about 1 which need to be refined. This uncertainty causes the refractive index dependence on the composition to also be unreliable. However,

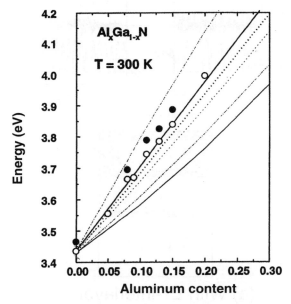

**Fig. 54.** Plot of the transition energies versus aluminum composition together with Band-gap composition dependencies previously reported in the literature. Solid symbols represent the bandgap determined from the position of the C line, open symbols indicte the energy position of the A-B lines. All the lines represent the data available in the literature. After Ref. 306.

when the refractive index is plotted against the bandgap, the data behave as expected.[309] More work is under way to narrow this variation in reported values.

Returning to growth related issues, Bremser *et al.*[310] and Ruffenach-Clur *et al.*[311] reported growth of AlGaN films over the entire range of compositions at substrate temperatures between 980 and 1150°C. Ruffenach-Clur *et al.* observed a high efficiency of Al incorporation in AlGaN films grown at 1000°C. The Al content at a given precursor input $Me_3Al/(.Me_3Al+Me_3Ga)$ ratio showed a super linear relationship for all compositions. Nevertheless, the structural quality of the AlGaN films observed by Ruffenach-Clur *et al.*[311] and Kung *et al.*[312] decreased with increasing Al content. The FWHM of the XRD (0002) rocking curve

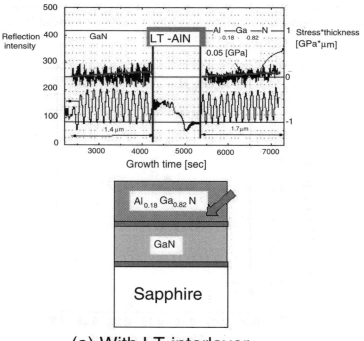

## (a) With LT-interlayer

**Fig. 55.** Schematic structure of AlGaN/GaN heterostructure grown (a) with and (b) without LT-interlayer. Upper figures show stress * thickness product and reflectivity measured *in situ*. The value of the grown-in tensile stress for each growth process is also shown. Courtesy of H. Amano.[313]

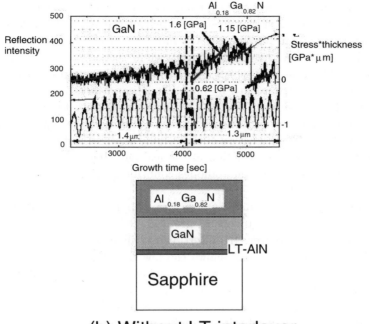

(b) Without LT-interlayer

**Fig. 55**   (*Continued*)

broadened up to 9 arcminute for $Al_{0.85}Ga_{0.15}N$. Bremser *et al.*[310] determined the optimum growth temperature for AlGaN films with $x$ above 0.5 to be around 1130°C. From the published data, one can speculate that the substrate temperature for the deposition of $Al_x Ga_{1-x}N$ films by MOCVD should be increased linearly with $x$ from 1000–1050°C at $x = 0$ to about 1150°C for $x = 1$.

As in the case of GaN growth, the quality of AlGaN also improved with LT buffer insertion as demonstrated by Amono *et al.*[313] Although the crystalline quality of AlGaN on sapphire improves by using an LT buffer layer, it progressively degrades with increasing AlN content. When the underlying GaN is improved, so did the AlGaN quality but a network of high-density cracks caused by the tensile stress which is induced by the lattice mismatch between AlGaN and GaN is generated when the thickness of AlGaN exceeds a critical value. This problem is alleviated by LT-AlN insertion. Amano *et al.*[313] inserted a LT-AlN between the underlying GaN layer and the upper AlGaN layer. The LT-interlayer reduces the tensile stress and threading dislocations which have screw components. Figure 55 shows the difference of the grown-in stress of the $Al_{0.18}Ga_{0.82}N$ on GaN (a) with and (b) without the LT-AlN interlayer. Nearly strain free AlGaN could be grown on the LT-interlayer, while relaxation occurs during a growth of $Al_{0.18}Ga_{0.82}N$ which is confirmed by the steep decrease of the stress*thickness product in Fig. 55(b). It is important to emphasize that the crystalline quality of this AlGaN is much superior to that grown on sapphire covered with only one LT buffer layer directly on sapphire, as confirmed by TEM. The reduction of the density of screw and mixed threading dislocations leads to a reduction in the leakage current in solar blind UV photoconductors and pin photodiodes confirming that these defects act as a current leakage path.

A technique similar to LEO has been applied to grow AlGaN with reduced defect density, by using grooved GaN templates on which to grow AlGaN.[313] However, LT insertion is still necessary to alleviate the cracking problem. Figure 56 shows the low temperature CL image of an $Al_{0.25}Ga_{0.75}N$ layer grown on grooved GaN covered with the LT-interlayer. As shown in Fig. 56, several dark spots are visible in the grooved region while it is entirely dark in the terraced regions which implies that edge dislocations act as nonradiative recombination centers. The upper limit for the density of the dark spots on the grooved region is around $10^7$ $cm^{-2}$. Therefore, a reduction in the threading dislocation density by as many as two orders of magnitude was achieved.

### 4.3.8.2. InGaN

The growth of GaInN alloys has proven to be relatively more difficult. The large difference in interatomic spacing between GaN and InN and high nitrogen pressure over InN due to N volatility are causes that give rise to a solid phase miscibility gap. Growth at high temperature (800°C) results in higher crystalline quality, but the amount of InN in the alloy is low and very high $NH_3$ flow rates are required, on the order of 20 l/min. Conversely, growth at lower temperatures (500°C) increases the InN concentration but at the expense of low crystalline quality.[314] Increasing In pressure in the vapor results in formation of In droplets on the surface.[315]

The first growth of single crystalline GaInN by MOVPE was realized by Nagatomo *et al.* in 1989[316] and followed by Yoshimoto *et al.* in 1991.[317] Since then, considerable work has been expended worldwide. Needless to say, high-efficiency blue and green LEDs utilizing InGaN active layers are

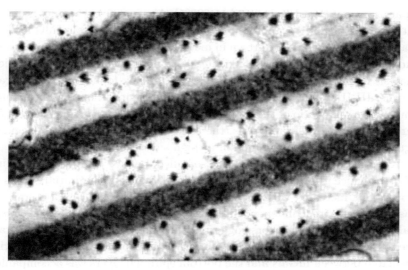

**Fig. 56.** CL image taken at cryogenic temperature of $Al_{0.25}Ga_{0.75}N$ grown on grooved GaN. From Amano.[313]

commercially available, and many reviews have been published already.[2] Matsuoka *et al.*[323] discovered that lowering the growth temperature to 500°C from nominal temperatures, such as 800°C, increased the In content in the layers, but at the expense of reduced quality. Efforts to increase the In concentration by raising the indium precursor temperature or the carrier gas flow rate resulted in the degradation of the structural and surface morphology so much so that In droplets were formed on the surface.[318]

The great disparity between Ga and In could lead to issues such as phase separation and instabilities. In this vein, Ho and Stringfellow[319] investigated the temperature dependence of the binodal and spinodal boundaries in the InGaN system with a modified valence force field model. The calculation of the extent of the miscibility gap yielded an equilibrium InN mole fraction in GaN of less than 6% at 800°C.[319] In the annealing experiments in an argon ambient, the phase separation of an $Ga_{1-x}In_xN$ alloy with $x \geq 0.1$ was observed at 600 and 700°C,[320] pointing to the large region of solid immiscibility of these alloys. However, under non-equilibrium growth conditions, $Ga_{1-x}In_xN$ layers were grown in the entire range of compositions. Decomposition into two phases upon annealing of the $Ga_{1-x}In_xN$ alloys ($x = 0.11$ and $x = 0.29$) at 600°C and 700°C was observed pointing to the existence of the miscibility gap. For some alloys with $x = 0.6$ the phase separation could not be observed at 600°C. Above 800°C the alloy samples with $x = 0.1$ actively evaporated from the substrate. These results suggest that the solid solutions are grown in metastable conditions and decomposed under annealing conditions. Koukitu *et al.*[321] performed a thermodynamical analysis of InGaN alloys grown by MOCVD. They found that in contrast to other III-III-V alloy systems where the solid composition is a linear function of the molar ratio of the Group III metalorganic precursors at constant partial pressure of Group V gas, the solid composition of InGaN deviates significantly from a linear function at high substrate temperatures.

Kawaguchi *et al.*[323] reported a so-called InGaN composition pulling effect in which the indium fraction is smaller during the initial stages of growth but increases with increasing growth thickness. This observation was to a first extent independent of the underlaying layer whether it was GaN or AlGaN. The authors suggested that this effect is caused by strain due to the lattice mismatch at the interface. It was found that a larger lattice mismatch between InGaN and the bottom epitaxial layers was accompanied by a larger change in the In content. What one can glean from this is that the indium distribution mechanism in InGaN alloy is caused by the lattice deformation due to the lattice mismatch. With increasing thickness, the lattice strain is relaxed due to the formation of structural defects, which weakens the composition pulling effect. Table 10 summarizes general trends between deposition parameters and structural qualities and the In incorporation in InGaN.

Other substrates were also used for InGaN growth. It was reported that the crystalline quality of InGaN was superior when grown with the composition that lattice matches ZnO substrate than when grown on bare (0001) sapphire substrate.[317,323] In the same investigations, it was observed that InGaN films grown on sapphire substrates using GaN as buffer layers exhibited much better optical properties than InGaN films grown directly on sapphire substrates.[324] For a given set of growth conditions an increase of InN in InGaN can be achieved by reducing the hydrogen flow.[325]

The bandgap dependence of $In_xGa_{1-x}N$ on In mole fraction has been studied by a number of researchers. The energy bandgap of $In_xGa_{1-x}N$ across $0 \leq x \leq 1$ can be expressed by

$$E_g(x) = (1 - x)E_g(\text{GaN}) + xE_g(\text{InN}) - bx(1 - x)$$

where $E_g(\text{GaN}) = 3.40$ eV, $E_g(\text{InN}) = 2.07$ eV, and $b = 1.0$ eV. The observed and calculated dependencies of InGaN energy bandgap with InN mole fraction are displayed in Fig. 57. The square data points are from

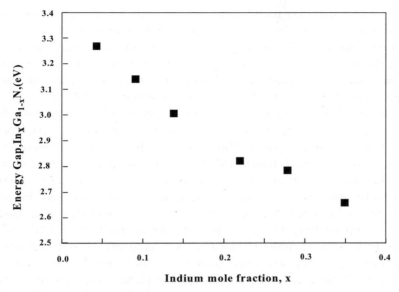

**Fig. 57.** Dependence of InGaN bandgap energy on composition. The composition is determined assuming that InGaN is homogeneous. Courtesy of S. Bedair.

Ref. 305. For calculations one gets a positive bowing parameter of 3.2 eV. However, the results of Nakamura *et al.*[326] support a bowing parameter of +1.0 eV which obviously do not agree with those of Ref. 305. Since the InGaN is grown on GaN, there are many complicating factors such as piezoelectric effect and inhomogeneities and thus non-uniform strain; the impact of the former can be made negligible by growing thick films. If the strain caused by the lattice mismatch were uniform, caused by the InN lattice constant being 11% larger than GaN, it would be compressive and optical transitions woud experience an accompanying blue shift. Herein lies the dilemma faced by the experimentalists. The extent of strain and piezoelectric effect could be minimized by growing thick films but compositional inhomogeneities and phase separation complicate matters. Additional complications involve the need for the relaxation value of the lattice constant and the origin of the optical transitions to be known accurately to determine the bandgap vs. composition dependence. Absorption and/or reflection measurements, providing that the absorption edge is sharp, are more useful in determining the bandgap but again require thick and/or good films which are lacking. Detailed X-ray reciprocal space mapping undertaken by Amano *et al.*[305] are purported to indicate that InGaN wells and even somewhat thicker InGaN layers grown on GaN buffer layers are coherently strained, a conclusion reached by the observation that the in-plane lattice constants of GaN and InGaN match. At the same time though, the layer thicknesses well exceeded the calculated critical values. This paradox remains unresolved, possibly until such time when heterostructures with much reduced structural defect concentrations can be attained.

Progress is being made, however, on the front of better understanding processes involved in the growth of In containing ternaries. Reaction pathways in effect during growth of In-based alloys have been investigated with good progress.[327] Such insight led to the attainment of InGaN films with high InN molar fraction, and improved structural and optical properties. It is to be noted that the phase separation is a real concern in this alloy particularly for high InN molar fraction. Using a unique susceptor, Bedair *et al.*[327] reported $In_{0.1}Ga_{0.9}N$ layers with symmetric X-ray FWHM values of as little as 6 arcminute. The PL was characterized as being intense bandedge type with occasional deep level transitions. In this particular approach, dubbed molecular stream epitaxy (MSE), the substrate is first exposed to a mixed gas composition inclusive of Groups III and V species. The sample is then rotated away from the impinging gas along the axis of the susceptor. When a complete rotation is made, the substrate is again exposed to reactant gases.

### 4.3.8.3. *AlInN*

The attraction to AlInN stems from the fact that it can be lattice matched to GaN and AlGaN in a wide range of compositions while engineering its bandgap. Doing so would lead to high quality heterointerfaces as well as the elimination of strain induced piezoelectric polarization. Despite this lure, only a very limited number of publications concerning the deposition of AlInN films is available. The difficulties associated with growing AlInN alloys are similar to or even greater than (higher differences in lattice spacing, thermal stability) those experienced in the deposition of InGaN films owing in part to the large difference of the epitaxial temperature of AlN and InN. Single crystalline and luminescent AlInN was first grown by Vartuli *et al.* and was found to contain both cubic and hexagonal form.[328] As-grown $In_{0.75}Al_{0.25}N$ alloy was *n*-type conducting ($n \approx 1019$ $cm^{-3}$) due to native defects. After annealing at 800°C, the sheet resistance of the samples decreased by two orders of magnitude. This alloy was synthesized by Yamaguchi *et al.* as well in 1998.[329] Photoluminescence obtained in samples with varying mole fractions point to a large bowing of the band gap in this system.[329]

Several groups have used AlInN or AlGaInN as the barrier layers in the MQWs[330,331] and reported the superior properties when they are applied to UV emitters and UV detectors. In addition to optoelectronics devices, applications to electronics devices are also promising because it is possible to control the lattice mismatch between GaN, either compressive or tensile strain. It is well established by now that large spontaneous polarization and piezoelectric polarization[332,333] induces high-density two-dimensional electron gas at the interface of GaN/AlGaN heterostructure with great electronic device implications. Use of the AlInN/GaN heterostructure would eliminate the strain and thus the strain induced polarization charge. Consequently, one can limit the interfacial charge to spontaneous polarization charge and could lower the sheet carrier concentration formed at the interface. In another report, Kim *et al.*[334] deposited thin AlInN films with X-ray rocking curve FWHM values between 10 and 20 arcminute. They observed an increase of In content in AlInN of up to 8% by lowering the substrate temperature to 600°C. A further reduction of substrate temperature during MOCVD is not useful because of the needed efficient pyrolysis of ammonia.

### 4.3.9. *Growth of quaternary alloys*

By alloying InN together with GaN and AlN, the band gap of the resulting alloy(s) can be increased from 1.9 eV to a value of 6.2 eV, which is critical for making high efficiency visible light sources and UV detectors. Quarternary alloys permit to widen the selection of materials that can simultaneously satisfy both the lattice match and band energy requirements. In quarternary alloys of nitrides the N atoms constitute anion sublattice while three groups elements (In, Ga, Al) constitute the cation sublattice. Quarternary alloys, $Ga_xAl_yIn_{1-x-y}N$ are expected to exist in the entire composition range $0 < x < 1$ and $0 < y < 1$. Unfortunately, as in the case of the InGaN alloy, incorporation of indium in these quaternary alloys is not easy. To prevent InN dissociation, InGaN crystal was originally grown at low temperatures (about 500°C).[335] The use of a high nitrogen flux rate allowed the high temperature (800°C) growth of high quality InGaN and InGaAlN films on (0001) sapphire substrates. It was noted that the incorporation of indium into InGaN film is strongly dependent on the flow rate, N/III ratio and growth temperature. The incorporation efficiency of indium decreases with increasing growth temperatures. Observations made in the case of InGaN should be applicable to In incorporation in quaternary nitrides.

## 5. Conclusions

Semiconductor III-Nitrides, primarily GaN, AlN, InN, and their ternaries have attracted considerable interest after an uneven path with ups and downs in effort around the first half of nineteen seventies followed by an intense activity in the latter part of the nineteen nineties and the following years. Substrates used for nitride growth, growth methods, growth of nitrides by MBE, HVPE and MOCVD have been reviewed. The inability to attain very high substrates temperatures and lack of reducing gases, such as H, in the MBE environment have prevented high quality GaN growth for some time by this technique. With the *ex situ* annealing technique described in the text for sapphire and SiC, issues with respect to the substrates were made moot. However, a viable method is yet to be developed for ZnO, but thermal annealing in $O_2$ appears to give smooth surfaces, though chemically clean surfaces are yet to be achieved on a regular basis. The nature of MBE growth lends itself to growth with both Ga and N-polarity with some control over polarity. Imbedded GaN quantum dots with AlN wetting layers were used to nucleate dots on defect sites and either loop or annihilate them to the surface of the dots. Using well-established chemical etching techniques to delineate defects, the defect concentration was reduced as much as 3 orders of magnitude, which is remarkable. Defect stoppage was confirmed by TEM.

High quality GaN layers and templates have been grown with HVPE. In fact, the freestanding GaN templates grown by this method has so far exhibited the best combined properties in terms of impurity concentration, extended defect concentration and electronic transport properties. The total impurity concentration away from the original epi-substrate interface is in the low $10^{16}$ $cm^{-3}$. The extended defect concentration is on the order of low $10^5$ $cm^{-3}$. The electron mobility approaches a remarkable 1400 $cm^2V^{-1}s^{-1}$ at room temperature. The layers grown on these templates by both MBE and MOCVD exhibited excellent properties. MOCVD with available lateral epitaxial overgrowth is everpresent in the realm of nitrides. The growth methods, the processes involved in the growth, LEO and its variants, low temperature buffer insertions, selective growth using metal masks, growth on grooved substrates, growth of binary and ternary alloys associated with the MOCVD process have been reviewed. The results presented indicate that layers grown with MBE and MOCVD on high quality templates look very promising.

## Acknowledgments

This work was funded by AFOSR (Dr. G. L. Witt, D. Johnstone, G. Pomrenke), NSF (Drs. L. Hess and U. Varshney), and ONR (Drs. C. E. C. Wood and Y. S. Park). The author would like to thank colleagues at VCU and members of the Wood Witt program dealing with defect characterization in GaN for discussions and data exchange. Much of the discussion dealing with the characterization of HVPE material is based on the effort conducted under the Wood Witt program. He specifically thanks Prof. R. Davis, Prof. S. Bedair, Prof. K. Hiramatsu, Dr. M. Koike, Dr. H. Amano, and Dr. D. C. Look for providing many reprints, preprints, and figures for this review. Prof. D. J. Smith of Ariziona State University produced the cross-sectional images of Qdot samples. The author thanks Dr. F. Yun for reading the manuscript.

## References

1. H. Morkoç, *Nitride Semiconductors and Devices* (Springer-Verlag, Heidelberg, 1999).
2. S. N. Mohammad and H. Morkoç, *Prog. Quantum Electron.* **20** (5–6), 361 (1996).
3. S. N. Mohammad, A. Salvador and H. Morkoç, *Proc. IEEE* **83**, 1306–1355 (1995).
4. H. Morkoç, S. Strite, G. B. Gao, M. E. Lin, B. Sverdlov and M. Burns, *J. Appl. Phys. Rev.* **76** (3), 1363 (1994).
5. S. T. Strite and H. Morkoç, *J. Vacuum Sci. Technol.* **B10**, 1237 (1992).
6. O. Ambacher, *J. Phys. D: Appl. Phys.* **31**, 2653 (1998).
7. S. J. Pearton, J. C. Zolper, R. J. Shul and F. Ren, *J. Appl. Phys.* **86**, 1 (1999).
8. S. C. Jain, M. Willander, J. Narayan and R. V. Overstraeten, *J. Appl. Phys.* **87**, 965 (2000).
9. M. T. Duffy, C. C. Wang, G'D. O'Clock, S. H. McFarlane III and P. J. Zanzucchi, *J. Electron. Mater.* **2**, 359 (1973).
10. M. Razeghi and A. Rogalski, *J. Appl. Phys.* **79**, 7433 (1996).
11. G. Y. Xu, A. Salvador, W. Kim, Z. Fan, C. Lu, H. Tang, H. Morkoç, G. Smith, M. Estes, B. Goldenberg, W. Yang and S. Krishnankutty, *Appl. Phys. Lett.* **71**, 2154 (1997).
12. S. Nakamura, T. Mukai and M. Senoh, *Appl. Phys. Lett.* **64**, 1687 (1994).
13. For a review see: H. Morkoç and S. N. Mohammad, *Science* **267**, 51 (1995).

14. For a review, see: H. Morkoç and S. N. Mohammad, *Wiley Encyclopedia of Electrical Engineering and Electronics Engineering*, Ed. J. Webster (John Wiley and Sons, New York 1999).

15. S. Nakamura, M. Senoh, N. Nagahama, N. Iwara, T. Yamada, T. Matsushita, H. Kiyoku, Y. Sugimoto, T. Kozaki, H. Umemoto, M. Sano and K. Chocho, *Jpn. J. Appl. Phys.* **38**, L1578 (1997).

16. H. Morkoç, *IEEE J. Selected Topics in Quantum Electronics*, Eds. R. Miles and I. Akasaki, **4**, 537 (1998).

17. J. Schetzina *et al.*, in *Proceedings of the International Conference on Silicon Carbide and Related Materials*, October 10–15, 1999, Research Triangle Park, NC, USA.

18. B. K. Ridley, *J. Appl. Phys.* **84**, 4020 (1998).

19. U. V. Bhapkar and M. S. Shur, *J. Appl. Phys.* **82**, 1649 (1997).

20. J. Kolnik, I. H. Oguzman, K. F. Brennan, R. Wang, P. P. Ruden and Y. Wang, *J. Appl. Phys.* **78**, 1033 (1995).

21. H. Morkoç, in *SiC Materials and Devices*, Ed. Y. S. Park, Academic Press, Willardson and Beer Series, Eds. Willardson and Weber, **52** (8), 307–394 (1998).

22. N. Nguyen and C. Nguyen of HRL Laboratories (private communication).

23. M. Micovic, A. Kurdoghlian, P. Janke, P. Hashimoto, D. W. S. Wong, J. S. Moon, L. McCray and C. Nguyen, *IEEE Trans. Electron. Devices* **48** (3), 591–592 (2001).

24. N. Nguyen and C. Nguyen, and of HRL Laboratories (private communication).

25. R. J. Molnar, W. Goetz, L. T. Romano and N. M. Johnson, *J. Cryst. Growth* **178**, 147 (1997).

26. S. Yamaguchi, M. Kariya, S. Nitta, T. Takeuchi, C. Wetzel, H. Amano and I. Akasaki, *J. Appl. Phys.* **85**, 7682 (1999).

27. D. C. Look, D. C. Reynolds, J. W. Hemsky, J. R. Sizelove, R. L. Jones and R. J. Molnar, *Phys. Rev. Lett.* **79**, 2273 (1997).

28. S. Nakamura, T. Mukai and M. Senoh, *J. Appl. Phys.* **71**, 5543 (1992).

29. M. E. Lin, B. Sverdlov, G. L. Zhou and H. Morkoç, *Appl. Phys. Lett.* **62**, 3479 (1993).

30. H. M. Ng, D. Doppalapudi, T. D. Moustakas, N. G. Weimann and L. F. Eastman, *Appl. Phys. Lett.* **73**, 821 (1998).

31. D. C. Look and J. R. Sizelove, *Phys. Rev. Lett.* **82**, 1237 (1999).

32. N. G. Weimann, L. F. Eastman, D. Doppalapudi, H. M. Ng and T. D. Moustakas, *J. Appl. Phys.* **83**, 3656 (1998).

33. Z. Q. Fang, D. C. Look, W. Kim, Z. Fan, A. Botchkarev and H. Morkoç, *Appl. Phys. Lett.* **72**, 2277 (1998).

34. K. Wook, A. E. Botohkarev, H. Morkoç, Z. Q. Fang, D. C. Look and D. J. Smith, *J. Appl. Phys.* **84**, 6680 (1998).

35. B. Heying, I. Smorchkova, C. Poblenz, C. Elsass, P. Fini, S. Den Baars, U. Mishra and J. S. Speck, *Appl. Phys. Lett.* **77**, 2885 (2000).

36. M. J. Manfra, L. N. Pfeiffer, K. W. West, H. L. Stormer, K. W. Baldwin, J. W. P. Hsu, D. V. Lang and R. J. Molnar, *Appl. Phys. Lett.* **77**, 2888 (2000).

37. J. Cui, A. Sun, M. Reshichkov, F. Yun, A. Baski and H. Morkoç, MRS Internet Journal — The URL for the front page is http://nsr.mij.mrs.org/5/7/.
38. J. A. Powell, D. J. Larkin and A. J. Trunek in *Silicon Carbide, III-Nitrides, and Related Materials*, Eds. G. Pensl, H. Morkoç, B. Monemar and E. Janzen (Trans Tech Publications, Sweden, 1998) **264–268**, pp. 421–424.
39. E. J. Tarsa, B. Heying, X. H. Wu, P. Fini, S. P. DenBaars and J. S. Speck, *J. Appl. Phys.* **82**, 5472 (1997).
40. B. Heying, R. Averbeck, L. F. Chen, E. Haus, H. Riechert and J. S. Speck, *J. Appl. Phys.* **88**, 1855 (2000).
41. N. Grandjean and J. Massies, CHREA-CNRS, Valbonne, France (private communication).
42. M. Leszczynski, I. Grzegory and M. Bockowski, *J. Cryst. Growth* **126**, 601 (1993); S. Porowski, *Acta Phys. Polon.* **A87**, 295 (1995); S. Porowski, I. Grzegory and J. Jun, in *High Pressure Chemical Synthesis*, Eds. J. Jurczak and B. Baranowski (Elsevier, Amsterdam, 1989) p. 21.
43. G. A. Slack and T. F. McNelly, *J. Cryst. Growth* **34**, 263 (1976).
44. H. P. Maruska and J. J. Tietjen, *Appl. Phys. Lett.* **15**, 327 (1969).
45. I. Akasaki and H. Amano, *Int. Symp. Blue Lasers and Light Emitting Diodes*, Chiba University, Japan (1996) pp. 11–16.
46. G. Popovici and H. Morkoç, *Deposition and Properties of III-Nitrides by Molecular Beam Epitaxy in iPhysics and Applications of Group III Nitride Semiconductors*, Ed. B. Gill, Oxford University (1998).
47. T. Detchprohm, H. Amano, K. Hiramatsu and I. Akasaki, *J. Cryst. Growth* **128**, 384 (1993).
48. D. Huang, F. Yun, P. Visconti, M. A. Reshchikov, D. Wang, H. Morkoç, D. L. Rode, L. A. Farina, Ç. Kurdak, K. T. Tsen, S. S. Park and K. Y. Lee, *Solid State Electron.* **45** (5), 711–715 (2001).
49. H. P. Maruska and J. J. Tietjen, *Appl. Phys. Lett.* **15**, 327 (1969).
50. Gene Cantwell, Eagle Picher Technologies, LLC (private communication).
51. J. A. Powell, D. J. Larkin, P. G. Neudeck, J. W. Yang and P. Pirouz, in *Silicon Carbide and Related Materials*, Eds. M. G. Spencer, R. P. Devaty, J. A. Edmond *et al.* (IOP Publishing, Bristol, 1994) pp. 161–164.
52. C. D. Lee, V. Ramachandran, A. Sagar, R. M. Feenstra, D. W. Greve, W. L. Sarney, L. Salamanca-Riba, D. C. Look, Bai Song Bai, W. J. Choyke and R. P. Devaty, *TMS*; *IEEE J. Electron. Mater.* **30** (3), 162–169 (2001).
53. T. Lei, K. F. Ludwig Jr. and T. Moustakas, *J. Appl. Phys.* **74**, 4430 (1993).
54. R. C. Powell, N. E. Lee, Y.-W. Kim and J. Green, *J. Appl. Phys.* **73**, 189 (1993).
55. J. Nause, F. Yun and H. Morkoç (unpublished).
56. G. A. Slack and T. F. McNelly, *J. Cryst. Growth* **34**, 263 (1976).
57. M. Leszczynski, I. Grzegory and M. Bockowski, *J. Cryst. Growth* **126**, 601 (1993); S. Porowski, *Acta Phys. Polon.* **A87**, 295 (1995); S. Porowski, I. Grzegory and J. Jun, in *High Pressure Chemical Synthesis*, Eds. J. Jurczak and B. Baranowski (Elsevier, Amsterdam, 1989) p. 21.

58. I. Grzegory, J. Jun, M. Bockowski, S. Krukowski, M. Wroblewski, B. Lucznik and S. Porowski, *J. Phys. Chem. Solids* **56**, 639 (1995).
59. G. Popovici, S. N. Mohammad and Hadis Morkoç, *Group III Nitride Semiconductor Compounds: Physics and Applications* Ed. B. Gil (Clarendon Press, 1998). Oxford ISBN 0-19-850159-5.
60. D. D. Koleske, A. E. Wiekenden, R. L. Henry, W. J. DeSisto and R. J. Gorman, *J. Appl. Phys.* **84** (4), 1998–2010 (1998).
61. S. Yoshida, S. Misawa and A. Itoh, *Appl. Phys. Lett.* **26**, 461 (1975).
62. S. Winsztal, B. Wauk, H. Majewska-Minor and T. Niemyski, *Thin Solid Films* **32**, 251 (1976).
63. K. R. Elliott and R. W. Grant, Rockwell Project Final Report MRDC41116.2FR (1984).
64. H. U. Baier and W. Mönch, *J. Appl. Phys.* **68**, 586 (1990).
65. S. Yoshida, S. Misawa and S. Gonda, *J. Vac. Sci. Technol.* **B1**, 250 (1983).
66. W. Kim, Ö. Aktas, A. Botchkarev and H. Morkoç, *J. Appl. Phys.* **79** (10), 7657–7666 (1996).
67. H. Gotoh, T. Suga, H. Suzuki and M. Kimata, *Jpn. J. Appl. Phys.* **20**, L545 (1981).
68. K. W. Boer, *Survey of Semiconductor Physics*, **1** (Van Nostrand Reinhold, New York, 1990).
69. A. Botchkarev, A. Salvador, B. Sverdlov, J. Myoung and H. Morkoç, *J. Appl. Phys.* **77**, 4455 (1995).
70. V. S. Ban, *J. Electrochem. Soc.* **119**, 761 (1972).
71. H. P. Maruska, W. C. Rhines and D. A. Stevenson, *Mater. Res. Bull.* **7**, 777 (1972).
72. D. K. Wickenden, K. R. Faulkner, R. W. Brander and B. J. Isherwood, *J. Cryst. Growth* **9**, 158 (1971); M. Ilegems, *J. Cryst. Growth* **13/14**, 360 (1972); G. Burns, F. Dacol, J. C. Marinace, B. A. Scott and E. Burstein, *Appl. Phys. Lett.* **22**, 356 (1973).
73. W. Seifert, G. Fitzl and E. Butter, *J. Cryst. Growth* **52**, 257 (1981); see also, W. Seifert and A. Tempel, *Phys. Status Solidi* **A23**, K39 (1974); A. Shintani and S. Minagawa, *J. Cryst. Growth* **22**, 1 (1974); Y. M. Suleimanov, I. G. Pichugin and L. A. Marasina, *Sov. Phys. Semicond.* **8**, 537 (1974); B. Monemar, *Phys. Rev.* **B10**, 676 (1974); T. Matsumoto, M. Sano and M. Aoki, *Jpn. J. Appl. Phys.* **13**, 373 (1974).
74. E. Ejder and P. O. Fagerström, *J. Phys. Chem. Solids* **36**, 289 (1975); R. K. Crouch, W. J. Debnam and A. L. Fripp, *J. Mater. Sci.* **13**, 2358 (1978); R. Fremunt, P. Cerny, J. Kohout, V. Rosicka and A. Bürger, *Cryst. Res. Tech.* **16**, 1257 (1981); A. S. Adonin, V. A. Evmenenko, L. N. Mikhailov and N. G. Ryabtsev, *Inorg. Mater.* **17**, 1187 (1982); A. V. Kuznetsov, V. G. Galstyan, V. I. Muratova and G. V. Chaplygin, *Sov. Microelectron.* **11**, 214 (1983); K. Naniwae, S. Itoh, H. Amano, K. Itoh, K. Hiramatsu and I. Akasaki, *J. Cryst. Growth* **99**, 381 (1990).
75. See, for example, T. L. Chu, D. W. Ing and A. J. Noreika, *Electrochem. Technol.* **6**, 56 (1968); see also, W. M. Yim, E. J. Stofko, P. J. Zanzucchi, J. I. Pankove, M. Ettenberg and S. L. Gilbert, *J. Appl. Phys.* **44**, 292 (1973); T. L. Chu and R. W. Kelm Jr., *J. Electrochem. Soc.* **122**, 995 (1975).

76. For gaseous GaCl approach, see, B. Baranov, L. Däweritz, V. B. Gutan, G. Jungk, H. Neumann and H. Raidt, *Phys. Status Solidi* **A49**, 629 (1978); S. S. Liu and D. A. Stevenson, *J. Electrochem. Soc.* **125**, 1161 (1978); E. Butter, G. Fitzl, D. Hirsch, G. Leonhardt, W. Seifert and G. Preschel, *Thin Solid Films* **59**, 25 (1979); W. Seifert, R. Franzheld, E. Butter, H. Subotta and V. Riede, *Cryst. Res. Tech.* **18**, 383 (1983); H. Neumann, W. Seifert, M. Staudte and A. Zehe, *Krist. Tech.* **9**, K69 (1974).

77. For gaseous $GaCl_2$, $GaCl_3$ and $Ga(C_2H_5)_2Cl$ sources, see, D. Troost, H. U. Baier, A. Berger and W. Mönch, *Surf. Sci.* **242**, 324 (1991); P. J. Born and D. S. Robertson, *J. Mater. Sci.* **15**, 3003 (1980); J. J. Nickl, W. Jus and R. Bertinger, *Mater. Res. Bull.* **10**, 1097 (1975).

78. For gaseous $GaCl_2NH_3$ sources, see, L. A. Marasina, A. N. Pikhtin, I. G. Pichugin and A. V. Solomonov, *Phys. Status Solidi* **A38**, 753 (1976).

79. For gaseous $AlCl_3$ sources, see, M. P. Callaghan, E. Patterson, B. P. Richards and C. A. Wallace, *J. Cryst. Growth* **22**, 85 (1974); I. Akasaki and H. Hashimoto, *Solid State Comm.* **5**, 851 (1967); T. L. Chu, D. W. Ing and A. J. Noreika, *Solid State Electron.* **10**, 1023 (1967); E. A. Irene, V. J. Silvestri and G. R. Woolhouse, *J. Electron. Mater.* **4**, 409 (1975); J. Bauer, L. Biste and D. Bolze, *Phys. Status Solidi* **A39**, 173 (1977).

80. For gaseous $AlBr_3$, $InCl_3$, and $GaBr_3$ sources, see, Y. Someno, M. Sasaki and T. Hirai, *Jpn. J. Appl. Phys.* **30**, 1792 (1991); L. A. Marasina, I. G. Pichugin and M. Tlaczala, *Krist. Tech.* **12**, 541 (1977); T. L. Chu, *J. Electrochem. Soc.* **118**, 1200 (1971); Y. Morimoto, K. Uchiho and S. Ushio, *J. Electrochem. Soc.* **120**, 1783 (1973); S. E. Aleksandrov, A. Y. Kovalgin and D. M. Krasovitskii, *Russ. J. Appl. Chem.* **67**, 663 (1994).

81. J. Pastrnak and L. Souckova, *Phys. Status Solidi* **11**, K71 (1963); J. Pastrnak and L. Roskovcova, *Phys. Status Solidi* **9**, K73 (1965).

82. P. M. Dryburgh, *J. Cryst. Growth* **94**, 23 (1989).

83. O. Igarashi and Y. Okada, *Jpn. J. Appl. Phys.* **24**, L792 (1985); O. Igarashi, *Jpn. J. Appl. Phys.* **27**, 790 (1988).

84. R. J. Molnar, W. Goetz, L. T. Romano and N. M. Johnson, *J. Cryst. Growth* **178** (1–2), 147–156 (1997).

85. H. M. Manasevit, F. M. Erdman and W. I. Simpson, *J. Electrochem. Soc.* **118**, 1864 (1971).

86. H. Amano, N. Sawaki, I. Akasaki and Y. Toyoda, *Appl. Phys. Lett.* **48**, 353 (1986).

87. C. Yuan, T. Salagaj, A. Gurary, P. Zavadski, C. S. Chen, W. Kroll, R. A. Stall, Y. Li, M. Schurman, C. Y. Hwang, W. E. Mayo, Y. Lu, S. J. Pearton, S. Krishnankutty and R. M. Kolbas, *J. Electrochem. Soc.* **142**, L163 (1995).

88. S. Nakamura, S. Mukai and T. Senoh, *J. Appl. Phys.* **76**, 8189 (1994).

89. S. Nakamura, M. Senoh, S. Magahama, N. Iwasha, T. Yamada, T. Matsushita, H. Kioku and Y. Sugimoto, *Jpn. J. Appl. Phys.* **L74**, 1998 (1996).

90. M. A. Khan, J. N. Kuznia, Bhattarai and D. T. Olson, *Appl. Phys. Lett.* **62**, 1786 (1993).

91. F. Binet, J. Y. Duboz, E. Rosencer, F. Scholz and V. Harle, *Appl. Phys. Lett.* **69**, 1002 (1996).

92. S. Nakamura, Y. Harada and M. Seno, *Appl. Phys. Lett.* **58**, 2021 (1991).
93. See, for example, D. K. Wickenden, J. A. Miragliotta, W. A. Bryden and T. J. Kistnmachr, *J. Appl. Phys.* **75**, 7585 (1994); J. C. Knights and R. A. Lujan, *J. Appl. Phys.* **49**, 1291 (1978); A. E. Wickenden, D. K. Wickenden and T. J. Kistenmacher, *J. Appl. Phys.* **75**, 5367 (1994).
94. M. Matloubian and M. Gershenzon, *J. Electron. Mater.* **40**, 633–644 (1985).
95. A. Wakahara and A. Yoshida, *Appl. Phys. Lett.* **54**, 709 (1989).
96. T. Y. Sheng, Z. Q. Yu and G. J. Collins, *Appl. Phys. Lett.* **52**, 576 (1988).
97. S. Fujieda, M. Mizuta and Y. Matsumoto, *Jpn. J. Appl. Phys.* **26**, 2067 (1987).
98. D. K. Gaskill, N. Bottka and M. C. Lin, *Appl. Phys. Lett.* **48**, 1949 (1986).
99. M. Mizuta, S. Fujieda, T. Jitsukawa and Y. Matsumoto, *Proc. Int. Symp. GaAs and Related Compounds* (Las Vegas, Nevada, 1986) appeared in 1987 (Adam Hilger, Bristol).
100. S. Miyoshi, K. Onabe, N. Ohkouchi, H. Yaguchi and R. Ito, *J. Cryst. Growth* **124**, 439 (1992).
101. M. Ishii, T. Minami, T. Miyata, S. Karaki and S. Takata, *Inst. Phys. Conf. Ser.* **142**, 899 (1996).
102. H. M. Manasevit, F. M. Erdmann and W. I. Simpson, *J. Electrochem. Soc.* **118**, 1864 (1971); see also, F. A. Pizzarello and J. E. Coker, *J. Electron. Mater.* **4**, 25 (1975).
103. G. D. O'Clock Jr. and M. T. Duffy, *Appl. Phys. Lett.* **23**, 55 (1973).
104. J. K. Liu, K. M. Lakin and K. L. Wang, *J. Appl. Phys.* **46**, 3703 (1975).
105. M. Morita, N. Uesugi, S. Isogai, K. Tsubouchi and N. Mikoshiba, *Jpn. J. Appl. Phys.* **20**, 17 (1981).
106. M. A. Khan, R. A. Skogman, R. G. Schulze and M. Gershenzon, *Appl. Phys. Lett.* **42**, 430 (1983).
107. M. Hashimoto, H. Amano, N. Sawaki and I. Akasaki, *J. Cryst. Growth* **68**, 163 (1984).
108. M. Matloubian and M. Gershenzon, *J. Electron. Mater.* **14**, 633 (1985); see also, T. Sasaki and T. Matsuoka, *J. Appl. Phys.* **64**, 4531 (1988); T. Nagatomo, T. Kuboyama, H. Minamino and O. Omoto, *Jpn. J. Appl. Phys.* **28**, L1334 (1989); M. A. Khan, J. N. Kuznia, J. M. Van Hove, D. T. Olsen, S. Krishnankutty and R. M. Kolbas, *Appl. Phys. Lett.* **58**, 526 (1991).
109. S. Nakamura, Y. Harada and M. Seno, *Appl. Phys. Lett.* **58**, 2021 (1991).
110. See, for example, D. K. Wickenden, J. A. Miragliotta, W. A. Bryden and T. J. Kistnmachr, *J. Appl. Phys.* **75**, 7585 (1994); J. C. Knights and R. A. Lujan, *J. Appl. Phys.* **49**, 1291 (1978); A. E. Wickenden, D. K. Wickenden and T. J. Kistenmacher, *J. Appl. Phys.* **75**, 5367 (1994).
111. M. Shiloh and J. Gutman, *Mater. Res. Bull.* **8**, 711 (1973).
112. R. Lappa, G. Glowacki and S. Galkowski, *Thin Solid Films* **32**, 73 (1976).
113. T. Y. Sheng, Z. Q. Yu and G. J. Collins, *Appl. Phys. Lett.* **52**, 576 (1988).
114. A. Wakahara and A. Yoshida, *Appl. Phys. Lett.* **54**, 709 (1989).
115. E. N. Eremin, L. I. Nekrasov, E. A. Rubtsova, V. M. Belova, V. L. Ivanter, L. N. Zakharov and L. N. Petukhov, *Russ. J. Phys. Chem.* **56**, 788 (1982).
116. M. Matloubian and M. Gershenzon, *J. Electron. Mater.* **40**, 633–644 (1985).

117. J. L. Dupuie and E. Gulari, *Appl. Phys. Lett.* **59**, 549 (1991).
118. H. Liu, D. C. Bertolet and J. W. Rogers, *Surf. Sci.* **320**, 145 (1994).
119. A. C. Jones, C. R. Whitehouse and J. S. Roberts, *Chem. Vap. Dep.* **1**, 65 (1995).
120. A. C. Jones, J. Auld, S. A. Rushworth, D. J. Houlton and G. W. Critchlow, *J. Mater. Chem.* **4**, 1591 (1994).
121. A. C. Jones, J. Auld, S. A. Rushworth, E. W. Williams, P. W. Haycock, C. C. Tang and Critchlow, *Adv. Mater.* **6**, 6 (1994).
122. B. Beaumont, P. Gibart and J. P Faurie, *J. Cryst. Growth* **156**, 140 (1995).
123. A. P. Purdy, *Inorg. Chem.* **33**, 282 (1994).
124. R. A. Fischer, A. Miehr, T. Metzger, E. Born, O. Ambacher, H. Angerer and R. Dimitrov, *Chem. Mater.* **8**, 1357 (1996).
125. F. Takeda, T. Mori and T. Takahashi, *J. Appl. Phys.* **20**, L169 (1981).
126. H. C. Lee, K. Y. Lee, Y. J. Yong, J. Y. Lee and H. Kim, *Thin Solid Films* **271**, 50 (1995).
127. M. Yoshida, H. Watanabe and F. Uesugi, *J. Electrochem. Soc.* **132**, 677 (1985).
128. F. Widmann, G. Feuillet, B. Daudin and J. L. Rouvie're, *J. Appl. Phys.* **85** (3), 1550–1555 (1999), and references therein.
129. B. Heying, R. Averbeck, L. F. Chen, E. Haus, H. Riechert and J. S. Speck, *J. Appl. Phys.* **88**, 1855 (2000).
130. H. Amano, I. Akasaki, K. Hiramatsu and N. Sawaki, *Thin Solid Films* **163**, 415 (1988).
131. Akasaki, H. Amano, Y. Koide, K. Hiramatsu and N. Sawaki, *J. Cryst. Growth* **98** (1989).
132. K. Uchida, W. Watanabe, F. Yano, M. Kougucci, T. Tanaka and S. Minegava, *Proc. Int. Symp. Blue Lasers and Light Emitting Diodes*, Chiba University, Japan (1996) p. 48.
133. S. Keller, B. P. Keller, Y.-F. Wu, B. Heying, D. Kaponek, J. S. Speck, U. K. Mishra and S. P. DenBaars, *Appl. Phys. Lett.* **68**, 1525 (1996).
134. N. Grandjean, J. Massies and M. Leroux, *Appl. Phys. Lett.* **69**, 2071 (1996).
135. D. Huang, P. Visconti, K. M. Jones, M. A. Reshchikov, F. Yun, A. A. Baski, T. King and H. Morkoç, *Appl. Phys. Lett.* (in press).
136. M. A. Khan, M. S. Shur, Q. C. Chen and J. N. Kuznia, *Electron. Lett.* **30**, 2175 (1994).
137. Ö. Aktas, W. Kim, Z. Fan, A. Bochkarev, A. Salvador, B. Sverdlov and H. Morkoç, *Electron. Lett.* **31**, 1389 (1995).
138. Ö. Aktas, W. Kim, Z. Fan, F. Strengel, A. Bochkarev, A. Salvador, B. Sverdlov and S. N. Mohammad, *IEEE Int. Electron Devices Meeting Technol.* (Washington D.C., 1995) pp. 205–208.
139. Z. Fan, S. N. Mohammad, O. Aktas, A. E. Botchkarev, A. Salvador and H. Morkoç, *Appl. Phys. Lett.* **69**, 1229 (1994).
140. M. Smith, J. Y. Lin, H. X. Jiang, A. Salvador, A. Botchkarev, W. Kim and H. Morkoç, *Appl. Phys. Lett.* **69**, 2453 (1996).
141. R. Singh, D. Doppalaudi and T. D. Moustakas, *Appl. Phys. Lett.* **69**, 2388 (1996).

142. H. H. Yang, T. J. Schmidt, W. Shan, J. J. Song and B. Goldenberg, *Appl. Phys. Lett.* **66**, 1 (1995).
143. R. K. Sink, S. Keller, B. P. Keller, D. I. Babic, A. L. Holmes, D. Kapolnek, S. P. deBaars, J. E. Bowers, X. H. Wu and J. S. Speck, *Appl. Phys. Lett.* **68**, 2147 (1996).
144. G. Popovici, S. N. Mohammad and Hadis Morkoç, Deposition and Properties of III-Nitrides, *Group III Nitride Semiconductor Compounds: Physics and Applications*, Ed. B. Gil (Clarendon Press. April 1998) Oxford ISBN 0-19-850159-5.
145. G. Popovici and H. Morkoç, *GaN and Related Materials II Optoelectronic Properties of Semiconductors and Superlattices* **7** 93–172, Ed. S. J. Pearton (Series Editor M. O. Manaresh, Gordon and Breach Science Publishers, Amsterdam 1999) ISSN 1023-6619.
146. E. S. Hellman, *MRS Internet J. Nitride Semicond. Res.* **3**, 11 (1998).
147. F. Bernardini, V. Fiorentini and D. Vanderbilt, *Phys. Rev.* **B56**, R10024 (1997).
148. H. Morkoç and R. Cingolani and B. Gil, *Solid State Electron.* **43**, 1909 (1999).
149. G. A. Martin, A. Botchkarev, A. Rockett and H. Morkoç, *Appl. Phys. Lett.* **68**, 2541 (1996).
150. B. Daudin, J. L. Rouviere and M. Arley, *Appl. Phys. Lett.* **69**, 2480 (1996).
151. J. L. Rouviere, J. L. Weyher, M/Seelmann-Eggebert and S. Porowski, *Appl. Phys. Lett.* **73**, 668 (1998).
152. M. Sumiya, M. Tanaka, K. Ohtsuka, S. Fuke, T. Ohnishi, I. Ohkuboi, M. Yoshimoto, H. Koinuma and M. Kawasaki, *Appl. Phys. Lett.* **75**, 674 (1999).
153. X.-Q. Shen, T. Ide, S.-H. Cho, M. Shimizu, S. Hara, H. Okumura, S. Sonoda and S. Shimizu, *Jpn. J. Appl. Phys.* **39**, L16–18 (2000).
154. R. Dmitrov, M. Murphy, J. Smart, W. Schaff, J. R. Shealy, L. F. Eastman, O. Ambacher and M. Stutzmann, *J. Appl. Phys.* **87**, 3375 (2000).
155. O. Ambacher, J. Smart, J. R. Shealy, N. G. Weimann, K. Chu, L. F. Eastman, R. Dimitrov, L. Wittmer, M. Stutzmann, W. Rieger and J. Hilsenbeck, *J. Appl. Phys.* **85**, 3222 (1999).
156. A. R. Smith, R. M. Feenstra, D. W. Greve, M.-S. Shin, M. Skowronski, J. Neugebauer and J. E. Northrup, *Appl. Phys. Lett.* **72**, 2114 (1998).
157. K. M. Jones, P. Visconti, F. Yun, A. A. Baski and H. Morkoç, *Appl. Phys. Lett.* **78**, 2497 (2001).
158. M. Suyima, K. Yoshimura, K. Ohtsuka and S. Fuke, *Appl. Phys. Lett.* **76**, 2098 (2000).
159. D. Huang, P. Visconti, K. M. Jones, M. A. Reshchikov, F. Yun, A. A. Baski, T. King and H. Morkoç, *Appl. Phys. Lett.* **78**, 4145 (2001).
160. F. Widmann, G. Feuillet, B. Daudin and J. L. Rouviere, *J. Appl. Phys.* **85** (3), 1550–1555 (1999).
161. X.-Q. Shen, T. Ide, S.-H. Cho, M. Shimizu, S. Hara, H. Okumura, S. Sonoda and S. Shimizu, *Jpn. J. Appl. Phys.* **39**, L16 (2000).

162. Q. Zhu, A. Botchkarev, W. Kim, Ö. Aktas, B. N. Sverdlov and H. Morkoç, *Appl. Phys. Lett.* **68** (8) 1141–1143 (1996).
163. D. J. Smith, B. N. Sverdlov and H. Morkoç (unpublished).
164. B. Heying, X. H. Wu, S. Keller, Y. Li, D. Kapolnek, B. P. Keller, S. P. DenBaars and J. S. Speck, *Appl. Phys. Lett.* **68**, 643 (1996).
165. D. Huang, C. W. Litton, M. A. Reshchikov, F. Yun, T. King, A. A. Baski and H. Morkoç, *Int. Symp. Compound Semiconductors* (Japan, 2001).
166. D. Huang, M. A. Reshchikov, F. Yun, T. King, A. A. Baski and H. Morkoç, *Appl. Phys. Lett.* (in press).
167. H. Morkoç, M. A. Reshchikov, A. A. Baski and M. I. Nathan, *Materials Science Forum*, Eds. C. H. Carter Jr., R. P. Devaty and G. S. Rohrer (Trans Tech Publications, 2000) **338–342** — Part 2, p. 1453.
168. M. A. Reshchikov, J. Cui, F. Yun, P. Visconti, M. I. Nathan, R. Molnar and H. Morkoç, *Proc. Fall MRS* **639** (2000).
169. P. Visconti, K. M. Jones, M. A. Reshchikov, R. Cingolani, H. Morkoç and R. Molnar, *Appl. Phys. Lett.* **77**, 3532 (2000).
170. P. Visconti, K. M. Jones, M. A. Reshchikov, F. Yun, R. Cingolani, H. Morkoç, S. S. Park and K. Y. Lee, *Appl. Phys. Lett.* **77**, 3743 (2000).
171. T. Lei, T. D. Moustakas, R. J. Graham, Y. He and S. J. Berkowitz, *J. Appl. Phys.* **71**, 4933 (1992).
172. K. Yokouchi, T. Araki, T. Nagatomo and O. Omoto, *Inst. Phys. Conf. Ser.* **142**, 867 (1996).
173. A. Ohtani, K. S. Stevens and R. Beresford, *Proc. MRS Symp.* **339** (MRS, Pittsburg, PA) pp. 471–476 (1994).
174. S. N. Basu, T. Lei and T. D. Moustakas, *J. Mater. Res.* **9**, 2370 (1994).
175. G. A. Martin, B. N. Sverdlov, A. Botchkarev, H. Morkoç, D. J. Smith, S.-C. Y. Tsen, W. H. Thompson and M. H. Nayfeh, *MRS Fall Meeting*, *Proc.* (1995).
176. M. Godlewski, J. P. Bergman, B. Monemar, U. Rossner and A. Barski, *Appl. Phys. Lett.* **69**, 2089 (1996).
177. B. Damilano, N. Grandjean, F. Semond, J. Massies and M. Leroux, *Appl. Phys. Lett.* **75**, 962 (1999).
178. H. Morkoç, M. A. Reshchikov, A. A. Baski and M. I. Nathan, *Materials Science Forum*, Eds. C. H. Carter Jr., R. P. Devaty and G. S. Rohrer (Trans Tech Publications, 2000) **338–342** — Part 2, pp. 1453–1455.
179. M. E. Lin, S. Strite, A. Agarwal, A. Salvador, G. L. Zhou, N. Teraguchi, A. Rocket and H. Morkoç, *Appl. Phys. Lett.* **62**, 702 (1993).
180. M. E. Lin, S. Strite, A. Agarwal, A. Salvador, G. L. Zhou, N. Teraguchi, A. Rockett and H. Morkoç, *Appl. Phys. Lett.* **62** (7), 702–704 (1993).
181. D. Byun, G. Kim, D. Lim, I.-H. Choi, D. Park and D.-W. Kum, *Proc. Int. Symp. Blue Lasers and Light Emitting Diodes*, Chiba University, Japan (1996) p. 380.
182. C. Kisielowski, J. Kruger, S. Ruvimov, T. Suski, J. W. Ager III, E. Jones, Z. Liliental-Weber, M. Rubin, E. R. Weber, M. D. Bremser and R. F. Davis, *Phys. Rev.* **B54** (24), 17745–17753 (1996).
183. D. J. Smith, D. Chandrasekhar, B. Sverdlov, A. Bochkarev, A. Salvador and H. Morkoç, *Appl. Phys. Lett.* **67**, 1830 (1995).

184. J. A. Powell, D. J. Larkin and A. J. Trunek, *Silicon Carbide, III-Nitrides, and Related Materials*, Eds. G. Pensl, H. Morkoç, B. Monemar and E. Janzen (Trans Tech Publications, Sweden) **264–268**, 421–424 (1998).

185. J. A. Powell, D. J. Larkin, P. G. Neudeck, J. W. Yang and P. Pirouz, *Silicon Carbide and Related Materials*, Eds. M. G. Spencer, R. P. Devaty, J. A. Edmond *et al.* (IOP Publishing, Bristol, 1994) pp. 161–164.

186. G. Melnychuk, S. E. Saddow, M. Mynbaeva, S. Rendakova and V. Dmitriev, *Mat. Res. Soc.* **T4.2**, San Francisco, CA (2000).

187. M. Mynbaeva, A. Titkov, A. Kryzhanovski, A. Zubrilov, V. Ratnikov, V. Davydov, N. Kuznetsov, K. Mynbaev, S. Stepanov, A. Cherenkov, I. Kotousova, D. Tsvetkov and V. Dmitriev, *GaN and AlN Layers Grown by Nano Epitaxial Lateral Overgrowth Technique on Porous Substrates* (to be published in Proc. MRS Fall 1999 meeting).

188. M. Micovic, A. Kurdoghlian, P. Janke, P. Hashimoto, D. W. S. Wong, J. S. Moon, L. McCray and C. Nguyen, *IEEE Trans. Electron. Devices* **48** (3), 591–596 (2001).

189. T. Matsuoka, N. Yoshimoto, T. Sasaki and A. Katsui, *J. Electron. Mater.* **21**, 157 (1992).

190. F. Hamdani, A. Botchkarev, W. Kim, A. Salvador and H. Morkoç, M. Yeadon, J. M. Gibson, S. C. T. Tsen, D. J. Smith, D. C. Reynolds, D. C. Look, K. Evans, C. W. Litton, W. C. Mitchel and P. Hemenger, *Appl. Phys. Lett.* **70**, 467 (1997).

191. T. Detchprohm, H. Amano, K. Hiramatsu and I. Akasaki, *J. Cryst. Growth* **128**, 384 (1993).

192. M. A. L. Johnson, S. Fujita, W. H. Rowland, W. C. Hughes, J. W. Cook and J. F. Schetina, *J. Electron. Mater.* **25**, 855 (1996).

193. T. Matsuoka, N. Yoshimoto, T. Sasaki and A. Katsui, *J. Electron. Mater.* **21**, 157 (1992).

194. F. Hamdani, M. Yeadon, D. J. Smith, H. Tang, W. Kim, A. Salvador, A. E. Botchkarev, J. M. Gibson, A. Y. Polyakov, M. Skowronski and H. Morkoç, *J. Appl. Phys.* **83**, 983–990 (1998).

195. W. A. Doolittle, S. Kang and A. Brown, *Solid-State Electron.* **44** (2), 229–238 (2000).

196. B. Heying, E. Tarsa, C. Elsass, P. Fini, S. Denbaars and J. Speck, *J. Appl. Phys.* **85**, 6470 (1999).

197. M. Leszczynski, B. Beaumont, E. Frayssinet, W. Knap, P. Prystawko, T. Suski, I. Grzegory and S. Porowski, *Appl. Phys. Lett.* **75** (9), 1276–1278 (1999).

198. F. A. Ponce, D. P. Bour, W. Gotz, N. M. Johnson, H. I. Helava, I. Grzegory, J. Jun and S. Porowski, *Appl. Phys. Lett.* **68**, 917 (1996).

199. F. A. Ponce, D. P. Bour, W. T. Young, M. Sounders and J. W. Steeds, *Appl. Phys. Lett.* **69**, 337 (1996).

200. A. Gassmann, T. Susski, N. Newmann, C. Kiselowski, E. Jones, E. R. Weber, Z. Liliental-Weber, M. D. Rubin, H. I. Helava, I. Grezegory, M. Bockovski, J. Jun and S. Porowski, *J. Appl. Phys.* **80**, 2195 (1996).

201. A. Pelzmann, C. Kirchner, M. Mayer, V. Schwegler, M. Schauler, M. Kamp, K. J. Ebeling, I. Grzegory, M. Leszczynski, G. Nowak and S. Porowski, *J. Cryst. Growth* **189–190**, 167 (1998).

202. C. R. Miskys, M. K. Kelly, O. Ambacher, G. Marti'nez-Criado and M. Stutzmann, *Appl. Phys. Lett.* **77** (12), 1868–1870 (2000).

203. P. Hacke, G. Feuillet, H. Okumura and S. Yoshida, *Appl. Phys. Lett.* **69**, 2507 (1996).

204. D. J. Smith, D. Chandrasekhar, B. Sverdlov, A. Botchkarev, A. Salvador and H. Morkoç, *Appl. Phys. Lett.* **67**, 1830 (1995).

205. B. N. Sverdlov, G. A. Martin, H. Morkoç and D. J. Smith, *Appl. Phys. Lett.* **67**, 2063 (1995).

206. D. J. Smith, D. Chandrasekhar, B. Sverdlov, A. Botchkarev, A. Salvador and H. Morkoç, *Appl. Phys. Lett.* **67**, 1830 (1995).

207. M. A. Reshchikov, D. Huang, F. Yun, L. He, D. C. Reynolds, S. S. Park, K. Y. Lee and H. Morkoç, *Appl. Phys. Lett.* **79**, 3779 (2001).

208. H. Morkoç, *Mater. Sci. Eng. Rep.* (MSE-R) **259** (R33/5–6), 1–73 (2001) and references therein.

209. K. Kornitzer, M. Grehl, K. Thonke, R. Sauer, C. Kirchner, V. Schwegler, M. Kamp, M. Leszczynski, I. Grzegory and S. Porowski, *Phys.* **B273–274**, 66 (1999).

210. B. K. Meyer, *Mat. Res. Symp. Proc.* **449**, 497 (1997).

211. V. Kirilyuk, A. R. A. Zauner, P. C. M. Christianen, J. L. Weyher, P. R. Hageman and P. K. Larsen, *Appl. Phys. Lett.* **76**, 2355 (2000).

212. J. M. Baranowski, Z. Liliental-Weber, K. Korona, K. Pakula, R. Stepniewski, A. Wysmolek, I. Grzegory, G. Nowak, S. Porowski, B. Monemar and P. Bergman, in *III-V Nitrides*, Eds. F. A. Ponce, T. D. Moustakas, I. Akasaki and B. Monemar (Materials Research Society, Pittsburgh, 1997) **449**, p. 393.

213. B. J. Scromme, J. Jayapalan, R. P. Vaudo and V. M. Phanse, *Appl. Phys. Lett.* **74**, 2358 (1999).

214. P. J. Dean, J. D. Cuthbert, D. G. Thomas and R. T. Lynch, *Phys. Rev. Lett.* **18**, 122 (1967).

215. B. Monemar, *J. Mater. Sci.: Materials in Electronics* **10**, 227 (1999).

216. G. Pozina, J. P. Bergman, T. Paskova and B. Monemar, *Appl. Phys. Lett.* **75**, 4124 (1999).

217. A. A. Yamaguchi, Y. Mochizuki, H. Sunakawa and A. Usui, *J. Appl. Phys.* **83**, 4542 (1998).

218. D. Volm, K. Oettinger, T. Streibl, D. Kovalev, M. Ben-Chorin, J. Diener, B. K. Meyer, J. Majewski, L. Eckey, A. Hoffmann, H. Amano, I. Akasaki, K. Hiramatsu and T. Detchprohm, *Phys. Rev.* **B53**, 16543 (1996).

219. We assumed that the donor concentration in the substrate and MBE-overgrown layer is the same and equals $2 \times 10^{16}$ cm$^{-3}$, as has been found previously for one of the similarly grown GaN template (see Ref. 243).

220. Y. J. Wang, R. Kaplan, H. K. Ng, K. Doverspike, D. K. Gaskill, T. Ikedo, I. Akasaki and H. Amono, *J. Appl. Phys.* **79**, 8007 (1996).

221. F. Mireles and S. E. Ulloa, *Appl. Phys. Lett.* **74**, 248 (1999).

222. J. Jayapalan, B. J. Skromme, R. P. Vaudo and V. M. Phanse, *Appl. Phys. Lett.* **73**, 1158 (1998).
223. W. Goetz, L. T. Romano, J. Walker, N. M. Johnson and R. J. Molnar, *Appl. Phys. Lett.* **72**, 1214 (1998).
224. K. Naniwae, S. K. Ithoh, H. Amano, K. Ithoh, K. Hiramatsu and I. Akasaki, *J. Cryst. Growth* **99**, 381 (1990).
225. C. Wetzel, D. Volm, B. K. Meyer, K. Pressel, S. Nilsson, E. N. Mokhov and P. G. Baranov, *Appl. Phys. Lett.* **65**, 1033 (1994).
226. S. Fischer, C. Wetzel, W. L. Hansen, E. D. Bourret-Chourchesne, B. K. Meyer and E. E. Haller, *Appl. Phys. Lett.* **69**, 2716 (1996).
227. Y. A. Vodakov, M. I. Karklina, E. N. Mokhov and A. D. Roenkov, *Inorg. Mater.* **17**, 537 (1980).
228. T. Detchprohm, K. Hiramatsu, H. Amano and I. Akasaki, *Appl. Phys. Lett.* **61**, 2688 (1992).
229. Z.-Q. Fang, D. C. Look, J. Jasinski, M. Benamara, Z. Liliental-Weber, K. Saarinen and R. J. Molnar, *Appl. Phys. Lett.* **78**, 332 (2000).
230. J. Jasinski, W. Swider, Z. Liliental-Weber, P. Visconti, K. M. Jones, M. A. Reshchikov, F. Yun, H. Morkoç, S. S. Park and K. Y. Lee, *Appl. Phys. Lett.* **78** 2297–2299 (2001).
231. M. K. Kelly, R. P. Vaudo, V. M. Phanse, L. Gorgens. O. Ambacher and M. Stutzmann, *Jpn. J. Appl. Phys.* **38** (3A), L217–219 (1999).
232. Z.-Q. Fang, D. C. Look, P. Visconti, D.-F. Wang, C.-Z. Lu, F. Yun, H. Morkoç, S. S. Park and K. Y. Lee, *Appl. Phys. Lett.* **78** (15), 2178–2181 (2001).
233. L. Chernyak, A. Osinsky, G. Nootz, A. Schulte, J. Jasinski, M. Benamara, Z. Liliental-Weber, D. C. Look and R. J. Molnar, *Appl. Phys. Lett.* **77**, 875 (2000).
234. F. Yun, M. A. Reshchikov, K. M. Jones, P. Visconti, H. Morkoç, S. S. Park and K. Y. Lee, *Solid State Electron.* **44**, 2225 (2000).
235. B. Heying, X. H. Wu, S. Keller, Y. Li, B. Keller, S. P. DenBaars and J. S. Speck, *Appl. Phys. Lett.* **68**, 643 (1996).
236. T. Metzger, P. Hopler, E. Born, O. Ambacher, M. Stutzmann, R. Stommer, M. Schuster, H. Gobel, S. Christiansen, M. Albrecht and H. P. Strunk, *Philosophy* **A77**, 1013 (1998).
237. L. T. Romano, B. S. Krusor and R. J. Molnar, *Appl. Phys. Lett.* **71**, 2283 (1997).
238. W. S. Wong, T. Sands and N. W. Cheung, *Appl. Phys. Lett.* **72**, 599 (1998).
239. M. Leszczynski, T. Suski, H. Teisseyre, P. Perlin, I. Grzegory, J. Jun, S. Porowski and T. D. Moustakas, *J. Appl. Phys.* **76**, 4909 (1994).
240. M. K. Kelly, R. P. Vaudo, V. M. Phanse, L. Gorgens, O. Ambacher and M. Stutzmann, *Jpn. J. Appl. Phys.* **38** (3A), L217–219 (1999).
241. D. C. Look and R. J. Molnar, *Appl. Phys. Lett.* **70**, 3377 (1997).
242. C. E. Stutz, M. Mack, M. D. Bremser, O. H. Nam, R. F. Davis and D. C. Look, *J. Electron. Mater.* **27**, L26 (1998).

243. D. Huang, F. Yun, P. Visconti, M. A. Reshchikov, K. M. Jones, D. Wang, H. Morkoç, D. L. Rode, L. A. Farina, Ç. Kurdak, K. T. Tsen, S. S. Park and K. Y. Lee, *Solid State Electron.* **45** (5), 711–715 (2001).
244. D. L. Rode, *Semicond. Semimetals* **10**, 1–89 (Academic, New York, 1975).
245. H. Brooks, *Phys. Rev.* **83**, 879 (1951).
246. D. C. Look, *Electrical Characterization of GaAs Materials, Devices* (Wiley, New York, 1989).
247. H. Ehrenreich, *J. Phys. Chem. Solids* **8**, 130 (1959).
248. D. A. Anderson and N. Aspley, *Semicond. Sci. Technol.* **1**, 187 (1986).
249. L. Hsu and W. Walukiewicz, *Phys. Rev.* **B56**, 1520 (1997).
250. C. Erginsoy, *Phys. Rev.* **79**, 1013 (1950).
251. J. Bauer, L. Biste and D. Bolze, *Phys. Status Solidi* **A39**, 173 (1977).
252. B. Baranov, L. Daweritz, V. B. Gutan, G. Jungk, H. Neumann and H. Raidt, *Phys. Status Solidi* **A49**, 629 (1978).
253. H. Sato, H. Tokashi, A. Watanabe and H. Ota, *Appl. Phys. Lett.* **68**, 3617 (1996).
254. S. Fujieda, M. Mizuta and Y. Matsumoto, *Jpn. J. Appl. Phys.* **26**, 2067 (1987).
255. H. Okumura, S. Misawa and S. Yoshida, *Appl. Phys. Lett.* **59**, 1058 (1991).
256. J. E. Andrews and M. A. Littlejohn, *J. Electrochem. Soc.* **122**, 1273 (1975).
257. H. M. Manasevit, F. M. Erdmann and W. I. Simpson, *J. Electrochem. Soc.* **118**, 1864 (1971); see also, F. A. Pizzarello and J. E. Coker, *J. Electron. Mater.* **4**, 25 (1975).
258. G. D. O'Clock Jr. and M. T. Duffy, *Appl. Phys. Lett.* **23**, 55 (1973).
259. J. K. Liu, K. M. Lakin and K. L. Wang, *J. Appl. Phys.* **46**, 3703 (1975).
260. M. Morita, N. Uesugi, S. Isogai, K. Tsubouchi and N. Mikoshiba, *Jpn. J. Appl. Phys.* **20**, 17 (1981).
261. M. A. Khan, R. A. Skogman, R. G. Schulze and M. Gershenzon, *Appl. Phys. Lett.* **42**, 430 (1983).
262. M. Hashimoto, H. Amano, N. Sawaki and I. Akasaki, *J. Cryst. Growth* **68**, 163 (1984).
263. M. Matloubian and M. Gershenzon, *J. Electron. Mater.* **14**, 633 (1985); see also, T. Sasaki and T. Matsuoka, *J. Appl. Phys.* **64**, 4531 (1988); T. Nagatomo, T. Kuboyama, H. Minamino and O. Omoto, *Jpn. J. Appl. Phys.* **28**, L1334 (1989); M. A. Khan, J. N. Kuznia, J. M. Van Hove, D. T. Olsen, S. Krishnankutty and R. M. Kolbas, *Appl. Phys. Lett.* **58**, 526 (1991).
264. S. Nakamura, Y. Harada and M. Seno, *Appl. Phys. Lett.* **58**, 2021 (1991).
265. H. Amano, N. Sawaki, I. Akasaki and Y. Toyoda, *Appl. Phys. Lett.* **48**, 353 (1986).
266. I. Akasaki, H. Amano, Y. Koide, K. Hiramatsu and N. Sawaki, *J. Cryst. Growth* **98**, 209 (1989).
267. K. Hiramatsu, S. Itoh, H. Amano, I. Akasaki, N. Kuwano, T. Shiraishi and K. Oki, *J. Cryst. Growth* **115**, 628 (1991).
268. S. D. Hersee, J. C. Ramer and K. J. Malloy, *MRS Bulletin* (1997).
269. S. Nakamura, *Jpn. J. Appl. Phys.* **30**, L1705 (1991).

270. A. E. Wickenden, D. K. Wickenden and T. J. Kistenmacher, *J. Appl. Phys.* **75**, 5367 (1994).

271. B. D. Joyce and J. A. Baldrey, *Nature*, **195**, 485 (1962).

272. For a review, see M. R. Goulding, *Materials Science and Engineering* **B17**, 47–67 (1993).

273. R. P. Zingg, G. W. Neudeck, B. Hoefflinger and S. T. Liu, *J. Electrochem. Soc.* **133** (6), 1274–1275 (1986).

274. B.-Y. Tsaur, R. W. McClelland, J. C. C. Fan, R. P. Gale, J. P. Salerno, B. J. Vojak and C. O. Bozler, *Appl. Phys. Lett.* **41**, 347 (1982).

275. A. D. Morrison and T. Daud, *United States Patent 4, 522, 661* (1985).

276. K. Kato, T. Kusunoki, C. Takenaka, T. Tanahashi and K. Nakajima, *J. Cryst. Growth* **115**, 174 (1991).

277. O. Parillaud, E. Gil-Lafon, B. Gerard, P. Etienne and D. Pribat, *Appl. Phys. Lett.* **68**, 2654 (1996).

278. Y. Kato, S. Kitamura, K. Hiramatsu and S. Sawaki, *J. Cryst. Growth* **144**, 133 (1994).

279. O. H. Nam, M. D. Bremser, T. S. Zheleva and R. F. Davis, *Appl. Phys. Lett.* **71**, 2638 (1997).

280. T. S. Zheleva, O.-H. Nam, D. M. Bremser and R. F. Davis, *Appl. Phys. Lett.* **71**, 2472 (1997).

281. S. Nakamura, M. Senoh, S. Nagahama, N. Iwasa, T. Yamanda, T. Matsushita, H. Kiyoku, Y. Sugimoto, T. Kozaki, H. Umemoto, M. Sano and K. Chocho, *Appl. Phys. Lett.* **72**, 211 (1998).

282. D. Kapolnek, S. Keller, R. Vetury, R. D. Underwood, P. Kozodoy, S. P. DenBaars and U. K. Mishra, *Appl. Phys. Lett.* **71**, 1204 (1997).

283. K. J. Linthicum, T. Gehrke, D. Thomson, E. Carlson, P. Rajagopal, T. Smith and R. Davis, *Appl. Phys. Lett.* **75**, 196 (1999).

284. R. F. Davis, T. Gehrke, K. J. Linthicum, T. S. Zheleva, P. Rajagopal, C. A. Zorman and M. Mehregany, *Proc. Fall 1999 MRS meeting.*

285. P. Fini *et al.*, *Appl. Phys. Lett.* **75**, 1706 (1999).

286. Y. Kawaguchi, S. Nambu, H. Sone, M. Yamaguchi, H. Miyake, K. Hiramatsu, N. Sawaki, Y. Iyechika and T. Maeda, *Proc. Mater. Res. Soc.* **4** (S.1), G 4.1 (1999).

287. K. Hiramatsu, M. Haino, M. Yamaguchi, H. Miyake, A. Motogaito, N. Sawaki, Y. Iyechika and T. Maeda, *Mater. Sci. Eng.* **B82**, 62–64 (2001).

288. H. Amano, M. Iwaya, T. Kashima, M. Katsuragawa, I. Akasaki, J. Han, S. Hearne, J. A. Floro, E. Chason and J. Figiel, *Jpn. J. Appl. Phys.* **37**, L1540 (1998).

289. M. Koike (private communication).

290. D. D. Koleske, M. E. Twigg, A. E. Wiekenden, R. L. Henry, R. J. Gorman, J. A. Freitas Jr. and M. Fatemi, *Appl. Phys. Lett.* **75** (20), 3141–3143 (1999).

291. S. Fischer, A. Gisbertz, B. K. Meyer, M. Topf, S. Koynov, I. Dirnstorfer, D. Volm, R. Uecker, P. Reiche, S. Ganschow and Z. Liliental-Weber, *Symp. Proc. EGN-1* (Rigi, Switzerland, 1996).

292. P. Kung, A. Saxler, X. Zhang, D. Walker, R. Lavado and M. Razeghi, *Appl. Phys. Lett.* **69**, 2116 (1996).

293. T. Detchprohm, K. Hiramatsu, N. Sawaki and I. Akasaki, *J. Cryst. Growth* **137**, 170 (1994); see also, T. Detchprohm, H. Amano, K. Hiramatsu and I. Akasaki, *Appl. Phys. Lett.* **61**, 2688 (1992); T. Detchprohm, K. Hiramatsu, K. Itoh and I. Akasaki, *Jpn. J. Appl. Phys.* **31**, L1454 (1992).

294. M. A. L. Johnson, S. Fujita, W. H. Rowland Jr., W. C. Hughes, Y. W. He, N. A. El-Masry, J. W. Cook Jr., J. F. Schetzina, J. Ren and J. A. Edmond, *J. Electron. Mater.* **25**, 793 (1996).

295. M. Leszczynski, B. Beaumont, E. Frayssinet, W. Knap, P. Prystawko, T. Suski, I. Grzegory and S. Porowski, *Appl. Phys. Lett.* **75** (9), 1276–1278 (1999).

296. C. R. Miskys, M. K. Kelly, O. Ambacher, G. Marti'nez-Criado and M. Stutzmann, *Appl. Phys. Lett.* **77** (12), 1858–1860 (2000).

297. C. Kirchner, V. Schwegler, F. Eberhard, M. Kamp, K. J. Ebeling, P. Prystawko, M. Leszczynski, I. Grzegory and S. Porowski, *Prog. Crystal Growth & Characterization of Materials* **41** (1–4), 57–83 (2000).

298. K. Komitzer, T. Ebner, K. Thonke, R. Sauer, C. Kirchner, M. Kamp, V. Schwegler, M. Leszczynski, I. Grzegory and S. Porowski, *Phys. Rev. Condensed Matter* **B60** (3), 1471–1473 (1999).

299. C. Kirchner, V. Schwegler, F. Eberhard, M. Kamp, K. J. Ebeling, K. Kornitzer, T. Ebner, K. Thonke, R. Sauer, P. Prystawko, M. Leszczynski, I. Grzegory and S. Porowski, *Appl. Phys. Lett.* **75** (8), 1098–1100 (1999).

300. J. Neugebauer and C. G. Van de Walle, *Appl. Phys. Lett.* **69**, 503 (1996).

301. M. A. Reshchikov, P. Visconti and H. Morkoç, *Appl. Phys. Lett.* **78**, 177 (2001).

302. M. A. Reshchikov, F. Shahedipour, R. Y. Korotkov, B. W. Wessels and M. P. Ulmer, *J. Appl. Phys.* **87**, 3351 (2000).

303. D. K. Wickenden, C. B. Bargeron, W. A. Bryden, J. Miragliova and T. J. Kistenmacher, *Appl. Phys. Lett.* **65**, 2024 (1994).

304. Y. Koide, H. Itoh, M. R. H. Khan, K. Hiramatsu, N. Sawaki and I. Akasaki, *J. Appl. Phys.* **61**, 4540 (1987).

305. H. Amano, T. Takeuchi, S. Sota, H. Sakai and I. Aksaski, *Mat. Res. Soc. Symp. Proc.* Eds. F. A. Ponce, T. D. Moustakas, I. Akasaki and B. A. Monemar, **449**, 1143–1150 (1997). MRS Fall Meeting, Boston 1996.

306. T. J. Ochalski, B. Gil, P. Lefebvre, N. Grandjean, M. Leroux, J. Massies, S. Nakamura and H. Morkoç, *Appl. Phys. Lett.* **74** (22), 3353–3355 (1999).

307. M. D. Mc Cluskey, C. G. Van de Walle, C. P. Master, L. T. Romano and N. M. Johnson, *Appl. Phys. Lett.* **72**, 2725 (1998).

308. A. F. Wright and J. S. Nelson, *Appl. Phys. Lett.* **66**, 3051 (1995).

309. Ü. Özgur, G. Webb-Wood, H. O. Everitt, F. Yun and H. Morkoç, *Appl. Phys. Lett.* (2001).

310. M. D. Bremser, W. G. Perry, T. Zheleva, N. V. Edwards, O. H. Nam, N. Parikh, D. E. Aspnes and R. F. Davis, *Mater. Res. Soc. Int. J. Nitr. Semi. Res.* **1**, 8 (1996).

311. S. Ruffenach-Clur, O. Briot, B. Gil, R.-L. Aulombard and J. L. Rouviere, *Mater. Res. Soc. Int. J. Nitr. Semi. Res.* **2**, 27 (1997).

312. P. Kung, A. Saxler, D. Walker, X. Zhang, R. Lavado, K. S. Kim and M. Razeghi, *Mater. Res. Soc. Symp. Proc.* **449**, 79; *Japan* **87**, 62 (1997).
313. H. Amano and I. Akasaki, *J. Phys. Condensed Matter* **13**, 6935–6944 (2001).
314. T. Matsuoka, N. Yoshimoto, T. Sasaki and A. Katsui, *J. Electron. Mater.* **21**, 157 (1992).
315. M. Shimizu, K. Hiramatsu and N. Sawaki, *J. Cryst. Growth* **145**, 209 (1994).
316. T. Nagatomo, T. Kuboyama, H. Minamino and O. Omoto, *Jpn. J. Appl. Phys.* **28**, L1334 (1989).
317. N. Yoshimoto, T. Matsuoka, T. Sasaki and A. Katsui, *Appl. Phys. Lett.* **59**, 2251 (1991).
318. T. Matsuoka, N. Yoshimoto, T. Sasaki and A. Katsui, *J. Electron. Mater.* **21**, 157 (1992).
319. M. Shimizu, K. Hiramatsu and N. Sawaki, *J. Cryst. Growth* **145**, 209 (1994).
320. I.-H. Ho and G. B. Stringfellow, *Appl. Phys. Lett.* **69**, 2701 (1996).
321. K. Osamura, S. Naka and Y. Murakami, *J. Appl. Phys.* **46**, 3432 (1975).
322. A. Koukitu, N. Takahashi, T. Taki and H. Seki, *Jpn. J. Appl. Phys.* **35**, L673 (1996).
323. Y. Kawaguchi, M. Shimizu, K. Hiramatsu and N. Sawaki, *Mater. Res. Soc. Symp. Proc.* **449**, 89 (1997).
324. S. Nakamura and T. Mukai, *Jpn. J. Appl. Phys.* **31**, L1457 (1992).
325. E. L. Piner, M. K. Behbehani, N. A. El-Mastry, F. G. McIntosh, J. C. Roberts, K. S. Boutros and S. M. Bedair, *Appl. Phys. Lett.* **70**, 461 (1997).
326. S. Nakamura and T. Mukai, *J. Vac. Sci. Technol.* **A13**, 6844 (1995).
327. S. M. Bedair. F. G. McIntosh, J. C. Roberts, E. L. Piner, K. S. Boutros and N. A. El-Masry, *J. Cryst. Growth* **178**, 32–44 (1997).
328. C. B. Vartuli, S. J. Pearton, C. R. Abernaty, J. D. MacKenzie, J. C. Zolper and E. S. Lambers, *MRS Proc.* **423** (MRS, Pittsburg, PA, 1996) pp. 569–574.
329. S. Yamaguchi, M. Kariya, S. Nitta, T. Takeuchi, C. Wetzel, H. Amano and I. Akasaki, *Appl. Phys. Lett.* **73**, 830 (1998).
330. J. Han, J. J. Figiel, G. A. Petersen, S. M. Myers, M. H. Crawford and M. A. Banas, *Jpn. J. Appl. Phys.* **39**, 2372 (2000).
331. M. Kosaki, S. Mochizuki, T. Nakamura, Y. Yukawa, S. Nitta, S. Yamaguchi, H. Amano and I. Akasaki, *Jpn. J. Appl. Phys.* **40**, L420 (2001).
332. F. Bernardini, V. Fiorentini and D. Vanderbilt, *Mater. Res. Soc. Symp. Proc.* **449**, 923 (1997).
333. T. Takeuchi, C. Wetzel, S. Yamaguchi, H. Sakai, H. Amano, I. Akasaki, Y. Kaneko, S. Nakagawa, Y. Yamaoka and N. Yamada, *Appl. Phys. Lett.* **73**, 1691 (1998).
334. K. S. Kim, A. Saxler, P. Kung, R. Razeghi and K. Y. Lim, *Appl. Phys. Lett.* **71**, 800 (1997).
335. T. Nagamoto, T. Kuboyama, H. Minamino and O. Omoto, *Jpn. J. Appl. Phys.* **28**, L1334 (1989).

# CHAPTER 2

# GaN AND AlGaN HIGH VOLTAGE POWER RECTIFIERS

An-Ping Zhang

*GE Corporate R&D Center, Niskayuna, NY, USA*

Fan Ren

*Department of Chemical Engineering, University of Florida,*
*Gainesville, FL, USA*

Jung Han

*Department of Chemical Engineering, Yale University, New Haven, CT, USA*

Stephen J. Pearton

*Department of MS&E, University of Florida, Gainesville, FL, USA*

Sung Soo Park and Yong Jo Park

*Samsung Advanced Institute of Technology, Suwon, South Korea*

Jen-Inn Chyi

*Department of Electrical Engineering*
*National Central University, Chung-Li, Taiwan*

## Contents

## 1. Introduction

There is a strong interest in developing wide bandgap power devices for use in the electric power utility industry.[1-3] With the onset of deregulation in the industry, there will be increasing numbers of transactions on the power grid in the US, with different companies buying and selling power. The main applications are in the primary distribution system (100 ~ 2000 kVA) and in subsidiary transmission systems (1 ~ 50 MVA). A major problem in the current grid is commentary voltage sags, which affect motor drives, computers and digital controls. Therefore, a system for eliminating power sags and switching transients would dramatically improve power quality. For example it is estimated that a 2-second outage at a large computer center can cost US\$ 600,000 or more, and an outage of less than one cycle, or a voltage sag of 25% for two cycles, can cause a microprocessor to malfunction. In particular, computerized technologies have led to strong consumer demands for less expensive electricity, premium quality power and uninterruptable power.

The basic power electronics hierarchy would include the use of widegap devices such as Gate Turn-Off Thyristors (GTOs), MOS-Controlled Thyristors (MCT) or Insulated Gate Bipolar Transistors (IGBTs) combined with appropriate packaging and thermal management techniques to make subsystems (such as switches, rectifiers or adjustable speed devices) which then comprise a system such as Flexible AC Transmissions (FACTS). Common power electronics systems, which are inserted between the incoming power and the electrical load include uninterruptable power supplies, advanced motors, adjustable speed drives and motor controls, switching power supplies, solid-state circuit breakers and power conditioning equipment. About 50% of the electricity in the US is consumed by motors. Motor repairs cost ~US\$ 5 billion each year and could be dramatically reduced by high power electronic devices that permit smoother switching and control. Moreover, control electronics could dramatically improve motor efficiency. Other end uses include lighting, computers, heating and air-conditioning.

Some desirable attributes of next generation, widegap power electronics include the ability to withstand currents in excess of 5 kA and voltages in excess of 50 kV, provide rapid switching, maintain good thermal stability while operating at temperatures above 250°C, have small size and light-weight, and be able to function without bulky heat-dissipating systems.

The primary limits of Si-based power electronics are:

(1) Maximum voltage ratings < 7 kV

   • Multiple devices must be placed in series for high-voltage systems.

(2) Insufficient current-carrying capacity

   • Multiple devices must be placed in parallel for typical power grid applications.

(3) Conductivity in one direction only

   • Identical pairs of devices must be installed in anti-parallel for switchable circuits.

(4) Inadequate thermal management

   • Heat damage is a primary cause of failure and expense.

(5) High initial cost

   • Applications are limited to the highest-value settings.

(6) Large and heavy components

   • Costs are high for installation and servicing, and equipment is unsuitable for many customers.

For these reasons, there is a strong development effort on widegap power devices, predominantly SiC, with lesser efforts in GaN and diamond, which should have benefits that Si-based or electromechanical power electronics cannot attain. The higher standoff voltages should eliminate the need for series stacking of devices and the associated packaging difficulties. In addition these widegap devices should have higher switching frequency in pulse-width-modulated rectifiers and inverters.

The absence of Si devices capable of application to 13.8 kV distribution lines (a common primary distribution mode) opens a major opportunity for widegap electronics. However, cost will be an issue, with values of US$ 200 ∼ 2000 per kVA necessary to have an impact. It is virtually certain that SiC switches will become commercially available within 3 ∼ 5 years, and begin to be applied to the 13.8 kV lines. MOS Turn-Off-Thyristors involving a SiC GTO and SiC MOSFET are a promising approach.[4] An inverter module can be constructed from an MOS turn-off thyristor (MTO) and a SiC power diode.

Packaging and thermal management will be a key part of future power devices. For current Si IGBTs, there are two basic package types — the first is a standard attached die, wire bond package utilizing soft-solder and wire-bonds as contacts, while the second is the presspack, which employs dry-pressed contacts for both electrical and thermal paths.[5,6] In the classical package the IGBTs and control diodes are soldered onto ceramic substrates, such as AlN, which provide electrical insulation, and this in turn is mounted to a heat sink (typically Cu). Thick Al wires (500 mm) are used for electrical connections, while silicone gel fills the package.[5] In the newer presspack style, the IGBT and diode are clamped between Cu electrodes, buffered by materials such as molybdenum or composites,[6] whose purpose is to account for the thermal expansion coefficient differences between Si and Cu. The package is again filled with gel for electrical insulation and corrosion resistance.

## 2. GaN Schottky Rectifiers with 3.1 kV Reverse Breakdown Voltage

The GaN materials system is attractive from the viewpoint of fabricating unipolar power devices because of its large bandgap and relatively high electron mobility.[7-10] An example is the use of Schottky diodes as high-voltage rectifiers in power switching applications.[7-9] These diodes will have lower blocking voltages than $p$-$i$-$n$ rectifiers, but have advantages in terms of switching speed and lower forward voltage drop. Edge termination techniques such as field rings on filed plates, bevels or surface ion implantation are relatively well-developed for Si and SiC, and they maximize the high voltage blocking capability by avoiding sharp field distributions within the device. However, in the few GaN Schottky diode rectifiers reported to date,[7,8] there has been little effort made on developing edge termination techniques. Proper design of the edge termination is critical both for obtaining a high breakdown voltage and reducing the on-state voltage drop and switching time.

Based on the punch-through model, Fig. 1 shows a plot of avalanche and punch-through breakdown of GaN Schottky diodes calculated as a function of doping concentration and standoff layer thickness. It can be seen that 20 kV device may be obtained with $\sim 100$ $\mu$m thick GaN layer with doping concentration $< 10^{15}$ cm$^{-3}$.

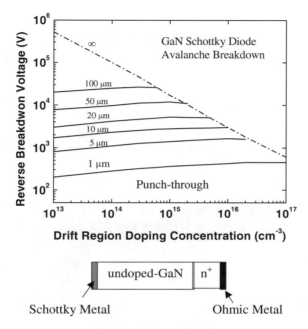

**Fig. 1.** The calculation of reverse breakdown voltage as a function of doping concentration and standoff region thickness based on a punch-through model.

In this chapter we report on the effect of various edge termination techniques on the reverse breakdown voltage, $V_B$, of planar GaN Schottky diodes which deplete in the lateral direction. A maximum $V_B$ of 3.1 kV at 25°C was achieved with optimized edge termination, which is a record for GaN devices. We also examined the temperature dependence of $V_B$ in mesa diodes and found a negative temperature coefficient of this parameter in these structures.

The GaN was grown on $c$-plane $Al_2O_3$ substrates by MOCVD using trimethylgallium and ammonia as the precursors. To create a Schottky rectifier with high breakdown voltage, one needs a thick, very pure GaN depletion layer. Figure 2 shows SIMS profile of H and other background impurities in a 2 $\mu$m thick, high resistivity ($10^7$ $\Omega$ cm) GaN layer grown by MOCVD. The reverse breakdown voltage of simple Schottky rectifiers fabricated on this material was > 2 kV, a record for GaN. Notice that in this material the hydrogen concentration is at the detection sensitivity of the SIMS apparatus. The amount of hydrogen present in GaN after cooldown from the growth temperature will depend on the number of sites to which

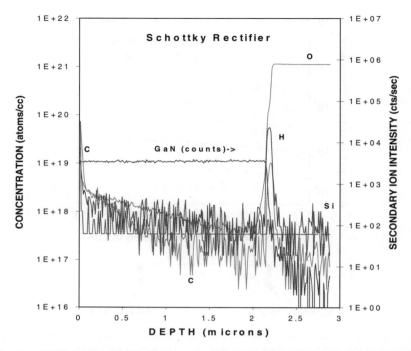

**Fig. 2.** SIMS profiles of H and other background impurities in as-grown, MOCVD Schottky rectifier structure.

it can bond, including dopants and point and line defects. In the absence of p-type doping, it is clear that the number of these sites is $\leq 8 \times 10^{17}$ cm$^{-3}$ under our growth conditions.

For vertically-depleting devices, the structure consisted of a 1 $\mu$m $n^+$ ($3 \times 10^{18}$ cm$^{-3}$, Si-doped) contact layer, followed by undoped ($n \sim 2.5 \times 10^{16}$ cm$^{-3}$) blocking layers which ranged from 3 to 11 $\mu$m thick. These samples were formed into mesa diodes using ICP etching with Cl$_2$/Ar discharges (300 W source power, 40 W rf chuck power). The d.c. self-bias during etching was $-85$ V. To remove residual dry etch damage, the samples were annealed under N$_2$ at 800°C for 30 seconds. Ohmic contacts were formed by lift-off of e-beam evaporated Ti/Al, annealed at 700°C for 30 seconds under N$_2$ to minimize the contact resistance. Finally, the rectifying contacts were formed by lift-off of e-beam evaporated Pt/Au. Contact diameters of 60–1100 mm were examined.

For laterally depleting devices, the structure consisted of $\sim 3$ micron of resistive ($10^7$ $\Omega/\square$) GaN. To form Ohmic contacts, Si$^+$ was implanted at $5 \times 10^{14}$ cm$^{-2}$, 50 keV into the contact region and activated by annealing at 150°C for 10 seconds under N$_2$. The Ohmic and rectifying contact metallization was the same as described above.

Three different edge termination techniques were investigated for the planar diode:

(1) Use of a $p$-guard ring formed by Mg$^+$ implantation at the edge of the Schottky barrier metal. In these diodes the rectifying contact diameter was held constant at 124 $\mu$m, while the distance of the edge of this contact from the edge of the Ohmic contact was 30 $\mu$m in all cases.
(2) Use of $p$-floating field rings of width 5 mm to extend the depletion boundary along the surface of the SiO$_2$ dielectric, which reduces the electric field crowding at the edge of this boundary. In these structures

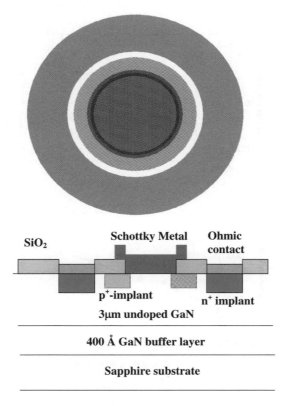

**Fig. 3.** GaN power rectifiers with $p$-guard ring for edge terminations.

a 10 $\mu$m wide $p$-guard ring was used, and one to three floating field rings employed.

(3) Use of junction barrier controlled Schottky (JBS) rectifiers, i.e. a Schottky rectifier structure with a $p$-$n$ junction grid integrated into its drift region.

In all of the edge-terminated devices the Schottky barrier metal was extended over an oxide layer at the edge to further minimize field crowding, and the guard and field rings formed by Mg$^+$ implantation and 1100°C annealing.

Figure 3 shows a schematic of the planar diodes fabricated with the $p$-guard rings, while the lower portion of the figure shows the influence of guard ring width on $V_\text{B}$ at 25°C. Without any edge termination, $V_\text{B}$ is $\sim$ 2300 V for these diodes. The forward turn-on voltage was in the range 15 $\sim$ 50 V, with a best on-resistance of 0.8 $\Omega$ cm$^2$. The figure-of-merit

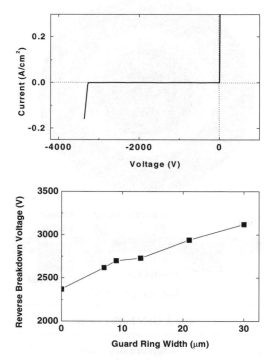

**Fig. 4.** Current-Voltage characteristics of GaN power rectifiers with $p$-guard ring for edge terminations (top), and effect of $p$-guard ring on the reverse breakdown voltage of GaN power rectifiers (bottom).

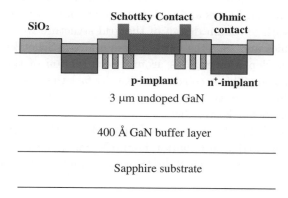

**Fig. 5.** GaN power rectifiers with floating field ring for edge termination.

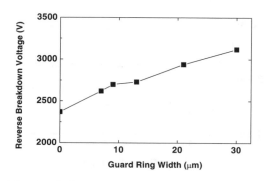

**Fig. 6.** Effect of floating field ring on the reverse breakdown voltage of GaN power rectifiers.

$(V_B)^2/R_{ON}$ was 6.8 MW/cm$^2$. As the guard-ring width was increased, we observed a monotonic increase in $V_B$, reaching a value of $\sim$ 3100 V for 30 $\mu$m wide rings (Fig. 4). The figure-of-merit was 15.5 MW/cm$^2$ under these conditions. The reverse leakage current of the diodes was still in the nA range at voltages up to 90% of the breakdown value.

Figure 5 shows a schematic of the floating field ring structures, while Fig. 6 shows the effect of different edge termination combinations on the resulting $V_B$ at 25°C. Note that the addition of the floating field rings to a guard ring structure further improves $V_B$, with the improvement saturating for a three-floating field ring geometry.

Figure 7 shows the effect of the junction barrier control on $V_B$, together with a schematic of the *p-n* junction grid in Fig. 8. In our particular structure we found that junction barrier control slightly degraded $V_B$ relative to devices with guard rings and various numbers of floating field rings. We believe that with optimum design of the grid structure we should achieve higher $V_B$ values and that the current design allows Schottky barrier lowering since the depletion regions around each section of the grid do not completely overlap. This is consistent with the fact that we did not observe the decrease in forward turn-on voltage expected for JBS rectifiers relative to conventional Schottky rectifiers.

The results of Figs. 3 to 6 are convincing evidence that proper design and implementation of edge termination methods can significantly increase reverse breakdown voltage in GaN diode rectifiers and will play an important

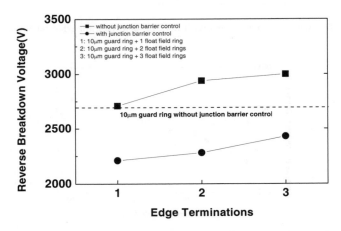

**Fig. 7.** Effect of junction barrier control on the reverse breakdown voltage of GaN power rectifiers.

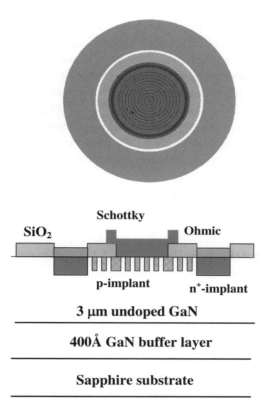

**Fig. 8.** GaN power rectifiers with junction barrier control.

role in applications at the very highest power levels. For example, the target goals for devices, intended to be used for transmission and distribution of electric power or in single-pulse switching in the subsystem of hybrid-electric contact vehicles are 25 kV standoff voltage, 2 kA conducting current and a forward voltage drop < 2% of the standoff voltage. At these power levels, it is expected that edge termination techniques will be essential for reproducible operation.

The devices designed for vertical depletion had lower on-state voltages than the lateral diodes, due to the fact that a highly-doped $n^+$ contact layer can be included in the epitaxial structure, obviating the need for implantation. However, we have not yet perfected the ability to grow resistive GaN on top of conducting GaN and, therefore, the depletion layers in the vertical devices typically had lightly $n$-type conductivity ($2 \times 10^{16}$ to $5 \times 10^{16}$ cm$^{-3}$).

The typical on-state resistances were $6 \sim 10$ m$\Omega$ cm$^2$, with reverse break-down voltages at 25°C of $200 \sim 550$ V (depending on doping level and layer thickness). The maximum figure-of-merit in these devices was higher than for the planar diodes, reaching values as high as 48 MW/cm$^2$.

In summary, GaN Schottky diodes with vertical and lateral geometries were fabricated. A reverse breakdown voltage of 3.1 kV was achieved on a lateral device incorporating $p$-type guard rings. Several types of edge termination were examined, with floating field rings and guard rings found to increase $V_B$. The best on-state resistance obtained in these lateral devices was 0.8 $\Omega$ cm$^2$. In mesa diodes incorporating $n^+$ contact layers, the best on-state resistance was 6 m$\Omega$ cm$^2$, while $V_B$ values were in the range $200 \sim 550$ V. These GaN rectifiers show promise for high power electronics applications.

## 3. AlGaN Schottky Rectifiers with 4.1 kV Reverse Breakdown Voltage

There is a strong interest in developing high current, high voltage switches in the AlGaN materials system for applications in the transmission and distribution of electric power and in the electrical subsystems of emerging vehicle, ship, and aircraft technology.[9,11,12] It is expected that packaged switches made from AlGaN may operate at temperatures in excess of 250°C without liquid cooling, therefore reducing system complexity, weight, and cost. In terms of voltage requirements, there is a strong need for power quality enhancement in the 13.8 kV class, while it is estimated that availability of 20–25 kV switches in a single unit would cause a sharp drop in the cost of power flow control circuits. Schottky and $p$-$i$-$n$ rectifiers are an attractive vehicle for demonstrating the high-voltage performance of different materials systems, and blocking voltages from 3–5.9 kV have been reported in SiC devices.[13-15] The reverse leakage current in Schottky rectifiers is generally far higher than expected from thermionic emission, most likely due to defect states around the contact periphery.[13] To reduce this leakage current and prevent breakdown by surface flashover, edge termination techniques such as guard rings, field plates, beveling, or surface ion implantation are necessary.[16,17] However, in the GaN rectifiers reported so far, there has been little effort in employing edge termination methods and no investigation of the effect of increasing the band gap by use of AlGaN.

We report in this section on the reverse breakdown voltage ($V_{RB}$) of AlGaN Schottky rectifiers for different Al compositions (0–0.25) and on the effect of various edge termination techniques in suppressing premature edge breakdown. A maximum $V_{RB}$ of 4.3 kV was achieved for $Al_{0.25}Ga_{0.75}N$ diodes, with very low reverse current densities. At low reverse biases the rectifiers typically show currents which are proportional to the contact perimeter, whereas at higher biases the current is proportional to contact area. The forward current characteristics show ideality factors of 2 at low bias (Shockley–Read–Hall recombination) and 1.5 at higher voltage (diffusion current).

The undoped $Al_xGa_{1-x}N$ layers were grown by atmospheric pressure metalorganic chemical vapor deposition at 1040°C (pure GaN) or 1100°C (AlGaN) on (0001) oriented sapphire substrates. The precursors were trimethylgallium, trimethylgaluminum, and ammonia, with $H_2$ used as a carrier gas. The growth was performed on either GaN (in the case of GaN active layers) or AlN (in the case of AlGaN active layers) low temperature buffers with nominal thicknesses of 200 Å. The active layer thickness was $\sim$ 2.5 $\mu$m in all cases and the resistivity of these films was of order $10^7$ $\Omega$ cm.[18] To form ohmic contacts in some cases, $Si^+$ was implanted at $5 \times 10^{14}$ cm$^{-2}$, 50 keV into the contact region and activated by annealing at 1150°C for 10 seconds under $N_2$. The contacts were then formed by lift off of *e*-beam evaporated Ti/Al/Pt/Au annealed at 700°C for 30 seconds under $N_2$. The rectifying contacts were formed by lift off of *e*-beam evaporated Pt/Ti/Au (diameter 60–1100 $\mu$m). A schematic of the planar diodes is shown in Fig. 9. The devices were tested at room temperature under a Fluorinert® ambient.

On the GaN diodes, we also examined the use of three different edge termination methods, namely *p*-guard rings formed by $Mg^+$ implantation at the edge of the rectifying contact, use of *p*-type floating field rings of width 5 $\mu$m to extend the depletion boundary along the edge of a $SiO_2$

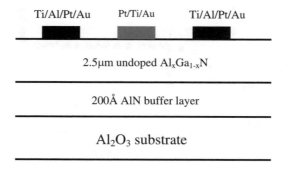

**Fig. 9.**   Schematic of AlGaN power rectifiers without edge termination.

passivation layer and finally, use of junction barrier controlled Schottky rectifiers (a rectifier with integrated $p$-$n$ junction grid in its drift region). In all of these edge-terminated diodes the Schottky metal was extended over a $SiO_2$ layer at the edge to minimize field crowding.

Figure 10 shows current-voltage (I-V) characteristics from two different diodes. The GaN device employed 30 $\mu$m wide $p$-guard rings. This was found to be the most effective edge termination method for these structures, producing an increase in $V_{RB}$ of $\sim$ 800 V over devices without any passivation or edge termination, i.e. breakdown occurred at 2.3 kV in the control diodes and 3.1 kV in devices with guard rings. The use of guard rings or floating field rings each produced improvements in $V_{RB}$ over the control diodes, with increases in the range 200–800 V. By sharp contrast, junction barrier control was unsuccessful in our structures, leading to decreases in $V_{RB}$ of 300–400 V. We believe this is due to Schottky barrier lowering because of the depletion regions around each section of the grid not completely overlapping in our initial design. The best on resistance ($R_{ON}$) achieved for GaN diodes was 0.8 $\Omega$ cm$^2$, producing a figure-of-merit $(V_{RB})^2/R_{ON}$ of 15.5 MW cm$^{-2}$. Figures 3–10 also shows an I-V characteristic from an $Al_{0.25}Ga_{0.75}N$ rectifier, without any edge termination or surface passivation. In this case $V_{RB}$ was 4.3 kV, which is far in excess of the values reported previously for GaN rectifiers, i.e. 350–450 V.[16,17] The on resistance of the AlGaN diodes was higher than for pure GaN, due to higher ohmic contract resistance. The lowest $R_{ON}$ achieved was 3.2 $\Omega$ cm$^2$, leading to a figure-of-merit of $\sim$ 5.5 MW cm$^{-2}$.

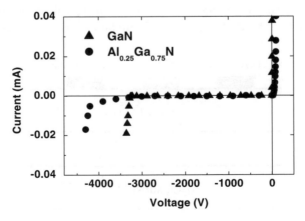

**Fig. 10.**  Room temperature I-V characteristics from an $Al_{0.25}Ga_{0.75}N$ rectifier.

Figure 11 shows the variation of $V_{RB}$ with Al percentage in the AlGaN active layers of the rectifiers. In this case we are using the $V_{RB}$ values from diodes without any edge termination or surface passivation. The calculated band gaps as a function of Al composition are also shown, and were obtained from the relation:

$$E_g(x) = E_{g,\text{GaN}}(1-x) + E_{g,\text{AlN}} \cdot x - bx(1-x)$$

where $x$ is the AlN mole fraction and $b$ is the bowing parameter with value 0.96 eV.[19] Note that $V_{RB}$ does not increase in linear fashion with band gap. In a simple theory, $V_{RB}$ should increase as $(E_g)^{1.5}$, but it has been empirically established that factors such as impact ionization coefficients and other transport parameters need to be considered and that consideration of $E_g$ alone is not sufficient to explain measured $V_{RB}$ behavior. The fact that $V_{RB}$ increases less rapidly with $E_g$ at higher AlN mole fractions may indicate increasing concentrations of defects that influence the critical field for breakdown.

The reverse I-V characteristics of all of the rectifiers showed I $\propto$ V$^{0.5}$ over a broad range of voltage (50–2000 V), indicating that Shockley–Read–Hall recombination is the dominant transport mechanism. The current density in all devices was in the range $5$–$10 \times 10^{-6}$ A cm$^{-2}$ at 2 kV. At low biases (25 V) the reverse current was proportional to the perimeter of the rectifying contact, suggesting that surface contributions are the most important in this voltage range. For higher biases, the current was proportional to the area of the rectifying contact. Under these conditions, the main contribution to the reverse current is from under this contact, i.e. from the bulk of the

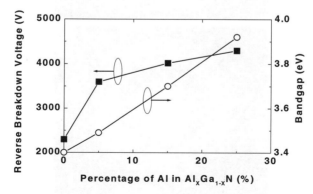

**Fig. 11.** Variation of $V_{RB}$ in Al$_x$Ga$_{1-x}$N rectifiers without edge termination, as a function of Al concentration. The band gaps for the AlGaN alloys are also shown.

material. It is likely that the high defect density in heteroepitaxial GaN is a primary cause of this current. The forward I-V characteristics showed that the current density was proportional to $\exp\left(-eV/2kT\right)$ at lowest voltages (up to current densities of $\sim 5 \times 10^{-4}$ A cm$^{-2}$) and to $\exp\left(-eV/1.5kT\right)$ at higher voltages (current densities in the range $10^{-3} - 1.5 \times 10^{2}$ A cm$^{-2}$). These results are consistent with Shockley–Read–Hall recombination as the dominant mechanism at low bias, followed by diffusion current at higher voltage. Qualitatively similar behavior has been reported previously for SiC Schottky rectifiers.

When pushed beyond breakdown, the diodes invariably failed at the edges of the rectifying contact, as shown in Fig. 12. As described earlier, the use of metal field plate contact geometries with SiO$_2$ as the insulator and either guard rings or floating field rings significantly increased $V_B$. These rectifiers generally did not suffer irreversible damage to the contact upon reaching breakdown and could be re-measured many times.

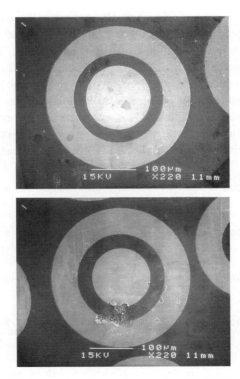

**Fig. 12.** Scanning electron microscopy micrographs of AlGaN rectifiers before (top) and after (bottom) pushing the applied bias beyond the value for breakdown.

In summary, Schottky rectifiers on high resistivity $Al_xGa_{1-x}N$ epi layers produced reverse breakdown voltages up to 4.3 kV for $Al_{0.25}Ga_{0.75}N$ diodes without edge termination. The current transport mechanisms were investigated as a function of bias voltage, with Shockley–Read–Hall recombination being dominant over a broad range of conditions. Minimizing electric field crowding at the corners of the rectifying contact was effective in increasing the breakdown voltage. The AlGaN materials system appears promising for high voltage applications.

## 4. Temperature Dependence and Current Transport Mechanisms in $Al_xGa_{1-x}N$ Schottky Rectifiers

*p-i-n* rectifiers are expected to have larger reverse blocking voltages than Schottky rectifiers, but inferior switching speeds and higher forward turn-on voltages. GaN Schottky rectifiers with reverse breakdown voltage ($V_{RB}$) to 3.1 kV have been demonstrated when $p^+$ guard rings and metal overlap onto a dielectric are employed as edge termination techniques. Use of $Al_{0.25}Ga_{0.75}N$ instead of GaN produced $V_{RB}$ values up to 4.3 kV.

Since this type of device is intended for elevated temperature operation, there is a need to understand the current transport mechanisms, the origin of the reverse leakage current and the magnitude and sign of the temperature coefficient for $V_{RB}$. In this section all of these properties are investigated. Over a broad range of voltages, the reverse leakage current is proportional to the diameter of the rectifying contact indicating that surface periphery leakage is the dominant contributor. The temperature coefficient for $V_{RB}$ was found to be negative for both GaN and AlGaN, even in edge-terminated devices.

The GaN and $Al_{0.25}Ga_{0.75}N$ layers were found to be resistive ($\sim 10^7 \ \Omega$ cm). Each was grown on $c$-plane $Al_2O_3$ substrates by metal organic chemical vapor deposition using conventional precursors and growth temperatures of 1040 (GaN) or 1100°C ($Al_{0.25}Ga_{0.75}N$). The layer thicknesses were 2.5–3 $\mu$m. Schematics of the completed rectifiers are shown in Fig. 13. The GaN devices employed $p^+$ guard rings formed (7 $\mu$m wide) by $Mg^+/P^+$ implantation, $n^+$ source/drain region formed by $Si^+$ implantation (annealing was performed at 1150°C for 10 seconds under $N_2$) and overlap of the rectifying contact onto a $SiO_2$ passivation layer. The AlGaN devices did not use any edge termination techniques. The contacts on all rectifiers were formed by lift-off, with the ohmic metallization annealed at 700°C for 30 seconds under $N_2$. The rectifying contact diameters were 45–125 $\mu$m with a separation of 124 $\mu$m between these contacts and the ohmic contacts.

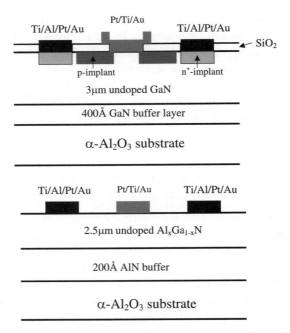

**Fig. 13.** Schematic of GaN (top) and AlGaN (bottom) rectifiers. The GaN devices employ several edge termination techniques.

Current-voltage (I-V) characteristics from both types of rectifiers are shown in Fig. 14 as a function of measurement temperature. The most obvious feature of the data is that there is a negative temperature coefficient for $V_{RB}$. The only previous information for GaN-based devices comes from GaN/AlGaN heterostructure field effect transistors in which a value of $+0.33$ V $K^{-1}$ was found,[19] and from linearly graded GaN $p^{+}pn^{+}$ junctions, in which a value of $+0.02$ V $K^{-1}$ was determined.[9] In both cases the $V_{RB}$ values were more than an order of magnitude lower than in the present diodes.

Figure 15 shows the variation of $V_{RB}$ with temperature. The data can be represented by a relation of the form:

$$V_{RB} = V_{RB0}[1 + \beta(T - T_0)]$$

where $\beta = -6.0 \pm 0.4$ V $K^{-1}$ for both types of rectifiers. However, in Schottky and $p$-$i$-$n$ rectifiers we have fabricated on more conducting GaN, with $V_{RB}$ values in the 400–500 V range, the values were consistently around $-0.34$ V $K^{-1}$. Therefore, in present state-of-the-art GaN rectifiers, the

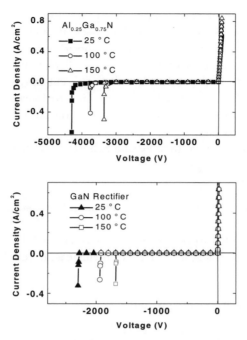

**Fig. 14.** I-V characteristics as a function of temperature for GaN (top) and AlGaN (bottom) rectifiers.

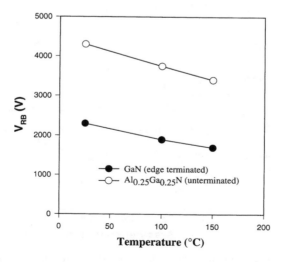

**Fig. 15.** Temperature dependence of $V_{RB}$ for GaN and AlGaN rectifiers.

temperature coefficient of $V_{RB}$ appears to be a function of the magnitude of $V_{RB}$. Regardless of the origin of this effect, it is clearly a disadvantage for GaN. While SiC is reported to have a positive temperature coefficient for $V_{RB}$ there are reports of rectifiers that display negative $\beta$ values.[21] One may speculate that particular defects present may dominate the sign and magnitude of $\beta$, and it will be interesting to fabricate GaN rectifiers on bulk or quasibulk substrates with defect densities far lower than in heteroepitaxial material.

The forward turn-on voltage $V_F$ of a Schottky rectifier can be written as

$$V_F = \frac{nkT}{e} \ln\left(\frac{J_F}{A^{**}T^2}\right) + n\phi_B + R_{ON} \cdot J_F$$

where $n$ is the ideality factor, $k$ is Boltzmann's constant, $T$ is the absolute sample temperature, $e$ the electronic charge, $J_F$ the forward current density (usually taken to be 100 A cm$^{-2}$) at $V_F$, $A^{**}$ the Richardson constant, $\phi_B$ the barrier height ($\sim 1.1$ eV in this case), and $R_{ON}$ the on-state

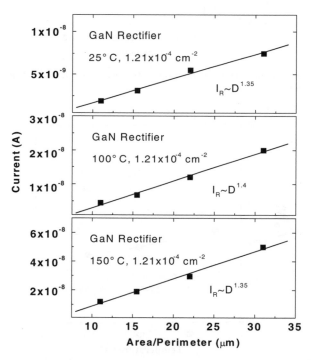

**Fig. 16.** Reverse current at $-100$ V bias for AlGaN rectifiers measured at three different temperatures.

resistance. The typical best $V_F$ values were $\sim$ 5 V for GaN and $\sim$ 7.5 V for $Al_{0.25}Ga_{0.75}N$, with best $R_{ON}$ values of 50 and 75 m$\Omega$ cm$^2$, respectively. The ideality factors derived from the forward I-V characteristic were typically $\sim$ 2 for both GaN and $Al_{0.25}Ga_{0.75}N$ for biases up to $\sim$ 2/3 of $V_F$. This is consistent with recombination being the dominant current transport in this bias range. At high voltages, $n$ was typically $\sim$ 1.5 for both types of rectifiers, indicating that diffusion currents were dominant. Beyond $\sim$ 2 $\times$ $V_F$, series resistance effects controlled the current. This behavior is often reported for SiC junction rectifiers, while Schottky rectifiers in that materials system show ideality factors of 1.1–1.4. In our GaN devics, the higher ideality factors may reflect the high compensation levels in the material.

Figure 17 shows the reverse current ($I_R$) at $-100$ V reverse bias for GaN and AlGaN rectifiers of different contact diameter, for three different measurement temperatures. Since $I_R \propto$ contact diameter, this indicates that

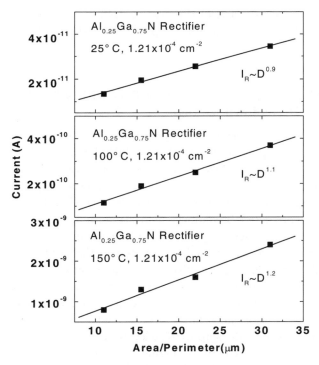

**Fig. 17.** Reverse current at $-100$ V bias for AlGaN rectifiers measured at three different temperatures.

under these conditions the reverse current originates from surface periphery leakage. Similar results were obtained for the GaN rectifiers as shown in Fig. 16. The activation energy for this periphery leakage was $\sim 0.13$ eV, which may represent the most prominent surface state giving rise to the current. At voltages approximately 90% of the breakdown values, the reverse current was proportional to contact area, indicating that bulk leakage is dominant under these conditions.

In conclusion, the temperature dependence of $V_{RB}$ has been measured in high breakdown GaN and AlGaN Schottky rectifiers. The temperature coefficient is negative, which is a significant disadvantage for devices intended for high temperature operation, and there are indications that it is a function of $V_{RB}$. The forward current conduction makes a transition from recombination to diffusion currents. The reverse leakage current originates from surface components around the rectifying contact at modest voltages. This current is thermally activated with an energy of 0.13 eV. The yield of acceptable devices (i.e. with $V_{RB}$ at least 90% of the maximum found on a wafer and $R_{ON}$ within 50% of the best values obtained) was rather small ($\sim 15\%$), so there is still much development needed on both materials and processing.

## 5. Lateral $Al_x Ga_{1-x}N$ Power Rectifiers with 9.7 kV Reverse Breakdown Voltage

There have been advances in developing GaN and AlGaN power rectifiers, which are key components of inverter modules for power flow control circuits. Vertical geometry GaN Schottky rectifiers fabricated on conducting materials typically show reverse breakdown voltages ($V_B$) 750 V whereas lateral devices on insulating GaN and AlGaN have $V_B$ values up to 4.3 kV.

Since the predicted breakdown field strength in GaN is of order 2–$3 \times 10^6$ V cm$^{-1}$,[22] there appears to be much room for improvement in rectifier performance and a need to understand the origin of reverse leakage currents, breakdown mechanisms, and the effect of contact spacing on $V_B$. In this section we report on the variation of $V_B$ with Schottky-to-ohmic contact gap spacing in $Al_x Ga_{1-x}N$ diodes ($x = 0$–0.25) employing $p$-guard rings and extension of the Schottky contact edge over an oxide layer for edge termination. $V_B$ values up to 9700 kV were achieved for $Al_{0.25}Ga_{0.75}N$ rectifiers, with breakdown still occurring at the edges of the Schottky contact. The reverse leakage current just before breakdown is dominated by bulk contributions, scaling with the area of the rectifying contact.

The rectifiers were fabricated on resistive ($\sim 10^7$ $\Omega$ cm) layers of 2.5–3 $\mu$m thick GaN or AlGaN grown on $c$-plane $Al_2O_3$ substrates at 1040–1100°C by metalorganic chemical vapor deposition. To create $n^+$ regions for ohmic contacts, $Si^+$ ions were implanted at $5 \times 10^{14}$ cm$^{-2}$, 50 keV, and activated by annealing at 1150°C for 10 seconds under $N_2$. It is important to control both the heating and cooling rates to avoid cracking of the AlGaN layer. $Mg^+$ implantation at $5 \times 10^{14}$ cm$^{-2}$, 50 keV was used to create 30 $\mu$m diameter $p$-guard rings at the edge of the Schottky barrier metal. The rectifying contact diameter was 124 $\mu$m in most cases, while the distance of this contact from the edge of the ohmic contact was varied from 30–100 $\mu$m. The Schottky metal was extended over a $SiO_2$ layer deposited by plasma-enhanced chemical vapor deposition in order to minimize field crowding. Ohmic contacts were created by lift off of $e$-beam evaporated Ti/Al/Pt/Au annealed at 750°C for 30 seconds under $N_2$. The Schottky contacts were formed by lift off of $e$-beam evaporated Pt/Ti/Au. A schematic of the completed rectifiers is shown in Fig. 18. Current-voltage (I-V) characteristics were recorded on a HP 4145 parameter analyzer, with all testing performed at room temperature under a Fluorinert® ambient.

Figure 19 shows the measured $V_B$ values for GaN and $Al_{0.25}Ga_{0.75}N$ rectifiers as a function of the gap spacing between the rectifying and ohmic contacts. For gaps between 40 and 100 $\mu$m, $V_B$ is essentially linearly dependent on the spacing, with slopes of where $W_B$ is the depletion width at breakdown. In our laterally depleting devices the surface quality will dominate the onset of breakdown, which is reflected in the lower breakdown field observed. However, given the current state of defect densities in epitaxial GaN, the lateral geometry seems the most promising, for the time being,

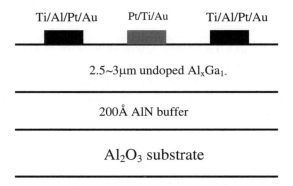

**Fig. 18.** Schematic of lateral geometry AlGaN rectifiers employing edge termination.

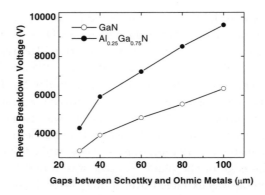

**Fig. 19.** Effect of Schottky-ohmic contact gap spacing on $V_B$ for GaN and $Al_{0.25}Ga_{0.75}N$ rectifiers.

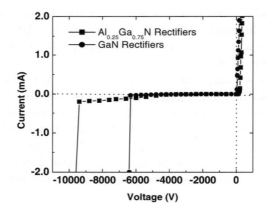

**Fig. 20.** Current-voltage characteristics of GaN and $Al_{0.25}Ga_{0.75}N$ rectifiers.

for achieving very high $V_B$ values. Quasi-substrates of GaN, produced by thick epi-growth on mismatched substrates and subsequent removal of this template, are soon to be commercially available. In some cases the background doping in these is as low as $7.9 \times 10^{15}$ cm$^{-3}$ which makes feasible the use of these thick (200 $\mu$m) freestanding GaN films for vertically depleting rectifiers.

Figure 20 shows some I-V characteristics from the 100 $\mu$m gap spacing GaN and $Al_{0.25}Ga_{0.75}N$ rectifiers. The best forward turn-on voltages, $V_F$ (defined as the forward voltage at a current density of 100 A cm$^{-2}$) was $\sim 15$ V for GaN and $\sim 33$ V for $Al_{0.25}Ga_{0.75}N$. These are much higher than

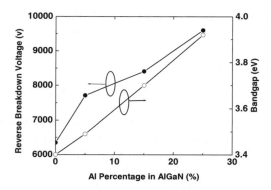

**Fig. 21.** Variation of $V_B$ with Al percentage in the AlGaN layer of the rectifiers.

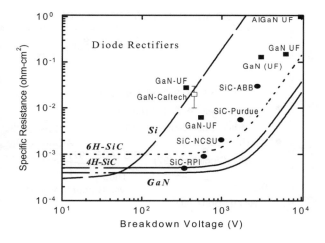

$$\bullet R_{ON} = \frac{4(V_{RB})^2}{\mu \varepsilon \, E_C^{\ 2}} + P_{sub} W_{sub} + R_C$$

Where:
$\qquad V_{RB}$ = breakdown voltage
$\qquad \mu \quad$ = carrier mobility
$\qquad \qquad$ = permittivity
$\qquad E_C \quad$ = critical field for breakdown
$\qquad P_{sub} W_{sub}$ = resistivity/thickness of substrate

**Fig. 22.** On-state resistance vs $V_B$ for wide bandgap Schottky rectifiers. The theoretical performance limits of Si (solid line), SiC (dashed line for 6H and dot-dashed line for 4H) and GaN (bottom line) devices are shown.

the values obtained on more conducting GaN films, where $V_F$ is typically 5–8 V. Note, however, that the ratio $V_B/V_F$ is still very high for the resistive diodes, with values ranging from 294 to 423. The specific on-state resistance for a rectifier is given by

$$R_{ON} = (4V_B^2/\varepsilon \cdot \mu \cdot E_M^3) + \rho \cdot s \cdot W_S + R_C$$

where $\varepsilon$ is the GaN permittivity, $\mu$ the carrier mobility, $S$ and $W_S$ are substrate resistivity and thickness, and $R_C$ is the contact resistance. The best on-state resistances we achieved were 0.15 $\Omega$ cm$^2$ for GaN and 1 $\Omega$ cm$^2$ for Al$_{0.25}$Ga$_{0.75}$N, leading to figure-of-merits $V_B^2/R_{ON}$ of 268 MW cm$^{-2}$ and 94 MW cm$^{-2}$, respectively. At low reverse voltages (2000 V), the magnitude of the reverse current was proportional to contact diameter. As the diodes

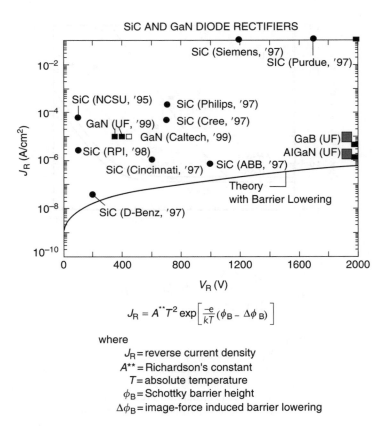

**Fig. 23.**  Reverse leakage current $J_R$ vs $V_B$ for wide band gap Schottky rectifiers.

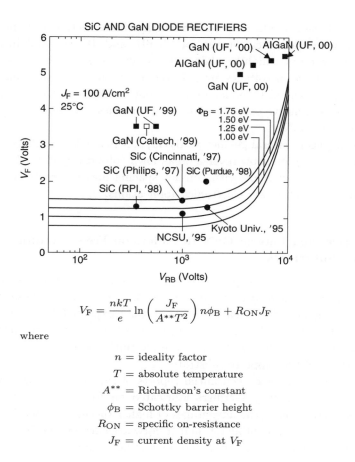

$$V_F = \frac{nkT}{e} \ln \left( \frac{J_F}{A^{**}T^2} \right) n\phi_B + R_{ON}J_F$$

where

$$n = \text{ideality factor}$$
$$T = \text{absolute temperature}$$
$$A^{**} = \text{Richardson's constant}$$
$$\phi_B = \text{Schottky barrier height}$$
$$R_{ON} = \text{specific on-resistance}$$
$$J_F = \text{current density at } V_F$$

**Fig. 24.** Forward voltage drop $V_F$ vs $V_B$ for wide band gap Schottky rectifiers.

approached breakdown the reverse current was proportional to contact area, suggesting bulk leakage became dominant.

The variation of $V_B$ with Al percentage in the AlGaN layer of the rectifiers is shown in Fig. 21, along with the calculated bandgaps. $V_B$ does increase with increasing bandgap $E_g$, but is not proportional to $(E_g)^{1.5}$ as expected from a simple theory. The presence of bulk and surface defects will have a strong influence on $V_B$, and these are not well controlled at this stage of AlGaN rectifier technology.

To place our results in context, Figs. 22, 23 and 24 show a compilation of $R_{ON}$, reverse leakage current and forward turn-on voltages versus $V_B$ data for state-of-the-art SiC and GaN Schottky diode rectifiers, respectively, together with theoretical curves for Si, 6H, and 4H–SiC and hexagonal GaN. Our results for high breakdown GaN devices show the on resistances and forward turn-on voltages are still well above the theoretical values and more work is needed to understand current conduction mechanisms, the role of residual native oxides on contact properties, and impact ionization coefficients in GaN.

In conclusion, lateral geometry $Al_xGa_{1-x}N$ Schottky rectifiers employing edge termination show reverse breakdown voltages up to 9.7 kV. These breakdown voltages scale with contact spacing and the rectifiers appear promising for high power electronics applications.

## 6. Vertical and Lateral GaN Rectifiers on Free-Standing GaN Substrates

Although the GaN-based power rectifiers on sapphire substrate show impressive results, there are still numerous short-comings in these devices, including higher reverse leakage current than expected from thermionic emission, high forward turn-on voltages, negative temperature coefficients for reverse breakdown voltage, non-uniformities and the low thermal conductivity of the sapphire substrate.

Recently there have been initial reports of reverse recovery characteristics of GaN Schottky rectifiers fabricated on free-standing substrates. Those substrates have the advantages of higher thermal conductivity than sapphire and the potential for higher forward current densities and reverse breakdown voltages than lateral rectifiers fabricated in insulating substrates.

We investigated the effect of contact dimension and current flow direction (lateral versus vertical) on the on-state resistance and breakdown voltage of GaN Schottky rectifiers fabricated on free-standing GaN substrates. There was a dramatic effect of contact diameter on $V_B$, with the latter ranging from 6 to 700°C as the diameter was decreased from 7 mm to 75 $\mu$m. At the lower end of this range the on-state resistance $(R_{ON})$ are exceptionally low $(1.71 \sim 3.01$ m$\Omega$ cm$^{-2})$, producing maximum figure-of-merit $(V_B{}^2/R_{ON})$ above 100 MW cm$^{-2}$.

The 200 $\mu$m thick GaN quasi-substrates were grown by hydride vapor phase epitaxy on sapphire substrate, lifted-off by laser heating and then etched and polished as shown in Fig. 25. The measured $n$-type doping concentration was $\sim 10^{17}$ cm$^{-3}$. Mg$^+$ implantation at $5 \times 10^{14}$ cm$^{-2}$, 50 keV,

**Fig. 25.** Free-standing GaN substrate grown by HVPE.

followed by annealing was used to create 30 $\mu$m diameter $p$-guard rings at the edge of the Schottky contacts. The rectifying contact diameter was 75 $\mu$m for the small-area device and 7 mm for the large-area devices. On these latter structures the Schottky metal was extended over a $SiO_2$ layer deposited by rf (13.56 MHz) plasma enhanced chemical vapor deposition using $SiH_4$ and $N_2O$ as the precursors. Full-area back ohmic contacts were placed on the $N$-face using $e$-beam evaporation of Ti/Al/Pt/Au. On the small-area devices we also placed ohmic contacts on the top (Ga-face) surface so that we could compare results from the lateral and vertical geometries. The top Schottky contacts were $e$-beam deposited Pt/Ti/Au in both large and small area devices. In the latter case, the Schottky-ohmic metal spacing was 30 $\mu$m. Schematics of the completed structures are shown in Fig. 27. Current-voltage (I-V) characteristics were recorded on an HP 4145B parameter analyzer at 25°C for the forward part of the characteristics, while Tektronix 370A curve tracer was used for the reverse characteristics measurement.

Figure 28 shows the I-V characteristic from the large area rectifiers. The reverse breakdown voltage ($V_B$) is only $\sim$ 6 V and is obviously far below anything of practical use. The on-state resistance ($R_{ON}$) was 3.4 $\Omega$ cm$^2$ for these devices. The low $V_B$ is in stark contrast to the values achieved in smaller devices, as described below. Since the defect density in the quasi-substrate was $\sim 10^5$ cm$^{-2}$ as measured by combined photochemical etching and atomic force microscopy, the large area rectifiers are highly likely to include one or more defects. Hsu *et al.* found that reverse bias leakage in GaN Schottky diodes occurred primarily at defects and dislocations. The

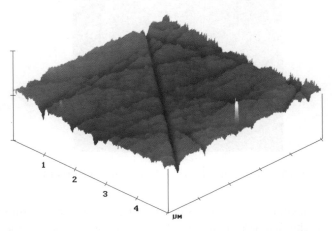

X   1.000 µм/div
Z   10.000 nм/div

allganto.001

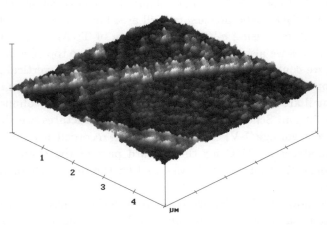

X   1.000 µм/div
Z   300.000 nм/div

allganbo.002

**Fig. 26.** AFM images showing Ga- (front surface, top) and *N*- (backside surface, bottom) terminated surfaces.

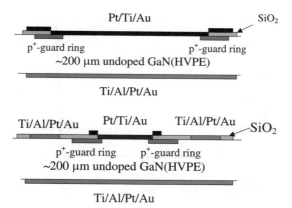

**Fig. 27.** Schematic of 7 mm contact diameter (top) and 75 $\mu$m contact diameter (bottom) rectifiers.

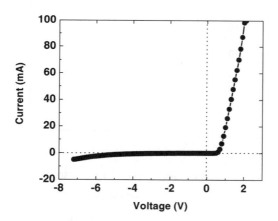

**Fig. 28.** I-V characteristics from 7 mm contact diameter GaN rectifier.

figure-of-merit $V_B{}^2/R_{ON}$ had a value of 10.7 W cm$^{-2}$ for the large area rectifiers while maximum current of $\sim$ 500 mA could be achieved before sample heating became a problem.

I-V characteristics from the small-area rectifiers, measured in the lateral geometry are shown in Fig. 29. The $V_B$ was $\sim$ 250 V, with an excellent $R_{ON}$ of 1.7 m$\Omega$ cm$^2$. This on-state resistance is the lowest reported for any GaN rectifiers and shows that continued improvements in surface cleaning and contact technologies for this materials system have led to a rapid maturation of our understanding of how to process these devices. The value of

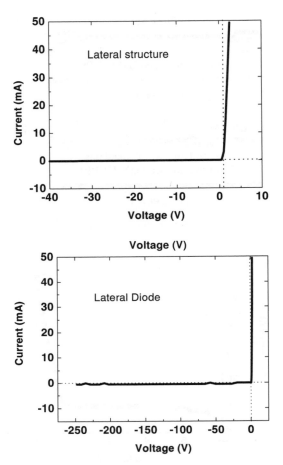

**Fig. 29.** I-V characteristics from 75 $\mu$m contact diameter GaN rectifiers measured in lateral geometry.

$V_B{}^2/R_{ON}$ was 36.5 MW cm$^{-2}$. Note that remarkable improvement in the electrical characteristics in the small-area rectifiers relative to the large devices fabricated on the same material. The forward turn-on voltage, $V_F$, defined as the forward bias at which the current density was 100 A cm$^{-2}$, was 1.8 V. This is roughly half of what has been reported previously for GaN Schottky rectifiers on heteroepitaxial layers. The forward turn-on voltage for Schottky rectifiers is given by

$$V_F = \frac{nkT}{e} \ln\left[\frac{J_F}{A^{**}T^2}\right] + n\phi_B + R_{ON} \cdot I_F$$

where $n$ is the ideality factor, $k$ is Boltzmann's constant, $T$ the absolute diode temperature, $e$ the electronic charge, $I_F$ the forward current density at $V_F$, $A^{**}$ is the Richardson constant, $\phi_B$ the barrier height ($\sim 1.1$ eV in this case for Pt on $n$-GaN) and $R_{ON}$ the on-state resistance. One of the reasons for the much lower $V_F$ in these quasi-bulk rectifiers is the small $R_{ON}$ value to heteroepitaxial devices. The ideality factors were $\sim 2$ in the large area rectifiers, indicating that recombination was the dominant current transport mechanism. In the small-area rectifiers, $n$ values was $\sim 1.5$, which is consistent with diffusion currents being dominant. These results can be explained by the relative probabilities for having defects in the active region of the rectifiers for the different contact diameters.

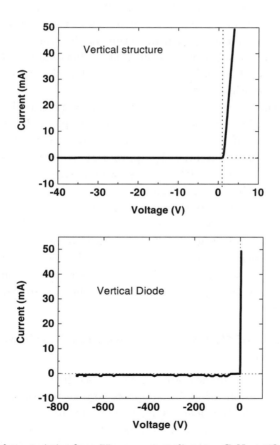

**Fig. 30.** I-V characteristics from 75 $\mu$m contact diameter GaN rectifiers measured in vertical geometry.

Figure 30 shows the I-V characteristics from the small-area diode measured from top-to-bottom, i.e. through the GaN substrate, rather than in the lateral geometry employed for the data of Fig. 29. The $V_B$ in the vertical geometry was $\sim 700$ V, while $R_{on}$ was 3.01 m$\Omega$ cm$^2$, leading to a figure-of-merit of 162.8 MW cm$^2$. This $V_B$ is close to the expected maximum for the drift region doping concentration of $\sim 10^{17}$ cm$^{-3}$. The forward turn-on voltage was still $\sim 1.8$ V, which is close to the minimum expected for a GaN rectifier with a $V_B$ of 700 V and assuming a barrier height of 1.1 eV. The ratio $V_B/V_F$ is $\sim 389$, a record for GaN rectifiers, and the forward current density could be pushed above 1000 A cm$^{-2}$. A plausible explanation for the large improvement in $V_B$ in the vertical geometry may not be only in the larger thickness in this direction (200 $\mu$m) compared to the 30 $\mu$m spacing between Schottky and ohmic contacts in the lateral direction, but also in the fact that the vertical depletion mode would minimize surface breakdown problems. The reverse current at bias value close to $V_B$ was proportional to contact diameter, suggesting that the surface is playing a strong role in the origin of the leakage current.

One can expect major improvements in $V_B$ in quasi-bulk GaN rectifiers as the background doping is decreased. For example, a 200 $\mu$m thick sample with a doping of $10^{15}$ cm$^{-3}$ (which is quite feasible by reducing the background Si and O content, or by appropriate compensation) would have a predicted $V_B$ of $\sim 1.5 \times 10^4$ V. These would have application for power control system in the 13.8 kV class.

In summary, the size and geometry dependence of GaN Schottky rectifiers on quasi-bulk substrate has been investigated. The reverse breakdown voltage increases dramatically as contact size is decreased and is also much larger for vertically-depleting devices. The low on-state resistances produce high figure-of-merits for the rectifiers and show their potential for applications involving high power electronic control systems.

## 7. Comparison of GaN *p-i-n* and Schottky Rectifiers Performance

Schottky and *p-i-n* diodes are employed as high-voltage rectifiers in power switching application. To suppress voltage transients when current is switched to inductive loads such as electric motors, these diodes are placed across the switching transistors. The advantage of simple metal-semiconductor diodes relative to *p-n* junction diodes is the faster turn-off because of the absence of minority carrier storage effects and lower power dissipation during switching. Wide bandgap semiconductors such as GaN offer additional advantages for fabrication of diode rectifiers, including much higher breakdown voltages and operating temperatures. There is much interest in developing advanced switching devices and control circuits for CW and pulsed electrical sub-systems in emerging hybrid-electric and all-electric vehicles, more-electric airplanes and naval ships and for improved transmission, distribution and quality of electric power in the utilities industry. Eventually one would like to reach target goals of 25 kV stand-off voltage, 2 kA or higher conducting current, forward drop less than 2% of the rated voltage and maximum operating frequency of 50 kHz.

Figure 31 shows a SIMS profile of H and other background impurities (along with intentional Si doping) in an MOCVD-grown $p$-$i$-$n$ diode structure (left), together with the Mg profile in the structure (right). Notice once again that the H decorates the Mg due to formation of the neutral $(Mg-H)^\circ$ complexes. About 70–80% of the Mg atoms have hydrogen attached.

The structures were grown by MOCVD at $1050°C$ on $c$-plane $Al^2O^3$ substrates. Growth commenced with a low temperature ($530°C$) AlN template $\sim 300$ Å thick, followed by a 1.2 mm thick, Si-doped ($n = 3 \times 10^{18}$ cm$^{-3}$) GaN layer, 4 $\mu$m of undoped ($n \times 10^{16}$ cm$^{-3}$) GaN and 0.5 $\mu$m of Mg-doped ($N_A \sim 10^{18}$ cm$^{-3}$) GaN. Both $p$-$i$-$n$ and Schottky diodes were fabricated on the same wafer. This was achieved by dry etch removal of the $p^+$-GaN layer

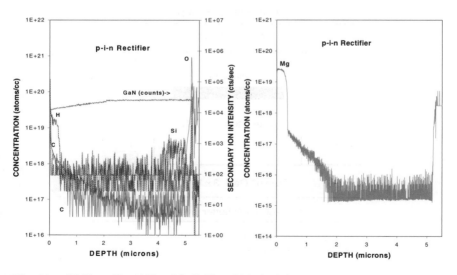

**Fig. 31.** SIMS profiles of H and Si (left) and Mg (right) in an as-grown, MOCVD $p$-$i$-$n$ rectifier structure.

in some regions, followed by a wet etch (NaOH, 0.1 M, 80°C) clean-up of 600 Å of material to remove residual lattice damage prior to deposition of the rectifying contact. In some cases, various amounts of material were removed by wet etching after plasma exposure and deposition of the rectifying contact.

For both *p-i-n* and Schottky diode structures, mesas were fabricated by dry etching down to the $n^+$ layer using $Cl_2/Ar$ ICP etching, followed by annealing at 750°C under $N_2$ to remove sidewall damage. Ohmic contacts to the $n^+$ layer were prepared by lift-off of Ti/Al/Pt/Au alloyed at 700°C, while rectifying contacts to the undoped GaN were formed by lift-off of Ni/Pt/Au. To form an Ohmic contact to the $p^+$ layer, we used the same

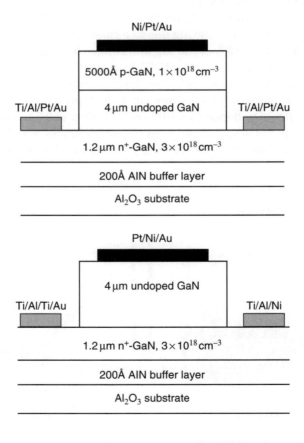

**Fig. 32.** Schematic of *p-i-n* (top) an Schottky (bottom) rectifiers.

metallization. Schematics of the *p-i-n* and Schottky rectifier structures are shown in Fig. 32. A scanning electron micrograph of a typical diode is shown in Fig. 33.

Current-voltage (I-V) measurements were recorded on a HP 4145B parameter analyzer. Figure 34 shows the I-V characteristics at 25°C for *p-i-n* and Schottky rectifiers fabricated side-by-side on the same wafer. There is a clear difference in the $V_{RB}$ of the two devices ($\sim$ 490 V for the *p-i-n* versus $\sim$ 347 V for the Schottky). For very high blocking voltages (typically in excess of $\sim$ 3 kV) or forward current densities ($>$ 100 A cm$^{-2}$) the *p-i-n* is expected to have the advantage because of the prohibitive leakage and

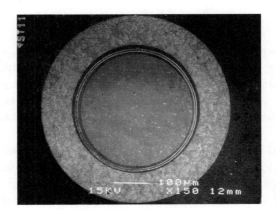

**Fig. 33.** Schematic of *p-i-n* (top) an Schottky (bottom) rectifiers.

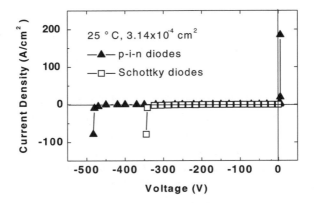

**Fig. 34.** I-V characteristics from *p-i-n* and Schottky rectifiers.

resistance of the drift region in a Schottky diode. However, for high frequency operation the Schottky has an advantage in switching speed due to the absence of minority carriers. The on-state characteristics of the Schottky and $V_{\mathrm{RB}}$ characteristics of the $p$-$n$ junction can be achieved in junction barrier controlled Schottky rectifiers.

The forward voltage drop ($V_{\mathrm{F}}$) of the two types of rectifiers can be written as

$$V_{\mathrm{F}} = \frac{nkT}{e} \ln\left(\frac{J_{\mathrm{F}}}{A^{**}T^2}\right) + n\phi_{\mathrm{B}} + R_{\mathrm{ON}} \cdot J_{\mathrm{F}} \quad \text{(Schottky)}$$

and

$$V_{\mathrm{F}} = \frac{kT}{e} \ln\left(\frac{n_- n_+}{n_i^2}\right) + V_m \quad \text{($p$-$i$-$n$)}$$

where $n$ is the ideality factor, $k$ the Boltzmann's constant, $T$ the absolute temperature, $e$ the electronic charge, $J_{\mathrm{F}}$ the forward current density at $V_{\mathrm{F}}$, $A^{**}$ the Richardson constant, $\phi_{\mathrm{B}}$ the barrier height, $R_{\mathrm{ON}}$ the on-state resistance, $n_-$ and $n_+$ the electron concentrations in the two end regions of the $p$-$i$-$n$ (the $p^+$-$n$ and $n^+$-$n$ regions) and $V_m$ is the voltage drop across the $i$-region.

The typical values of $V_{\mathrm{F}}$ were $\sim 5$ V for the $p$-$i$-$n$ rectifiers and $\sim 3.5$ V for the Schottky rectifiers, both measured at 25°C and defining $V_{\mathrm{F}}$ as the forward voltage at which the current density was 100 A/cm$^2$. Both of these values are still well above the theoretical minima, which should be of the order of the barrier height for the Schottky metal (between 1 and 1.5 eV in our case) or the bandgap for the $p \pm i \pm n$ diode (3.4 eV for GaN). A similar situation occurs in SiC rectifiers, although there have been reports of $V_{\mathrm{F}}$ values relatively close to the theoretical values. In general it is expected that $V_{\mathrm{F}}$ will remain fairly constant for GaN $p$-$i$-$n$'s and Schottky rectifiers to breakdown voltages in the $3 \sim 5$ kV range, at which point there is a sharp increase due to the increase in on-state resistance.

Forward I-V characteristics for $p$-$i$-$n$ rectifiers of different contact areas are shown in Fig. 35. It is often found for SiC $p$-$i$-$n$ rectifiers that the Sah–Noyce–Shockley model for forward current conduction cannot be applied due to the presence of a multiple number of deep and shallow impurity levels in the bandgap, which can act as recombination sites. The I-V characteristic of Fig. 34 corresponds well to the four different exponential regimes predicted by this model, with ideality factors close to 1 at low bias ($< 0.6$ V), $\sim 2$ at higher bias (0.6–5 V), dependent on the number of deep and shallow levels (around 5 V) and then $\sim 2$ at higher bias. In the latter

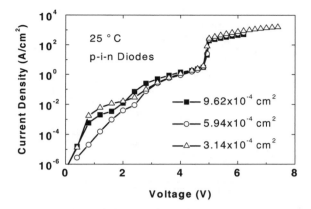

**Fig. 35.** Forward I-V characteristics at 25°C from *p-i-n* rectifiers of different contact areas.

case our characteristic becomes dominated by series resistance effects. In the multi-recombination center model, the forward current density $J_F$ can be written as

$$J_F = J_{01} \exp\left(\frac{eV}{2kT}\right) + J_{02} \exp\left(\frac{eV}{nkT}\right) + J_{S2} \exp\left(\frac{eV}{kT}\right)$$

where the first two terms represent the recombination current components and the third represents the diffusion current component originating from the recombination of electrons and holes in neutral regions outside the space-charge region. When we measure the forward I-V characteristics at elevated temperatures ($150 \sim 250°C$) for the *p-i-n* diodes, the shape of the curves became more simplified and appeared to revert to the more common Sah–Noyce–Shockley form. This is probably a result of the fact that recombination through multiple deep and shallow levels becomes far less effective at elevated temperature. The Schottky diodes typically showed ideality factors in the range $1.3 \sim 1.6$ at $V_F = 2.5 \sim 4$ V and $n = 2$ at $V_F > 4$ V, consistent with recombination at low bias and diffusion current at higher bias.

Figure 36 shows the reverse current measured at 100 V bias for *p-i-n* diodes with different contact diameters. For all the temperatures employed in these measurements, the reverse current was directly proportional to the diameter of the contact, indicating the dominance of surface perimeter leakage. If bulk leakage were dominant then we would obtain a slope of 2 for the plot of reverse current versus contact diameter. Note again that all of our devices are unpassivated and unterminated, thus no effort was made

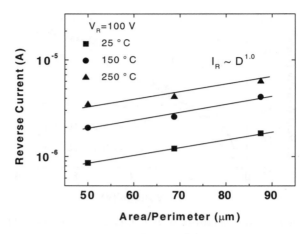

**Fig. 36.** Reverse current as a function of contact diameter for *p-i-n* rectifiers at different temperatures.

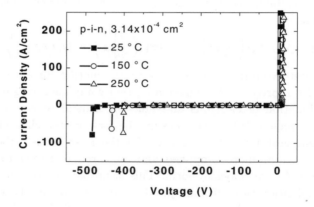

**Fig. 37.** I-V characteristics from *p-i-n* rectifiers as a function of temperature.

to minimize surface contributions to the leakage current. However, one of the expected attributes of GaN for electronic devices was a relative insensitivity to surface effects. We have consistently observed to the contrary that the GaN surface is easily disrupted during plasma processing or thermal annealing, usually through the preferential loss of nitrogen.

Figure 37 shows the I-V characteristics from a *p-i-n* diode at three different temperatures. It is clear that there is a negative temperature coefficient for the reverse breakdown voltage. We observed similar behavior for

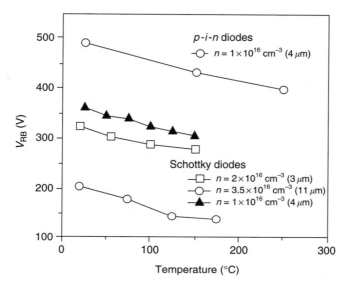

**Fig. 38.** Temperature dependence of $V_{RB}$ in *p-i-n* and Schottky rectifiers.

Schottky diodes and a compilation of such data is shown in Fig. 38. Here we have also included results from several other GaN Schottky rectifiers we have fabricated on *n*-type material. Note that in all cases the measured breakdown voltage decreases with temperature as

$$V_{RB} = V_{RBO}(1 + \beta \cdot (T - T_0))$$

where $\beta = -0.34 \pm 0.05$ V/K. Other reports have found positive temperature coefficients for AlGaN-GaN high electron mobility transistors (+0.33 V/K) and GaN $p^+$-$p$-$n^+$ linearly-graded junctions (+0.02 V/K). In separate experiments we have found that use of edge termination techniques (floating field rings, junction barrier control or metal overlap onto a dielectric on the surface) also produced negative temperature coefficients of $V_{RB}$ in GaN Schottky rectifiers.

A direct comparison of GaN *p-i-n* and Schottky rectifiers fabricated on the same GaN wafer showed higher reverse breakdown voltage for the former (490 V versus 347 V for the Schottky diodes), but lower forward turn-on voltages for the latter ($\sim$ 3.5 V versus $\sim$ 5 V for the *p-i-n* diodes). The forward I-V characteristics of the *p-i-n* rectifiers show behavior consistent with a multiple recombination center model. For the Schottky rectifiers, the forward I-V characteristics were consistent with Shockley–Read–Hall recombination at low bias and diffusion currents at higher bias. The reverse current in both types of rectifiers was dominated by surface perimeter leakage at moderate bias. Finally, all of the devices we fabricated showed negative temperature coefficients for reverse breakdown voltage, which is a clear disadvantage for elevated temperature operation.

Figure 39 (top) shows the $J_F \sim V_F$ characteristics at 25°C from lateral geometry GaN rectifiers with different $p^+$ guard-ring widths. The series

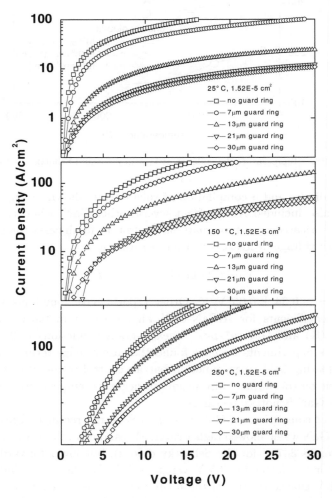

**Fig. 39.** Forward I-V characteristics at 25°C, 150°C and 250°C from GaN Schottky rectifiers employing $p^+$ guard rings of different widths.

resistance effects are more obvious when the guard rings are employed, especially for widths beyond 7 $\mu$m. When we also used floating field rings of 5 $\mu$m width or junction barrier control we also observed an increase in the series resistance in the rectifiers. Even though edge termination techniques improved $V_{RB}$ (from 2.3 kV in unterminated diodes to 3.1 kV with optimized edge termination), they produced a deterioration of the forward current conduction. We believe that more work needs to be done to optimize the design and implementation of the implanted $p$-guard rings. In currently available GaN it is difficult to activate implanted acceptor dopants because of the presence of compensating defects and impurities and the difficulty in avoiding nitrogen loss from the surface during annealing.

Figure 39 (center) shows the forward $J_F$–$V_F$ characteristics at 150°C from the lateral GaN rectifiers with guard rings of different widths. The series resistance effect is less evident at this temperature and current densities of 100 A/cm$^2$ are reached at lower forward biases. Similarly, the use of blocking rings and junction barriers control deteriorated the current conduction, but in all cases the current densities were much higher than at 25°C.

Similar data are also shown in Fig. 39 for measurement at 250°C. The same trend holds, namely that lower forward voltages were needed to reach high current densities. The best on-resistances obtained at different temperatures were 0.13 $\Omega$ cm$^2$ at 25°C, 0.075 $\Omega$ cm$^2$ at 150°C and 0.067 $\Omega$ cm$^2$ at 250°C. The figure-of-merit $V_{RB}{}^2/R_{ON}$ was 73.9 W/cm$^2$ at 25°C, 64.5 MW/cm$^2$ at 150°C and 38.2 MW/cm$^2$ at 250°C. Thus, even though $R_{ON}$ decreased with temperature, so did $V_{RB}$ because of a negative temperature coefficient for this parameter of $\sim$ 6 V/K in these high breakdown rectifiers.

A compilation of data for $V_F$ versus $V_{RB}$ for GaN and SiC Schottky rectifiers is shown in Fig. 40. The solid lines are theoretical values for SiC, assuming different barrier heights. While SiC rectifiers have achieved $V_F$ values close to the theoretical values, the GaN devices are still well below their optimal performance. This clearly indicates that much more work is needed in the areas of thermally stable rectifying contacts, surface cleaning and preparation, and material quality. On the positive side, all of the GaN results have been reported only in the past year or so, and represent excellent progress in this field.

Figure 41 shows a compilation of published results for $R_{ON}$ versus $V_{RB}$ for SiC and GaN Schottky rectifiers. The lines on the plot represent theoretical values for Si, 4H–SiC, 6H–SiC and GaN. The first reports on GaN ($V_{RB}$ values of $\sim$ 500 V) show performance similar to that expected for

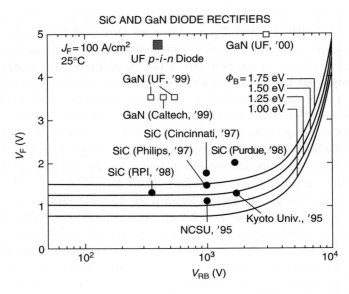

**Fig. 40.** $V_F$ versus $V_{RB}$ for GaN and SiC Schottky rectifiers in the literature.

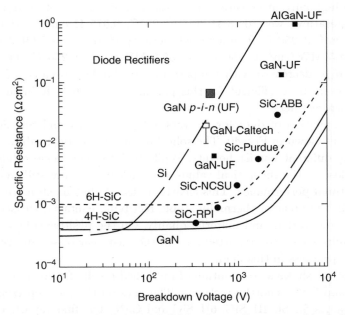

**Fig. 41.** $R_{ON}$ versus $V_{RB}$ for GaN and SiC Schottky rectifiers reported in the literature.

Si. The more recent results, with $V_{RB}$ values of $> 3$ kV at $R_{ON}$ values $< 1 \ \Omega \ cm^2$, show performance well in excess of that possible with Si. Note that SiC has achieved performance close to its theoretical values, emphasizing its more mature state of development.

The reverse current density for SiC and GaN Schottky rectifiers is shown in Fig. 42 as a function of reverse bias. Note that this is not reverse breakdown voltage, but simply reverse bias. The solid line represents values calculated for SiC, assuming barrier lowering due to image-force effects. Once again, these are some reports for SiC rectifiers with $J_R$ values close to the theoretical minimum. However, most wide bandgap rectifiers show reverse currents well above the predicted values, which indicates the role of crystal defects and surface leakage in influencing the device performance. In all of our GaN diodes we find $I_R \propto V_{rmR}^{0.5}$, which is indicative of Shockley–Read–Hall recombination being the dominant current mechanism.

To date there have not been any published reports on $p$-$i$-$n$ GaN rectifiers intended for high voltage applications, at least to our knowledge. Figure 43 shows the Mg profile in our $p$-$i$-$n$ structure where the turn-on of the acceptor begins well before the $p^+$ contact layer. However, the overwhelming majority of the dopant is sharply confined to the top 0.5 $\mu$m of the structure. It is not clear if the low concentration tail represents in-diffusion of interstitial Mg or a build-up of Mg prior to growing the $p^+$ layer.

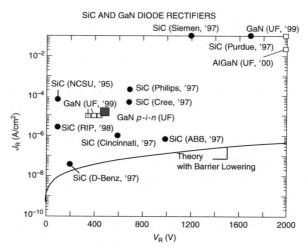

**Fig. 42.** $J_R$ versus $V_{RB}$ for GaN and SiC Schottky rectifiers reported in the literature.

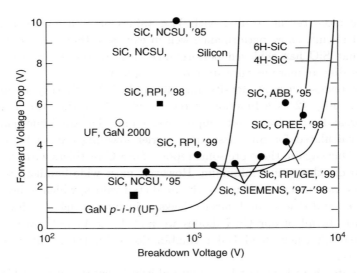

**Fig. 43.** $V_F$ versus $V_{RB}$ for GaN and SiC *p-i-n* rectifiers reported in the literature.

Figure 43 shows a compilation of $V_F$ versus $V_{RB}$ data for SiC *p-i-n* rectifiers, as well as our GaN result. The solid lines represent theoretical values for Si, 4H–SiC and 6H–SiC *p-i-n* rectifiers. As time has progressed, the SiC devices have produced performance relatively close to the theoretical values; by contrast, our GaN result is quite respectable, given the fact that the theoretical values will be slightly higher than for SiC because of the layer bandgap of GaN. It is expected that *p-i-n* diodes will have larger $V_{RB}$ values than do Schottky rectifiers, but higher $V_F$ values and slower switching speeds. The latter is a result of the presence of minority carriers in the device.

We have presented a summary of results for GaN Schottky and *p-i-n* diode rectifiers. In particular the forward turn-on voltages of these devices are ∼ 3.5 and 5 V, respectively. Reverse blocking voltages of 350–550 V are typical at present for rectifiers fabricated on conducting GaN, while lateral diodes fabricated on resistive GaN produce $V_{RB}$ values above 3 kV with proper edge termination. The results are still inferior to theoretically predicted values for this materials system, but they show the rapid progress over the last 2 years. The current transport in GaN Schottky rectifiers is still mainly by Shockley–Read–Hall recombination. The forward I-V characteristics of *p-i-n* rectifiers show a number of slope changes, which are consistent with a multiple-recombination-level model.

## Acknowledgments

The work at UF was partially supported by DARPA/EPRI, NSF 0101438 and CTS-991173 and ONR-DURIP N00014-99-1-0336.

## References

1. H. Akagi, *IEEE Trans. Power Electron.* **13**, 345 (1998).
2. G. T. Heydt and B. K. Skromme, *Mater. Res. Soc. Symp. Proc.* **483**, 3 (1998).
3. E. R. Brown, *Solid-State Electron.* **43**, 1918 (1998).
4. J. B. Casady, A. K. Agarwal, L. B. Rowland, S. Seshadri, R. R. Siergiej, S. S. Mani, D. C. Sheridan, P. A. Sanger and C. D. Brandt, *Mater. Res. Soc. Symp. Proc.* **483**, 27 (1998).
5. R. Zehringer, A. Stack and T. Lang, *Mater. Res. Soc. Symp. Proc.* **483**, 369 (1998); *J. Mater.* **50**, 46 (1998).
6. B. M. Green, K. K. Chu, E. M. Chumbes, J. A. Smart, J. R. Shealy and L. F. Eastman, *IEEE Electron. Dev. Lett.* **21**, 268 (2000).
7. J.-I. Chyi, C.-M. Lee, C.-C. Chuo, G. C. Chi, G. T. Dang, A. P. Zhang, F. Ren, X. A. Cao, S. J. Pearton, S. N. G. Chu and R. G. Wilson, *MRS Internet J. Nitride Semicond. Res.* **4**, 8 (1999).
8. Z. Z. Bandic, D. M. Bridger, E. C. Piquette, T. C. McGill, R. P. Vaudo, V. M. Phanse and J. M. Redwing, *Appl. Phys. Lett.* **74**, 1266 (1999).
9. M. Trivedi and K. Shenai, *J. Appl. Phys.* **85**, 6889 (1999).
10. V. A. Dmitriev, K. G. Irvine, C. H. Carter Jr., N. I. Kuznetsov and E. V. Kalinina, *Appl. Phys. Lett.* **68**, 229 (1996).
11. M. S. Shur, *Solid-State Electron.* **82**, 2131 (1998).
12. R. Hickman, J. M. Van Hove, P. P. Chow, J. J. Klaassen, A. M. Wowchack and K. Shenai, *J. Appl. Phys.* **85**, 6898.
13. C. I. Harris and A. O. Konstantinov, *Phys. Scr.* **T79**, 27 (1999).
14. K. G. Irvine, R. Singh, M. J. Paisley, J. W. Palmour, O. Kordina and C. H. Carter Jr., *Mater. Res. Soc. Symp. Proc.* **512**, 119 (1998).
15. F. Dahlquist, C.-M. Zetterling, M. Ostling and K. Rottner, *Mater. Sci. Forum* **264–268**, 1061 (1998).
16. M. Adler, V. Temple, A. Ferro and E. Rustay, *IEEE Trans. Electron. Devices* **24**, 107 (1997).
17. E. Stefanov, G. Charitat and L. Bailon, *Solid-State Electron.* **42**, 2251 (1998).
18. A. Y. Polyakov, N. B. Smirnov, A. V. Govorkov and J. M. Redwing, *Solid-State Electron.* **42**, 831 (1998).
19. H. Amano and I. Akasaki, in *Properties, Processing and Application of GaN and Related Semiconductors*, EMIS Data Review No. 23, eds. J. H. Edgar, S. Strite, I. Akasaki, H. Amano and C. Wetzel (IEE, London, 1999).
20. N. Dyakonova, A. Dickens, M. S. Shur, R. Gaska and J. W. Yang, *Appl. Phys. Lett.* **72**, 2562 (1998).
21. V. Khemkar, P. Patel, T. P. Chow and R. J. Gutman, *Solid-State Electron.* **43**, 1945 (1999).
22. J. C. Zolper, *Mater. Res. Symp. Proc.* **622**, T2.4.1 (2000).

# CHAPTER 3

# GaN-BASED POWER HIGH ELECTRON MOBILITY TRANSISTORS

Shreepad Karmalkar

*EE Department, IIT-Madras, Chennai 600 036, India*

Michael S. Shur* and Remis Gaska[†]

*ECSE Department, RPI, 8th Street Troy, New York 12180, USA
[†]Sensor Electronic Technology, Inc., Columbia, SC 29209, USA*

## Contents

# 1. Introduction

High Electron Mobility Transistors (HEMTs) have emerged as a promising candidate for microwave ($f > 1$ GHz) power amplification,[1-3] with applications ranging from satellite links to wireless communications, from highways to electronic warfare. Also, they have a potential for low frequency ($f < 100$ MHz) high voltage (up to 1 kV) switching power control.[4-6] Until recently, power HEMTs were primarily based on AlGaAs/GaAs, AlGaAs/InGaAs, AlInAs/InGaAs and related epitaxial films grown on GaAs or InP substrates. In late 1990s, AlGaN/GaN, AlGaInN/GaN, and AlGaInN/InGaN power HEMTs grown on sapphire, insulating 4H–SiC, conducting SiC, and even bulk GaN have demonstrated much larger output powers and have become promising contenders for a variety of high power amplification and switching applications. Moreover, the use of wide bandgap semiconductors in power amplifiers not only increases the output power, but also extends the temperature tolerance and the radiation hardness of the circuits. The latter is corroborated by recently demonstrated operation of GaN-based HFETs at 750°C.[7] In this chapter, we review the various aspects of these devices, namely — history, current status, technology, characteristics, modeling and circuits.

The HEMT is also known as MODFET (Modulation-doped FET), TEGFET (Two-dimensional Electron Gas FET), SDHT (Selectively Doped Heterostructure Transistor) or simply, HFET (Heterojunction FET). The unique feature of the HEMT is channel formation from carriers accumulated along a grossly asymmetric heterojunction,[8] i.e. a junction between a heavily doped high bandgap and a lightly doped low bandgap region. In HEMTs based on GaN substrates, this carrier accumulation is mainly due to polarization charges developed along the heterojunction in the high bandgap AlGaN side.[4] This is in contrast to the situation in other HEMTs, such as those on GaAs or InP substrates. Here, the accumulation is a result of carrier diffusion from the heavily doped to the lightly doped region, the diffusion being enhanced significantly by the bandgap difference between the two regions. Whatever the physical origin of carrier accumulation, the accumulated carriers might have high mobility due to their separation from their heavily doped source region, and their location in the low-doped region where impurity scattering is absent. A discussion of the basic principles of HEMT operation is readily available in several books, e.g. Refs. 9 and 10.

The physics of carrier transport parallel to a heterojunction was first considered in 1969.[11] The development of MOCVD and MBE technologies in the 1970s made heterojunctions practical. Although substantial research on GaN growth was initiated in early 1960s, the technological spin-offs came late because of lack of ideal substrates. The enhanced mobility effect was first demonstrated in AlGaAs/GaAs heterojunctions in 1979,[12] and applied to demonstrate a HEMT in 1980.[13,14] Electron mobility enhancement at AlGaN/GaN heterojunction was first reported in 1991.[15] Later, this enhancement was attributed to the 2D nature of the electrons, based on observations of mobility increase with lowering of the temperature down to 77 K and Shubnikov-de-Haas oscillations. The potential of AlGaN/GaN HEMTs for microwave electronics was demonstrated in 1994.[16] With advancements in material quality, heterostructure design, and ohmic contact formation, the excellent power capability of these HEMTs was established by 1997.[17,18] Recently, some AlGaN/GaN structures useful for several hundred volts power switching have been proposed.[6,19]

Two factors, namely — new markets and technological advances[3,5] have made the HEMT into a viable industrial product. Until 1980, most applications of RF transistors had been military or exotic scientific projects, e.g. electronic warfare, missile guidance, smart ammunition etc. In the 1980s, satellite television using low-noise transistors operating at 12 GHz in the receiver front-ends was the first civil application of RF transistors with a market volume worth mentioning. Currently, we witness far-reaching developments in civil communication technology. Mass consumer markets have been created for RF systems by the advent of mobile phones, whose production surpassed that of personal computers in 1998. There have been several advances in HEMT technology. Electron Beam Lithography (EBL) has enabled realization of a short ($\sim 0.1$ $\mu$m) T-shaped gate in a repeatable manner. Improvements in device modeling/simulation have permitted better prediction of experimental characteristics. Progress in MBE and MOCVD growth techniques have made it possible to control the material composition and thickness of the heterojunction layers.

## 1.1. *Power versus Small-Signal HEMT*

There are some important differences between HEMTs used in power and small-signal applications. A power device has to withstand a larger voltage and current amplitude as compared to a small-signal device, to provide high output power. High drain voltage operation is necessary for two reasons. First, increasing the device output power by increasing $I_{\max}$ alone involves increasing the device area, resulting in the input and output impedances too low for a good match with the surrounding circuitry. Second, to achieve a high power added efficiency, class AB or class B operation, requiring a large quiescent drain voltage, is used, resulting in a high drain voltage when the gate-source voltage is swung below pinch-off.

Consider the role of small-signal and power HEMTs in wireless communication, which occupies a significant RF market.[1] The wireless communication hardware consists of the infrastructure (base station), and the user part (handset) having receiver and transmitter sections. In the handset, supply voltage as low as 3 V is used to reduce power consumption. Small-signal low noise transistors are required to amplify incoming signals in receiver front-ends, and power transistors with low on-resistance and high on-current are required in the transmitter section. Today, the small-signal low-noise applications of a HEMT include collision avoidance radar at 77 GHz and military radar at 94 GHz. On the other hand, power applications, which are the subject of the present chapter, include base station applications at ~ 40 GHz, mobile communication at ~ 0.9/1.9 GHz, and switching power supplies at < 1 MHz.

Small-signal devices typically have a gate length below 0.15 μm and breakdown voltage of ~ 5 V, since they have to be optimized mainly with respect to their RF performance. However, power HEMTs typically have uncritical gate lengths in the range 0.5–1 μm and breakdown voltages > 10 V. Gate lengths > 1 μm may be used for several hundred volts breakdown voltage required in power switching applications. Note that gate definition accounts for a significant fraction of the total fabrication cost. Thus, the use of uncritical gate-lengths in power HEMTs facilitates their cheap large volume production, so that they compete successfully with various other technologies in power applications, such as Si/Ge-HBTs, III-V HBTs, and GaAs-MESFETs.

## 2. Device Structures and Fabrication

### 2.1. *Basic Device Structure*

The cross section of a basic AlGaN/GaN power HEMT is given in Fig. 1. We have fabricated AlGaN/GaN HFETs with the source-to-drain spacing from 2 μm to 7 μm, the gate length from 0.25 μm to 5 μm and the total gate width from 50 μm to 150 μm (2 × 25 μm to 2 × 75 μm). Although not shown in the figure, plated via holes (~ 20 μm × 40 μm) cut through the substrate are used to connect the source metallization to the back

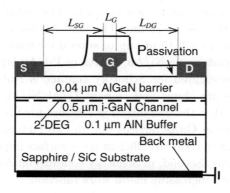

**Fig. 1.**   The basic AlGaN/GaN power HEMT structure.

metal, which is grounded. This reduces the parasitic source inductance and improves thermal dissipation.

### 2.1.1. *The substrate*

GaN substrates with large enough diameter do not exist for growing GaN channels; the largest GaN substrates obtained so far are 1.7 cm × 1 cm.[20] Sapphire ($Al_2O_3$) and SiC are the most popular substrate materials used currently. Sapphire substrates are cheaper for GaN growth than SiC. However, the low thermal conductivity of sapphire presents a serious challenge for packaging of high power devices. Long term reliability may be compromised due to thermally induced stress on the contacting pads, which also serve as the heat conducting path. On the other hand, SiC has lower lattice mismatch with GaN or AlN (3.5% against 13% of sapphire), and 10 times higher thermal conductivity.[21] Other substrates, such as $LiAlO_2$,[22] $LiGaO_2$,[23] Si,[24] and AlN[92,93] are also being investigated. The Si substrates are the cheapest, and have the potential of integrating GaN optoelectronics with silicon devices. However, the growth of GaN on these substrates is still a new area. Properties of various substrate materials appear in Table 1.

### 2.1.2. *The contacts*

The shape of the gate contact is crucial to device performance. T- or Y-shaped cross-sections are employed (Fig. 1), wherein the stem provides the short channel length, while the wide "hat" reduces the gate resistance of the connection to the bonding pads. Multilayer metallization schemes are used for the gate, source and drain contacts. The initial and barrier layers are usually as thin as 20–50 nm. The final Au layer is thick, with the thickness ranging from $\sim 0.15$ $\mu$m up to several microns. The gate contact is Schottky type, made using Ni/Au, Pt/Au or Pd/Au layers.[25,26] Another material being investigated for Schottky contact is $RuO_2$.[27] The source and drain contacts are ohmic and employ Ti/Al/Ni/Au or Ti/Al/Ti/Au layers.[25,26,28] Note that the metallization schemes used in GaN devices are different from those in GaAs devices due to two factors.[26] First, in GaN devices, it is difficult to create a sufficient $n++$ surface impurity doping during contact annealing. Second, in Schottky contacts on AlGaN, the barrier height increases with increasing metal work function.

**Table 1.** Properties of some substrate materials for GaN growth.

| Substrate | $E_g$ (eV) | $\chi$ (W/cm°C) | $E_c$ (MV/cm) | $\varepsilon$ | $\mu$ (cm²/Vs) | $v_s$ (cm/s) | $\dfrac{\chi E_c{}^2 \varepsilon \mu v_s{}^*}{(\chi E_c{}^2 \varepsilon \mu v_s)_{\mathrm{Si}}}$ |
|---|---|---|---|---|---|---|---|
| Si | 1.1 | 1.31 | 0.3 | 11.8 | 1350 | $1 \times 10^7$ | 1 |
| GaN | 3.39 | 1.3 | 3.3 | 9 | 1200 | $2.5 \times 10^7$ | 153 |
| SiC | 2.86 | 4.9 | 2.0 | 10 | 650 | $2 \times 10^7$ | 136 |
| Sapphire | — | 0.25 | — | — | — | — | — |
| AlN | 6.2 | 3.4 | — | 8.5 | 300 | $1.5 \times 10^7$ | — |
| GaAs | 1.43 | 0.53 | 0.4 | 12.8 | 8500 | $2 \times 10^7$ | 10 |

*Combined figure of merit for high temperature/high power/high frequency applications.

### 2.1.3. *Passivation*

Passivation of the gate-drain and gate-source surfaces by depositing $SiO_2$ and $Si_3N_4$[29,30] arrests the degradation in maximum current and transconductance due to the surface charges and traps present in these regions. These traps are a result of disorder in surface bonds of the III-V compound semiconductor material and cause Fermi-level pinning.

### 2.1.4. *Packaging*

An AlGaN/GaN power HEMT is encapsulated in packages similar to those developed for silicon power devices. However, the choice of packaging materials plays a more critical role in compound semiconductors than in silicon, due to differences in the coefficient of thermal expansion, and because compound semiconductors are more fragile and may exhibit mechanical stresses causing device degradation and failure.[31] Low power ($< 1$ W) packages include hermetic metal can packages (TO-39), plastic encapsulations (TO-92), plastic surface mount packages, and metal-ceramic packages for severe environmental conditions. For powers $> 1$ W, the RF die is mounted in metal ceramic package. Hermeticity is seldom warranted for commercial applications, and is required for special applications such as military and space, where environmental conditions are severe. The difficulty in dissipating heat from conventionally die bonded/wire bonded power HEMT chips forces thinning of finished wafers. Heat dissipation is also a problem with substrates having low thermal conductivity such as sapphire. In such cases, flip-chip mounting of the die on the package base is a solution (see Fig. 2).

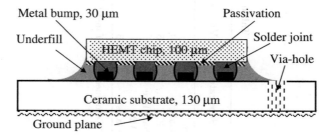

**Fig. 2.** Cross-section of a flip-chip assembly. The planar dimensions of the chip and the substrate are about 0.5 cm and 1 cm respectively. The diameters of the bump and the plated via hole is is 80 $\mu$m, and 200 $\mu$m respectively (after Ref. 32).

## 2.2. *Improved Device Structures*

Some improved device structures proposed by us are shown in Fig. 3. In Fig. 3(a), 50–100 nm doped GaN channel with a donor concentration of $5 \times 10^{17}$–$1 \times 10^{18}$ cm$^{-3}$ has been incorporated between the nominally doped GaN and the AlGaN layer.[33] This increases the density of two-dimensional (2D) electron gas near the heterointerface up to $n_s = 2 \times 10^{13}$ cm$^{-2}$, and thus raises the current carrying capability, $I_{\max}$, of the device. It has also led us to the idea of highly doped channel GaN MESFETs[34] that has many advantages in terms of the ease of fabrication and stability. Note that, the current capability can also be increased simply by increasing the device periphery by laying out multiple gate fingers and interdigitation (see Fig. 4). However, following this approach beyond a point creates problems in matching the device impedance to the surrounding circuitry.[1] Another feature of the improved structure is the use of a conducting 6H–SiC substrate.[35] A high-quality epitaxial insulating AlN buffer layer allows us to use both

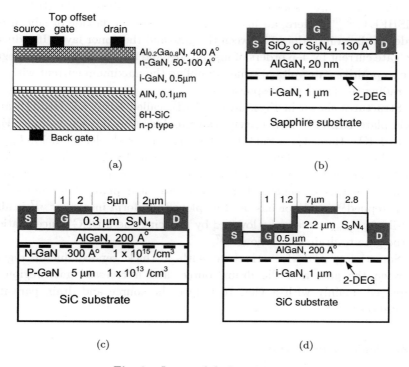

**Fig. 3.** Improved device structures.

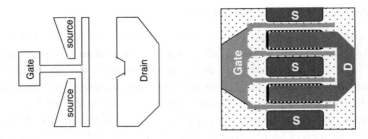

**Fig. 4.**   Top views of (a) offset gate and, (b) multiple finger gate, designs.

$n$-type and $p$-type SiC substrates as second (buried) gates, which strongly affect the device performance. Finally, in order to increase the breakdown voltage, $V_{br}$, an offset gate design has been used (also see Fig. 4).[94] We have studied the effect of varying the ratio of the source-to-gate distance, $L_{SG}$, and the gate-to-drain distance, $L_{GD}$, from 0.2 to 5. These separations as well as the gate lengths were determined from the SEM pictures.

The structure shown in Fig. 3(b) is called MOSHFET or MISHFET.[28,36,91] Here, an insulator, such as silicon dioxide or silicon nitride, has been introduced between the gate and the donor layer. It reduces the gate current by 4–6 orders of magnitude, and also permits a large negative to positive gate voltage swing, doubling the maximum current without degrading the frequency response.

Structures shown in Figs. 3(c) and (d), called RESURF HEMT[6] and Field-plate HEMT[19] respectively, can provide a breakdown voltage, $V_{br}$, up to 1 KV. Here, $V_{br}$ is raised by lowering the peak electric field, $E_{peak}$, near the drain edge. This has been achieved in Fig. 3(c), using a novel combination of field plates and a buried $p$-$n$ junction. Figure 3(d) is a variation of this arrangement, where a gate field plate alone, deposited on a stepped[19] insulator has been employed (uniform insulator[25,37] is also possible). $E_{peak}$ can also be lowered by reducing the 2-DEG concentration in the gate-to-drain separation.[19,37]

Sometimes, a device structure with a thick semiconductor region between the gate and the drain/source, and a recessed gate region is employed, mainly with a view to reduce the source and drain parasitic resistances.[38,38a]

## 2.3.  *Device Fabrication*

AlGaN/GaN HEMTs are fabricated in the following steps:

1. Growth of the semiconductor epitaxial and insulator layers.
2. Photolithography for ohmic contact openings.
3. Ohmic contact metallization.
4. Rapid Thermal Annealing (RTA) of ohmic contacts.
5. Photolithography for device isolation level.
6. Reactive ion etching or ion implantation for device isolation.
7. Contact or *e*-beam lithography for gate openings.
8. Gate metallization.
9. Substrate thinning and slot via formation.
10. Backside metallization.
11. Dicing and packaging.


### 2.3.1. *Material growth*

To-date, most AlGaN/GaN and AlInGaN/GaN heterostructures are grown by Metal Organic Chemical Vapor Deposition (MOCVD). MOCVD systems are capable of operating in both conventional and atomic layer deposition regimes. The conventional deposition regime is that in which precursors entering the growth chamber are simultaneously used to deposit GaN layers. Triethylgallium and ammonia are used as the precursor gases. $Al_x Ga_{1-x} N$ layers are deposited when precursors enter the chamber in a cyclic fashion. Triethylgallium, triethylaluminum and ammonia are used as precursors. The precursors are introduced into the chamber using hydrogen or nitrogen as a carrier gas. Epilayers are deposited on substrates placed on a graphite susceptor, which is heated to the growth temperature by RF-induction or resistive heating. Schematics of AlGaN HFET, MOSHFET, and MISH-FET epilayers without and with doped channel are shown in Fig. 5. The epilayers are typically grown on basal (0001) sapphire, conducting 6H–SiC and semi-insulating 4H–SiC substrates. The substrates are slightly off-axis (1–2°) production grade 6H and 4H silicon carbide, available from Cree Research. The doping levels of both *n*- and *p*-type 6H–SiC are on the order of $2 \times 10^{18}$ $cm^{-3}$. The micropipe density is approximately 30 $cm^{-2}$. The deposition of an approximately 50 nm thick AlN buffer is followed by the growth of insulating/modulation doped GaN, which is finally capped with an AlGaN barrier layer. The major difference between the AlGaN/GaN growth on sapphire and SiC substrates is the thickness of the insulating GaN layer. The details of MOCVD epitaxial growth are given in Refs. 39 and 40. An important and somewhat unique feature used in Refs. 39 and 40 and related work show that the use of trace amounts of indium throughout

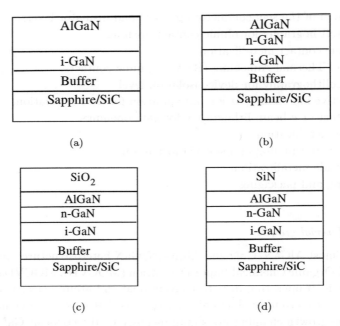

(a)                                    (b)

(c)                                    (d)

**Fig. 5.**  Schematics of AlGaN HFET, MOSHFET, and MISHFET epilayers without and with doped channel.

the entire epitaxial structure, which dramatically improves materials and surface quality.

Cross-sectional TEM analysis of GaN grown on sapphire and SiC reveals a strong dependence of growth defect distribution along the growth direction on the substrate material. A significant reduction in the number of threading dislocations in GaN on sapphire is observed for layer thicknesses above 1.5–2 $\mu$m. A similar improvement in material quality for GaN grown on SiC was achieved at thicknesses as low as 0.5 $\mu$m or even lower. Also, preliminary data obtained at Lawrence Berkeley National Lab point to a significantly lower number of domains with inverted crystal face polarity in GaN layers on SiC. This is in good agreement with our electron transport studies, which show better quality AlGaN/GaN heterointerfaces in the structures grown on SiC, in particular, on 6H–SiC. Thus, the performance of the devices with higher levels of dissipated power can be improved by effective heat sinking through the SiC substrate.

Epitaxial films for GaN-based HEMTs have also been grown by Molecular Beam Epitaxy on 4H–SiC substrates using RF-assisted gas source molecular beam epitaxy technology. The details of this technique can be found in Refs. 41–44.

### 2.3.2. Gate lithography

Realization of the gate electrode is one of the most crucial steps in device fabrication. Either optical[38,45,46] or Electron Beam Lithography (EBL)[47,48] may be employed for this purpose. Generally, EBL approach is preferred for sub-half micron gate lengths. In the EBL approach, a T-shaped gate (see Fig. 1) is created in a multi-layer *e*-beam resist deposited on the device, by a direct write exposure system. Tri-layer and four-layer resist processes are available.[48] However, devices fabricated using the four-layer resist process show improved values of breakdown voltage and $f_{\max}$, the latter due to lower values of gate-drain feedback capacitance and output conductance, with only slight reduction of drain current and transconductance.[48]

### 2.3.3. Annealing

Rapid Thermal Anneal (RTA) is used for the formation of ohmic contacts to source and drain, for radiation hardening, and for recovering the electrical properties of the devices damaged by plasma-assisted deposition and etching processes. For ohmic-contacts, the annealing is carried out in $N_2/H_2$ ambient for 30–60 seconds. It reduces the contact resistance to $\sim 10^{-6}$ $\Omega cm^2$ due to alloying. The RTA temperature is $\sim 850°C$ for Ti/Al/Ni/Au. It has been reported that, an RTA for 40 seconds at 800°C could recover the serious degradation in $I_{\max}$ and $g_m$ of AlGaN/GaN HEMTs, induced by proton irradiation,[49] and by plasma assisted processes.[50]

## 3. Characteristics

The power HEMT characteristics can be classified into DC and RF. Their critical parameters, reliability and uniformity are addressed below.

### 3.1. DC Characteristics

#### 3.1.1. Drain current

The measured DC characteristics of a power HEMT are shown in Fig. 6. The critical parameters related to these characteristics are the maximum drain current ($I_{\max}$), the threshold voltage ($V_T$), the peak DC transconductance ($g_{m(peak)}$), the breakdown voltage ($V_{br}$) and the output conductance

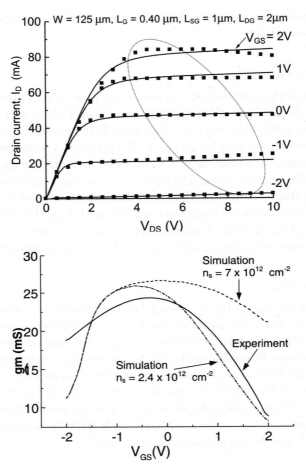

**Fig. 6.** Measured $I_D$–$V_{DS}$ and $g_m$–$V_{GS}$ curves (experimental data from Ref. 51). The dashed oval superimposed over the $I_D$–$V_{DS}$ curves represents a possible load line at microware frequencies. On the $I_D$–$V_{DS}$ map, points represent experimental data, and lines represent AIM-spice simulated curves.

($g_o$). $I_{max}$ is the maximum current that a device can allow under any bias condition. It is measured for the device biased in saturation, which is the condition used for the measurement of $g_m$ and $g_o$ as well. This is because the active region of a power transistor operation for amplification covers is restricted to the saturation region of the $I_D$–$V_{DS}$ curves. $V_T$ can be defined in two ways. It may be found as the $V_G$ intercept of the linear extrapolation of the $I_D$–$V_{GS}$ curve to the $V_{GS}$ axis. Alternately, it can be taken as the value of $V_{GS}$ corresponding to a given low value of $I_D$, e.g. 0.01 mA/mm; this is physical pinch-off condition. The value of $V_T$ in these two definitions can differ significantly because the $I_D$–$V_{GS}$ curve of the power HEMTs approaches cut-off gradually.

### 3.1.2. *Breakdown voltage*

The device supports the maximum voltage when turned off, i.e. when $V_{GS} < V_T$. Hence, the breakdown voltage for this condition, $V_{br(off)}$, is important. For devices such as the HEMTs, which operate at microwave frequencies, the breakdown in the on-state, i.e. for high $V_{GS}$, also becomes important for two reasons (see Fig. 6). First, $V_{br(on)}$ can be significantly less than $V_{br(off)}$; e.g. in a study,[51a] the measured values for these were 60 V and 100 V respectively. Second, the shape of the load line for RF applications is a closed curve and not a straight line of the DC case. The curve area increases not only with the applied input voltage swing but also with increasing frequency. Thus, a much larger area of the output characteristics is covered than in the DC case, and higher voltages are reached even at higher currents. It is then possible that the load line may trespass the $V_{br(on)}$ locus. The reason for the change in shape of the load-line with frequency is understood with the help of the basic amplifier circuit shown in Fig. 7. The transistor is biased with a DC voltage $V_{DD}$ and $V_{GG}$ is varied. When the frequency of $V_{GG}$ variation is low, the slope of the load line is defined by the load resistance, so that the load line is straight. However, as the frequency increases, the capacitive and inductive elements of the transistor and the matching networks eventually become significant. In this case, the transistor voltages and currents can be determined by circuit simulation, and these translate to the non-linear load-line.

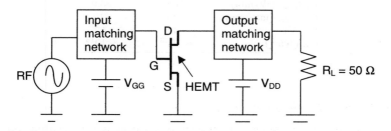

**Fig. 7.** An amplifier circuit with a HEMT biased at $V_{DD}$ and $V_{GG}$.

### 3.1.3. *Passivation effects*

Refer to the measured DC parameters of an AlGaN/GaN device shown in Table 2. Passivation, which suppresses surface traps and charges, improves $I_{max}$ by 24%, $V_{br}$ by 26%, and $g_{m(peak)}$ by 8%, and shifts $V_T$ by $-0.25$ V. Accordingly, a significant increase in output power capacity and reasonable increase in gain are predicted. Another study of GaN devices[30] supports these trends. The improvement in $I_{max}$ and $g_m$ should be attributed partly to the elimination of surface depletion leading to reduction of the sheet resistivity of the recess region, and partly to the $V_T$ shift. It is of interest to contrast the effects of passivation in GaN and GaAs devices. Unlike in GaN devices, in GaAs devices,[52] passivation degrades $V_{br}$ and makes it independent of the gate-drain separation, $L_{GD}$. This points to different origins of breakdown in unpassivated GaAs and GaN HEMTs. In GaN devices, the conduction and ionization processes associated with this breakdown occur in the surface layer, while in GaAs devices, they occur in the channel layer. In the absence of passivation, surface charges deplete the channel over $L_{GD}$ in GaAs HEMTs, spreading the electric field over $L_{GD}$, and thus reducing the $E_{peak}$ at the drain edge of the gate. This is not possible after passivation. However, the electric field can spread over several microns of $L_{GD}$ in GaN devices even after passivation, mainly due to their much higher breakdown field than GaAs. It is for this reason that $V_{br}$ increases with $L_{GD}$ in GaN devices even after passivation.

**Table 2.** Measured DC parameters[29]; $V_{br}I_{max}$ is a figure of merit for maximum power output capacity. The device has $L_G = 0.5$ $\mu$m and $L_{GD} \sim 1$ $\mu$m.

| Parameter | Unpassivated | Passivated |
|---|---|---|
| $I_{max}$ (mA/mm) | 520 | 640 |
| $V_{br(off)}$ (V) | 42 | 53 |
| $V_{br}I_{max}$ (W/mm) | 21.8 | 33.9 |
| $g_{m(peak)}$ (mS/mm) | 195 | 210 |
| $V_T$ (V) | $-4.5$ | $-4.75$ |
| $n_{so}$ (/cm$^2$) | $\sim 10^{13}$ | $\sim 10^{13}$ |

### 3.1.4. *Geometry scaling effects*

It is important to consider the impact of gate geometry scaling on DC characteristics, since smaller $L_G$ and larger width are preferred for better performance. In a study,[53] as $L_G$ was reduced in the micrometer range from

6 $\mu$m to 2 $\mu$m, the extrinsic $g_m$, measured at 7 V drain bias, rose linearly from 97 mS/mm to 107 mS/mm. However, the $g_m$ reached a maximum of about 130 mS/mm for $L_G = 0.85$ $\mu$m, and fell for further reduction of $L_G$ in the sub-micrometer range to 100 mS/mm at $L_G = 0.1$ $\mu$m. This could be attributed to short-channel effects such as drain induced barrier lowering and gate-fringing. Other consequences of such short-channel effects observed, such as $V_T$ and $V_{br}$ reduction, are shown in Fig. 8. The effect of increasing the width has also been studied.[28] The $g_m$ increased to 410 mS for a rise in width up to 3 mm, but saturated at 440 mS, on further width increments to $W = 6$ mm. This saturation could be attributed to the random variation in $V_T$ over the large gate area. Thus, the law of diminishing returns starts operating for increase in $W$ and decrease in $L_G$ beyond a certain point.

### 3.1.5. *Temperature effects*

Several researchers have studied the effect of temperature variation on device characteristics.[54,55] Results of one such study are given in Table 3. While the device $V_T$ ($= -3.6$ V) showed little change with temperature variation, the $g_m$ and $I_{max}$ fell by 35% over the temperature range of 23–187°C. This was attributed mainly to the fall in $v_s$ and mobility with temperature.

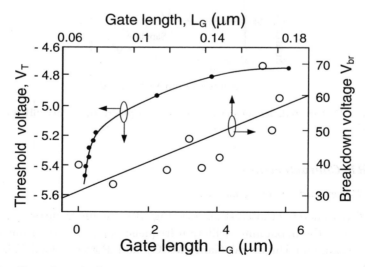

**Fig. 8.** Short-channel effects, namely $V_T$ and $V_{br}$ reduction, in GaN HEMTs (after Ref. 53).

**Table 3.**   Effect of temperature variation.[54]

| $T$ (°C) | $g_{m(peak)}$ (mS/mm) | $I_{max}$ (mA/mm) | $f_T$ (GHz) | $v_s$ ($10^7$ cm/s) |
|----------|----------------------|-------------------|-------------|---------------------|
| 23 | 110 | 480 | 13.7 | 1.2 |
| 187 | 72 | 320 | 8.7 | 0.8 |

### 3.1.6. *Negative resistance*

An anomaly observed in the $I_D$–$V_{DS}$ characteristics of GaN HEMTs on sapphire substrates is shown in Fig. 9. These devices show negative resistance for high currents and voltages. This has been attributed to self-heating of the device due to the poor thermal conductivity of sapphire.[56] Detailed explanation for this anomaly will be given in the modeling Sec. 4.3.3. This effect can be minimized by measuring the characteristics under pulsed condition to reduce power consumption, by using higher thermal conductivity substrate, and by proper cooling.

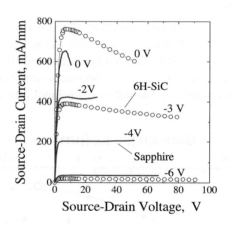

**Fig. 9.**   I-V characteristics for devices grown on sapphire (dots) and 6H–SiC (open circles).[56]

## 3.2. *RF Characteristics*

### 3.2.1. *Power and cut-off frequency*

The RF parameters of interest are (see Fig. 10), the unity current gain frequency ($f_T$), the maximum oscillation frequency ($f_{max}$), the output power ($P_{out}$), Maximum Unilateral Gain (MUG) and Power Added Efficiency (PAE).

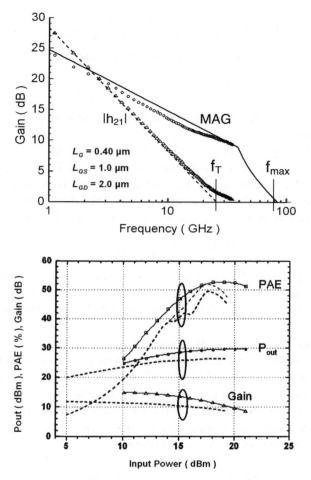

**Fig. 10.** RF characteristics, namely $P_{\text{out}}$, PAE, gain, $f_T$ and $f_{\text{max}}$ for the device of Fig. 6. The gain is obtained from the $P_{\text{out}}$ versus $P_{\text{in}}$ curve. Points represent experimental data, while dotted lines represent simulation.[95]

In general, MUG is the power gain of a two-port network having no output-to-input feedback, but with input and output conjugately impedance-matched to signal source and load, respectively. Because any RF transistor has a non-zero feedback from output to input, a lossless network must be added to cancel the feedback. Evidently, the resulting circuit will not oscillate unintentionally, so that MUG is defined over the whole frequency range. It is given by

$$\text{MUG} = \frac{|y_{21} - y_{12}|^2}{2[\text{Re}(y_{11})\text{Re}(y_{22}) - \text{Re}(y_{12})\text{Re}(y_{21})]} \tag{1}$$

or, in terms of the more commonly used $S$-parameters,

$$\text{MUG} = \frac{|S_{12}|^2}{(1 - |S_{11}|^2)(1 - |S_{22}|^2)} . \tag{2}$$

Note that, MUG (dB) = 10 log MUG. This gain rolls off at $-20$ dB/dec at high frequencies.

The output power, $P_{\text{out}}$, is the power delivered to the load, and depends upon the amplifier circuit, e.g. class A, AB, or B, for a given device. However, it is common to find $P_{\text{out}}$ being expressed as W/mm gate width, which is not exactly the total output power of the device. Since heat dissipation or battery power is of concern, we define the efficiency,

$$\text{PAE} = \frac{P_{\text{out}}(\text{ac}) - P_{\text{in}}(\text{ac})}{P_{\text{in}}(\text{dc})} . \tag{3}$$

Note that $P_{\text{in}}(\text{dc})$ is delivered by the DC supply, and is transformed into heat in the device. The variation of the PAE and $P_{\text{out}}$ with $P_{\text{in}}$ represent the large-signal behavior, of which, the saturated $P_{\text{out}}$ is an important figure of merit.

The frequency at which MUG reaches unity (or 0 dB) is called $f_{\text{max}}$. Its somewhat misleading designation, maximum frequency of oscillation, stems from the fact that it is also the highest frequency at which an ideal amplifier would still be expected to operate. On the other hand, $f_T$, also referred to as the gain-bandwidth product, is the frequency at which the small-signal current gain under output short circuit conditions, $h_{21}$, decreases to unity. In terms of $S$-parameters, $h_{21}$ is calculated as

$$h_{21} = \frac{2S_{12}}{(1 - S_{11})(1 + S_{22}) + S_{12}S_{21}} ; \tag{4}$$

$|h_{21}|$ rolls off at $-20$ dB/dec at high frequencies. Note that transistors with $f_{\text{max}} > f_T$ can have useful power gains for $f > f_T$ and up to $f_{\text{max}}$, since a current gain $< 1$ may be compensated by a voltage gain $> 1$ for $f_T < f < f_{\text{max}}$. There is no unequivocal answer to the question, which of the two frequencies, $f_T$ and $f_{\text{max}}$, is more important, although generally it is assumed that $f_T$ is important for digital and $f_{\text{max}}$ for analog applications. Manufacturers strive to achieve $f_T \approx f_{\text{max}}$ so that the devices have a wide application range. The thumb rules are, $f_T \geq f_{\text{op}}$ (the operating frequency) and $f_{\text{max}} \geq 3f_{\text{op}}$ for power transistors, and $f_T \geq 2f_{\text{op}}$ for small-signal devices. Sometimes in literature, $f_{\text{max}}$ is referred to as the frequency at which the maximum available power gain (MAG) rather than the gain MUG, decreases to unity. This is not entirely correct, but the results of the two definitions are not significantly different.

The resistivity of the buffer/substrate layer has an important effect on the frequency response. Use of an actively compensated buffer doubled the $f_{\max}$ to 21 GHz from 10 GHz for a conventional buffer.[26] Similarly, the device temperature also affects the frequency behavior. As shown in Table 3, the $f_T$ decreases significantly with temperature, on account of the decrease in saturation velocity. The $f_T$ is a function of $V_{DS}$, e.g. Ref. 54. It increases rapidly with $V_{DS}$ for small $V_{DS}$ values (below about 10 V), due to increase in the electric field along the channel and hence electron velocity. However, it falls gradually after reaching a peak due to extension of the gate depletion region towards the source.

### 3.2.2. Passivation effects

Measured RF power parameters provided in Table 4 show that passivation improves $P_{out}$ and PAE greatly; 100% increase in $P_{out}$ and 28% increase in PAE are seen. Another study showed these values to be 145% and 11% respectively. (Note that the higher percentage increase in $P_{out}$ is associated with a lower percentage increase in PAE, in keeping with the trade-off between $P_{out}$ and PAE). These improvements should be attributed to the fact that passivation prevents channel depletion due to traps and surface charges, so that the gate can modulate the channel completely. Consequently, $I_{max}$ increases and source and drain resistances decrease. There is a 30% reduction in gain, however, because of the increase in $C_{GD}$ (gate-drain capacitance) due to two factors. First, the surface dielectric constant increases due to the passivating layer ($Si_3N_4$ in this example), and second, there is a reduction in gate-drain depletion layer because of the elimination of surface charges/traps which deplete the channel. These effects are also responsible for a reduction in $f_T$ and $f_{\max}$ (of $\sim 9\%$).

**Table 4.** Measured RF power parameters of the GaN HEMT having $L_G = 0.5\ \mu$m, at $f = 4$ GHz.[30]

| Parameter | $P_{out}$ (W/mm) | PAE (%) | Power gain (dB) |
|-----------|-----------|---------|-----------------|
| Unpassivated | 1.0 | 36 | 20 |
| Passivated | 2.0 | 46 | 18.5 |

### 3.2.3. *Geometry scaling effects*

It is important to note the effects of scaling the gate geometry, $W$ and $L_G$. Due to the random variation in parameters, e.g. $V_T$, over larger device areas, the power tends to saturate with increase in $W$, particularly at higher frequencies. Decreasing the channel length significantly raises $P_{out}$ and gain, mainly due to an increase in $f_T$. However, for short channel lengths of $L_G < 0.1$ $\mu$m, $f_T$ may saturate, as parasitics from gate fringing and gate bonding pad capacitances become comparable to the gate capacitance.[1]

## 3.3. *Reliability*

Since the technology of power HEMTs is less mature than that of silicon power devices, many failure mechanisms limit their useful life, i.e. mean time between failure (MTTF). Some failure mechanisms of silicon devices, such as electromigration, apply to HEMTs as well,[57] but there are many, which are different. Moreover, failure mechanisms of GaAs and InP devices are not necessarily transferable to GaN devices, whose technology is in infancy.

The technique used to determine MTTF of HEMTs is also a subject of discussion. A measurement based rather than theoretical technique is preferred.[57] The usual extrapolation of the MTTF corresponding to the required temperature (usually 125°C) from high temperature accelerated stress experiments may yield unrealistically high MTTF estimates, because the failure mechanisms of compound semiconductor devices frequently show high values of activation energy at temperatures $> 200°C$.[58] Hence, electrically accelerated stress methods are also being considered.[59]

Failures are classified into catastrophic and degradation failures. Their exact mechanism depends on device bias, resultant channel temperature, passivation, and interaction among materials including those used for packaging. Below, we provide an account of the various failure mechanisms. Some of these are unique to power HEMTs, and may not affect small-signal HEMTs. High power and efficiency come at the cost of reduced reliability due to several reasons. First, the close proximity of the channel to the gate, essential for a high transconductance, makes the device susceptible to surface effects. Second, the plasma treatments and/or wet etching associated with gate recessing, employed for reducing parasitic source/drain resistances, can cause surface damage. Third, high efficiency power amplification involves large quiescent drain voltage condition (see Sec. 1.2), resulting in high electric fields at the drain end of the gate.

The reliability considerations of GaN devices revolve around epitaxial growth, contact technology, implantation and dry etching.[26] Presently, the epitaxial buffer layer has high defect concentration and low resistivity. Refined epi-growth using Mg-compensation and LEO may provide defect free, high resistivity buffer layers. Ga-faced rather than N-faced buffers should be deposited to provide higher channel 2-DEG concentration. The standard Ti/Al ohmic contact system shows weak edge definition and needs to be modified to avoid occurrence of liquid phases during annealing. Pt-based Schottky contacts pose problems at temperatures $> 300°$C attained by power devices. Their replacement by Re-based or WSiN-barrier metal system is being considered. Another problem is related to the implantation process used for device isolation. The high activation temperature ($1100°$C) of implanted dopants affects the integrity of semiconductor layers because of self-diffusion. Finally, dry-etching processes induce nitrogen vacancies enhancing surface conductivity. The hydrogen used in RIE gas compositions may compensate semiconductor layers.

Other failure mechanisms are related to trapping of hot electrons in the buffer and donor layers adjacent to the channel, or on the surface. Such trapping has been shown[60,61] to be the origin of transconductance frequency dispersion, drain current collapse, gate and drain lag transients, and restricted microwave $P_{out}$. Current collapse is the reduction in $I_D$ at low $V_{DS}$ after the application of a high $V_{DS}$ (see Fig. 11). This has a characteristic time dependence of minutes, after which the current is restored through release of trapped charge by thermal emission or illumination. This phenomenon is due to hot electron injection and trapping in the buffer layer. The trapped charge reduces 2-DEG concentration in the active channel, thereby reducing $I_D$. Since surface is not involved, this effect cannot be controlled by passivation. But lower resistivity, i.e. lesser deep level concentration, in the buffer layer lowers this effect. On the other hand, gate lag, which is the delayed response of $I_D$ to a $V_{GS}$ change (see Fig. 11), is sensitive to surface states in the gate-to-drain/source access regions. The trapped charge in these states reduces 2-DEG concentration in the access regions, raising parasitic source and drain resistances. Thus passivation can control this effect. Both buffer and surface layer trapping effects restrict the practical microwave $P_{out}$ to a value significantly below the estimations from the DC drain current and operating point. This is because the gate lag due to surface traps enhances large signal response time, while buffer layer traps reduce $I_{max}$ and increase the knee voltage.

Recently, Simin *et al.* reported on a novel, current collapse free, Double Heterostructure AlGaN/InGaN/GaN Field Effect Transistors (DHFETs)

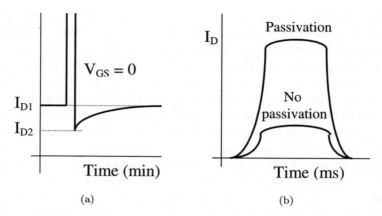

**Fig. 11.** Schematics illustrating drain current response to short voltage pulses, due to trapping effects. (a) Drain collapse: $V_{DS}$ is pulsed between a low value (10–100 mV) to a high value (15–20 V). (b) Gate lag: $V_{GS}$ pulsed from $V_T$ to 0 V, keeping $V_{DS}$ low (10–100 mV).

fabricated on the insulating SiC substrates.[90] Their simulations showed that a combined effect of the bandgap offsets and polarization charges provides an excellent 2D carrier confinement. These devices demonstrated output RF powers as high as 4.3 W/mm in CW mode and 7.5 W/mm in the pulsed mode, with the gain compression as low as 4 dB.

### 3.4. *Uniformity*

Manufacturing uniformity control is a critical issue, because high cost represents a significant road block in the development of RF systems. For cost-effective manufacturing, continuous process yield diagnostics and improvement are essential. Understanding of this issue requires a statistical study of characteristics of a large number of devices on an experimental wafer. For this purpose, simple DC measurements on actual transistors without any specialized test structures may suffice. Uniformity of device parameters is strongly process related.

Reproducibility and yield can be enhanced using $C$-doping for the GaN insulating buffer, and precision temperature control during growth of undoped GaN channel by reactive ammonia MBE. This is confirmed by an analysis of 2-DEG mobility values obtained in 50 experiments on 2 inch diameter substrates.[62] About 75% of the experiments yielded values higher than 800 cm$^2$/Vs. Reproducibility measurements from five consecutive growth runs showed that the variation in the electron density values was $< 7.5\%$, and in electron mobility values was $< 5.7\%$; the average values were $1.54 \times 10^{13}/\text{cm}^2$ and $987.6 \text{ cm}^2/\text{Vs}$ respectively. $C$-doping provides both a high resistivity buffer and a low defect density channel. Note that, a thick insulating buffer is essential for device isolation and good RF performance.

Figure 12 gives the distributions of $I_{\max}$, $f_{\mathrm{T}}$ and $f_{\max}$, and Fig. 13 gives the variation of the latter two, across a 2 inch diameter wafer.[63] The $f_{\mathrm{T}}$ and $f_{\max}$ have maximum values of 8.3 and 10.2 GHz, and average values of 7.7

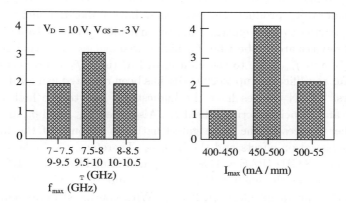

**Fig. 12.** The spread in critical frequencies and $I_{\max}$ distributions of AlGaN/GaN HEMTs on 2 in diameter sapphire at $T = 23°\text{C}$ (after Ref. 63).

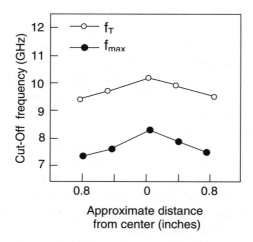

**Fig. 13.** Spatial distribution across 2 inch wafer of cut-off frequencies for AlGaN/GaN HEMTs at $23°\text{C}$ and $V_{\mathrm{DS}} = 10$ V, $V_{\mathrm{G}} = -3$ V (after Ref. 63).

and 9.8 GHz. The $I_{max}$ has minimum and maximum values of 473 mA/mm and 505 mA/mm, and an average of 473 mA/mm.

### 3.5. *GaN versus GaAs and InP Power HEMTs*

We now compare GaN HEMTs with GaAs and InP HEMTs, based on their characteristics. Some important details appear in Table 5. GaAs pHEMTs have been a commercial reality for a few years, and cover an extremely wide frequency range. On the other hand, GaN and InP HEMTs are just more than a laboratory product. However, they have important niche markets and are striving to fully emerge from technological infancy. Table 5 shows that InP devices are the best bet for higher frequencies, since they have the highest $f_T$ and $f_{max}$ due to the high $v_s$ of InP (note $f_T$ is proportional to $v_s$). Useful amplification up to 200 GHz has been reported using InP MMIC amplifiers.[64] GaN devices have the highest $V_{br}$ due to the high channel bandgap and associated critical field $E_c$. Also, their $I_{max}$ and $g_m$ are higher than other devices of the same gate-length, on account of the high $n_{so}$ due to polarization effects and the high $v_s$. Therefore, the highest $P_{out} = 9.8$ W/mm at 8 GHz (PAE = 47%) has been obtained from GaN HEMTs.[65] This is $\sim 6$ times the highest $P_{out} = 1.6$ W/mm at 2 GHz (PAE = 62%) of the widely prevalent GaAs pHEMT.[5] With continued improvements in material quality and device design, a power density of 12 W/mm or higher appears practical. Recently, it has also been shown[66] that these devices can achieve low microwave added Noise Figures (NF = 0.6 dB at 10 GHz) while maintaining a large breakdown voltage ($> 60$ V) and hence a large dynamic range. These results imply that AlGaN HEMTs can be used to perform the active transmit and receive functions in more robust, higher dynamic range modules. Thus for X-band applications, GaN HEMTs are a powerful alternative to GaAs pHEMTs. As mentioned in Sec. 2.2, very high voltage low frequency ($f < 100$ MHz) GaN HEMTs for switching power control are also possible.

To make full use of the high voltage capability, the transistor must not be thermally limited. GaN also has an advantage over GaAs in this regard, with thermal conductivities up to 2.0 W/cmK for GaN versus 0.46 W/cmK for GaAs. Furthermore, GaN HEMTs can be grown on semi-insulating SiC with a thermal conductivity of 3.3 W/cmK. The high breakdown field and good thermal conductivity allow these devices to be used in high efficiency (theoretical efficiency of 78.5%) class B push/pull amplifiers at full power

**Table 5.** Typical values of parameters in various power HEMTs.

| Device | Hetero-junction | $\Delta\phi_c$ (eV) | $n_{so}$ (cm$^{-2}$) | $\mu$ (cm$^2$/Vs) | $V_{br}$ (V) | $f_T$ (GHz) | $f_{max}$ (GHz) |
|---|---|---|---|---|---|---|---|
| GaN HEMT | Al$_{0.3}$Ga$_{0.7}$N/GaN | 0.45 | $1.5 \times 10^{13}$ | 1180 | 40 | 65 | 180 |
| GaAs PHEMT | Al$_{0.3}$Ga$_{0.7}$As/GaAs | 0.24 | $1.0 \times 10^{12}$ | 8500 | 7 | 130 | 250 |
| InP HEMT | Al$_{0.5}$In$_{0.5}$As/In$_{0.6}$Ga$_{0.4}$As | 0.64 | $3.4 \times 10^{12}$ | 13000 | 5 | 290 | 480 |

rating. GaAs microwave devices, on the other hand, are usually implemented in a class B push/pull amplifier by backing off the voltage bias (and hence the power level) to accommodate the higher voltage swing in this configuration as compared to class A, single-ended operation.

Other advantages of GaN are that the intrinsic temperature is much higher (due to wide bandgap), and the $v_s$ less temperature sensitive, than in GaAs. So, GaN power devices are more suitable for high temperature applications than GaAs devices, and can operate with less cooling and with fewer high cost processing steps associated with complicated structures designed to maximize heat extraction. Also, electrons in GaN exhibit much more pronounced overshoot effects than in GaAs but at much higher electric fields. This result illustrates the potential of GaN for ballistic power devices.[4]

### 3.6. *HEMTs versus MESFETs and HBTs*

It is of interest to compare HEMTs *vis-à-vis* their other compound semiconductor counterparts, namely MESFETs and HBTs. The advantages of HBTs are a higher current and power density, due to vertical current transport, which offers better utilization of wafer area.[67] This allows compact devices with smaller periphery, and therefore, easy matching with the surrounding circuit. Also, the device dimensions critical for HBT speed are not planar but vertical and thus independent of the photolithography process. These devices also have better linearity and more uniform threshold voltage. However, the HEMT has lower noise and better performance at high frequencies. Further, as compared to MESFET it has a higher transconductance, stemming from two factors. First is the close confinement of channel carriers to the gate, and second is the high mobility of the carriers due to diminished impurity scattering. The carrier confinement close to the gate also results in lower short-channel effects.

### 4. Modeling

The various models for calculating the DC and RF characteristics are now described. These may be employed to design a power HEMT, or an amplifier circuit based on this device. The model equations are presented following a discussion of the basic phenomena, namely gate-controlled charges, carrier transport, hot electrons, and polarization effects.

### 4.1. Basic Phenomena

#### 4.1.1. Gate-controlled 2-DEG

The DC and RF characteristics, i.e. $I_D$ or $g_m$ and the capacitance, are controlled by the behavior of 2-DEG $(n_s)$ as a function of the $V_{GS}$ (see Fig. 14). The $n_s$–$V_{GS}$ behavior is non-linear at low $V_{GS}$ due to 2-DEG movement. A unified expression for the subthreshold and above threshold behavior has been given as[68]

$$V_{GS} - V_T = \left(1 + \frac{d}{W_d}\frac{\varepsilon_d}{\varepsilon_b}\right)\frac{kT}{q}\ln\left(\frac{n_s}{n_0}\right) + q\left(\frac{d}{\varepsilon_d} + \frac{\Delta d}{\varepsilon_b}\right)n_s,\quad (5)$$

and

$$W_d = \sqrt{\frac{\varepsilon_b\phi_G}{(qN_b)}}.\quad (6)$$

Here, $d$ is the depth of the heterojunction from the gate, $\Delta d$ ($\sim$ 4.5 nm) is the location of the 2-DEG from the heterojunction, $n_0 = 5 \times 10^{10}/\text{cm}^2$, $W_d$ is the channel depletion width, $N_b$ is the channel doping.

On the other hand, following approach Refs. 69 and 70 gives a simple and explicit expression for the nonlinear behavior, including the gradual saturation at high $V_{GS}$ (see Fig. 14), over the complete bias range of interest. First, a simple function is chosen to represent the detailed shape of the charge–voltage curve. For example, above threshold, a function that represents the $n_s$–$V_{GS}$ behavior including gradual saturation and gradual

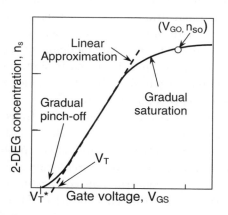
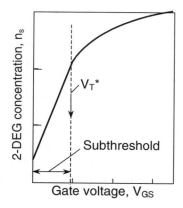

**Fig. 14.** Gate controlled charges in a HEMT. (a) Above threshold behavior; (b) Subthreshold behavior of the 2-DEG charge, (after Ref. 69).

pinch-off can be written as[69,71]

$$n_s = n_c \left[ \alpha + (1 - \alpha) \tanh \left( \frac{V_{GS} - V_{GM}}{V_1} \right) \right] \quad V_T^* < V_{GS} < V_{GO} . \quad (7)$$

Here, $V_T^*$ is the threshold voltage representing the pinch-off condition and is less than the linearly extrapolated $V_T$, due to the gradual pinch-off nature of the $I_{DS}$–$V_{GS}$ curve. $V_{GO}$ is a critical voltage[8] representing the onset of gate control on the 2-DEG, and $n_c$, $\alpha$, $V_{GM}$ and $V_1$ are the constants to be determined in terms of device parameters. Next, the constants are expressed in terms of device parameters, by forcing the function to pass through (intersect) readily derivable critical points (lines) on the accurate curve. These provide the required number of simultaneous equations to solve for the constants. For example, we need four such equations for the constants in $n_s$–$V_{GS}$ function (Eq. (7)). Two of these are obtained by forcing the function to pass through points $(V_T^*, 0)$ and $(V_{GO}, n_{so})$; the other two are obtained by forcing the function to intersect the well known linear $n_s$–$V_{GS}$ approximation,

$$n_s = \theta(V_{GS} - V_T) \quad V_T \le V_{GS}, n_s \le n_{so} . \quad (8)$$

Note that $V_T^*$, $V_T$, $V_{GO}$, $n_{so}$, and $\theta$ are readily derived in terms of device parameters.[69] Current (capacitance) expressions may be derived by integrating (differentiating) Eq. (7) respectively.

### 4.1.2. *Carrier transport*

Since power HEMTs are compound semiconductor devices, the carriers flowing in the bulk between the source and drain contacts through high field regions, experience not only velocity saturation, but also velocity overshoot preceding velocity saturation. Therefore for accurate modeling of the current, a simple Drift-Diffusion (DD) transport model, which can simulate velocity saturation alone, is not adequate. A Hydrodynamic (HD) model is required for capturing velocity overshoot effects on the current.[72]

Current through the contacts requires accurate modeling of the transport across interfaces. The Schottky contact governs $I_G$. The metal-semiconductor interface associated with this contact is clean and sharp, so that its modeling is straightforward. However, the ohmic source/drain contact, which governs the all-important $I_D$, is a concern. This contact interface is rough due to metal alloy penetration into the semiconductor. There are two possible models for transport across contact interfaces: Thermionic Emission (TE) without tunneling, and Thermionic Field Emission (TFE), which includes tunneling. For Schottky contacts, these apply for low and high reverse bias, respectively. For ohmic contacts, the latter has been found to be more appropriate.[72]

It is also necessary to include effects of temperature. This is because, power HEMTs have to dissipate large amount of power, so that heat sinking is a significant matter. A substrate with low thermal conductivity, such as sapphire, which is, presently, the preferred choice for growing AlGaN/GaN power HEMT, may accentuate self-heating effects, and these need to be modeled.

So far, analytical modeling of the velocity overshoot, TFE and temperature effects has been difficult, and numerical simulation techniques have been employed.[56,72,73]

### 4.1.3. *Hot electrons*

High-mobility electrons are easily "heated" up by the channel field. This effect is exacerbated by the use of high $V_{DS}$ for higher $P_{out}$, and is intense near the gate edge close to the drain where the electrical field is the highest. A few electrons can accumulate sufficient energy to overcome the heterojunction barrier, resulting in the reversible charging/discharging of deep levels in the donor and/or interface layers. Similarly, a few hot electrons may overcome the channel potential well, and get injected and trapped in the buffer layer beneath. In principle, Monte Carlo simulation approach employed to model hot electron effects in GaAs and InP HEMTs,[74,75] may be employed for GaN HEMTs as well. In GaN HEMTs, these phenomena may affect the DC characteristics through transconductance frequency dispersion, current collapse, and the RF performance through gate and drain lags.[60,61] However, so far, modeling of these phenomena has mostly remained qualitative.[60,61]

### 4.1.4. *Polarization effects*

Group-III nitrides such as AlN, GaN and InN, exhibit large polarization effects at heterointerfaces.[4] The origin of polarization is two-fold: piezoelectric effects due to stress along the $c$-direction of these ionic and non-centrosymmetric materials, and the difference in spontaneous polarization between these. The polarization dipole in the AlGaN layer is responsible for the high $\sim 10^{13}/cm^2$ 2-DEG concentration in the AlGaN/GaN HEMTs. Although this effect is complex, and numerical simulation has been employed for accurate results,[76] in cases such as

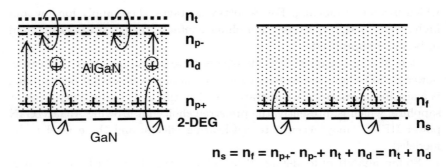

$$n_s = n_f = n_{p+} - n_{p-} + n_t + n_d = n_t + n_d$$

**Fig. 15.** Polarization effects in AlGaN/GaN heterojunction. Actual picture (left); simplified model (right), (after Ref. 37).

simulation of $V_{br}$ enhancement using field plate techniques, the simple model shown in Fig. 15 has worked well.[6,19,37]

### 4.2. *Analytical Modeling*

#### 4.2.1. *Critical DC and RF parameters*

The maximum 2-DEG concentration, $n_{so}$, equals the charge transferred across the heterojunction under equilibrium. In the case of AlGaN/GaN heterojunctions, $n_{so} \sim 10^{13}/cm^2$ is decided primarily by polarization effects in AlGaN, and is almost independent of donor doping.

Based on the threshold voltage expressions given in Refs. 10 and 68, the threshold voltage corresponding to the linearly extrapolated $I_D$–$V_{GS}$ characteristics to the $V_{GS}$ axis, at low $V_{DS}$, can be written as[10,68]

$$V_T = \phi_b - \Delta\phi_c - \frac{q}{\varepsilon}\left[\frac{N_d d_d{}^2}{2} + n_f d_d - N_b W_d\right], \qquad (9)$$

where $\phi_b$ is the gate Schottky-barrier height, $n_f$ is the effective positive polarization charge in AlGaN (can be regarded equal to $n_{so}$) defined in Fig. 14, and $d_d$ is the distance between the gate and the donor/spacer interface in the gate region. The expression for the threshold voltage representing pinch-off condition, $V_T{}^*$, can be written as,[69]

$$V_T{}^* = \phi_b - \Delta\phi_c + \delta_3 - \frac{q}{\varepsilon}\left[\frac{N_d d_d{}^2}{2} + n_f d_d - \left(N_b W_d + \frac{n_{so}}{100}\right)(d_d + d_i)\right], \qquad (10)$$

where $\delta_3$ is a doping dependent potential. The difference $(V_T - V_T{}^*)$ can exceed 100 mV.

In short-channel devices ($L_G < 0.25$ $\mu$m), $V_T$ (or $V_T^*$) will reduce (see Fig. 8) below the above value of the long channel HEMTs, and this reduction is a function of $V_{DS}$. The movement of the 2-DEG location accentuates this $V_T$ reduction. To obtain short-channel $V_T$, numerical simulation involving a simultaneous 2D solution of Schrödinger's and Poisson's equations is necessary. This procedure may be simplified using an analogy between the HEMT and a Buried-Channel (BC) MOSFET.[77] Here, the HEMT is replaced by an equivalent BC-MOSFET, whose oxide thickness and BC doping profile are adjusted so that the BC depletion edge simulates the movement of the 2-DEG location in the HEMT. The $V_T$ of the BC-MOSFET is obtained by solving the Poisson's equation alone, and corresponds to that of the HEMT. For circuit simulation purposes, the following simple equation is adequate

$$V_T(\text{or } V_T^*) = V_{T0}(\text{or } V_{T0}^*) - \sigma V_{DS}. \tag{11}$$

The parameter $\sigma$ may be extracted experimentally.

Simple and approximate expressions for $I_{\max}$ and intrinsic $g_{m(\text{peak})}$ (per unit width) are given by[4,78]

$$I_{\max} \approx q n_{so} \nu_s \tag{12}$$

$$g_{m(\text{peak})} \approx \frac{q n_{so} \mu / L_G}{\sqrt{1 + \left(\frac{\theta n_{so} \mu (d + \Delta d)}{\varepsilon_d \nu_s L_G}\right)^2}}. \tag{13}$$

The complete $I_D$–$V_{DS}$ characteristics including all effects can only be derived numerically. For purposes of hand calculation and circuit simulation, however, analytical models are necessary. Below, we discuss analytical models followed by numerical models.

The frequencies $f_T$ and $f_{\max}$ can be expressed in terms of the small-signal device equivalent circuit parameters (see Fig. 16 later) as[79]

$$f_T = \frac{g_m}{2\pi(C_{GS} + C_{GD})} \tag{14}$$

$$f_{\max} = \frac{f_T}{\sqrt{4\{1 + [R_S + R_G]g_o\} + 2\left(\frac{C_{GD}}{C_{GS}}\right)\left[\frac{C_{GD}}{C_{GS}} + g_m\left(R_S + \frac{1}{g_o}\right)\right]}}. \tag{15}$$

Extrinsic values of $g_m$ and $C_{GS}$ are given by $g_m^* = g_m/(1 + g_m R_S)$ and $C_{GS}^* = C_{GS}(1 + g_m R_S)$. In the active bias region of a HEMT ($V_{DS} > 0.5$ V and $V_{GS} > V_T$) $C_{GD}$ is only about 15% of $C_{GS}$. Therefore $f_T$ can also be

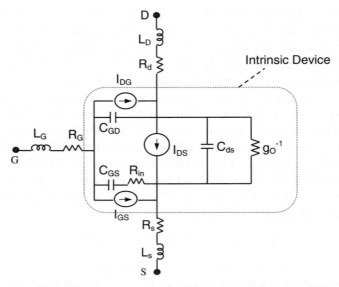

**Fig. 16.** RF large signal equivalent circuit model. In the small-signal model, the current sources $I_{DG}$ and $I_{GS}$ are removed (after Ref. 87).

approximated to be $\approx g_m{}^*/2\pi C_G$, where $C_G = C_{GS}{}^* + C_{GD}$ is the extrinsic total gate capacitance.

### 4.2.2. *Drain current and power*

An $I_D$–$V_{DS}$ expression can be derived by integrating analytical $n_s$–$V_{GS}$ and $n_f$–$V_{GS}$ expressions, e.g. those given in Sec. 4.1.1, assuming gradual channel approximation and simple saturating velocity-field curve ignoring velocity overshoot. Such expressions have been shown to represent the experimental results adequately (e.g. see Ref. 69). We shall discuss detailed simulations of an experimental GaN HEMT using AIM-Spice HFET model.[80–82] The goal of our simulation is to establish a device model, valid at both DC and microwave frequencies. To this end, we compare the results of such a simulation with the experimental data from GaN/AlGaN HFETs from Cree Research, Inc.[83]

The HFET model implemented in AIM-Spice uses a relatively small number of scalable parameters and allows for easy parameter extraction from the experimental data.

From DC measurements[83] on the 0.4 and 0.45 $\mu$m gated channel GaN/AlGaN HFETs, we have extracted the following parameter values: gate-to-channel separation 27 nm, output conductance parameter $0.005$ $V^{-1}$, low-field mobility $1900$ $cm^2$ $V^{-1}s^{-1}$, maximum sheet charge density in the channel $2.4 \times 10^{12}$ $cm^{-2}$, Schottky barrier height $1.1$ eV, drain series resistance $13$ $\Omega$, gate series resistance $12$ $\Omega$, source series resistance $13$ $\Omega$, saturation velocity $1.9 \times 10^7$ cm/s, knee voltage parameter 4, threshold voltage $-2.1$ V, charge partitioning parameter 1.2. The automatic extraction program described in Ref. 84 and parameter adjustment procedure were employed for this purpose. The agreement between measured and simulated results is quite reasonable (see Fig. 6). However, at $V_{GS} = 2$ V, the measured $I_D$ decreases with increased $V_{DS}$ due to self-heating. The extracted value of the saturation drift velocity, $v_s = 1.9 \times 10^7$ cm/s, is higher than the values extracted from the I-V characteristics and microwave data for other GaN-based HEMTs ($v_s = 1$–$1.5 \times 10^7$ cm/s).[85]

The extracted value of the maximum 2D-electron density, $n_{max} = 2.4 \times 10^{12}$ $cm^{-2}$ is lower than the Hall measurement value of over $7 \times 10^{12}$ $cm^{-2}$.[83] This might reflect a considerable transfer of hot electrons into the three dimensional states in GaN or into the AlGaN barrier layer at high $V_{GS}$ as evidenced by a drop in $g_m$ at high $V_{GS}$ (see Fig. 6). The thermal impedance is $\sim 2.5$ Kmm/W,[85] and the power dissipation is around 5–15 W/mm. Hence, the operating temperature might $\sim 380$–$390$ K, and the deduced parameters might be somewhat different from those expected at $T = 300$ K.

Microwave simulations of the device should allow us to better understand microwave device performance, and to establish the expected performance limitations of such devices. So far, simulations of GaN/AlGaN HFETs have been mostly limited to DC fitting, although Trew *et al.* reported on microwave simulations of GaN/AlGaN HFETs using the harmonic balance method.[86] The parameters $P_{out}$, gain and PAE depend on the entire circuit configuration including the device equivalent circuit, and are obtained by circuit simulation. An equivalent circuit model useful for this purpose is given in Fig. 16. Here, the current $I_{DS}$ is the large signal drain-to-source current. $I_{DG}$ and $I_{GS}$ represent the drain to gate to source currents, respectively. At large signal operation if $V_{DG}$ exceeds the breakdown voltage ($V_B$) or/and $V_{GS}$ exceeds the Schottky cut-in voltage ($V_{bi}$), respectively; simple linear relationships, $I_{DG} = (V_{DG} - V_B)/R_1$ and $I_{GS} = (V_{GS} - V_{bi})/R_F$, may be used.[87] The currents $I_{DG}$ and $I_{GS}$ may be represented by the two diodes used to model the gate current, as shown in Fig. 17 and discussed later.

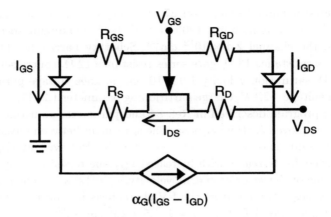

**Fig. 17.** Power HEMT equivalent circuit accounting for gate leakage (after Ref. 81).

The equivalent circuit parameters may be calculated using formulae based on device analysis. Alternately, they may be extracted from measured $S$-parameters. For this purpose, initially, a reasonable value is assumed for each parameter, and for this assumption, the $S$-parameters of the circuit are calculated. This procedure is repeated by modifying the assumption until the differences between the calculated and measured $S$-parameters is minimum.

We implemented the AIM-Spice small-signal model in the microwave simulator Libra (see Fig. 10). The device is biased under class-A operation in the large signal simulation ($V_{\mathrm{gs}} = 0.7$ V and $= 21.2$ V). The largest simulated output power is close to 4.5 W/mm, while the experimental value is 6.8 W/mm. Some of this discrepancy can be attributed to self-heating, which might play a smaller role at microwave frequencies than at DC, since power is low during a substantial part of the microwave cycle.

### 4.2.3. *Gate current*

For values of $V_{\mathrm{DS}}$ low enough so that breakdown does not occur, an equivalent circuit showing the effect of $I_{\mathrm{G}}$ is given in Fig. 17.[81] Each of the two diodes in this figure obeys the conventional Schottky diode law

$$I_{\mathrm{G}} \approx J_{\mathrm{SS}} L_g W \left[ \exp\left( \frac{qV}{mkT} \right) - 1 \right]. \qquad (16)$$

Here, $J_{\mathrm{SS}}$ is the reverse saturation current density calculated using the TE or TFE model. $V$ and $m$ for the diodes may differ. Note that $I_{\mathrm{G}}$ affects $I_{\mathrm{D}}$.

This is because $I_G$ is distributed along the channel, with the largest density near the source side of the channel. This redistributes the channel electric field, increasing it near the source, and decreasing $I_D$.

### 4.3. Numerical Modeling

#### 4.3.1. Drain current

Several reports discussing numerical $I_D$–$V_{DS}$ models are available.[6,19,37,56,72,73,76] These show that device geometry used for simulation is extremely important. The ohmic source/drain contacts should be assumed to be on the top of the cap layer. Results of any other assumption, such as contacts directly on the channel, or on the cap with heavy doping between the contacts and the channel, deviate significantly from experiment. Furthermore, the choice of transport models is very important. TFE model should be assumed for the transport across the ohmic contact interface; a TE model gives currents much lower than measured. Assumption of a DD model for transport parallel to the hetero-interface grossly underestimates the $I_D$ and $g_{m(peak)}$, and overestimates the width of the flat portion of $g_m$–$V_{GS}$ curve. An HD model in the channel (and to accurately describe the parasitic MESFET behavior occurring for high $V_{GS}$, a HD model in the donor layer as well) is required. A simultaneous fitting of the currents and capacitances is necessary to separate the effects of the carrier concentration and the velocity. The carrier concentration and surface potential within the ungated source-gate and gate-drain regions vary widely, so modeling of these regions as linear resistors is inaccurate. Some other issues will be considered in the context of breakdown simulation below.

#### 4.3.2. Breakdown voltage

The breakdown voltage in AlGaN/GaN HEMTs is at least several tens of volts, so that $V_{br(on)}$ measurement is difficult due to excessive self-heating, and $V_{br(off)}$ is the relevant parameter. So far, impact ionization has been assumed to be the cause of breakdown. Modeling of polarization charge is an important issue. Generally, numerical simulation has been employed to predict $V_{br(off)}$.

Because the intrinsic $V_{br}$ and $I_{max}$ of GaN devices can be $> 50$ V and $> 600$ mA/mm [see Table 2, Eq. (12)], respectively, some numerical simulation studies have explored the possibility of raising the $V_{br}$ to as high as 600–1000 V using field plate structures $V$ (see Figs. 3(c), (d), and Refs. 6,

19 and 37). This was with a view to consider high voltage power switching applications for this device. In these simulations, DD transport with impact ionization was used. The $V_{br}$ was directly extracted from the simulated $I_D$–$V_{DS}$ curve for $V_{GS} = V_T$. This approach is better than the indirect inference of $V_{br}$ from the peak electric field condition. The peak field is sensitive to the simulation mesh while the area under the field at breakdown is not, and the impact ionization depends not only on the peak but also on the shape of the field distribution near the peak. The simplified polarization charge model of Fig. 15 was employed, as the spatial separation of the polarization and other charges associated with the AlGaN layer was irrelevant due to the presence of a thick insulator between the AlGaN layer and the field plate. A dielectric medium above the device was found to raise the $V_{br}$ by 6–20%. These simulations showed that an optimized gate field plate over uniform insulator configuration can enhance the $V_{br}$ by as much as 5 times, depending upon the dielectric constant of the insulator below the field plate, and the value of $n_s$.[37] By way of example, the simulated reduction in peak electric field for this configuration is illustrated in Fig. 18. More than 10 times enhancement in $V_{br}$ is predicted, either by stepping the insulator thickness along the channel (see Fig. 3(d) and Ref. 19), or by introducing a *p-n* junction below the channel along with a drain field plate (see Fig. 3(c) and Ref. 6).

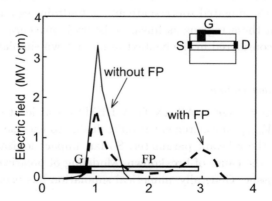

**Fig. 18.** The simulated distribution of the electric field along the 2-DEG, directed from drain to source, with and without a field plate (FP).[29] The FP extends 2 $\mu$m from the gate ($L_G = 0.4$ $\mu$m), and is deposited on a 0.3 $\mu$m silicon nitride insulator; $n_s = 1 \times 10^{13}$/cm$^2$; $V_{DS}$, $V_{GS} = 123$ V, $-2.8$ V, respectively. The device without FP is on the verge of breakdown.

### 4.3.3. Negative resistance

Mobility reduction with rise in temperature is the main cause of the downward sloping current curves in Fig. 9. The device temperature increases at high $I_D$ — high $V_{DS}$ due to enhanced power dissipation. Proper modeling of this effect requires accurate determination of channel temperature distribution from the Laplace Heat Conduction Equation, which is possible by numerical simulation. Here, definition of the heat generation source is a crucial issue. Single[73] and dual[56] heat source models have been employed. Alternative assumptions for the heat source length are: the gate, the gate-drain and the source-drain regions. Predictions for these assumptions can vary by as much as 30%. The dual heat source model assumes the gate length plus the drain depletion depth as one source, and the ohmic contacts together with the source/drain access regions as the other source. The total power is divided between the two sources in the ratio of their voltage drops. Temperature rises caused by the two heat sources are superimposed to get the overall temperature distribution. In all the models, the strong temperature dependence of thermal conductivity of substrate is accounted for. Simulations with the dual heat source model showed a channel temperature of 360°C for 4 W/mm dissipation on sapphire substrates.[56] For such high temperatures, the mobility can degrade by a factor of three as compared to room temperature.

## 5. Circuits

Here, we shall discuss the circuit configuration of a broadband power amplifier for X-band (0.2–6 GHz) phased array-radar and instrumentation. Simple lumped broadband amplifiers have gain-bandwidth products limited by $f_T$. Distributed or Traveling Wave Amplifiers (TWAs) can provide gain-bandwidth products up to $f_{max}$. However, a lumped amplifier based on $f_T$-doubler topology[88] can provide higher PAE and occupy smaller die area than the distributed amplifiers.

The basic $f_T$-doubler configuration is shown in Fig. 19(a). Essentially, it is a modified Darlington stage where suitable source loading of $Q_1$ splits the input voltage equally between $Q_1$ and $Q_2$, so that $I_{D1} = I_{D2}$ at all frequencies. Note that the two devices have half the width of a common-source stage of equal $P_{out}$. The current gain of the $f_T$-doubler stage becomes unity at $2f_T$. Thus, the bandwidth limitation due to $C_{GS}$ is alleviated. To absorb the effect of $C_{DS}$ and $C_{GD}$, broadband $L$-$C$ $\pi$-sections may be used between the amplifier output and the load. The $f_T$-doubler has been implemented

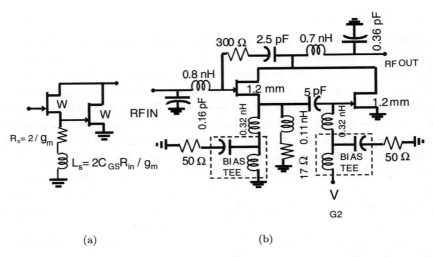

(a)                                                    (b)

**Fig. 19.**  $f_T$-doubler resistive feed back amplifier. (a) Basic configuration; (b) Complete circuit diagram.[89]

using 0.7 $\mu$m gate-length AlGaN/GaN HEMTs.[89] The complete circuit diagram is given in Fig. 19(b). Resistive feedback has been used to match the input and output to 50 $\Omega$. The die size is 1.1 × 1.45 mm². Following parameters were measured: gain = 11 dB over 0.2–7.5 GHz, $P_{out}$ > 1 W and PAE > 10% over 1 − 7 GHz.

## 6. Summary

In this chapter, we have presented an account of the evolution, technology, DC and RF characteristics, modeling and circuits of GaN based power HEMTs. These devices are the focus of intense research, as they are a powerful alternative to GaAs pHEMTs for use in transmitter and receiver sections of mobile communication sets operating in the X-band. A power density of 12 W/mm or higher appears practical for these devices, which can also achieve low noise figures (< 0.6 dB at 10 GHz) while maintaining a large breakdown voltage (> 60 V) and hence a large dynamic range. Very high voltage (∼ 1 KV) low frequency ($f$ < 100 MHz) switching power control is also possible using this device. However, the technology of these devices is still in infancy, so that a number of processes and hot electron effect related reliability issues afflict this device. These issues have been pointed out. The discussion on technology considered various device structures, and critical fabrication steps including the all important material growth. The effects of passivation, device geometry scaling and temperature on drain current and RF power were presented. It was pointed out that some of these effects, particularly those caused by thermal and electron heating, need numerical techniques for accurate modeling. However, simple analytical models were discussed, which are useful for preliminary design calculations of the device current and frequency parameters, and for simulating microwave power performance in circuit applications. Finally, a circuit configuration of this device for X-band amplification was considered by way of example.

## References

1. F. Schwierz and J. J. Liou, *Microelectron. Reliab.* **41**, 145–168 (2001).
2. S. Arai and H. Tokuda, *Solid-State Electron.* **41**, 1575–1579 (1997).
3. S. Takamiya *et al.*, *Solid-State Electron.* **38**, 1581–1588 (1995).
4. M. S. Shur, *Solid-state Electron.* **42**, 2131–2138 (1998).
5. S. J. Pearton *et al.*, *Materials Science and Engineering: Reports* **30**, 55–212 (2000).
6. S. Karmalkar *et al.*, *IEEE Electron Device Lett.* **22**, 373–375 (2001).
7. I. Daumiller, C. Kirchner, M. Kamp, K. J. Ebeling, L. Pond, D. E. Weitzel and E. Kohn (Device Research Conference Digest, 1998) pp. 114–115.
8. S. Karmalkar, *IEEE Trans. Electron. Devices* **45**, 2187–2195 (1998).
9. H. Morkoc *et al.*, *Principles and Technology of MODFETs* Vols. 1 and 2 (Wiley, New York, 1991).
10. M. Shur, *GaAs Devices and Circuits* (Plenum press, New York, 1997).
11. L. Esaki and R. Tsu, *IBM Research*, RC 2418 (1969).
12. R. Dingle *et al.*, *Appl. Phys. Lett.* **33**, 665 (1978).
13. T. Mimura *et al.*, *Jap. J. Appl. Phys.* **19**, L225–L227 (1980).
14. D. Delagebeaudeuf *et al.*, *Electron Lett.* **16**, 667–668 (1980).
15. M. A. Khan *et al.*, *Appl. Phys. Lett.* **58**, 2408–2410 (1991).
16. M. A. Khan *et al.*, *Appl. Phys. Lett.* **65**, 1121–1123 (1994).
17. Q. Chen *et al.*, *IEEE Electron Device Lett.* **19**, 44–46 (1998).
18. Y.-F. Wu *et al.*, *Electron. Lett.* **33**, 1742–1743 (1997).
19. S. Karmalkar and U. K. Mishra, *Solid-State Electron.* **45**, 1645–1652 (2001).
20. Akira Usui *et al.*, *Jap. J. Appl. Phys.* Part 2 **36**, L899–L901 (1997).
21. Q. Chen, *Materials Science and Engineering* **B59**, 395–400 (1999).
22. S. Fischer *et al.*, *Symp. Proc.* EGN-1, 1996.
23. P. Kung *et al.*, *Appl. Phys. Lett.* **69**, 2116–2118, (1996).
24. F. Semond *et al.*, *Appl. Phys. Lett.* **78**, 335–337 (2001).
25. N. Q. Zhang *et al.*, *IEEE Electron Device Lett.* **21**, 421–423 (2000).
26. J. Würfl, *Microelectron. Reliab.* **39**, 1737–1757 (1999).
27. S.-H. Lee *et al.*, *IEEE Electron Device Lett.* **21**, 261–263 (2000).
28. G. Simin *et al.*, *IEEE Electron Device Lett.* **22**, 53–55 (2001).
29. M. Bruce Green *et al.*, *IEEE Electron Device Lett.* **21**, 268–270 (2000).
30. J.-S. Lee *et al.*, *Electron. Lett.* **37**, 130–132 (2001).
31. S. Kayali, *Microelectron. Reliab.* **39**, 1723–1736 (1999).

32. Y. Arai *et al.*, *IEEE Trans. Microwave Theory and Tech.* **45**, 2261–2266 (1997).
33. S. Michael Shur and M. Asif Khan, *Phys. Scripta* **T69**, 103–107 (1997).
34. R. Gaska, M. S. Shur, X. Hu, A. Khan, J. W. Yang, A. Taraki, G. Simin, J. Deng, T. Werner, S. Rumyantsev and N. Pala, *Appl. Phys. Lett.* **78** (6), 769–771 (2001).
35. R. Gaska, M. S. Shur, J. W. Yang, A. Osinsky, A. O. Orlov and G. L. Snider, *Materials Science Forum* **264–268**, 1445–1448 (Trans. Tech. Publ.; Switzerland; 1998).
36. G. Simin *et al.*, *Electron. Lett.* **36**, 2043–2044 (2000).
37. S. Karmalkar and U. K. Mishra, *IEEE Trans. Electron. Devices* **48**, 1515–1521 (2001).
38. Oliver Breitschadel *et al.*, *J. Electron. Mater.* **28**, 1420–1423 (1999).
38a. T. Egawa *et al.*, *IEEE Trans. Electron. Devices* **48**, 603–608 (2001).
39. W. Knap *et al.*, *Mater. Res. Soc. Symp. Proc.* **639**, G7.3, eds. G. Wetzel, M. S. Shur, U. K. Mishra, B. Gil and K. Kishino (2001).
40. E. Frayssinet *et al.*, *Appl. Phys. Lett.* **77**, 2551–2553 (2000).
41. N. X. Nguyen *et al.*, *Electron. Lett.* **34**, 811–812 (1998).
42. C. Nguyen *et al.*, *Electron. Lett.* **34**, 309–311 (1998).
43. N. X. Nguyen *et al.*, *Electron. Lett.* **33**, 334–335 (1997).
44. C. Nguyen *et al.*, *Electron. Lett.* **35**, 1380–1381 (1999).
45. J. Hu *et al.*, *Sensors and Actuators A: Physical* **86**, 122–126 (2000).
46. J.-L. Lee *et al.*, *Solid-State Electron* **42**, 2063–2068 (1998).
47. P. M. Frijlink, *Microelectronic Engineering* **35**, 313–316 (1997).
48. R. Grundbacher *et al.*, *IEEE Trans. Electron. Devices* **44**, 2136–2142, (1997).
49. S. J. Cai *et al.*, *IEEE Trans. Electron. Devices* **47**, 304–307 (2000).
50. R. Dimitrov *et al.*, *Solid-State Electron.* **44**, 1361–1365 (2000).
51. S. T. Sheppard, K. Doverspike, W. L. Pribble, S. T. Allen and J. W. Palmour, *56th Device Research Conference*, June 1998.
51a. K. Kunihoro *et al.*, *IEEE Electron Device Lett.* **20**, 608–610 (1999).
52. Naoya Okamoto *et al.*, *IEEE Trans. Electron. Devices* **47**, 2284–2289 (2000).
53. O. Breitschädel *et al.*, *Materials Science and Engineering* **B82**, 238–240 (2001).
54. M. Akita *et al.*, *IEEE Electron. Device Lett.* **22**, 376–377 (2001).
55. M. A. Khan *et al.*, *Appl. Phys. Lett.* **66**, 1083–1085 (1995).
56. R. Gaska, Q. Chen, J. Yang, A. Osinsky, M. A. Khan and M. S. Shur, *IEEE Electron. Device Lett.* **18** (10), 492–494 (1997).
57. S. Kayali, *Microelectron. Reliab.* **39**, 1723–1736 (1999).
58. F. Fantini and F. Magistrali, *Microelectron. Reliab.* **32**, 1559–1569 (1992).
59. K. van der Zanden *et al.*, *IEEE Trans. Electron. Devices* **46**, 1570–1576 (1999).
60. K. Ikossi *et al.*, *IEEE Trans. Electron. Devices* **48**, 465–471 (2001).
61. A. Tarakji *et al.*, *Appl. Phys. Lett.* **78**, 2169–2171 (2001).
62. H. Tang *et al.*, *Solid-State Electron.* **44**, 2177–2182 (2000).
63. R. Hickman *et al.*, *Solid-State Electron.* **42**, 2183–2185, (1998).

64. S. Weinreb *et al.*, *IEEE Microwave and Guided Wave Lett.* **9**, 282–284 (1999).
65. Y.-F. Wu *et al.*, *IEEE Trans. Electron. Devices* **48**, 586–590 (2001).
66. J. W. Johnson *et al.*, *Solid-State Electron.* **45**, 1979–1985 (2001).
67. E. Alekseev and D. Pavlidis, *Solid-State Electron.* **44**, 245–252 (2000).
68. Y. H. Byun *et al.*, *IEEE Electron. Device Lett.* **11**, 50–52 (1990).
69. S. Karmalkar and G. Ramesh, *IEEE Trans. Electron. Devices* **47**, 11–23 (2000).
70. S. Karmalkar and R. R. Rao, *IEEE Trans. Electron. Devices* **47**, 667–676 (2000).
71. H. Rohdin and P. Roblin, *IEEE Trans. Electron. Devices* **33**, 664 (1986).
72. H. Brech *et al.*, *IEEE Trans. Electron. Devices* **47**, 1957–1964 (2000).
73. J. D. Albrecht *et al.*, *IEEE Trans. Electron. Devices* **47**, 2031–2036 (2000).
74. E. Zanoni *et al.*, *Int. J. High Speed Electron. and Systems* **10**, 119–128 (2000).
75. K. A. Christianson *et al.*, *Solid-State Electron.* **38**, 1623–1626 (1995).
76. Yuji Ando *et al.*, *IEEE Trans. Electron. Devices* **47**, 1965–1970 (2000).
77. S. Karmalkar, *IEEE Trans. Electron. Devices* **44**, 862–867 (1997).
78. K. Lee *et al.*, *J. Appl. Phys.* **54**, 2093–2096 (1983).
79. M. B. Das, *IEEE Trans. Electron. Devices* **32** (1), 11–17 (1985).
80. T. Fjeldly *et al.*, *Introduction to Device and Circuit Modeling for VLSI* (John Wiley and Sons, New York, ISBN 0-471-15778-3, 1998).
81. K. Lee *et al.*, *Semiconductor device modeling for VLSI* (Prentice Hall, New Jersey, 1993).
82. H. Shen *et al.*, *Advanced Workshop on Frontiers in Electronics* (May/June 1999) p. 97.
83. S. T. Sheppard *et al.*, *56th Device Research Conference*, June 1998.
84. M. A. Khan *et al.*, *Appl. Phys. Lett.* **65**, 1121–1123 (1994).
85. R. Gaska *et al.* (IEDM-97 Technical Digest, December 1997) pp. 565–568.
86. R. J. Trew, D. Park, Y. Wu, U. K. Mishra and S. DenBaars, *Proc. 1997 IEEE/Cornell Conf. Advanced Concepts in High Speed Semiconductor Devices and Circuits*, August 1997.
87. W. R. Curtice and M. Ettenberg, *IEEE Trans. Microwave Theory Tech.* **33**, 1383–1394 (1985).
88. Tektronix Inc., O. R. Beaverton, US Patent No. 4, 236, 119 (1980).
89. K. Krishnamurthy *et al.*, *IEEE J. Solid-State Circuits* **35**, 1285–1292 (2000).
90. G. Simin, X. Hu, A. Tarakji, J. P. Zhang, A. Koudymov, S. Saygi, J. W. Yang, M. A. Khan, M. S. Shur and R. Gaska, *Jpn. J. Appl. Phys.* **40** (11A), L1142–L1144 (2001).
91. X. Hu, A. Koudymov, G. Simin, J. Yang, M. Asif Khan, A. Tarakji, M. S. Shur and R. Gaska, *Appl. Phys. Lett.* **79**, 2832–2834 (2001).
92. J. C. Rojo, L. J. Schowalter, K. Morgan, D. I. Florescu, F. H. Pollak, B. Raghothamachar and M. Dudley, *Mat. Res. Soc. Symp. Proc.* **680E**, 221 (2001).
93. J. C. Rojo, L. J. Schowalter, R. Gaska, M. S. Shur, A. Khan, J. Yang and D. Koleske, Growth and characterization of epitaxial layers on single-crystal aluminum nitride substrates, submitted to J. Crystal Growth.

94. R. Gaska, Q. Chen, J. Yang, A. Osinsky, M. A. Khan and M. S. Shur, *Electro. Lett.* **33** (14), 1255–1257 (1997).
95. J. Deng, B. Iñiguez, M. S. Shur, R. Gaska, M. A. Khan and J. W. Yang, *Physica Status Solidi* **176** (1), 205–208 (1999).

## CHAPTER 4

## FABRICATION AND PERFORMANCE OF GaN
## MOSFETs AND MOSHFETs

Cammy R. Abernathy and Brent P. Gila

*Department of Materials Science and Engineering*
*University of Florida, Gainesville, FL 32606, USA*

## Contents

## 1. Introduction

III-V materials, in particular GaAs, have been quite successful in the area of high speed, high power applications because of their combination of high saturation velocity and moderately large bandgaps. As power requirements continue to increase, however, performance is becoming limited by device breakdown, suggesting larger bandgap materials are required. Recently, the Group III nitrides and SiC have received a great deal of attention as wide bandgap alternatives.[1-4] These materials are expected to be excellent candidates for use in high temperature, high power electronic applications because of their large thermal conductivities and saturation velocities in addition to the larger bandgaps.[1]

Advances in SiC and GaN have led to power components based on several device configurations including Metal-Semiconductor Field Effect Transistors (MESFETs), Heterojunction Field Effect Transistors (HFETs), thyristors and Heterojunction Bipolar Transistors (HBTs).[5-10] While these devices have shown promise for a number of applications, the metal oxide (or insulator) semiconductor field effect transistor (MOSFET or MISFET) is also a desirable structure. The MOS(MIS)FET structure has a number of advantages over the heterojunction type transistor such as a relative insensitivity to temperature during operation. The MOSFET structure can be made with either $n$-type or $p$-type material under the gate thus also allowing for fabrication of complementary device structures, known as C-MOS. The same dielectrics used in MOSFET devices may also be used in combination with HFETs to form MOSHFETs. This device combines the attractive features of the MOSFET, such as low gate leakage, with the high speeds and high current density of the HFET. For either MOS(MIS)FETs or MOSHFETs to be realized, a high quality dielectric material must be created for the gate insulator. Even in conventional HFET devices, where surface leakage limits RF power performance, development of a low interface state density dielectric which could be used as a surface passivation layer would significantly improve the device performance. The following sections outline the status of and the prospects for development of dielectrics for GaN MOS(MIS)FET and MOSHFET devices.

## 2. Candidate Dielectrics

Ideal dielectric materials are perfect insulators in which no mobile charged particles are present. The total effect of an electric field on a dielectric material is known as polarization. The ability of a material to resist the polarization of charge is described as the dielectric constant, $\kappa$, which is the ratio of the permittivity of the material, $\varepsilon_i$, to the permittivity of vacuum, $\varepsilon_v$. The dielectric constant can also be related to the internal field created within the material and the external applied field: $E_{\text{internal}} = E_{\text{applied}}/\tilde{\kappa}$.

The polarization, $P$, of the material is related to the dielectric constant by

$$P = (\kappa - 1)\varepsilon_o \xi\,,$$

where $\xi$ is the strength of the electric field in V/cm. It can be assumed from this relation that the polarization increases as the electric field strength increases, until all the dipoles are aligned.

There are several applications for dielectric materials in GaN device technology. Passivation of high voltage junctions, isolation of devices and interconnects, and gate insulation of field effect transistors are the most important. A successful dielectric must possess a number of characteristics including chemical stability over the life of the device and thermal stability, since many GaN applications require operation at elevated temperatures. For good electrical operation, the material should have immobile charge traps (to avoid shorting), low defect densities (to avoid breakdown at low electric fields), and a dielectric constant higher than that of the semiconductor (to avoid generation of high electric fields in the dielectric). This last requirement is the major disadvantage of $SiO_2$, as shown in Table 1. The dielectric should also have large band offsets to both the conduction and valence bands in order to provide adequate confinement of carriers. $SiO_2$ fills this requirement quite well, as shown in Fig. 1, while several of the other candidates to be discussed in this chapter are less than ideal.

In addition to the bulk properties, the dielectric/semiconductor interface is also critically important, as leakage at the interface will severely compromise device performance. In general, the interface state density of carrier traps must be $<$ mid-$10^{11}$ eV$^{-1}$cm$^{-2}$ for a successful device. It is this criteria which normally precludes the use of poly-crystalline dielectrics. In order to achieve low interface state densities, amorphous dielectrics are often chosen as these typically produce the lowest number of defects at the dielectric/semiconductor interface, e.g. $SiO_2$/Si or GGG/GaAs. In the last few years, however, success has been achieved by using single crystal oxide layers, e.g. $Gd_2O_3$ on GaAs. In the case of GaN, the number of crystalline oxides which can be grown epitaxially are limited due to the rather small GaN lattice constant. As shown in Table 1, the most likely candidates include the Bixbyite oxides $Gd_2O_3$ and $Sc_2O_3$ and MgO which has the rock salt structure. Even for these materials the lattice mismatch to GaN is still quite large. However, as will be discussed in later sections, in spite of this mismatch single crystal oxide material has been deposited on GaN, though in some cases only a few monolayers thick.

Table 1. Material properties for GaN and various dielectrics. W = Wurtzite, A = amorphous, B = Bixbyite, N = NaCl.

| | GaN[11] | SiO$_2$[12-14] | SiN$_x$[14] | AlN[15-17] | GGG[18] | Gd$_2$O$_3$[19-21] | Sc$_2$O$_3$[22] | MgO[23,24] |
|---|---|---|---|---|---|---|---|---|
| Structure | W | A | A | W or A | A | B | B | N |
| Lattice Constant | 3.186 | — | — | 3.113 | — | 10.813 | 9.845 | 4.2112 |
| Spacing in the (111) plane | — | — | — | — | — | 3.828 | 3.4807 | 2.978 |
| Mismatch to GaN (%) | — | — | — | 2.3 | — | 20.1 | 9.2 | -6.5 |
| T$_{MP}$ (K) | 2800 | 1900 | 2173 | 3500 | 2023 | 2668 | 2678 | 3073 |
| Bandgap (eV) | 3.4 | 9 | 5 | 6.2 | 4.7 | 5.3 | 6.3 | 8 |
| Electron Affinity (eV) | 3.4 | 0.9 | 0-2.9 | — | 0.63 | — | 0.7 | — |
| Work Function (eV) | — | — | — | 0.9-1.2 | — | 2.1-3.3 | 4 | 3.1-4.4 |
| Dielectric Constant | 9.5 | 3.9 | 7.5 | 8.5 | 14.2 | 11.4 | 14 | 9.8 |

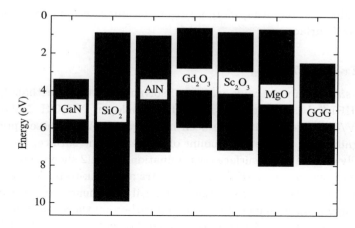

**Fig. 1.** Band alignments for GaN and various dielectrics. The top of the bar corresponds to the conduction band edge while the bottom corresponds to the valence band edge. The vacuum level is set at 0. Values for AlN are based on an average of the reported electron affinity values. The conduction band edge for $Sc_2O_3$ was calculated from the work function. Band offsets for GGG were taken from Ref. 25.

## 3. Surface Cleaning

Oxygen atoms are actively adsorbed on the surface of III-V semiconductors, resulting in the production of a native oxide. A native oxide layer is also expected to grow upon air exposure after growth of GaN. Ishikawa *et al.*[26] used X-ray Photoelectron Spectroscopy (XPS) to examine the GaN surface and found a contamination layer $\sim 2$ nm thick consisting of gallium oxide and adsorbed carbon. This oxide was shown by Prabhakran *et al.*[27] to be primarily $Ga_2O_3$. Even though the native oxide does not grow as rapidly on the GaN surface as compared to other III-V semiconductors such as GaAs, the presence of oxides on the GaN surface can cause effects which are detrimental to device performance. The surface oxide layer acts as a barrier for carrier transport from metal to semiconductor, and subsequently increases the Ohmic contact resistivity. Also the oxide plays a role in increasing the Schottky barrier height at the surface. Furthermore, device fabrication processes involve several layers of photolithography and various etching processes to define active device areas. These processing steps result in carbon contamination and chemical residuals on the surface. The removal of the native oxide and generation of a contamination free surface is required to grow a high quality gate dielectric layer. In particular, carbon and oxygen are more strongly bonded to GaN than to other III-V semiconductors, requiring more stringent cleaning techniques to obtain a clean GaN surface.

### 3.1. *Ex-situ Cleaning*

For *ex-situ* cleaning, a wide variety of solvents (acetone, methanol), acids (HCl, HF),[28] basic solutions (AZ 400 K developer, $NH_4OH$, NaOH)[27-30] and $UV/O_3$[31] treatments have been investigated. While these approaches can significantly reduce the amount of oxygen on the surface, they do not generally eliminate the surface contamination. Table 2 shows typical Auger Electron Spectroscopy (AES) survey spectra and peak-to-peak height ratios of GaN after various *ex-situ* surface chemical treatments. As-received GaN shows significant amounts of oxygen and carbon on the surface, which further increases after exposure to photoresist. Solvent cleaning has little effect on the level of carbon at the surface. A 1 : 1 HCl : DI solution was found to be very effective for removing the native oxide. However, while this acid treatment was found to produce the lowest O/N ratio, significant amounts of carbon and chlorine remained. $UV/O_3$ treatment was found be an excellent surface cleaning method to remove surface carbon contamination. An HCl acid treatment followed by $UV/O_3$ exposure for 25 minutes removed carbon and chemical residual (chlorine) completely from the GaN surface even for the case of photoresist exposed material. Preliminary evidence suggests that exposure to ozone does oxidize some of the GaN material. The combination of $UV/O_3$ oxidation and dipping in acid was also used to determine the best chemical cleaning method for removing carbon and oxygen from GaN surfaces. $UV/O_3$ exposure for 25 minutes followed by an HCl acid treatment was very effective for oxide removal but left significant amounts of carbon. To eliminate the carbon it is found to be necessary to also treat the surface with acid prior to exposure to the ozone.

**Table 2.** AES peak-to-peak height ratios for the $UV/O_3$ and various wet chemically treated GaN surfaces. After Ref. 31.

|  | C/N | O/N | Ga/N | Cl/N |
|---|---|---|---|---|
| As-received | 0.45 | 0.97 | 1.08 | ND |
| PR/deposit/strip | 1.66 | 0.95 | 1.05 | ND |
| Solvents (acetone/methanol) | 0.55 | 0.80 | 0.97 | ND |
| 1 : 1 HCl : DI (3 min.) | 0.40 | 0.24 | 0.64 | 0.14 |
| HCl (3 min.)/$O_3$ (25 min.) | ND | 1.47 | 1.00 | ND |
| $O_3$ (25 min.)/HCl (3 min.) | 0.13 | 0.26 | 0.68 | ND |
| HCl/$O_3$ (25 min.)/HCl | 0.06 | 0.45 | 0.79 | 0.09 |

AES Ga to N peak ratios from the GaN surfaces were investigated to study the surface stoichiometry changes after *ex-situ* chemical treatments as shown in Table 2. Similar to results in the AlN case, the Ga to N ratio was reduced $\sim 40\%$ compared to that of the as-received sample after acid HCl treatment. Since the oxide on as-received or UV/$O_3$ exposed sample surfaces is composed mostly of gallium oxide, the removal of this oxide layer with HCl can reduce the Ga/N ratio on the GaN surface. The UV/$O_3$ exposure again restored the Ga/N ratio to the levels of as-received material. AFM results showed that neither wet chemical treatments to remove oxide nor UV/$O_3$ exposure to restore the oxide layer increased the RMS roughness value beyond those of the as-received sample, though HCl treatment did appear to slightly planarize the GaN surface.

RHEED has been used to study the removal of oxides and the surface roughness after various wet chemical treatments. Figure 2 shows RHEED patterns taken along the $\langle 1\,1\,\bar{2}\,0 \rangle$ azimuth on the GaN surfaces. No diffraction pattern was detected for the as-received and UV/$O_3$ exposed samples. This indicates that the native oxide and UV/$O_3$ generated oxide surfaces are likely to be amorphous. GaN surfaces treated in UV/$O_3$ oxidation followed by dipping in 1 : 1 HF : DI or AZ 400 K developer displayed $(1 \times 1)$ streaky RHEED patterns indicating removal of the UV/$O_3$ generated oxide or the native oxide and the generation of an atomically flat surface. However, these RHEED results do not indicate that HF and AZ 400 K treated samples are completely free of oxygen or carbon contamination, as oxygen and carbon have been shown to be present on chemically cleaned GaN surfaces by AES and SIMS. RHEED results only indicate that the cleaned GaN surface is smooth and relatively free of oxides and other contaminants but significant amounts of oxygen and carbon contamination still

**Fig. 2.**   RHEED patterns taken along the azimuth on the $\langle 1\,1\,\bar{2}\,0 \rangle$ MOCVD-grown GaN surfaces after various chemical treatments (a) as-received GaN (b) HCl/$O_3$/HF-treated GaN and (c) HCl/$O_3$/AZ 400 K-treated GaN.

exist. Also RHEED results indicate that native oxides do not grow rapidly on the GaN surface during the time ($<$ 30 minutes) required to load the samples in the vacuum chamber. In summary, the lowest oxygen impurity levels on GaN surfaces have been achieved after HCl or HF based acid treatments, while $UV/O_3$ cleaning is very effective for removing carbon. However, complete removal of carbon and oxygen from the GaN surface generally cannot be achieved by *ex-situ* chemical treatments alone. *In-situ* chemical treatments are required for further removal of carbon and oxygen from the GaN surface.

## 3.2. *In-situ Cleaning*

### 3.2.1. *Thermal/Plasma Cleaning*

For further removal of surface carbon and oxygen, *in-situ* surface treatments including thermal treatments with or without $N_2$ or $H_2/N_2$ plasmas have been investigated. Smith *et. al*[32] and King *et. al*[33] have reported cleaning of GaN surfaces using various wet-chemical treatments followed by *in-situ* thermal desorption to remove carbon and oxygen. However, complete thermal desorption of all contaminants was not achieved even at a temperature of 900°C. The effects of combining *ex-situ* wet chemical and *in-situ* thermal treatments in an MBE chamber were investigated by capping the cleaned surfaces *in-situ* with GaN and subsequently characterizing the layer surface morphology (AFM) and the interfacial purity (SIMS). Table 3 shows the atomic concentration of carbon and oxygen as measured by AES after various combinations of wet chemical and thermal surface treatment. Virtually any thermal treatment was successful in obtaining a clean GaN surface within the detection limits of AES. From AFM, it was found that none of the *in-situ* thermal cleaning processes significantly affected the roughness of the treated and regrown GaN surface.

**Table 3.** Comparison of surface atomic concentration of C and O between AES and SIMS analysis on GaN after various combination of wet chemical and thermal surface treatment. After Ref. 31.

| Method | AES | | SIMS | |
|---|---|---|---|---|
| | C | O | C ($cm^{-3}$) | O ($cm^{-3}$) |
| As-received with 750°C $N_2$ Plasma | ND | ND | $1.5 \times 10^{21}$ | $5.1 \times 10^{21}$ |
| $HCl/O_3$ with 750°C $N_2$ Plasma Exposure | ND | ND | $3.2 \times 10^{20}$ | $2.0 \times 10^{22}$ |
| $HCl/O_3$ with 750°C $H_2/N_2$ Plasma Exposure | ND | ND | $2.9 \times 10^{20}$ | $2.0 \times 10^{22}$ |
| $HCl/O_3/HF$ with 700°C vacuum anneal without plasma | ND | ND | $2 \times 10^{20}$ | $5.5 \times 10^{20}$ |

At lower temperatures, AES analysis on the sample which did not receive a plasma exposure after the $PR/HCl/O_3$ treatment showed significant amounts of carbon and oxygen contamination at the interface. The surface carbon could be removed after $H_2/N_2$ plasma exposure only at 500°C. By contrast, the complete removal of oxygen was only achieved after $H_2/N_2$ plasma exposure at a much higher temperature of 750°C. However, the sample which was exposed to the $H_2/N_2$ plasma at 750°C shows a degraded surface. This can be attributed to preferential loss of surface nitrogen by formation of volatile $NH_X$ compounds or desorption of thermodynamically unstable GaN at high temperature and in high vacuum. Further, contrary to AES data, as shown in Table 3, SIMS data reveals that a significant concentration of surface carbon and oxygen still exists after thermal or plasma treatments. It is quite possible that the oxide generated by the $UV/O_3$

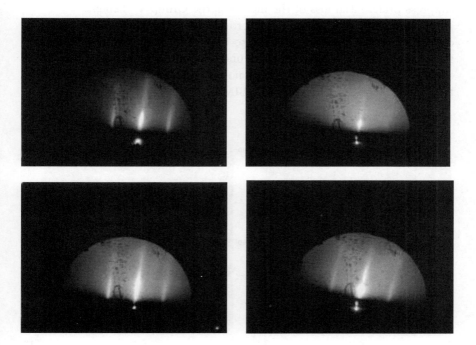

**Fig. 3.** RHEED patterns taken along the $\langle 1\,1\,\bar{2}\,0 \rangle$ azimuth on the MOCVD-grown GaN surfaces after various chemical and thermal treatments (upper left) $HCl/O_3/HF$ treated GaN, (upper right) $HCl/O_3/AZ$ 400 K-treated GaN, (lower left) $HCl/O_3/HF$ treatment followed by thermal annealing at 700°C and (lower right) $HCl/O_3/AZ$ 400 K followed by thermal annealing at 700°C.

treatment is too thermally stable to be removed simply by *in-situ* cleaning. Rinsing the GaN surface in HF or AZ 400 K immediately prior to loading into the reactor thermal desorption did lower the interfacial contamination substantially. This enhanced cleaning effect can also be seen in the brightness and sharpness of the RHEED patterns, shown in Fig. 3, which increased for both $HCl/O_3/HF$ and $HCl/O_3/AZ$ 400 K pretreated samples. From the RHEED, the acid rinse appears to be more effective than the base rinse. However, even in this case significant contamination remains.

Figure 4 shows an AFM image of the regrown MBE GaN layer at growth temperature of 650°C with or without $N_2$ plasma exposure during substrate heating to growth temperatures. The regrown GaN layer without $N_2$ plasma exposure during substrate heating shows a rougher surface than the GaN regrown with nitrogen plasma exposure. This may possibly be the result of high temperature heating without providing an overpressure of nitrogen to suppress preferential loss of the more volatile Group V element.

In summary, the best approach to cleaning the GaN surface prior to deposition of the dielectric appears to be *ex-situ* oxidation and oxide strip followed by an *in-situ* thermal clean under a stabilizing nitrogen flux. However, even using this approach does not completely remove all contamination on the surface as measured by SIMS.

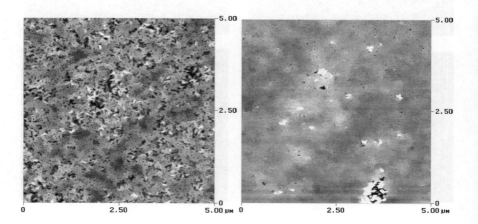

**Fig. 4.** AFM images of the MBE-grown GaN layers on MOCVD-grown GaN substrates (at left) overgrowth at 650°C without prior $N_2$ plasma exposure and (at right) overgrowth with prior $N_2$ plasma exposure during substrate heating to growth temperature. The RMS roughness values are 7.6 and 2.0 nm, respectively. After Ref. 31.

### 3.3. Gate Dielectrics on GaN

While GaN is a fairly new electronics technology, initial results using either MESFET or HEMT structures have been encouraging.[35-44] However, several problems remain. In particular these devices exhibit low Power Added Efficiencies (PAEs) and nonlinearity. This is most likely due to high parasitic resistances and knee voltages which arise from the high contact resistivities and high sheet resistances between the source contact and the gate. Conventional $n^+/n$ layer structures for GaAs technology cannot be applied in the nitride based material system since no adequate gate recess technology is presently available. Both of these problems may be overcome by using a MOSFET approach. MOSFET devices are expected to have a wide gate modulation range, usually resulting in better linearity than either MESFETs or HEMTs. This wide operating range is due to the fact that the device turn-on is dependent on the dielectric thickness and is not limited to low turn-on voltages as in case of a Schottky contact ($\sim 0.7$ V). Also, current FET processing does not have a gate fabrication technology that can tolerate the high temperature implant activation needed to form the low resistance source and drain region in the self-aligned MESFET configuration. By contrast, MOSFETs can use implantation to form a highly doped source and drain without compromising the gate contact. The temperature sensitivity of MOSFETs will also be superior since they do not suffer from the severe leakage encountered in Schottky based devices. Further, circuit design can be simplified since enhancement-mode MOSFETs can be used to form single supply voltage control circuits for power transistors. The use of MOSFETs also allows the use of complementary devices, thus producing less power consumption and simpler circuit design. The challenge for this technology is identifying a suitable gate dielectric. The following sections discuss the various candidates that have been explored thus far.

### 3.4. $SiO_2$

Most work in the GaN MOSFET area has focused on the amorphous dielectrics, and in particular on $SiO_2$. $SiO_2$ has a large bandgap and a small electron affinity. This makes an excellent band configuration with GaN, as shown in Fig. 1. However, the dielectric constant is low at 3.9. A number of methods have been used to deposit $SiO_2$ on GaN for use as a gate dielectric. Plasma Enhanced Chemical Vapor Deposition (PECVD)[45-48] at substrate temperatures of 300°C has been reported to give $SiO_2/GaN$ interface state densities using the Terman method on the order of low

$1 \times 10^{11}$ eV$^{-1}$cm$^{-2}$, while silicon oxide deposited by Electron Beam (EB) evaporation has shown an interface state density of $5.3 \times 10^{11}$ eV$^{-1}$cm$^{-2}$, again calculated using the Terman method.[49] Breakdown fields range from 1.8 to 2.5 MV/cm for $e$-beam and PECVD respectively.[48,49] The most successful deposition method appears to be a high temperature CVD process using dichlorosilane and N$_2$O, followed by a high temperature 30 seconds anneal at 1050°C in nitrogen. This approach has produced the best SiO$_2$/GaN $D_{it}$ at $8 \times 10^{10}$ eV$^{-1}$cm$^{-2}$ near the conduction band edge, as measured by the conductance method.[50]

The exact nature of the SiO$_2$/GaN interface is still under investigation. Therrien *et. al*[51] used remote plasma assisted oxidation to probe the effect of producing a few monolayers of Ga$_2$O$_3$ by oxidizing the GaN surface prior to depositing SiO$_2$ from plasma excited O$_2$ and SiH$_4$ at 300°C. This initial oxidation appeared to reduce the interfacial trap density by roughly an order of magnitude. This suggests that the best SiO$_2$/GaN interfaces may in fact have an unintentionally formed Ga$_2$O$_3$ interlayer which improves the electrical characteristics. Further work is needed to confirm this. While the SiO$_2$ results are promising, the low dielectric constant of SiO$_2$ and the absence of modulation at positive gate voltages (forward bias) above 3–4 V in SiO$_2$/GaN devices suggests that other alternatives should be considered.

### 3.5. *Silicon Nitride*

Silicon nitride is another amorphous dielectric which has been explored. Deposition by PECVD[47,52] on GaN resulted in an interface state density of $6.5 \times 10^{11}$ eV$^{-1}$cm$^{-2}$. Electrical measurements showed the MISFET structure had a large flat band voltage shift (3.07 V) and a low breakdown voltage (1.5 MV/cm) of the dielectric. A unique dielectric structure of SiO$_2$/Si$_3$N$_4$/SiO$_2$ (ONO) was reported to have a breakdown field strength of 12.5 MV/cm for temperatures as high as 300°C.[53] The ONO structure was deposited by jet vapor deposition from N$_2$O, N$_2$ and SiH$_4$ to a thickness of 10 nm/20 nm/10 nm. Interface state densities for this structure at room temperature were calculated using the AC conductance method to be $5 \times 10^{11}$ eV$^{-1}$cm$^{-2}$ near the band edge ($E_c - E = 0.2$ eV). In optimized structures this value has been estimated to be $5 \times 10^{10}$ eV$^{-1}$cm$^{-2}$ at 0.8 eV below the conduction band edge. This multilayer structure allows for unique engineering of the dielectric, and will most likely receive more attention in the future. However, though the high breakdown field will allow operation at fairly high powers, the dielectric constant of the stack may still limit the performance range somewhat.

### 3.6. *Aluminum Nitride*

In addition to the amorphous dielectrics, crystalline AlN has also been considered. Aluminum nitride deposited by MBE, MOMBE and MOCVD has been used to create MISFET devices and IG-HFET devices.[54,55] Using MOCVD at high substrate temperatures, 1150°C, AlN/GaN MISFETs have been fabricated.[56] The resulting single crystal AlN has produced $D_{it}$ values of $10^{11}$ eV$^{-1}$cm$^{-2}$ as calculated using the Terman method. Devices fabricated using this approach showed a peak transconductance of 135 mS/mm for gate lengths of 2 microns, with $I_{dss} \sim 600$ mA/mm. Gate modulation up to 1 V forward bias was demonstrated.

Lower temperature deposition by MOCVD at 550°C has also been explored, with the lower temperatures employed to suppress cracking of the AlN layer.[57] Using the Terman method, interface state densities varied from $10^{13}$ at the band edge ($E_c - E = 0.2$ eV) to $9.5 \times 10^{10}$ eV$^{-1}$cm$^{-2}$ at $E_c - E = 0.8$ eV. It is not known if the dielectric in this case was single crystal or poly-crystalline. The AlN gate in the MISFET structure grown by MOMBE at 400°C was polycrystalline and resulted in a dielectric breakdown field of only 1.2 MV/cm.[54] Unlike the amorphous dielectrics, single crystal AlN and polycrystalline AlN films are expected to suffer from defects and grain boundaries that reduce the breakdown field sustainable in the material.

While crystalline AlN shows some promise as a gate dielectric, the low breakdown fields and resulting low forward gate voltages are most likely due to dislocations and other defects. Presumably these arise from the small but finite lattice mismatch which exists between AlN and GaN and from propagation of defects from the GaN to the AlN. An alternative approach is the use of amorphous AlN. It is expected that reducing the growth temperature should suppress the surface mobility sufficiently to prevent nucleation of a crystalline phase. Kidder *et. al*[58] have in fact shown that by using a growth temperature of 200°C in the CVD deposition of AlN on Si it is possible to reduce the leakage and improve the breakdown relative to crystalline AlN grown on the same substrate material. The absence of XRD peaks in their material suggested that the deposited material was amorphous. Partially amorphous AlN has also been deposited using a gaseous Al precursor and an RF nitrogen plasma in a Metalorganic Molecular Beam Epitaxy (MOMBE)

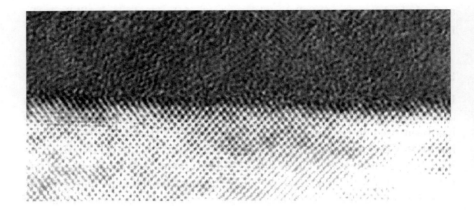

**Fig. 5.** XTEM of AlN deposited on Si at 325°C showing crystalline pockets dispersed in an amorphous matrix. After Ref. 59.

system.[59] As shown in Fig. 5, material was deposited at 325°C in which approximately 30% of the volume was amorphous. On SiC, this approach produced MIS capacitors with a forward gate breakdown of 6.8 MV/cm, and C-V analysis which showed little or no evidence of hysteresis. Similar experiments on GaN were not as successful, however, with breakdowns of only 1 MV/cm attainable. This was most likely due to enhanced crystallinity from the template effect of the GaN substrate. In order to improve the electrical behavior of the AlN/GaN structures it will be necessary to further reduce the crystallinity of the material. This can most likely be accomplished by further reducing the growth temperature. In order to accomplish this, it will be necessary to switch from the gaseous Al precursor, which has a lower thermal decomposition temperature of ∼ 325°C, to an elemental Al effusion oven which can in principle be used at much lower substrate temperatures.

### 3.7. *Gallium Oxide*

Gallium oxide is yet another material which has been investigated. Thermal oxidation of the GaN surface to form $Ga_2O_3$ has been performed in dry[60,61] and wet[62] atmospheres. Dry oxidation of GaN epilayers at temperatures below 900°C showed minimal oxidation. At temperatures above 900°C, a polycrystalline monoclinic $Ga_2O_3$ forms at a rate of 5.0 nm/h. This oxidation rate is too slow to be viable as a processing step. Wet oxidation of GaN also forms polycrystalline monoclinic $Ga_2O_3$, but at a rate of 50.0 nm/h at

900°C. From Cross-Sectional Transmission Electron Microscopy (XTEM), the interface between the oxide and the GaN is found to be non-uniform. Scanning Electron Microscopy (SEM) shows that both films are rough and facetted. Electrical characterization of the oxide films shows the dry oxide dielectric has a breakdown field strength of 0.2 MV/cm and the wet oxide dielectric field strength of 0.05 to 0.1 MV/cm. The oxidation rate can be increased to $> 200$ nm/h through the use of photoelectrochemical etching in KOH solutions.[63,64] Without annealing, $D_{it}$ values measured using the Terman method are typically in the $10^{12}$ eV$^{-1}$cm$^{-2}$ range. After annealing at 500°C for 20 minutes. this value is reduced to $3 \times 10^{11}$ eV$^{-1}$cm$^{-2}$. The best breakdown field reported for $Ga_2O_3$ thus far is 3.85 MV/cm obtained from dry oxidation at 850°C.[65]

ECR-MBE using elemental Ga and an ECR oxygen plasma has also been employed to deposit $Ga_2O_3$ on GaN.[66] Auger Electron Spectroscopy (AES)

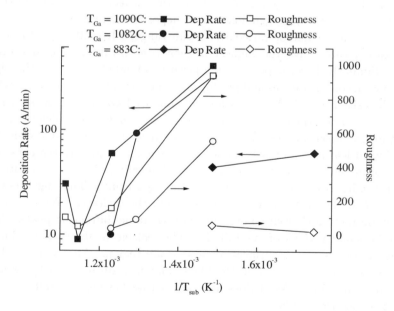

**Fig. 6.** GaO$_x$ deposition rate and roughness as a function of substrate temperature. As mentioned above, decreasing the substrate temperature enhances the deposition rate. These increased rates result in rougher surface morphologies. However, if the Ga cell temperature is reduced to offset the increase in deposition rate at low substrate temperatures, smooth surfaces can be obtained. In fact the smoothest films have been obtained at the lowest temperature tested, 300°C.

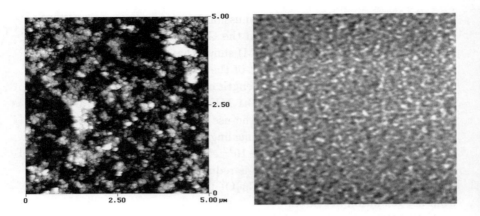

**Fig. 7.**   AFM (left) and SEM (right) of $Ga_2O_3$/GaN surface. After Ref. 66.

indicated that the Oxygen/Gallium ratio was a sensitive function of (a) Ga cell temperature, (b) oxygen flux and (c) deposition temperature. While the deposition rates could be varied over a wide range, $0.18-3.0$ $\mu m/h$, the lower deposition rates gave the highest oxygen concentrations and smoothest films. Similarly, lower deposition temperatures tended to provide the highest oxygen concentrations and produced the best surface morphologies when the deposition rate is corrected for the enhanced Ga uptake, as shown in Fig. 6. Even the smoothest films, however, still showed significant surface texture as shown in Fig. 7, $GaO_x$ deposited at $400°C$ on $p$-GaN showed significant leakage and low breakdown voltage. Lower growth temperatures did not appear to alleviate this problem. Annealing of these structures in hydrogen improved the C-V characteristics significantly. This suggests that much of the leakage is due to defects within the dielectric. While powder X-ray diffraction showed no evidence of crystalline diffraction peaks, there may be small regions of crystallinity. The leakage might then be due to dangling bonds at the surfaces of these small regions. Alternatively there may be insufficient oxygen in the $GaO_x$ matrix resulting in defect states. As was found for GaAs, it does not appear that $Ga_2O_3$ by itself will be a viable dielectric for GaN.

### 3.8. Gadolinium Gallium Garnet

Due to the recent success of $Ga_5Gd_3O_{12}$, Gallium Gadolinium Garnet (GGG), as a dielectric in GaAs MOSFETs,[67-70] attention has turned toward this as a dielectric material for GaN. This material is amorphous when deposited by $e$-beam evaporation. The large dielectric constant of this material is particularly attractive and the bandgap, though significantly lower than that of $SiO_2$, is expected to be adequate for $n$-type material.

The oxide thickness and the RMS roughness of the GGG/GaN interface formed by $e$-beam evaporation in an MBE chamber were studied by X-ray reflectivity.[54] The interfaces were found to be quite smooth with interfacial roughness as low as 0.3 nm. The roughness is believed to come from the GaN surfaces, which are known to be quite rough even when grown under optimum MOCVD growth conditions. X-ray reflectivity studies were performed on 8.5 nm GGG films before and after rapid thermal annealing at 825°C or 950°C, and are shown in Fig. 8. By analyzing the periodic spacing of the oscillatory curves, it is estimated that the thickness of the oxide films does not change significantly with annealing, even for a temperature

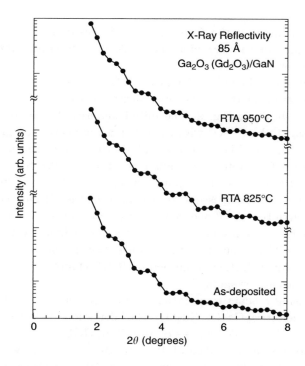

**Fig. 8.** X-ray reflectivity scans for 85 nm GGG films before and after annealing at either 825°C or 950°C. The reflectivity is a reciprocal space scan taken along the surface normal of GaN (0001) direction on very small angles. After Ref. 54.

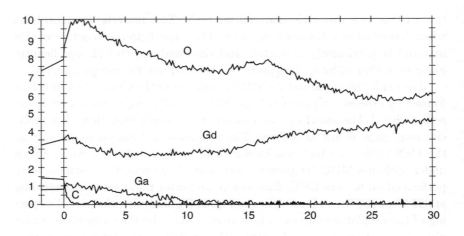

**Fig. 9.**   AES depth profile of GGG/GaN MOSFET structure. After Ref. 71.

of 950°C. Judging from the similarity of the intensity decay with increasing $2\theta$, it is clear that the air/oxide and oxide/GaN interfaces remain intact after annealing.

AES depth profiling of a GGG/GaN MOSFET structure indicates that the concentration of the GGG layer varies with distance from the GaN interface, as shown in Fig. 9. The initial material deposited on the substrate appears to be $Gd_2O_3$ with the Ga signal appearing roughly half way through the sample. The oxygen signal also increases with increasing distance from the GaN interface. Similar profiles have also been observed in GGG deposited by *e*-beam evaporation on GaAs. It is not known if the variation in composition is due to a change in the sticking coefficient of Ga with growth time, or if one element or compound is being preferentially driven from the *e*-beam source resulting in a change in flux with time. TEM analysis of GGG deposited by this manner has shown the material to be amorphous. This is further confirmed by RHEED analysis which shows no evidence of streaks or spots.

Again using *e*-beam evaporation in an MBE environment for deposition of the dielectric, $Ga_2O_3(Gd_2O_3)$/GaN depletion mode MOSFETs have been fabricated.[71,72] I-V analysis of an 8.5 nm GGG/GaN structure shows leakage currents in the range of $10^{-5}$ to $10^{-9} A/cm^2$. However, the I-V curve is not symmetrical, as shown in Fig. 10, and shows much higher leakage than is normally obtained for similar structures on GaAs. This suggests a problem with the dielectric/semiconductor interface. This is further

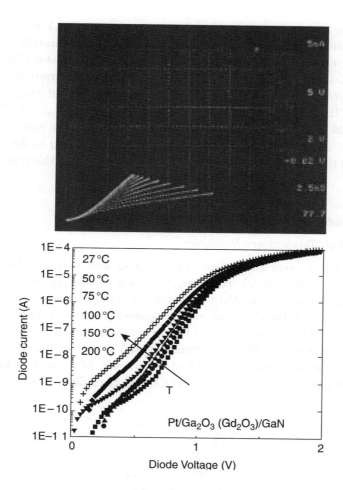

**Fig. 10.** Current–voltage characteristics of GGG/GaN MOSFET at room temperature (top), and gate characteristics (bottom) of GGG/GaN MOSFET as a function of temperature. After Ref. 72.

confirmed in the behavior of the MOSFET in that modulation of the device could not be achieved for voltages above 3 V due to gate leakage. By contrast, the breakdown field of the oxide used in the MOSFET structure was found to be excellent at 12.5 MV/cm. This shows that while the GGG/GaN interface typically contains a rather large number of traps, the dielectric itself is of high quality and contains a much lower concentration of traps.

Under reverse bias, gate breakdown voltages up to 39 V were obtained which are significantly higher than the breakdown normally obtained from a Pt/GaN Metal Semiconductor Field Effect Transistor (MESFET). The GGG/GaN device showed a maximum extrinsic transconductance of 15 mS/mm ($V_{ds}$ = 30 V), a unit current gain cut-off frequency, $f_T$, of 3.1 GHz and a maximum frequency of oscillation, $f_{max}$, of 10.3 GHz ($V_{ds}$ = 25 V and $V_{gs}$ = $-20$ V). As shown in Fig. 10, the gate characteristics are not strongly affected by temperature at least up to 200°C. This suggests that the conduction band offset is adequate over this temperature regime.

### 3.9. *Gadolinium Oxide*

#### 3.9.1. *High Temperature Deposition*

An attractive oxide alternative to both GGG and $SiO_2$ is $Gd_2O_3$ which also has a high dielectric constant, 11.4, and a bandgap of 5.3 eV.[19-21] This oxide exists in the $Mn_2O_3$ (Bixbyite) crystal structure,[73] shown in Fig. 11, which exhibits similar atomic symmetry in the (111) plane as the GaN (0001) basal plane. This similarity in symmetry offers the possibility to grow an epitaxial dielectric. Single crystal growth of $Gd_2O_3$ on GaAs has, for example, been demonstrated.[70] However, in the case of GaAs the expected orientation between the $Gd_2O_3$ and the substrate is (100) rather

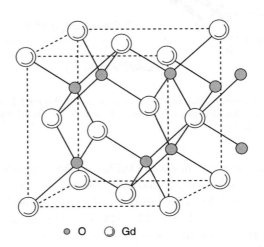

● O     ○ Gd

**Fig. 11.**   Bixbyite structure of $Gd_2O_3$. After Ref. 73.

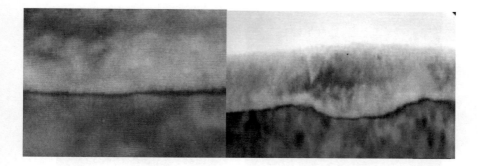

**Fig. 12.**    TEM images of $Gd_2O_3$ grown on GaN at 650°C using elemental sources. The GaN is on bottom and the $Gd_2O_3$ is on top. The dark layer between the $Gd_2O_3$ and the GaN is a thickness effect from the ion milling process used to fabricate the TEM sample. The bar length is 5.0 nm. After Refs. 66, 71.

than (111). Further, the bond length mismatch between the $Gd_2O_3$ and GaAs (100) planes is smaller than between the $Gd_2O_3$ (111) and GaN (0001), 4.2% vs. 20%. In spite of this mismatch, it has been shown that $Gd_2O_3$ can be deposited epitaxially on GaN using elemental Gd and an ECR oxygen plasma in GSMBE.[74] This is clearly demonstrated in the high resolution XTEM images shown in Fig. 12. There are dislocations visible in the image indicating that the film is highly defective. However, large areas of dislocation free material were also observed. From this image, it can be seen that the $Gd_2O_3$/GaN interface was clean. On the left side of the image, the $Gd_2O_3$ lattice can be seen in contact with the GaN lattice. A lower resolution TEM image (at right in Fig. 12) indicated that the $Gd_2O_3$ planarized the GaN quite well. In fact, the $Gd_2O_3$ filled in the void in the GaN surface with the same registry as the entire film. Thus there was no polycrystalline morphology in the void. There were however some pockets of crystal lattice that were tilted and rotated as seen in the image. This can be attributed to the dislocations arising from the single crystal material in the void meeting single crystal material on the surface. Unlike the $Ga_2O_3$, the $Gd_2O_3$ surfaces are smooth and relatively featureless, as shown in Fig. 13.

As expected, it has been shown that the growth initiation sequence is a critical step in the formation of epitaxial dielectric films on GaN. Four different procedures for $Gd_2O_3$ initiation have been tested on the $(1 \times 3)$ GaN surface produced by the *ex-situ* chemical treatment described in Sec. 2 followed by heating in vacuum to 700°C.[71] First a low temperature 300°C

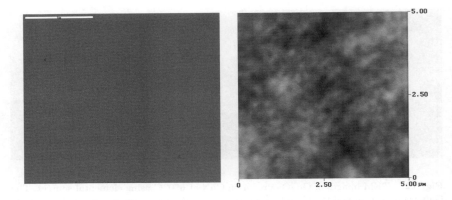

**Fig. 13.**   At left, SEM image of $Gd_2O_3$ on GaN grown at 650°C using a low temperature oxygen plasma exposure to initiate growth of the oxide. At right, AFM images of same sample. RMS roughness was approximately 0.3 nm. After Ref. 66.

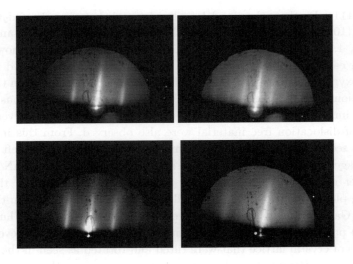

**Fig. 14.**   RHEED images of (top) a $(1 \times 3)$ GaN surface at 700°C and (bottom) the same surface at 300°C under an O plasma showing a $(1 \times 2)$.

exposure to the oxygen plasma was studied. As shown in Fig. 14, the surface changed from a sharp $(1 \times 3)$ to a hazy $(1 \times 2)$. If the $(1 \times 3)$ GaN surface was exposed to the oxygen plasma at 650°C, the surface also immediately changed to the $(1 \times 2)$ surface, then after an additional 5 minute exposure to the oxygen plasma the surface became further reconstructed to a $(4 \times 4)$

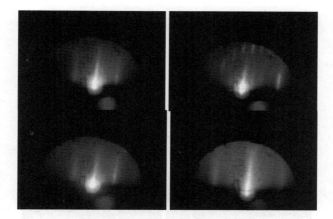

**Fig. 15.** RHEED images of (top) a GaN surface after receiving an O plasma exposure for 5 minutes at $700°$C showing a faint $(4 \times 4)$, and (bottom) a $(1 \times 1)$ $Gd_2O_3$ surface grown at $610°$C on the $(4 \times 4)$ oxygen treated surface.

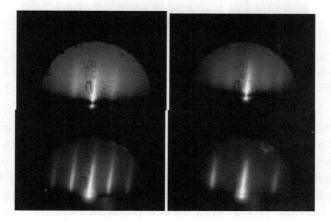

**Fig. 16.** RHEED image of (top) the GaN surface after exposure to Gd for 1 minute at $650°$C, and (bottom) a $(1 \times 1)$ $Gd_2O_3$ surface from a layer grown at $610°$C.

as shown in Fig. 15. This $(4 \times 4)$ surface remained after the oxygen plasma was turned off. When the $(1 \times 3)$ surface was exposed to a Gd atomic beam, the RHEED pattern, seen in Fig. 16, also became a $(1 \times 2)$, however the spacing of the main diffracted lines changed indicating that the spacing between the atoms changed and that a different crystal structure formed.

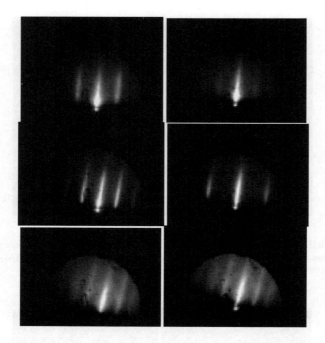

**Fig. 17.** RHEED images of (top) a $(1 \times 2)$ $300°$C oxygen treated GaN surface, (middle) a $(1 \times 1)$ $Gd_2O_3$ surface after 10 minutes of growth at $650°$C, and (bottom) a symmetric pattern after 90 minutes of growth. The $\langle 11-20 \rangle$ is at left and the $\langle 1-100 \rangle$ is at right.

Growth on the low temperature oxygen plasma exposed surface initially indicated a good $Gd_2O_3$ crystal structure as indicated by RHEED, but after 10 minutes of growth, a symmetric RHEED pattern emerged, as seen in Fig. 17, indicating that the $Gd_2O_3$ crystal quality had degraded. Reducing the growth rate produced the same symmetric pattern. Growth was also performed on the $(1 \times 2)$ GaN surface produced form the short oxygen exposure at $650°$C. Here the initial $Gd_2O_3$ RHEED pattern remained for the duration of the growth, however extra spots were seen in the RHEED images (see Fig. 18) indicating that the layer was partly polycrystalline. X-ray diffraction has shown only one peak for the $Gd_2O_3$ that coincided with the (111) peak from published JCPDS cards. This peak had a shoulder to the left side and had a full-width-at-half-maximum (FWHM) of 883 arc-seconds. The origin of the shoulder peak is unknown. X-ray diffraction on the gadolinium initiated sample and the 5 minute high temperature oxygen plasma initiated sample, seen in Fig. 19 shows a striking difference

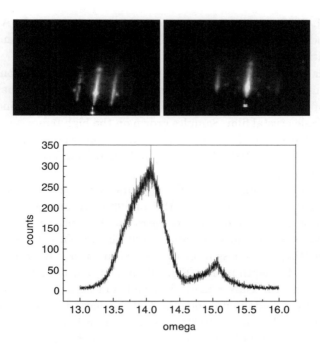

**Fig. 18.** RHEED image (top) of $Gd_2O_3$ ($1 \times 1$) grown at 650°C at 0.9 nm/min indicating extra spots due to defects. XRD pattern (bottom) from same film showing FWHM of 1623 arcseconds.

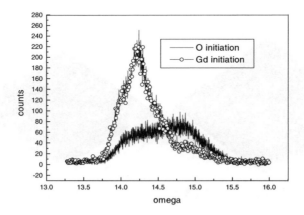

**Fig. 19.** XRD plot of samples grown on a Gd exposed GaN surface or a 5 minute high temperature O plasma exposed GaN surface. FWHM of the Gd initiated sample is 1058 arcseconds.

in FWHM, 1023 arcseconds. vs. several thousand arcseconds. As a result of this study it was determined that Gd-initiation is the best method for growth initiation.

Further growth experiments with reduced gadolinium effusion cell temperatures and a reduced substrate temperature of 605°C gave final RHEED patterns of the $Gd_2O_3$ crystal surface without the extra spots, indicating a uniform single crystal film. Samples grown on the high temperature oxygen plasma treated surface ($4 \times 4$) and the gadolinium exposed surface showed sharp $Gd_2O_3$ crystal RHEED patterns and smooth surfaces on the order of 0.5 nm RMS roughness.

The thermal stability of $Gd_2O_3$ appears to be quite good. A sample which showed an initial X-ray FWHM of 883 arcseconds was annealed at 1000°C for 30 seconds under flowing nitrogen. The surface RMS roughness of the annealed sample was 0.60 nm seen in Fig. 20. This is less than a 10% increase in RMS roughness from the as grown surface RMS roughness of 0.56 nm. An AES depth profile (Fig. 21) showed that the interface between the $Gd_2O_3$ and the GaN remained abrupt and there was no diffusion of either oxygen or gadolinium into the GaN surface. This showed that the interface was stable up to 1000°C. X-ray diffraction from this annealed sample gave a FWHM of 789 arcseconds as seen in Fig. 22. This was almost 100 arcseconds less than the as grown sample. This indicated that the 1000°C anneal improved the crystal quality without degrading the interface or surface. Thus it appears that, like the GGG, the $Gd_2O_3$/GaN interface is

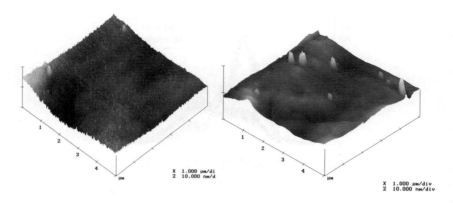

**Fig. 20.** AFM of $Gd_2O_3$ on GaN grown at 650°C before and after receiving a 1000°C anneal under nitrogen. RMS roughnesses were 0.56 and 0.60 nm for the as-grown and annealed layers respectively.

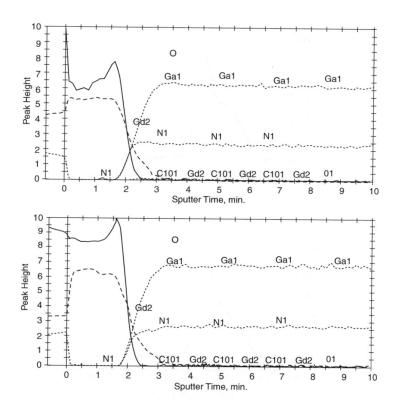

**Fig. 21.** AES plot of $Gd_2O_3$ grown on GaN at 650°C from elemental sources before (top) and after (bottom) annealing at 1000°C under nitrogen.

thermally stable and should be able to withstand even the most demanding device processing.

Electrical characterization of the $Gd_2O_3$/GaN showed evidence of severe leakage with a breakdown field of $\sim 0.1$ MV/cm. This is most likely due to the dislocations and tilt boundaries which can act as leakage paths and cause the breakdown voltage to be low. In order to improve the breakdown, it was necessary to deposit amorphous $SiO_2$ on top of the $Gd_2O_3$. This improved the breakdown field of the oxide to 0.8 MV/cm. With the additional 300 Å of $SiO_2$ on the $Gd_2O_3$/GaN MOSFET structure, the gate reverse leakage current was $\sim 10$ pA at a gate-source bias ($V_{GS}$) of $-10$ V, as shown in Fig. 23.[72] The reverse leakage current remained below 10 nA past $V_{GS} = -70$ V, clearly demonstrating the benefit of the insulated gate MOS structure. A comparison of the GGG/GaN and $SiO_2$/$Gd_2O_3$/GaN

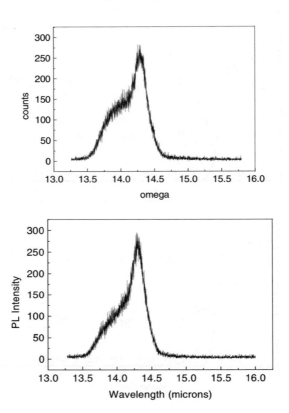

**Fig. 22.** X-ray diffraction plot of $Gd_2O_3$ grown at $650°C$ before (top) and after (bottom) a 30 second $1000°C$ anneal in nitrogen. The peak FWHM are 883 (top) and 789 (bottom) arcseconds.

leakage currents can be seen in Fig. 23. The $D_{it}$ of the $Gd_2O_3/SiO_2$ stack was measured to be $3 \times 10^{11}$ cm$^{-2}$ eV$^{-1}$ using the Terman method. The $Gd_2O_3/GaN$ interface appears to contain fewer traps, and thus exhibits less leakage, than the GGG/GaN device suggesting that the crystalline dielectric provides a better interface to the GaN. However, the superior breakdown field of the GGG clearly indicates that the defects in the crystalline material are deleterious to the performance of the dielectric.

Further evidence of the quality of the $Gd_2O_3/GaN$ interface and of the improvement in device performance provided by the $SiO_2$ overlayer can be seen in the DC output characteristics of a $1 \times 200$ $\mu m^2$ gate dimension $SiO_2/Gd_2O_3/GaN$ MOSFET,[72] shown in Fig. 24. Without the $SiO_2$,

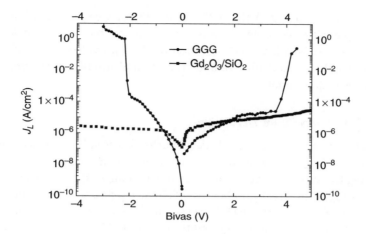

**Fig. 23.** Gate leakage current density for GGG/GaN and $SiO_2$/$Gd_2O_3$/GaN structures. After Ref. 71.

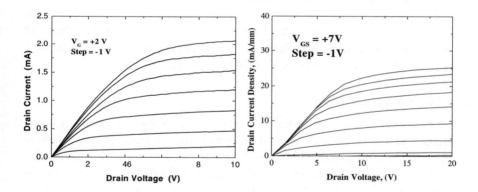

**Fig. 24.** IV of $Gd_2O_3$/GaN MOSFET without (at left) and with (at right) $SiO_2$. After Ref. 72.

modulation could only be achieved at forward voltages up to 2 V. However, the addition of $SiO_2$ allowed for modulation at gate voltages up to +7 V for the common-source MOSFET, clearly indicating the excellent quality of $Gd_2O_3$/GaN interface. A maximum intrinsic transconductance of 61 mS/mm was measured at $V_{GS} = -0.5$ V and $V_{DS} = 20$ V. Similar $1 \times 200$ $\mu$m devices from the same wafer showed source-drain breakdown greater than 80 V. The slightly high output conductance ($\sim 0.25$ mS/mm) of the drain IV in Fig. 24 was caused by short channel effects due to mid-$10^{15}$ cm$^{-3}$ doping in the buffer layer. The device had high knee voltage and parasitic resistance because the ohmic contact was annealed only to $\sim 300°$C. Although the O$^+$ isolation implant provided a good sheet resistivity for the as-implanted sample, the thermal stability of the implanted samples was quite poor.

### 3.9.2. *Low Temperature Deposition*

In an attempt to eliminate the leakage paths observed in the defective high temperature Gd$_2$O$_3$, low deposition temperatures have been investigated.[76] Reducing the substrate temperature to 100°C changes the RHEED patterns dramatically from streaky, indicative of single crystal, to hazy, indicative of amorphous. Post-growth XRD, however, shows that the layers are not amorphous but in fact consist of crystallites oriented along the (111) and (321) Gd$_2$O$_3$ planes. This microstructure resulted in significant improvement in the breakdown field to 3 MV/cm. Interface state density as determined from C-V analysis using the Terman method indicates a $D_{it}$ of $6 \times 10^{11}$ eV$^{-1}$cm$^{-2}$ at $E_c - 0.5$ eV. Unfortunately, the thermal stability of this material is not acceptable. Annealing at 1000°C for 30 seconds shows strong evidence of recrystallization. After annealing, the electrical properties are quite similar to those of the Gd$_2$O$_3$ deposited at high temperature.

### 3.10. *Scandium Oxide*

Sc$_2$O$_3$ exists in the same Bixbyite structure as Gd$_2$O$_3$ but has a much smaller lattice constant, see Table 1, which should make it less defective when deposited on GaN.[22,24] Like Gd$_2$O$_3$ it has a high melting point suggesting good thermal stability.[22] The material is often used as an optical coating for high power photonic devices and has a bandgap of 6.3 eV.[22] Based upon the reported work function of 4 eV[22] it is likely that the electron affinity is $\sim 0.85$ eV. If this estimate is correct then, unlike Gd$_2$O$_3$, the band offsets for Sc$_2$O$_3$ should allow fabrication of devices on both $n$- and $p$-type material. The dielectric constant is also favorable as it is slightly larger than that of Gd$_2$O$_3$, 14 vs. 11.

Similar to the Gd$_2$O$_3$, Sc$_2$O$_3$ deposited by GSMBE using elemental Sc and an ECR oxygen plasma also produces extremely smooth surfaces as indicated from AFM and SEM.[75,76] RMS roughnesses of 0.5 to 0.8 nm were seen for substrate temperatures of 100°C or 600°C, respectively. Using AES depth profiling and surface scans, it was shown that the MBE derived Sc$_2$O$_3$ was of uniform concentration throughout the film, as shown in Fig. 25. The

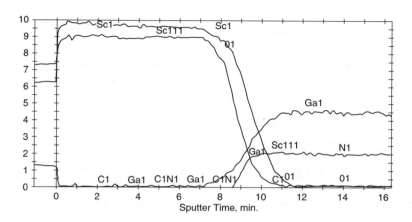

**Fig. 25.** AES depth profile of $Sc_2O_3$ on GaN. The conditions of the oxide growth were $T_{Sc} = 1130°C$ and $T_{sub} = 100°C$. After Ref. 75.

O : Sc ratio, as measured from comparing Auger peak-to-peak heights, was 0.85 to 0.90 and was not strongly dependent on either substrate or Sc cell temperature. The growth rate of the scandium oxide was 0.67 nm/min for $T_{Sc} = 1130°C$ and 1.25 nm/min for $T_{Sc} = 1170°C$, and again was found to be independent of substrate temperature.

From HXRD results, it was found that for the higher growth temperatures, the scandium oxide (111) plane grew parallel to the surface (0001) GaN as was also seen in the high temperature $Gd_2O_3$ films grown on GaN.[75] The 20% mismatch between the (111) $Gd_2O_3$ and the GaN (0001), leads to a highly defective single crystal layer, resulting in a full-width-at-half-maximum (FWHM) of over 800 arcseconds in HXRD scans. The smaller mismatch between the $Sc_2O_3$ and the GaN (0001) should lead to a lower defect density and in fact the FWHM is substantially reduced at 514 arcseconds as compared to the 883 arcseconds found in the best $Gd_2O_3$, shown in Fig. 26. From the RHEED data, it appears that reducing the substrate temperature changes the film microstructure from defective single crystal to amorphous or poly-crystalline after the first few minutes of growth, just as was the case for the $Gd_2O_3$. However, it is not known if the initial $Sc_2O_3$ material deposited at low temperature is single crystal.

Diodes fabricated from 40 nm of $Sc_2O_3$ grown on $n$-GaN at 600°C show a forward breakdown field of only 0.7 MV/cm, as shown in Table 4.[76] While this represents an improvement over the $Gd_2O_3$, the leakage is still so

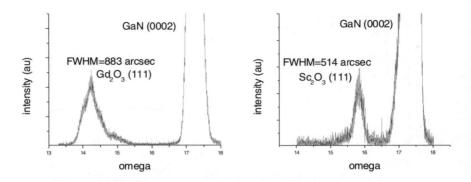

**Fig. 26.** HXRD scans of dielectrics grown at 600°C on GaN: (at left) $Gd_2O_3$, (at right) $Sc_2O_3$. After Ref. 77.

**Table 4.** Interface state density, $D_{it}$, calculated using the Terman method, and breakdown field at 5 mA/cm$^2$, $V_{BD}$, for various dielectrics grown on $n$-GaN by GSMBE using an ECR oxygen plasma.

| Dielectric | $D_{it}$ (eV$^{-1}$cm$^{-2}$) | $V_{BD}$ (MV/cm) |
|---|---|---|
| HT-$Gd_2O_3$ | $3 \times 10^{11}$ | 0.1 |
| LT-$Gd_2O_3$ | $6 \times 10^{11}$ | 3.0 |
| HT-$Sc_2O_3$ | NA | 0.7 |
| LT-$Sc_2O_3$ | $4 \times 10^{11}$ | 1.5 |
| LT-MgO | $4 \times 10^{11}$ | 1.0 |

severe that further electrical characterization is not possible without the use of some form of cap. This leakage can be reduced by annealing at 1000°C in nitrogen, which improves the breakdown field to 1.4 MV/cm. However, C-V analysis shows that the anneal severely degrades the interface, producing $D_{it}$ values of $\sim 10^{13}$ cm$^{-2}$ eV$^{-1}$. As with the $Gd_2O_3$, low deposition temperatures did improve the breakdown field to $\sim 1.5$ MV/cm. The $D_{it}$ for the low temperature material was found using the Terman method to be $\sim 4 \times 10^{11}$ eV$^{-1}$cm$^{-2}$ near the conduction band edge, again similar to that observed for $Gd_2O_3$.

While ECR plasmas are expected to produce the lowest ion energies and hence should produce the least amount of damage to the GaN surface, RF plasmas are expected to produce much more intense plasmas in terms of active oxygen species. Using the RF approach at low deposition temperatures, the amorphous/poly-crystalline $Sc_2O_3$ shows slightly less leakage as deposited, with a breakdown field of $\sim 1.9$ MV/cm, as

**Table 5.** Forward breakdown field and interface state density calculated using the AC conductance method ($E_c - E_t = 0.7 - 0.8$ eV) for dielectrics deposited at $100°$C on $n$-GaN using either an ECR or an RF plasma.

| | ECR | | RF | |
|---|---|---|---|---|
| | $D_{\text{it}}$ (eV$^{-1}$cm$^{-2}$) | $V_{\text{BD}}$ (MV/cm) | $D_{\text{it}}$ (eV$^{-1}$cm$^{-2}$) | $V_{\text{BD}}$ (MV/cm) |
| LT-Sc$_2$O$_3$ | $8 \times 10^{11}$ | 1.5 | $9 - 11 \times 10^{11}$ | 1.9 |
| LT-MgO | $2 \times 10^{11}$ | 1.2 | $2 - 3 \times 10^{11}$ | 2.3 |

shown in Table 5. C-V analysis showed modulation from accumulation to deep depletion. Deep depletion with no inversion capacitance is typical for wide-gap semiconductor MIS or MOS structures due to the slow generation rate of the minority carrier at room temperature.[78,79] A significant flat band voltage shift was observed, which indicated the existence of fixed charges in the oxide layer. From C-V analysis the interface state density was calculated using the Terman and AC conductance methods.[80,81] The values determined were $5 \times 10^{11}$ eV$^{-1}$ cm$^{-2}$ at $E_c - E_t = 0.2$ eV (Terman) and $8 \times 10^{11}$ eV$^{-1}$ cm$^{-2}$ at 0.7 eV (AC conductance). Additional AC conductance measurements conducted at $300°$C showed an interface state density of $1.11 \times 10^{12}$/eV cm$^2$ at $E_c - E_t = 0.42$ eV. No information is yet available on the thermal stability of low temperature Sc$_2$O$_3$.

### 3.11. *MgO*

MgO is a rock salt dielectric which has been explored as an intermediate buffer layer for growth of ferroelectric materials on semiconductors[82,83] or as a potential gate dielectric for GaAs[84-85] or Si.[83] While MgO deposition by MBE has been successfully demonstrated by a number of groups, the crystal quality of the films deposited on GaAs and Si has been poor due to the large lattice mismatch between the MgO and the semiconductor substrate. GaN has a smaller lattice constant than GaAs and is thus a much closer match to MgO.[23,24] An additional advantage of this system is the large bandgap and thus large band offsets that are expected relative to either an $n$- or $p$-type semiconductor. Further, the dielectric constant for MgO, 9.8, is substantially higher than for SiO$_2$.

Because of the high Mg vapor pressure, MgO cannot be grown at the same MBE growth temperatures used for the Bixbyite oxides. Above $\sim 350°$C the growth rate drops precipitously with growth temperature.[86] At a substrate temperature of $100°$C, the RHEED pattern remains streaky for

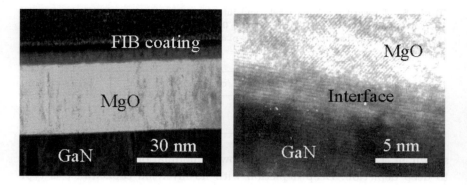

**Fig. 27.** XTEM images of MgO/$n$-GaN structure grown using an RF oxygen plasma.

a few monolayers then changes to a pattern similar to that observed for low temperature growth of $Gd_2O_3$. At 350°C the same transition is observed, though the final RHEED pattern appears to be more poly-crystalline than at 100°C, and the film seems to be more textured with an apparent [111] axis perpendicular to the substrate. XTEM of the MgO/GaN interface shows evidence of a thin single crystal layer at the interface in agreement with the RHEED, as shown in Fig. 27. In some areas this epitaxial region appears to be as thick as 40–50 monolayers, while in other areas it is substantially thinner.

Diodes with 100 nm MgO gate dielectrics grown using an ECR plasma show reverse breakdown and forward turn-on voltages of $> 40$ V and $> 10$ V, where these parameters are defined as the values at 100 nA and 5 mA/cm$^2$, respectively). A forward breakdown field of 1.2 MV/cm for the MgO was calculated from this data. The measured C-V curve at a frequency of 1 MHz and sweep-rate of 100 mV/s showed a clear deep depletion behavior for negative bias voltage and no measurable hysteresis was observed. The interface state density was calculated using the Terman method to be $4 \times 10^{11}$ eV$^{-1}$cm$^{-2}$ at 0.3 eV below the conduction band edge. The conductance technique was also used to characterize the MgO/GaN MOS capacitors, and an interface state density of $2 \times 10^{11}$/eVcm$^2$ at $E_c - E_t = 0.7$ eV was obtained. The AC conductance technique is not able to measure interface traps with a time constant much longer than the period of the applied AC signal, because these traps cannot respond to the AC perturbation. Therefore, high temperature AC conductance analysis was carried out over a temperature range of 25 to 300°C to investigate the traps with longer

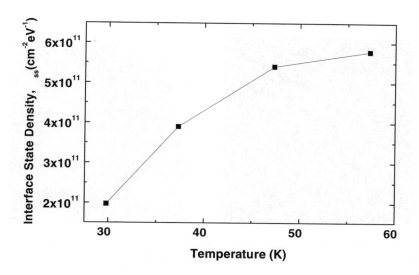

**Fig. 28.** The interface trap density in an ECR grown MgO/$n$-GaN diode measured at different temperatures using the AC conductance method. After Ref. 87.

time constants. Using this approach, the interface trap density was found to increase linearly from room temperature to 200°C and begin to level off at 300°C, as shown in Fig. 28. The interface trap density was almost three times larger at 300°C as compared to that at room temperature. This may have significant impact on the MOSFET performance at elevated temperatures. In general, however, the performance of the MgO dielectric appears to be superior to that of the Bixbyite oxides. This is most likely due to the reduced mismatch between the MgO and the GaN relative to the other oxides and to the larger bandgap of this material.

As with the $Sc_2O_3$, both ECR and RF plasmas have been explored for the growth of MgO dielectrics. In the case of the MgO, the RF appears to produce substantial improvement in both morphology, as shown in Fig. 29, and in the breakdown field, as shown in Table 6. The interface state density is only slightly higher with the RF-grown material, suggesting that ion damage from the RF is not occurring. The higher breakdown field is probably related to the improved morphology. The reason for the improvement in morphology is still under investigation, however it most likely stems from a higher concentration of active oxygen species at the growth surface.

As discussed previously, a clear demonstration of surface inversion has proven elusive in GaN due to the very low minority carrier generation rate

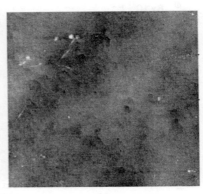

**Fig. 29.**  AFM images of MgO grown on $n$-GaN using ECR (at left) or RF (at right) oxygen plasma sources. The scale is 10 $\mu$m. After Ref. 88.

in GaN at room temperature. Even at 300°C in conventional GaN MOS devices, the generation rate is still too low to observe inversion. To overcome a similar problem in SiC MOS devices, the $n^+p$ junction of a MOS gate-controlled devices was employed as an external source of inversion charge.[89] A similar approach has been used to observe inversion in MgO/$p$-GaN gate-controlled MOS diodes in which $n^+$ gated contact regions were created by Si$^+$ implantation and subsequent activation annealing with the MgO gate oxide in place. The MgO was deposited by MBE at 100°C on a $p$-GaN($p \sim 3 \times 10^{17}$ cm$^{-3}$ at 25°C) layer grown by metal organic chemical vapor deposition on sapphire. The MgO was grown using elemental Mg and RF plasma-activated oxygen to a thickness of $\sim 80$ nm. For diode fabrication, implantation of 70, 195, 380 keV $^{29}$Si$^+$ ions at a dose of $2 \times 10^{13}$, $6 \times 10^{13}$, $1.8 \times 10^{14}$ cm$^{-2}$, followed by high temperature($\sim 950$°C) activation annealing, was used to create the $n^+$ gate-contact regions. The reverse breakdown of the diode with as-deposited dielectric was $\sim 12$ V, with a forward turn-on voltage of $\sim 10$ V. Figure 30 shows C-V characteristics of the MgO/GaN MOS-controlled diodes at 25°C in the dark as a function of the measurement frequency. In each case, $-20$ V was applied at the gated contact to provide a source of minority carriers. The frequency dispersion observed in inversion is due to the resistance of the inversion channel, as reported for $n^+p$ SiC MOS gate-controlled diodes.[89] At 5 KHz measurement frequency, only deep depletion is seen since the characteristics are dominated by majority carriers.[90] As the frequency is decreased, a

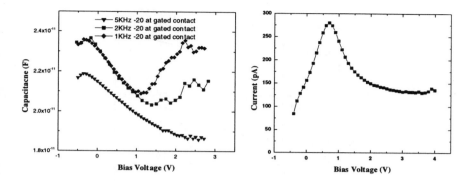

**Fig. 30.** C-V and I-V characteristics of MgO/$n$-GaN gate controlled diode. After Ref. 92.

clear inversion behavior is observed due to charge flow into and from the n+ regions external to the gate. Figure 30 also shows I-V characteristics from the diodes at 25°C in the dark. For the accumulation region at negative bias, the reverse current is mainly due to diffusion and generation in the depletion region. As the bias is moved to positive values, the current increases due to the presence of two additional components, namely the additional depletion region under the gate provides more gate current and the surface generation current increases. As the bias is further increased to the inversion region, this surface generation component is suppressed, leaving only current due to generation in the depletion region. This is the classical behavior for a gate-controlled diode.[91] In light of this result the prospect for realization of enhancement-mode GaN MOS transistors appears to be quite good.

## 4. MOSHFETs

Many of the same dielectrics studied for use in GaN MOS(MIS)FETs have also been tried in Insulated Gate Heterostructure Field Effect Transistors (IGHFETs) or Metal Oxide Semiconductor Heterostructure Field Effect Transistors (MOSHFETs). This approach combines the low gate leakage current feature of the MOSFET and the superior channel mobility of the HFET. In addition, if the gate dielectric is of high quality, this type of structure should allow the use of higher forward gate voltages to increase the carrier concentration in the channel and hence the current carrying capability of the device. AlN, $GaO_x$ and $SiO_2$ have all been investigated for this application. The AlN IG-HFET structure was grown at 990°C, forming

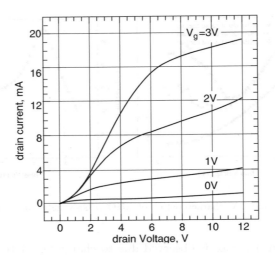

**Fig. 31.** Drain current vs. drain voltage for various gate voltages ($L_g$ = 1.4 $\mu$m, $W_g$ = 40 $\mu$m, and source-drain regrown layer separation is 3.6 $\mu$m). After Ref. 93.

a single crystal film of 4.0 nm.[93] In this device, the AlGaN/GaN structure is inverted so that the AlN gate can sit directly on the GaN channel. This device operated in enhancement mode and had a pinch-off voltage of 0 V as shown in Fig. 31. The gate leakage was found to be quite high, and to decrease with decreasing gate length. In spite of severe leakage in this device, the transconductance reached 235 mS/mm for a 1.4 micron gate at a gate voltage of 2 V.

To date the most extensive work on MOSHFETs has been done by Khan *et. al.* using $SiO_2$ as the dielectric.[94-97] In this structure, the $SiO_2$ is deposited to a thickness of 7 − 10 nm using PECVD. As shown in Fig. 32, this allows modulation of the gate at forward voltages up to 9 V.[95] The gate leakage in this device was ∼six orders of magnitude lower than for comparable HFET devices. Even at elevated temperatures, the performance advantage of the MOSHFET can still be clearly seen, as the leakage current remains roughly four orders of magnitude lower than for HFETs.[97]

Photoelectrochemical formation of $Al_xGa_{2-x}O_3$ from AlGaN has also recently been used to fabricate MOSHFET devices.[98] The oxide formed from partially oxidizing the $Al_{0.2}Ga_{0.8}N$ cap layer is found by AES to be richer in Al than the underlying semiconductor from which it is made. Since $Al_2O_3$ has a much larger bandgap than $Ga_2O_3$ (7.5 eV vs.4.4 eV) this enrichment is expected to produce better confinement than observed in

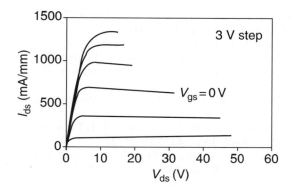

**Fig. 32.** Room temperature I-V characteristics of a MOSHFET at room temperature, 2 $\mu$m gate length. After Ref. 95.

the $Ga_2O_3$ MOS diodes discussed earlier. When compared to conventional HFETs prepared from the same base material, the oxidized $Al_xGa_{2-x}O_3$ devices showed higher transconductance and lower gate currents as expected. When compared to MOSHFETs fabricated using an optimized PECVD $SiO_2$ deposition process on the same base material, the two approaches show similar values for transconductance and gate leakage though the $SiO_2$ has a much higher turn-on voltage due at least in part to the low $SiO_2$ dielectric constant. If the PEC process can be made compatible with optimized HFET processing, it may prove to be an attractive approach for MOSHFET fabrication.

## References

1. S. Strite and H. Morkoc, *J. Vac. Sci. Technol.* **B10**, 1237 (1992) and references therein.
2. R. F. Davis, *Proc. IEEE* **79**, 702 (1991), and references therein.
3. H. Okumura, S. Misawa and S. Yoshida, *Appl. Phys. Lett.* **59**, 1058 (1991).
4. S. Strite, J. Ruan, Z. Li, A. Salvador, H. Chen, D. J. Smith, W. J. Choyke and H. Morkoc, *J. Vac. Technol.* **B9**, 1924 (1991).
5. B. J. Baliga, *IEEE Electron. Device Lett.* **10**, 455 (1989).
6. T. P. Chow and R. Tyagi, *IEEE Trans. Electron. Devices* **41**, 1481 (1994).
7. R. J. Trew, M. W. Shin and V. Gatto, *Solid-State Electron.* **41**, 1561 (1997).
8. M. N. Yoder, *IEEE Trans. Electron. Devices* **43**, 1633 (1996).
9. A. D. Bykhovski, B. I. Gelmont and M. S. Shur, *J. Appl. Phys.* **81**, 6332 (1997).
10. P. M. Asbeck, E. T. Yu, S. S. Lau, G. J. Sullivan, J. Van Hove and J. Redwing, *Electron. Lett.* **33**, 1230 (1997).

11. *Properties of Group III Nitrides*, ed. J. H. Edgar (Inspec, the Institution of Electrical Engineers, London, UK, 1994).
12. G. W. Kaye and T. H. Laby, *Tables of Physical and Chemical Constants and Some Mathematical Functions*, 11th edition (Longmans, New York, 1956).
13. C. T. Sah, *Fundamentals of Solid-State Electronics* (World Scientific, New Jersey, 1991).
14. S. M. Sze, *Physics of Semiconductor Devices*, 2nd edition (Wiley, New York, 1981).
15. W. M. Yim, E. J. Stofko, P. J. Zanzucchi, J. I. Pankove, M. Ettenburg and S. L. Gilbert, *J. Appl. Phys.* **44**, 292 (1973).
16. W. L. Chin, T. L. Tansley and T. Osotchan, *J. Appl. Phys.* **75**, 7365 (1994).
17. M. S. Shur and M. A. Khan, *Mat. Res. Bull.* **22** (2), 44 (1997).
18. Q. Xiao and J. J. Derby, *J. Cryst. Growth* **139**, 147 (1994).
19. S. S. Derbeneva and S. S. Batsano, *Dokl. Chem.* **175**, 710 (1967).
20. S. S. Batsono and E. V. Dulepov, *Sov. Phys. — Solid State* **7** (4), 995 (1965).
21. K. A. Gschneider, *Izd. Mir.* (1965).
22. V. N. Abramov, A. N. Ermoshkin and A. I. Kuznetsov, *Sov. Phys. Sol. State* **25**, 981 (1983).
23. N. Daude, C. Jouanin and C. Gout, *Phys. Rev.* **B15**, 2399 (1977).
24. G. V. Samsonov, *The Oxide Handbook* (Plenum, New York, 1973) and references therein.
25. T. S. Lay, M. Hong, J. Kwo, J. P. Mannaerts, W. H. Hung and D. J. Huang, *Mat. Res. Soc. Symp. Proc.* **573**, 131 (1999).
26. H. Ishikawa, S. Kobayashi, Y. Koide, S. Yamasaki, S. Nagai, J. Umezaki and M. Koike, *J. Appl. Phys.* **81**, 1315 (1997).
27. K. Prabhakran, T. G. Anderson and K. Nozawa, *Appl. Phys. Lett.* **69**, 3212 (1996).
28. J. L. Lee, J. K. Kim, J. W. Lee, Y. J. Park and T. Kim, *Sol. State Elec.* **43**, 435 (1999).
29. V. M. Bermudez, *J. Appl. Phys.* **80**, 1190 (1996).
30. N. V. Edwards, M. D. Bremser, T. W. Weeks, R. S. Kern, R. F. Davis and D. E. Aspnes, *Appl. Phys. Lett.* **69**, 2065 (1996).
31. K. N. Lee, S. M. Donovan, B. Gila, M. Overberg, J. D. MacKenzie and C. R. Abernathy, *J. Electrochem. Soc.* **99**, 241 (1999).
32. L. L. Smith, S. W. King, R. J. Nemanich and R. F. Davis, *J. Electron. Mat.* **25**, 805 (1996).
33. S. W. King, L. L. Smith, J. P. Barnak, J. H. Ku, J. A. Christman, M. C. Benjamin, M. D. Bremser, R. J. Nemanich and R. F. Davis, *Mater. Res. Soc. Symp. Proc.* **395**, 739 (1996).
34. S. W. King, J. P. Barnak, M. D. Bremser, K. M. Tracy, C. Ronning, R. F. Davis and R. J. Nemanich, *J. Appl. Phys.* **84**, 5248 (1998).
35. P. M. Asbeck, E. T. Yu, S. S. Lau, G. J. Sullivan, J. Van Hove and J. M. Redwing, *Electron. Lett.* **33**, 1230 (1997).
36. S. C. Binari, W. Kruppe, H. B. Dietrich, G. Kelner, A. E. Wickenden and J. A. Freitas *Solid State Electron.* **41**, 1549 (1997).
37. M. S. Shur, *Mat. Res. Soc. Symp. Proc.* **483**, 15 (1998).

38. R. Gaska, Q. Chen, J. Yang, A. Osinsky, M. A. Khan and M. S. Shur, *IEEE Electron. Dev. Lett.* **18**, 492 (1997).

39. M. A. Khan, Q. Chen, C. J. Sun, M. S. Shur and B. Gelmark, *Appl. Phys. Lett.* **67**, 1429 (1995).

40. Y.-F. Wu, B. P. Keller, P. Fini, S. Keller, T. J. Jenkins, L. T. Kenias, S. P. DenBaars and U. K. Mishra, *IEEE Electron. Dev. Lett.* **19**, 50 (1998).

41. Y. F. Wu, B. P. Keller, S. Keller, D. Kapolneck, P. Kozodoy, S. P. DenBaars and U. K. Mishra, *Appl. Phys. Lett.* **69**, 1438 (1996).

42. G. J. Sullivan, M. Y. Chen, J. A. Higgins, J. W. Yang, Q. Chen, R. C. Pierson and B. T. McDermott, *IEEE Electron. Dev. Lett.* **19**, 198 (1998).

43. A. T. Ping, Q. Chen, J. W. Yang, M. A. Khan and I. Adesida, *IEEE Electron. Dev. Lett.* **19**, 54 (1998).

44. O. Aktas, Z. F. Fan, A. Botchkarev, S. N. Mohammad, M. Roth, T. Jenkins, L. Kehias and H. Morkoc, *IEEE Electron. Dev. Lett.* **18**, 293 (1997).

45. H. C. Casey Jr., G. G. Fountain, R. G. Alley, B. P. Keller and S. P. Denbaars, *Appl. Phys. Lett.* **68**, 1850 (1996).

46. M. Sawada, T. Sawada, Y. Yamagata, K. Imai, H. Kumura, M. Yoshino, K. Iizuka and H. Tomozawa, *Proc. Second Int. Conf. Nitride Semiconductors, Tokushima* 482 (1997).

47. S. Arulkumaran, T. Egawa, H. Ishikawa, T. Jimbo and M. Umeno, *Appl. Phys. Lett.* **73**, 809 (1998).

48. P. Chen, Y. G. Zhou, H. M. Bu, W. P. Li, Z. Z. Chen, B. Shen, R. Zhang and Y. D. Zhang, *Mat. Res. Soc. Symp. Proc.* **639**, G11.24 (2001).

49. M. Hong, *Proc. Mat. Res. Soc. Symp.* (2000).

50. K. Matocha, T. P. Chow and R. J. Gutmann, *IEEE Electron. Device Lett.* **23**, 79 (2002).

51. R. Therrien, G. Lucovsky and R. F. Davis, *Phys. Stat. Sol.* **A176**, 793 (1999).

52. S. C. Binari, K. Doverspike, G. Kelner, H. B. Dietrich and A. E. Wickenden, *Solid State Electron.* **41**, 177 (1997).

53. B. Gaffey, L. G. Guido, X. W. Wang and T. P. Ma, *IEEE Trans. Electron. Device* **48**, 458 (2001).

54. F. Ren, C. R. Abernathy, J. D. MacKenzie, B. P. Gila, S. J. Pearton, M. Hong, M. A. Marcus, M. J. Schurman, A. G. Baca and R. J. Shul, *Solid State Electron.* **42**, 2177 (1998).

55. H. Kawai, M. Hara, F. Nakamura, T. Asatsuma, T. Kobayashi and S. Imanaga, *J. Crystal Growth* **189/190**, 738 (1998).

56. E. Alekseev, A. Eisenbach and D. Pavlidis, *Electron. Lett.* **35**, (1999).

57. T. Hashizume, E. Alekseev, D. Pavlidis, K. S. Boutros and J. Redwing, *J. Appl. Phys.* **88**, 1983 (2000).

58. J. N. Kidder, Jr., J. S. Kuo, T. P. Pearsall and J. W. Rogers, Jr., *Mater. Res. Soc. Symp. Proc.* **395**, 249 (1996).

59. K. K. Harris, B. P. Gila, J. Deroaches, K. N. Lee, J. D. MacKenzie, C. R. Abernathy, F. Ren and S. J. Pearton, *J. Electrochem. Soc.* **149**, G128 (2002).

60. S. D. Wolter, B. P. Luther, D. L. Waltemyer, C. Onneby, S. E. Mohney and R. J. Molnar, *Appl. Phys. Lett.* **70**, 2156 (1997).

61. S. D. Wolter, S. E. Mohney, H. Venugopalan, A. E. Wickenden and D. D. Koleske, *J. Electrochem. Soc.* **145**, 629 (1998).

62. E. D. Readinger, S. D. Wolter D. L. Waltemyer, J. M. Delucca, S. E. Mohney, B. I. Prenitzer, L. A. Gainnuzzi and R. J. Molnar, *J. Electron. Mater.* **28**, 257 (1999).

63. T. Rotter, D. Mistele, J. Stemmer, F. Fedler, J. Aderhold, J. Graul, V. Schwegler, C. Kirchner, M. Kamp and M. Heuken, *Appl. Phys. Lett.* **76** (26), 3923 (2000).

64. D. J. Fu, Y. H. Kwon, T. W. Kang, C. J. Park, K. H. Baek, H. Y. Cho, D. H. Shin, C. H. Lee and K. S. Chung, *Appl. Phys. Lett.* **80** (3), 446 (2002).

65. H. Kim, S. J. Park and H. Hwang, *J. Vac. Sci. Technol.* **B19** (2), 579 (2001).

66. B. P. Gila, K. N. Lee, J. LaRoche, F. Ren, S. M. Donovan and J. Han, *Mat. Res. Soc. Symp. Proc.* **573**, 247 (1999).

67. Y. C. Wang, M. Hong, J. M. Kuo, J. P. Mannaerts, J. Kwo, H. S. Tsai, J. J. Krajewski, J. S. Weiner, Y. K. Chen and A. Y. Cho, *Mater. Res. Soc. Symp. Proc.* **573**, 219 (1999).

68. J. Kwo, M. Hong, A. R. Kortan, D. W. Murphy, J. P. Mannaerts, A. M. Sergent, Y. C. Wang and K. C. Hsieh, *Mater. Res. Soc. Symp. Proc.* **573**, 57 (1999).

69. T. S. Lay, M. Hong, J. Kwo, J. P. Mannaerts, W. H. Hung and D. J. Huang, *Mater. Res. Soc. Symp. Proc.* **573**, 131 (1999).

70. A. R. Kortan, M. Hong, J. Kwo, J. P. Mannaerts and N. Kopylov, *Mater. Res. Soc. Symp. Proc.* **573**, 21 (1999).

71. B. P. Gila, K. N. Lee, W. Johnson, F. Ren, C. R. Abernathy, S. J. Pearton, M. Hong, J. Kwo, J. P. Mannaerts and K. A. Anselm, *Proc. Cornell Conf.* (2000).

72. J. W. Johnson, B. Lou, F. Ren, B. P. Gila, W. Krishnamourthy, C. R. Abernathy, S. J. Pearton, J. I. Chyi, T. E. Nee, C. M. Lee and C. C. Chuo, *Appl. Phys. Lett.* (2000).

73. F. Ren, M. Hong, S. N. G. Chu, M. A. Marcus, M. J. Schurman, A. Baca, S. J. Pearton and C. R. Abernathy, *Appl. Phys. Lett.* **73**, 3893 (1998).

74. R. G. Haire and L. Eyring, *Handbook on the Physics and Chemistry of Rare Earths* **18**, 429 (1994).

75. B. P. Gila, J. W. Johnson, K. N. Lee, V. Krishnamoorthy, C. R. Abernathy, F. Ren and S. J. Pearton, *Proc. Electrochem. Soc.* **1**, 71 (2001).

76. B. P. Gila, J. W. Johnson, R. Mehandru, B. Luo, A. H. Onstine, K. K. Allums, V. Krishnamoorthy, S. Bates, C. R. Abernathy, F. Ren and S. J. Pearton, *Phys. Stat. Sol.* **188** (1), 239–242 (2001).

77. R. Mehandru, B. P. Gila, J. Kim, J. W. Johnson, K. P. Lee, B. Luo, A. H. Onstine, C. R. Abernathy, S. J. Pearton and F. Ren, *Mat. Res. Soc. Symp. Proc.* (2001).

78. T. Hashizume, E. Alekseev and D. Pavlidis, *J. Appl. Phys.* **88** (4), 1983 (2000).

79. S. Berberich, P. Godignon, M. L. Locatelli, J. Millan and H. L. Hartnagel, *Solid State Electron.* **42** (6), 915 (1998).

80. E. H. Nicollian and J. R. Brews, *MOS Physics and Technology* (John Wiley, New York, 1982) pp. 212, 325, 363.
81. E. J. Miller, X. Z. Dang, H. H. Wieder, P. M. Asbeck, E. T. Yu, G. J. Sullivan and J. M. Redwing, **87**, 8070 (2000).
82. E. S. Hellman and E. H. Hartford, *Appl. Phys. Lett.* **64**, 1341 (1994).
83. F. Niu, B. H. Hoerman and B. W. Wessels, *J. Vac. Sci. Technol.* **B18**, 2146 (2000).
84. E. J. Tarsa, X. H. Wu, J. P. Ibbetson, J. S. Speck and J. J. Zinck, *Appl. Phys. Lett.* **66**, 3588 (1995).
85. M. Hong, M. Passlack, J. P. Mannaerts, J. Kwo, S. N. G. Chu, N. Moriya, S. Y. Hou and V. Fratello, *J. Vac. Sci. Technol.* **B14**, 2297 (1996).
86. B. P. Gila, J. Kim, A. H. Onstine, K. Siebein, C. R. Abernathy, F. Ren and S. J. Pearton, to be published (2002).
87. J. Kim, B. P. Gila, R. Mehandru, J. W. Johnson, J. H. Shin, K. P. Lee, B. Luo, A. Onstine, C. R. Abernathy, S. J. Pearton and F. Ren, *Mat. Res. Soc. Symp. Proc.* (2002).
88. A. Onstine, B. P. Gila, J. Kim, R. Mehandru, C. R. Abernathy, F. Ren and S. J. Pearton, Presented at the Florida AVS Meeting, Orlando, March 2002.
89. S. T. Sheppard, J. A. Cooper, Jr. and M. R. Melloch, *J. Appl. Phys.* **75**, 3205 (1994).
90. A. Goetzberger and J. C. Irvin, *IEEE Trans. Electron. Device* **ED-15**, 1005 (1968).
91. D. K. Schroder, *Advanced MOS Diodes* (Addison Wesley Longman, NY, 1986).
92. J. Kim, R. Mehandru, B. Luo, F. Ren, B. P. Gila, A. H. Onstine, C. R. Abernathy, S. J. Pearton and Y. Irokawa, submitted for publication to *Appl. Phys. Lett.*
93. H. Kawai *et al.*, *Electron. Lett.* **34**, 592 (1998).
94. M. A. Khan, X. Hu, G. Sumin, A. Lunev, J. Yang, R. Gaska and M. S. Shur, *IEEE Electron. Device Lett.* **21**, 63 (2000).
95. M. A. Khan, X. Hu, A. Tarakji, G. Simin, J. Yang, R. Gaska and M. S. Shur, *Appl. Phys. Lett.* **77**, 1339 (2000).
96. A. Tarakji, X. Hu, A. Koudymov, G. Simin, J. Yang, M. A. Khan, M. S. Shur and R. Gaska, *Solid State Electron.* (2001).
97. X. Hu, X. Zhang, A. Koudymov, G. Simin, J. Yang, A. Khan, A. Tarakji, M. Shur and R. Gaska, *Phys. Status Solidi* **A188** (1), 219 (2001).
98. D. Mistele, T. Rotter, K. S. Rover, S. Paprotta, M. Seyboth, V. Schwegler, F. Fedler, H. Klausing, O. K. Semchinova, J. Stemmer, J. Aderhold and J. Graul, *Mat. Sci. Eng.* B (2002).

# CHAPTER 5

# SiC MATERIALS GROWTH AND CHARACTERIZATION

Marek Skowronski

*Department of Materials Science and Engineering*
*Carnegie Mellon University, Pittsburgh, PA 15213, USA*

## Contents

*"I like materials science!!! I enjoy watching crystals grow."*

*Anonymous*

## 1. Introduction

Silicon carbide is the only known compound of silicon and carbon. The relative simplicity of the Si–C phase diagram is compensated by the wide variety of polymorphs this material can crystallize in, with about 200 different known polytypes. All of these correspond to the same 1 : 1 stoichiometry and a very similar arrangement of nearest neighbors. As is common in all covalently bonded semiconductors, both silicon and carbon atoms are tetrahedrally coordinated.

The SiC structure can be considered as an assembly of corner-sharing tetrahedra composed of a carbon atom in a centroid position and four silicon neighbors (or vice-versa). The tetrahedra can occupy three different positions (A, B and C) and can occur in two different variants corresponding to rotation around the c-axis by 180°.[1] The second "twinned" variant will be denoted by a prime sign as in: A′, B′, and C′ in the remainder of this chapter. The two simplest types of SiC structure are zinc blende (frequently referred to as 3C–SiC) composed of only one variant of tetrahedra with the stacking sequence ABCABC ... , and wurtzite (called 2H–SiC) with the stacking sequence A′BA′BA′B .... The projection of both structures on (11$\bar{2}$0) plane is shown in Fig. 1. It is easy to notice that stacking of subsequent layers of the same variant produces a cubic-like sequence while switching from the standard to the twinned variant produces hexagonal-type stacking. All other 200 polytypes correspond to longer stacking sequences and can be viewed as a mixture of cubic and hexagonally stacked layers. Although all the polytypes share many properties in common such as large bandgap, high thermal conductivity, and high saturated electron velocity,

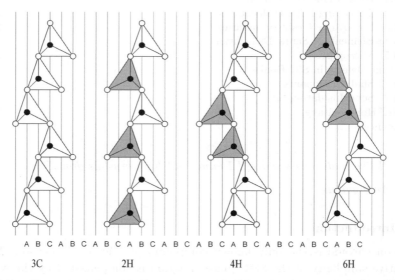

A B C A B C A B C A B C A B C A B C A B C A B C A B C A B C

3C          2H          4H          6H

**Fig. 1.** Projection of the SiC structure on (11$\bar{2}$0) plane. Standard variant of the tetrahedron is white; the twinned variant is grey.

only two have found applications in electronic devices. These are 6H–SiC with the stacking sequence ABCB′A′C′ABC ... and 4H–SiC corresponding to ABA′C′ABA′C′ .... They have primary advantage over all other forms of silicon carbide that we can grow high crystalline quality boules of these two polymorphs.

## 2. Growth of SiC Boules

The first silicon carbide crystals suitable for use in the fabrication of electronic devices were grown by Lely in 1955.[2] The method employed a closed graphite crucible packed with silicon carbide powder, filled with an inert gas at atmospheric pressure, and heated to temperatures between 2400 and 2600°C. The crystals nucleated spontaneously on the walls of the cavity in what was essentially an isothermal growth ambient. Lely's method produced material with very high structural perfection. The total extended defect density in some Lely crystals can be as low as 1–10 cm$^{-2}$, which is better than that of commercially available wafers. However, the crystals produced by this method take on the shape of irregular hexagonal platelets with an area of up to 1 cm$^{-2}$ and a thickness of about 1 mm. Due to this small size, Lely platelets are not suitable for device processing. The renewed interest in silicon carbide as a semiconductor material was generated by the invention of the Physical Vapor Transport growth method proposed by Tairov and Tsvetskov in 1978.[3] Their original approach with small changes is used to produce virtually all silicon carbide substrates used today.

### 2.1. *Physical Vapor Transport Growth*

A schematic drawing of a typical growth system and the hot zone is shown in Fig. 2. The hot zone is located in a double-wall water-cooled quartz chamber with heating provided by a high frequency field. The system does not have any moving parts with the exception of an adjustable coil, which can be moved up and down by a stepping motor. The chamber walls and hot zone are supported by a stainless steel support ring and graphite cylinder, respectively. In addition, the support ring provides several ports for vacuum gauges, gas inlet, exhaust, and a turbomolecular pump. The temperature of the growth crucible is monitored by two two-color pyrometers and two viewports located in the center of the top and bottom plates.

Two grades of graphite are used in the construction of the hot zone: a dense graphite used for the support rod, susceptor, and crucible, and carbon

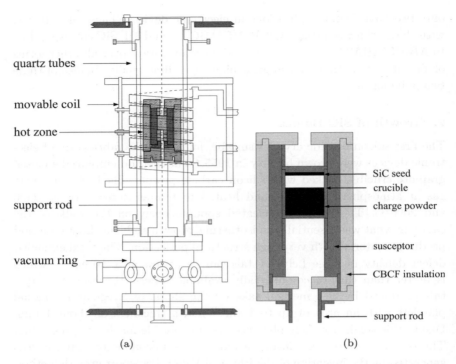

quartz tubes

movable coil

hot zone

support rod

vacuum ring

SiC seed
crucible
charge powder

susceptor

CBCF insulation

support rod

(a)                                                     (b)

**Fig. 2.** Schematic diagram of the SiC Physical Vapor Transport reactor, (a): and hot zone, (b) (after Skowronski[4]).

bonded carbon fiber insulation surrounding the susceptor. RF field with a frequency of 10 kHz couples mostly to the graphite susceptor cylinder and does not penetrate into the crucible, charge powder, or growing crystal. The desired temperature gradient is achieved by adjusting the position of the RF coil in respect to the hot zone and a pooper choice of the size of the opening in the top and bottom insulation discs. Radiation escaping through the top opening provides an effective heat extraction mechanism.

Silicon carbide wafers used as seeds are mounted on the crucible lid using either graphitized sucrose or a retaining ring. The crystal surface faces down during growth. The solid charge material fills most of the crucible volume. Growth runs are conducted at temperatures between 2000 and 2300°C, with a temperature difference between the top and bottom surfaces of the crucible of about 10–30°C. The growth rate is controlled by adjusting the temperature gradient and/or pressure inside the reactor.

PVT growth consists of three consecutive steps:

(i) evaporation of SiC charge material
(ii) transport of vapor species to the growth surface, and
(iii) adsorption, surface diffusion, and incorporation of atoms into the crystal.

The most abundant gaseous species in equilibrium with solid SiC are Si, $SiC_2$, and $Si_2C$ with several others detected in lower concentrations. Their partial pressures increase exponentially with temperature according to the formula

$$p = A \exp\left(-\frac{Q}{RT}\right)$$

where $Q$ is the activation energy of the particular vapor species, and $R$ is the gas constant. The values of activation energies and pre-exponential factors have been determined by Drowart[5] and refined by Lilov[6] and Glass *et al.*[7] The pressures are also affected by the polytype of the charge material and surface termination. The two limiting cases for silicon-rich charge and carbon-rich charge are shown in Fig. 3.

It is worth noting that for typical growth temperatures of 2000–2300°C, the vapor is always silicon-rich. The Si/C ratio at 2000°C ranges from 6 for Si-rich charge to 2.5 for carbon-rich source powder. This ratio decreases with increasing temperature and is between 3.5 and 1.5 at 2300°C. The vapor stoichiometry can change slightly during transport due to reactions with the graphite crucible walls. Nevertheless, the PVT-grown crystals can be expected to contain excess silicon in the form of carbon vacancies, silicon interstitials, and/or silicon antisites. If the temperature gradient is high, the silicon-rich vapor can condense, producing liquid silicon droplets on the growth surface and inclusions in the SiC crystal.

The growth rate-limiting step in PVT is the transport of vapor species from the charge powder to the growth surface. The convection contribution to transport is small and can be neglected[8]; the Raleigh number at typical growth conditions and crucible size is below 5 and much below the $R_a$ value corresponding to the onset of convection ($R_a \approx 230$).[9] The experimental data and modeling results performed by Mueller *et al.*[9] indicate that a thick Si-rich boundary layer forms near the growth surface with diffusion being the dominant transport mechanism. These authors found that radially averaged reciprocal flux of Si–C species can be expressed as

$$\frac{1}{J^{\text{TOT}}} = \frac{\Phi}{p^{\text{s},*}} + \frac{\Omega}{\Theta} + \frac{\Omega}{\Theta}\frac{p^{\text{Ar}}}{p^{\text{s},*}}$$

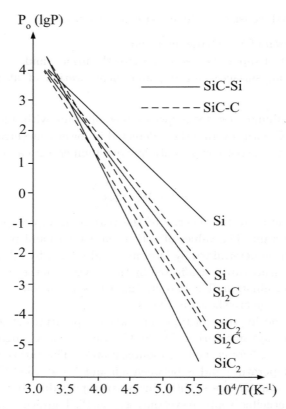

**Fig. 3.** Temperature dependence of partial pressures of components for silicon-rich (solid lines) and carbon-rich (dashed lines) systems (after Lilov[6]).

where $p^{s,*}$ is the saturated pressure of vapor species at charge temperature; $p^{Ar}$ is the total pressure of inert gas; and $\Phi$, $\Omega$, $\theta$ are temperature independent constants. The growth rate, therefore, is determined by the sublimation pressure and pressure of the inert gas in the reactor. The experimental data on growth rate agree well with the above formula (Fig. 4).

Silicon carbide boules at typical PVT growth conditions grow in the step flow mode. Adatoms diffuse on atomically flat terraces of the growth surface toward steps where they are incorporated in the crystal. The steps on (0001) silicon face are typically high. The smallest step size observed on 6H–SiC is one unit cell (1.5 nm), with many steps bunching up to produce macrosteps sometimes exceeding 100 nm. The steps on a $(000\bar{1})$ carbon face used for growth of 4H–SiC do not bunch up as frequently. In either case,

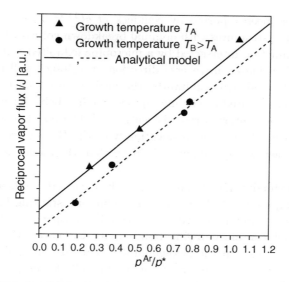

**Fig. 4.** Kinetics of SiC growth in general coordinates (after Mueller *et al.*[9]).

all steps originate at screw dislocations and/or micropipes intersecting the growth surface. There is no evidence of two-dimensional islands nucleating on terraces at later stages of growth. It leads to an interesting observation, namely that the polytype of the growing crystal is determined by the type of screw dislocations.

## 3. Extended Defects in SiC Boules

The interest in the crystalline quality of silicon carbide boules arises from the detrimental effect of extended defects on device performance and/or reliability. Extended defects originating from the substrate replicate into the epitaxial structures and thread the active layers of the devices. Once they are present in the layers subjected to high electric fields or current densities, extended defects can lead to a host of problems well documented in a variety of materials.

The defect that attracted most attention in the early days of SiC technology development is the micropipe. Koga *et al.* was the first to notice that the *p-n* junction intersected by a micropipe exhibits breakdown at very low voltage values.[10] They associated this phenomenon with the open core of a micropipe filling with metal during contact annealing and the shorting of the device. Similar correlation between reverse diode characteristics and micropipes has been reported by Neudeck *et al.*[11,12] These authors have investigated *p-n* structures with nominal breakdown voltage of 1000 V. Devices with a junction intersected by a micropipe exhibited sudden increase in current when biased in reverse direction at voltages well below nominal breakdown field. Concomitantly, the authors observed the formation of microplasmas located at the micropipes. Similar behavior was observed in the case of elementary screw dislocations.[13] The *p-n* diodes intersected by a single elementary screw dislocation with the Burgers vector $1c[0001]$ had a breakdown voltage about 20% lower than the diodes without screw dislocations.

Another extended defect that has recently attracted significant attention is the stacking fault.[14,15] In most semiconductors, stacking faults are not electrically active and do not affect the electrical properties of the material, as they do not contain dangling bonds (with the exception of bounding partial dislocations). In the hexagonal 4H– and 6H–SiC polytypes, however, all the stacking faults act as quantum wells leading to localization of electrons.[16–18] As such, they interfere with the electron transport along the *c*-axis with the result that the forward voltage drop in the SiC bipolar devices is increased. Lendenmann *et al.*[19] have documented that stacking faults in forward biased high voltage diodes nucleate at defects originating in the SiC substrate.

### 3.1. *Basal Plane Dislocations*

Probably the most numerous dislocations in SiC boules are basal plane dislocations with a Burgers vector of $a/3\langle11\bar{2}0\rangle$. The dislocation lines are confined to the close packed (0001) basal plane similar to 60° dislocations that usually lie in (111) planes in cubic semiconductors. In high quality SiC boules, basal plane dislocations form long straight segments extending along one of the three $\langle11\bar{2}0\rangle$ directions (Fig. 5). In such a case, their Burgers vector forms either a 60° or 0° angle with the dislocation line. Basal plane dislocation densities vary widely between different boules and manufacturers and fall usually in the $10^3$–$10^6$ cm$^{-2}$ range.

A convenient way of monitoring basal plane dislocations in off-cut wafers is etching in molten KOH.[10,20] Due to an oblique angle formed by the dislocation line with the wafer surface, the etch pits have a characteristic oval shape that is easily distinguished from other etch features (Fig. 6). Unfortunately, this method does not work on on-orientation wafers. An alternative technique is X-ray topography, which in addition to the density

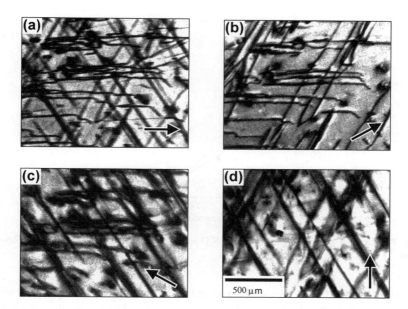

**Fig. 5.** X-ray topography image of basal plane dislocations in 4H–SiC wafer. Images were obtained using diffraction vectors (a) $[2\bar{1}\bar{1}0]$, (b) $[10\bar{1}0]$, (c) $[1\bar{1}00]$, and (d) $[01\bar{1}0]$ (courtesy of M. Dudley).

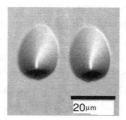

**Fig. 6.** Etch pits due to basal plane dislocations in 8° off cut 4H–SiC.

and dislocation line direction, can be used to determine the direction of the Burgers vector. In wafers with a dislocation density above $10^4$ cm$^{-2}$, X-ray topography does not resolve individual dislocations due to overlapping strain fields.

In general, dislocations can be produced by three different mechanisms: plastic deformation; misorientation between independent nuclei during growth; and coalescence of point defects. Plastic deformation is quite common in the growth of compound semiconductors.[21,22] In hexagonal SiC,

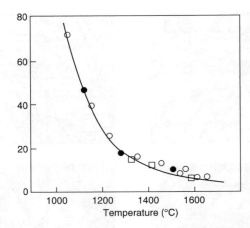

**Fig. 7.**   Critical resolved shear stress of SiC as a function of temperature (after Fujita *et al.*[23]).

deformation studies at elevated temperatures revealed that the primary slip system is $a/3\langle11\bar{2}0\rangle(0001)$.[23–25] Even in instances when the geometry of the sample was selected to minimize the resolved shear stress on the basal plane, the sample deformed by buckling and motion of basal plane dislocations. Also, although silicon carbide is extremely hard at room temperature, it becomes quite plastic at the growth temperatures. The results of the measurement of the Critical Resolved Shear Stress (CRSS) versus temperature are shown in Fig. 7. The data are available only up to 1600°C where CRSS value is about 5 MPa. Extrapolation of these data to 2200°C gives a CRSS value of between 1 and 2 MPa. This is below the corresponding value for GaAs at its melting point. Using an analogy between the growth of GaAs by the Liquid Encapsulated Czochralski method and SiC by PVT, one could expect that SiC boules can undergo plastic deformation during growth.

In most bulk crystal growth techniques, the latent heat of solidification is extracted through the crystal. This makes it necessary to impose the temperature gradient on the growing boule, which in turn can produce thermoelastic stresses in the solid. For the cylindrical boule, the stress magnitude is proportional to the second derivative of the temperature along the crystal growth axis and proportional to the radial thermal gradient. As discussed in the preceding section, the common SiC growth apparatus utilizes radiation leak in the form of the opening in the graphite insulation behind the growth crucible lid. The heat is supplied from a cylindrical susceptor/heater surrounding the crucible. Such a design is inherently susceptible to high radial temperature gradients. Several groups have calculated the temperature and stress distribution in growing SiC boules.[26-29] The results are strongly dependent on the system design and the assumed boundary conditions. Nevertheless, most results indicate that the shear stress resolved in the basal plane (assuming the on-axis growth) exceeds the extrapolated CRSS by about an order of magnitude. The $\sigma_{rz}$ stress values are highest halfway between the center and the periphery of the crystal. Von Mises stress, on the other hand, is highest in the center of the boule and at the periphery. Typical calculated von Mises stress level is higher by about an order of magnitude than the $\sigma_{rz}$ shear stress. The conclusion of several modeling efforts is that SiC boules are expected to plastically deform during growth.

The experimental data also indicate that the primary slip system is active during the growth of boules. A recent report has identified the slip bands in 4H–SiC boules consisting of arrays of etch pits on the surfaces of the off-cut SiC wafers (Fig. 8(a)).[30] The arrays are always perpendicular to the off-cut direction and are distributed in either a single basal plane or a narrow band of basal planes. They are usually distributed over most of the wafer area with the exception of the center, in agreement with the thermoelastic stress distribution. The length of some slip bands exceeds half the radius of the boule, indicating that stress is global rather than local in origin. Corresponding images of the dislocation arrays obtained by X-ray topography are shown in Fig. 8(b). An alternative method of assessing the degree of basal plane shear during growth was proposed by Ha *et al.*[26] Since some growth conditions result in negative $\sigma_{rz}$ stress throughout the boule, the expected deformation assumes the form of bending of the basal planes. This was measured by high resolution X-ray diffraction and

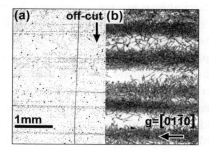

**Fig. 8.** Slip bands on the surface of the commercial 4H–SiC wafer off-cut 8° toward [1$\bar{1}$00] revealed by: (a) KOH etching; and (b) white beam synchrotron radiation topography (after Ha *et al.*[30]).

produced results in agreement in sign and magnitude with the typical temperature distribution in a growing crystal. The total basal plane reflection shift along the boule diameter was approximately 150 arcseconds.

The primary $\langle 11\bar{2}0 \rangle (0001)$ slip cannot relax the shear stress components along the c-axis, and at high growth temperatures other slip systems can be activated. One result that indicates that this does occur is the correlation between the total dislocation density and von Mises stress distribution.[27,29] Both normal and von Mises stresses are highest in the center of the boule and at the outer periphery and agree very well with the total dislocation density determined by KOH etching.[27]

The results discussed above provide evidence of plastic deformation through basal plane slip occurring during growth of SiC boules. It should be pointed out that the current SiC growth technology is far from mature and the temperature profiles in the hot zone are not optimized. It is possible to design the growth apparatus with greatly reduced level of thermoelastic stresses. Consequently, it is the belief of the author that the densities of dislocations related to plastic deformation of SiC boules can be reduced from the current average of $10^5$ cm$^{-2}$ to below $10^2$ cm$^{-2}$.

### 3.2. *Threading Edge Dislocations*

Threading edge dislocations have the same type of Burgers vector as the basal plane dislocation, i.e. $a/3\langle 11\bar{2}0 \rangle$. Their dislocation line extends along the c-axis. Since the Burgers vector is perpendicular to the line direction, threading edge dislocations have pure edge character (Fig. 9). Their typical density in wafers and homoepitaxial layers is between $10^4$ and $10^5$ cm$^{-2}$. This range is much narrower than for basal plane dislocations, indicating that the formation mechanisms are not as process dependent. The obvious source of threading edge dislocations is the seed which has approximately the same diameter as the boule. Since dislocations cannot terminate inside the crystal and are propagating in the general direction of the growth axis, all dislocations in the seed replicate into the crystal. Most threading edge dislocations are, therefore, of the grown-in type.

In addition to propagation from the seed, there are several other mechanisms for the introduction of threading edge dislocations. One is the conversion from basal plane dislocations as discussed in Sec. 5 for the case of SiC epitaxy. Another documented mechanism is the secondary slip system in which threading edge dislocations, such as that shown in Fig. 9(a), glide in the $(1\bar{1}00)$ prismatic plane determined by their Burgers vector and the

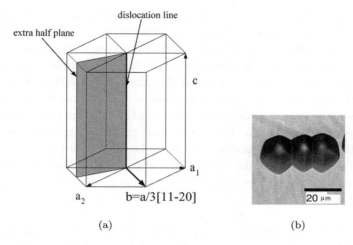

(a)

(b)

**Fig. 9.** Configuration of a threading edge dislocation in hexagonal SiC (a), and its characteristic etch pit on the Si (0001) wafer surface (b).

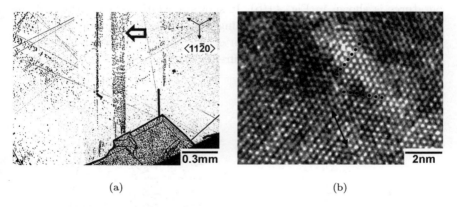

(a)

(b)

**Fig. 10.** (a) Prismatic slip band observed on the KOH etched wafer, and (b) plan view high resolution TEM image of an individual dislocation in the slip band. The black dots denote two extra $\langle 11\bar{2}0 \rangle$ half planes of a threading edge dislocation (after Ha *et al.*[31]).

dislocation line. This slip system is denoted by $a/3\langle 11\bar{2}0 \rangle \{1\bar{1}00\}$. The resulting slip bands due to repeated nucleation and glide have been observed by KOH etching and transmission electron microscopy.[31] An example is shown in Fig. 10, obtained by Nomarski contrast optical microscopy on the etched (0001) wafer surface. The etch pits form a band approximately 50 $\mu$m wide extending vertically along the $\langle 11\bar{2}0 \rangle$ direction and marked with an arrow.

The slip band appears to originate in a misoriented grain at the edge of the crystal. A narrower parallel band starts on the group of large hexagonal pits corresponding to several micropipes. Plan view high resolution TEM analysis confirmed that the dislocations are in fact threading edge dislocations with their Burgers vector parallel to the band direction (Fig. 10(b)). Prismatic slip bands are quite infrequent in SiC boules. Rarely, there is more than one source of them in a wafer and the total area covered is small compared to the wafer size. Because of these characteristics, the stress that causes this type of deformation appears to be local in origin.

The distribution of dislocations in a slip band corresponds to a high energy configuration. The largest stress component around an edge dislocation is $\sigma_{xx}$ as shown schematically in Fig. 11(a). In a slip band with Burgers vectors of dislocations parallel to the slip band direction, these stresses add up to produce a long range strain field and bending of the crystal. If the dislocations are mobile, at high enough temperatures, they rearrange themselves in such a way as to reduce the total strain energy. Direct experimental evidence of dislocation mobility at SiC growth temperatures is shown in Fig. 11(b). It shows a KOH etched surface of the 4H–SiC wafer with etch pits forming two vertical prismatic slip bands (marked with arrows at the bottom of the figure). Superimposed on the overall band shape is the fine structure: dislocations tend to group in short

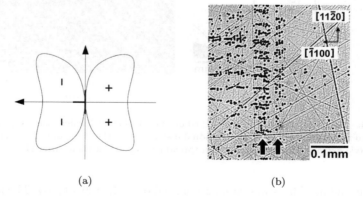

(a)　　　　　　　　　　　　　　(b)

**Fig. 11.** (a) Distribution of tensile (marked +) and compressive (marked −) stress regions around the edge dislocation line. The direction of the dislocation line is perpendicular to the plane of a drawing, with the extra half plane extending to the left. (b) Onset of polygonization in the prismatic slip band. The orientation of dislocations (Burgers vector, line direction, stress distribution, etc.) is identical to the schematic drawing in (a).

horizontal arrays. Such an arrangement allows the opposite strain fields of neighboring dislocations to overlap and annihilate each other. This process is referred to as polygonization and was observed in a variety of materials. After prolonged annealing, mobile dislocations are expected to fully polygonize, forming long linear arrays in direction perpendicular to the Burgers vector of individual dislocations.

### 3.3. *Low-Angle Grain Boundaries*

One of the characteristic structural features of SiC wafers is their domain structure reflected in the shape of high-resolution X-ray diffraction peaks. An example of the X-ray basal plane rocking curve obtained on PVT-grown wafer is shown in Fig. 12.[32] The reflection consists of about 7 peaks each with Full-Width-at-Half-Maximum of about 20 arcseconds. Individual peaks are shifted from each other by several tens of arcseconds. This characteristic structure has been interpreted as due to SiC boules consisting of high crystalline quality "domains" (hence the small width of individual peaks) that are misoriented in respect to each other.[32] The boundaries between domains were suggested to contain high defect densities accommodating the misorientation. Although initially the domain structure has been

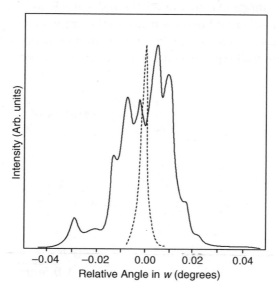

**Fig. 12.** High resolution X-ray rocking curves obtained on wafer grown by PVT (continuous line) and high quality Lely platelet (dashed line) (after Glass *et al.*[32]).

explained as a result of deposition on slightly misoriented growth centers, the recent interpretations lean toward domain wall formation due to interactions between threading edge and basal plane dislocations. Also, although the quality of SiC boules has improved dramatically since the publication of Fig. 12, formation of domains and low-angle grain boundaries still remains a quality issue in SiC technology.

Evidence of dislocations mobility and interaction during growth was presented in the preceding section. The onset of alignment of threading edge dislocations in Fig. 11 occurred during one growth run, while the full development of the domain structure in SiC boules occurs over many growth runs: the seeds cut from one boule replicate their microstructure into subsequent growth runs. This corresponds to very long high temperature annealing and resulting alignment can be much more extensive. An example of a threading dislocation distribution in a 2-inch diameter 4H–SiC commercial wafer is shown in Fig. 13. The characteristic feature is the highly non-uniform etch pit distribution. The etch pits are preferentially aligned in rows along $\langle 1-100 \rangle$ directions. These arrays are clear candidates for the high defect density domain walls with the areas between arrays exhibiting much higher crystal quality. TEM experiments showed that dislocation lines in such arrays lie along $[0001]$ and, therefore, the boundaries are $\{11\bar{2}0\}$ planes.[33] The Burgers vectors of all the dislocations in a single array are parallel to each other and perpendicular to the array direction. Such a configuration of dislocations results in a misorientation between two domains, and the domain wall represents a low-angle grain boundary.

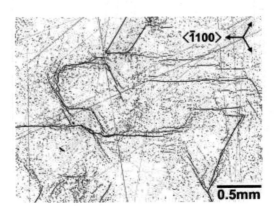

**Fig. 13.** Distribution of threading dislocations in the center of 4H–SiC wafer (after Ha *et al.*[33]).

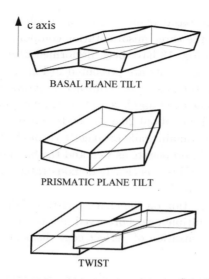

**Fig. 14.** Three types of misorientation across the domain boundary: (a) basal plane tilt; (b) prismatic plane tilt; and (c) twist type.

In general, there are three possible types of misorientation across a grain boundary: two tilts and a twist (Fig. 14). A twist-type misorientation is produced when one of the domains is rotated with respect to the other around the axis that is perpendicular to the boundary. Two tilt components correspond to rotation around two orthogonal axes in the plane of the boundary. In a hexagonal crystal, it is natural to select one of these axes along the [0001] direction. The rotation of a domain around [0001] results in misorientation between prismatic planes while leaving basal planes perfectly aligned. Such misorientation will be referred to as prism-plane-tilt. The second tilt axis that is perpendicular to the [0001] and lies in the $\{11\bar{2}0\}$ plane of the boundary, is along one of the $\langle 1\bar{1}00 \rangle$ directions. The rotation around it produces misorientation of basal planes and is called basal-plane-tilt. Each one of these boundaries is made up of a specific type of defects. Tilt boundaries are composed of edge dislocations with line direction along the misorientation rotation axis and Burgers vector perpendicular to the domain wall. A twist boundary has to be made up of two orthogonal arrays of screw dislocations.

According to above, the domain walls shown in Fig. 13 that consist of polygonized threading edge dislocations should be of the prism-tilt-type. The expected misorientation should be visible in prism plane X-ray reflections, while the basal plane reflection should not be affected. The measurements of all the three misorientation components across one of the boundaries shown in Fig. 13 have been performed using high resolution X-ray diffraction and the results are presented in Fig. 15. In agreement with the above discussion, the prism plane diffraction shows two peaks separated by 130 arcseconds. The twist and basal-plane-tilt components are below the value of the peak width of about 20 arcseconds. Since the threading dislocations making up the domain walls propagate through the entire boule, the misorientations should propagate as well and the SiC boules should exhibit a columnar structure. This was clearly demonstrated by X-ray topography using a $[11\bar{2}0]$ reflection.[34]

The remaining question on the characteristics of domain structure concerns the origin of multiple peaks observed in basal plane X-ray diffraction (Fig. 12). The polygonization of perfect edge dislocations cannot produce

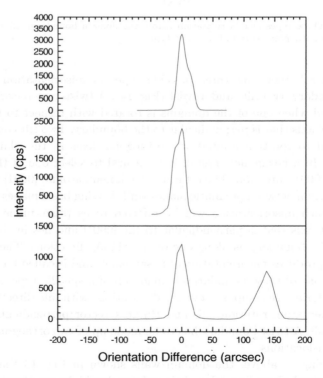

**Fig. 15.** High resolution X-ray diffraction $\omega$ scans on one of polygonized threading dislocation boundaries shown in Fig. 13 (after Ha *et al.*[33,35]).

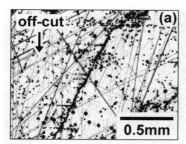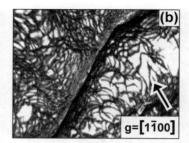

**Fig. 16.** Basal plane dislocation pile-ups against the grown-in domain wall: (a) optical micrograph of KOH etch surface of the off-cut wafer; and (b) white beam synchrotron radiation topograph of the same area (after Ha *et al.*[30]).

misorientation of the basal plane, and some other mechanism must be involved. Two possible effects have been suggested so far. Ha *et al.* have proposed that the in-grown prism-plane-tilt boundaries interact with the slip-induced basal plane dislocations leading to basal plane dislocation pile-ups.[30] This phenomenon has been observed in etch patterns on off-cut wafers and in X-ray topography and its results are shown in Fig. 16. The grown-in prismatic tilt boundary extends from the upper right to the lower left corner. On its left side, one can see short horizontal arrays of basal plane dislocations produced by primary slip. A corresponding basal dislocation pile-up is also visible in X-ray topography in (b). Such a configuration of dislocations is expected to produce a basal-plane-tilt component. High resolution X-ray diffraction confirmed this assertion. The density of dislocations in pile-ups was in good agreement with the value of basal plane misorientation. An alternative interpretation was suggested by Katsuno *et at.*[36] They proposed that the threading dislocations in a domain wall are inclined to the *c*-axis and, therefore, are of mixed character. In such a case, the misorientation across the domain wall should have a basal-plane-tilt component in addition to prism-plane-tilt. Both of the above groups have determined experimentally the degree of twist in low angle grain boundaries in SiC boules and found this misorientation component negligible.

### 3.4. *Elementary Screw Dislocations*

The only dislocations commonly observed in SiC crystals with the Burgers vector not in the basal plane are screw dislocations. Their Burgers vector is equal to the *c*-lattice parameter or is a multiple of it, with the dislocation line direction along the *c*-axis. The typical density of screw dislocations in

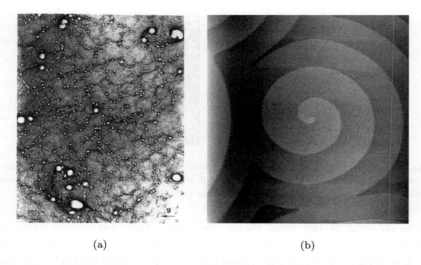

(a)                                              (b)

**Fig. 17.** (a) White-beam synchrotron radiation topography image of elementary screw dislocations in PVT SiC wafer (after Dudley *et al.*[38]) and the AFM image of a growth spiral starting on an emerging screw dislocation.

commercial SiC wafers is between $10^3$ and $10^4$ cm$^{-2}$. They are distributed almost uniformly across the wafers with the exception of slightly higher density in the center of the boule. In KOH etching, screw dislocations produce hexagonal pits similar in shape to, but larger in size than, that of threading edge dislocations.[20] The most reliable method of monitoring the screw dislocation density is back-reflection X-ray topography (Fig. 17). Screw dislocations produce characteristic white dots surrounded by a dark circle.[37] Large white spots visible in the figure correspond to micropipes which are discussed in the next section.

Screw dislocations in SiC play a very important role in growth of crystals along the [0001] direction. Examination of growth surfaces of PVT grown bulk SiC crystals reveals that growth always proceeds by the step flow mode (i.e. adatoms attach themselves at the step edges). All steps, in turn, originate at emerging screw dislocations (Fig. 17b). Because of this, the stacking sequence in the growing crystal is determined by the existing screw dislocations. It is interesting to note that $1c[0001]$ screw dislocations have not been observed in boules grown along directions other than [0001].[39] This defect is characteristic only to growth along the $c$-axis.

In 6H–SiC, the Burgers vector length of an elementary screw dislocation is 1.5 nm versus 0.307 nm for a basal plane or a threading edge dislocation.

Since the elastic energy of a dislocation is proportional to the square of the Burgers vector, it makes screw dislocations very energetically "expensive". As such, screw dislocations are unlikely to be produced by plastic deformation and, as a consequence, their origin has to be related to nucleation faults. The experimental evidence in support of this statement was obtained in studies of early stages of growth. In their seminal paper, Takahashi *et al.* have reported dislocation structure in SiC crystals grown on the silicon face of a low defect density Acheson seed.[34] X-ray topography of the crystal cross-section revealed a clean seed that remained essentially dislocation free. This eliminated the possibility of plastic deformation occurring during growth. The material deposited on the seed, however, exhibited contrast characteristic of elementary screw dislocations originating at the crystal/seed interface.

Similar results indicating nucleation of screw dislocations at the seed/crystal interface have been obtained by KOH etching and atomic force microscopy in an experiment involving simultaneous growth on two seeds.[40] One seed was polished on-axis, while the other was intentionally off-cut approximately 5° towards [11$\bar{2}$0]. After hydrogen etching, the vicinal surface of the off-cut seed should be covered by unit cell high steps separated by 17 nm wide terraces. Since the terrace width was much smaller than the adatom diffusion length at SiC growth temperatures, the growth proceeded by step flow throughout the short growth run. On the on-orientation seed, however, steps due to very small non-intentional misorientation were grown out and the growth had to proceed through the phase of 2D island nucleation. These two different growth modes resulted in a dramatic difference between screw dislocations. The layer growing through the step flow mechanism had a screw dislocation density of 50 cm$^{-2}$ while the layer that underwent the 2D island growth had screw dislocation density higher by a factor of 300.[40] It appears that nucleation of screw dislocations is related to the nucleation of 2D islands on basal plane facets of growing crystal.

One of the characteristics of forced 2D nucleation is the lack of "polytype memory": close packed layers are free to select any stacking sequence guided only by energy differences. These differences are exceedingly small at the growth temperatures as indicated by the existence of over 200 known SiC polytypes and the small energies of stacking faults. In a 4H–SiC, Shockley fault energy is 14.7 mJ/m$^2$, while that of a 6H–SiC is 2.9 mJ/m$^2$.[41] In a large number of small islands, it is likely that some layers will nucleate out of sequence of a polytype that is thermodynamically stable at given growth conditions. The evidence of such an occurrence is

**Fig. 18.** Stacking faults at the interface between Lely seed (lower right corner) and layer grown by forced 2D nucleation (upper left corner) (after Sanchez *et al.*[40]).

apparent in the cross-sectional TEM image of the on-orientation Lely seed/crystal interface (Fig. 18). The section of the Lely platelet seed is visible in the lower part of the image separated by the band of horizontal lines from the section of the deposited layer. The lines correspond to the band of basal plane stacking faults. Most of the faults are quite complex, involving multiple mis-stacked layers.[42]

It could be expected that stacking faults will expand together with islands, but will terminate whenever the neighboring islands coalesce. Depending on the bonding between layers at this point, some of the bounding dislocations can have Burgers vector along the *c*-axis and it is conceivable that they could give rise to the formation of screw dislocations. The exact processes involved still need to be identified.

### 3.5. *Micropipes*

Possibly the defect that attracted the most attention in silicon carbide and certainly the one that was investigated the longest is the micropipe. Micropipes are tubular voids extending along the *c*-axis in SiC crystals that frequently penetrate the entire boule. Micropipes in SiC wafers used as substrates for deposition of SiC CVD films are known to propagate into the epilayers. The densities of micropipes in commercially available wafers are between $10-100$ cm$^{-2}$, with the best 2-inch diameter wafers exhibiting densities approaching 1 cm$^{-2}$. Micropipes tend to form clusters in crystals, apparently due to some local disturbance of the growth process.

Numerous observations correlated the micropipes with screw dislocations. According to the model proposed by Frank,[43] the open core of a

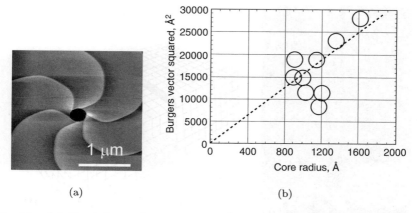

**Fig. 19.** (a) Atomic force microscopy image of a micropipe in 6H–SiC with the Burgers vector $b = 7c[0001]$; (b) Relationship between open core radius of a micropipe and the square of the Burgers vector. Frank model predicts a linear dependence (after Giocondi et al.[48]).

micropipe is stabilized by a strain energy of a dislocation with large Burgers vector. The elementary screw dislocations in both 4H– and 6H–SiC have closed cores.[44] With an increase of the Burgers vector length, the strain energy increases rapidly and at some critical value, it becomes energetically favorable to remove the material around the dislocation core, creating the additional free surface. The critical value of the Burgers vectors resulting in an open core appears to be $2c$ in 6H and $3c$ in 4H–SiC.[45,46] With further increase of the Burgers vector, the diameter of the void increases and can reach sizes in excess of 0.3 $\mu$m (Fig. 19).[44,47] Super-screw dislocations with Burgers vectors as large as $15c$ have been observed.

Although Frank's model appears to be widely accepted, it is still not clear how dislocations with large Burgers vectors can be created. Numerous mechanisms have been proposed over the last decade.[7,47,49–52] One possible route would be to form them by coalescence of several elementary screw dislocations with parallel Burgers vectors. This route is difficult due to repulsive forces acting between like-dislocations and is unlikely to occur in a solid. However, if several screw dislocations intersect the bottom of a local depression on the growth surface, caused for example by a second phase inclusion, then the elastic interaction is eliminated. During the overgrowth of such an inclusion, the lowest energy configuration for dislocations would be a single empty core micropipe with the Burgers vector corresponding to the net sum of Burgers vectors of all dislocations in the depression. This

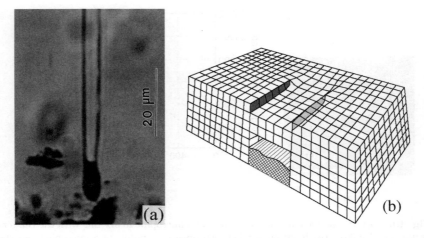

**Fig. 20.** Formation of pairs of micropipes. (a) Optical transmission image of an inclusion in SiC boule with two micropipes extending towards the growth surface. (b) Schematic model of crystal deformation leading to appearance of pairs of screw dislocations (after Dudley *et al.*[51]).

model, proposed originally by Giocondi *et al.*,[47] could explain the experimental observation of Balakrishna *et al.*[53] of micropipes originating on intentionally deposited metal dots on the seed surface. Similar observation of micropipes originating on non-intentional second phase inclusions has been reported by Dudley *et al.*[51] These authors interpret their observations as lateral overgrowth of inclusions on the growth surface. The cantilevered growth fronts can flex up and down under the influence of small forces and when the opposite fronts meet they can produce a pair of micropipes (Fig. 20).

A recent report by Kuhr *et al.*[52] presented evidence of an unusual mechanism of dislocation coalescence during growth. As described in Sec. 2.1, the seed crystal is affixed to the crucible with the help of high temperature adhesive (most often sucrose). All possible adhesives decompose during growth, producing volatile components and as a result, the attachment layer can contain voids. Since the Si, $SiC_2$, and $Si_2C$ vapor pressures are high at growth temperatures, the mass transport can occur across such voids. SiC material evaporates on the side of the void closer to the growth surface and deposits on the relatively colder crucible lid. The process continues and the void moves through the growing SiC boule, assuming a characteristic hexagonal shape reminiscent of a Lely platelet (Fig. 21(a)). The rate

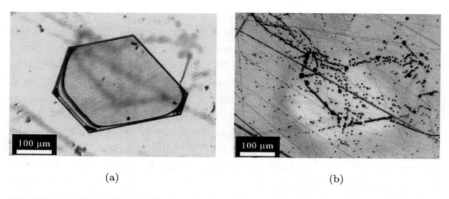

(a)                                                    (b)

**Fig. 21.** (a) Optical micrograph of a hexagonal void in SiC wafer. (b) Etched wafer under the void. Large etch pits at the void corners correspond to micropipes (after Kuhr et al.[52]).

of motion can be higher than the growth rate and bigger voids can traverse the entire boule and disappear at the growth surface. In the process of re-crystallization, the dislocations are re-distributed. Most of them are relocated to the last solidified part of the void bottom, which frequently forms a deep trench at the void edge. It is a situation very similar to the one discussed in the preceding paragraph. As the result, the path of the void is marked with micropipes at the void corners.

The models of micropipe formation discussed above do not exclude each other. Most likely, multiple mechanisms are operating in slightly different growth conditions which makes the task of eliminating this important defect so challenging.

## 4. Point Defects in SiC Boules

The interest in point defects in silicon carbide is driven by the necessity of controlling electrical properties of substrates and epilayers. The desired electron and hole concentrations range from as high as $10^{20}$ to as low as $10^0$ cm$^{-3}$. Such a wide range can be covered by manipulating the concentrations of intentional shallow dopants, residual impurities, and native point defects.

### 4.1. *Shallow Dopants*

The most commonly used intentional dopant in SiC is nitrogen. It occupies the carbon site forming a shallow donor level.[54] Mixed polytypes of silicon carbide have multiple inequivalent substitutional sites. 4H–SiC has two such sites: one hexagonal and the other cubic, while 6H–SiC has three: one hexagonal and two cubic. The ionization energies of the nitrogen donor in 4H are 52.1 meV and 91.8 meV,[55] respectively, while in 6H–SiC, the energies are 81.0, 137.6, and 142.4 meV[56] for hexagonal and cubic sites, respectively. These energies are small enough for the majority of shallow nitrogen donors to be ionized at room temperature. Electron concentration can be easily controlled in both bulk and epilayer growth, by adjusting the flow of gaseous nitrogen in the reactor. The solid solubility of nitrogen in SiC exceeds $10^{20}$ cm$^{-3}$ at bulk and epitaxial growth temperatures. However, the doping is limited to approximately $5 \times 10^{18}$ cm$^{-3}$ by the onset of $n$-type doping-induced crystalline instability. Wafers with electron concentrations above $1.5 \times 10^{19}$ cm$^{-3}$ were observed to warp during high temperature processing steps such as oxidation, contact annealing, and epigrowth.[57] This effect is due to spontaneous formation of double Shockley stacking faults shown in Fig. 22.[58] The driving force for the transformation appears to be the lowering of electron energy from the Fermi level high in the conduction band to the quantum well-like states associated with the double stacking faults.[17,18]

The most commonly used shallow acceptor-type dopant is aluminum. It substitutes for silicon in the SiC lattice and, similarly as nitrogen, forms either two or three inequivalent centers in 4H and 6H–SiC, respectively.[59] Higher effective masses in the valence band compared with the conduction band result in higher values of the ionization energies which fall between 190 and 250 meV for different centers and polytypes.[60] This results in

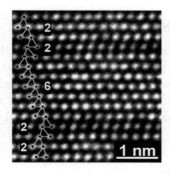

**Fig. 22.** Nitrogen-induced double Shockley stacking fault in 4H–SiC wafer (after Liu *et al.*[58]).

incomplete ionization of aluminum acceptors and higher resistivity of *p*-type material at room temperatures compared with the *n*-type, limiting its usefulness.

## 4.2. *Semi-Insulating SiC*

One of the very promising applications of silicon carbide is in high frequency devices. Possible structures include silicon carbide Static Induction Transistors and MESFETs, as well as nitride based devices with active Group III nitride layers deposited on silicon carbide substrates. These structures promise higher power densities at higher frequencies than possible with conventional semiconductors such as silicon and gallium arsenide. All of the structures intended for high frequency applications require high resistivity substrates with high thermal conductivity for efficient heat extraction. These requirements prompted research on residual impurities and native point defects in silicon carbide crystals.

Residual impurity content in SiC-boules grown by PVT and High Temperature CVD[61] has been investigated by mass spectrometry methods. The total impurity concentration in the cleanest PVT grown material is below $1 \times 10^{17}$ cm$^{-3}$, with the two dominant contaminants being nitrogen and boron (Table 1). Both were present in the low $10^{16}$ cm$^{-3}$ range.[62] The concentrations of all other elements (including transition metals which for a long time posed a significant contamination problem) were more than an order of magnitude lower and in all cases at the detection limit or a background level of the technique. The data available for HT CVD are not as complete but indicate similar transition metal concentrations and significantly lower boron content.

The obvious sources of residual impurities are the charge material and graphite used in the construction of the hot zone. Commercially available SiC powders produced by either the Acheson or CVD methods are

**Table 6.** Concentration of impurities observed in high purity PVT SiC and HT CVD wafers (per cm$^3$).

| Element | B | N | Na, Al | Ti | V, Cr | Fe, P | Ni, F | Cu | O* | S, Cl |
|---------|-----|-----|--------|------|-------|-------|-------|------|------|-------|
| PVT[62] | 1.5E16 | 5E16 | 5E13 | 1E14 | 5E13 | 2E14 | 5E14 | 3E14 | 4E16 | 1E15 |
| HTCVD[63] | 5E14 | 5E15# | 8E13 | 5E14 | 1E13 | X | X | X | X | X |

\* — SIMS background level, # — photoluminescence data, X — no data.

of relatively low purity with high concentration of boron and nitrogen. As a consequence, many manufacturers are producing the charge material in-house. Direct reaction between semiconductor grade silicon and carbon or a CVD-based process can be used and either method is capable of producing SiC with a purity high enough for growth of undoped semi-insulating SiC boules. The purity of source material for HT CVD (silane and propane) is high enough not to cause any problems. Graphite, however, is an on-going concern for both methods. It appears to be the primary source of boron in SiC wafers with the boron content in purified graphite between 0.1 and 1 ppm. Known boron-related centers are not deep enough to serve as efficient recombination centers, but play an important role in compensation mechanisms.

The first high resistivity SiC boules have been grown by Hobgood *et al.* by PVT[64] The undoped crystals grown at this time had boron as a major residual impurity in concentrations as high as $10^{18}$ cm$^{-3}$ and had the Fermi level pinned at shallow boron acceptor level 0.3 eV above the valence band.[65] In order to compensate boron, the boules had been intentionally doped with vanadium which substitutes for silicon and is known to produce two levels within the band gap. The lower donor level is located close to the middle of the gap.[66] The acceptor level lies 0.6 eV below the edge of the conduction band and about 0.9 eV above the donor level.[67] When vanadium is introduced in the p-type crystal, electrons from the donor level are transferred to the shallow acceptors. If the vanadium concentration is higher than the acceptor density (assuming that other impurities and native defects are present in negligible concentrations), all the shallow states will be filled and the Fermi level is pinned at the vanadium donor. The activation energy of resistivity versus temperature in vanadium-doped crystals agrees well with the known vanadium donor energy. Using an experimentally determined Fermi level position, one can estimate the resistivity at room temperature to be in the $10^{19}$ Ohm cm range. Intentional doping with vanadium has several drawbacks. The increased impurity concentration contributes to phonon scattering and reduces the thermal conductivity. Also, due to segregation of vanadium during growth, it is quite easy to exceed the solid solubility limit of vanadium. This can lead to formation of inclusions and degradation of crystalline quality. The most significant disadvantage of this compensation approach, however, is its detrimental effect on MESFET characteristics. Deep vanadium-induced traps have been reported to capture electrons injected from the channel during the MESFET operation.[63,68] As the negative charge in the substrate builds up, it restricts

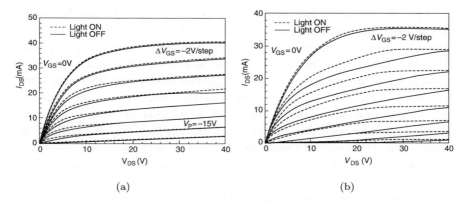

**Fig. 23.** Comparison of the DC I–V characteristics under 40 V $V_{DS}$ stress with and without illumination of $1 \times 100$ $\mu$m MESFETs processed on (a) low-doped HTCVD; and (b) commercial semi-insulating substrate (after Ellison *et al.*[63]).

the current flow between source and drain leading to the effect known as current collapse (Fig. 23).

Current research in this area is focused on the development of undoped high resistivity SiC with reduced density of electron traps. The approach relies on reduction of residual impurity concentrations and identification of an appropriate native defect that would play the role of a deep compensating center. Promising results have already been reported. In PVT-grown material, Deep Level Transient Spectroscopy measurements have detected three trap levels in the middle third of the band gap, with ionization energies of 0.68–0.74, 1.24–1.33, and 1.45–1.55 eV.[62] High temperature Hall effect data point to the Fermi energy pinned at five different levels: 0.9, 1.1, 1.2, 1.35, and 1.5 eV below conduction band in different crystals. The origin of discrepancies between these techniques is not clear at present, but both techniques indicate the presence of multiple deep levels in the middle third of the bandgap. In principle, any of the above traps can be used for compensation and fabrication of high resistivity material. The exact atomic structure of these centers is not known, although the presence of carbon vacancies in as-grown PVT boules has been reported.[62,69]

In HT CVD, two traps dominate the electrical and optical characteristics. The temperature dependent Hall effect measurements detected the Fermi level pinned at either 0.85 or 1.4 eV below the conduction band, while infrared spectroscopy detected two narrow absorption lines due to a silicon vacancy and to an unidentified defect referred to as UD-1.[63]

The above observations give important insight into the stoichiometry of the material produced by both methods. During sublimation growth, vapor is always silicon-rich and, as a consequence, the resulting crystals should contain excess silicon which should be accommodated in the form of point defects, such as a carbon vacancy, silicon interstitial, silicon antisite, and their complexes. CVD processes, on the other hand, are typically run with the carbon to silicon ratio $C/Si > 1$. This leads to carbon terminated surfaces with the growth rate controlled by the silicon flow. In such a case, the material should be carbon-rich with dominant point defects including a silicon vacancy, carbon interstitial, and carbon antisite. This assertion agrees with the observation of a carbon vacancy in PVT crystals and a silicon vacancy in CVD. As a net result, the point defect signatures in both types of material (CVD and PVT) are distinctly different. The compensation mechanisms leading to high resistivity are also expected to be different. At this point in time, we do not know the concentrations of any of these centers. However, assuming typical oscillator strength for optical transitions one can estimate vacancy concentration in as-grown materials at $10^{15}$ cm$^{-3}$. Electrically active point defects with densities in this range are clearly important in overall compensation. The ability to consistently produce semi-insulating material needed for RF device substrates hinges, therefore, on the ability to control vapor stoichiometry during growth.

## 5. SiC Homoepitaxy

All commercially available silicon carbide devices (Schottky diodes from Infineon AB and Cree, and MESFETs from Cree) are fabricated using epilayers deposited by chemical vapor deposition (CVD) on slightly misoriented (0001) SiC single crystal substrates. Barring an unexpected breakthrough in SiC hetero-epitaxy, CVD homo-epitaxial growth on basal plane will continue to dominate this field. Because of this, the discussion of SiC epitaxy will be limited here to homoepitaxy on vicinal [0001] surfaces.

Typically, the epitaxy is performed at temperatures much below bulk growth temperatures. In GaAs OMVPE and MBE deposition, the temperatures are between 550 and 650°C compared with the 1238°C melting point. Low temperatures result in low concentrations of residual impurities and native point defects. For example, the concentration of the arsenic antisite defect known as EL2 in bulk grown crystals is about $1 \times 10^{16}$ cm$^{-3}$ while it is only $5 \times 10^{13}$ cm$^{-3}$ in OMVPE epilayers. This difference is due to the much narrower existence range of GaAs compound at epitaxial growth temperatures compared to possible deviation from stoichiometry close to the melting point. Similar effects are expected in silicon carbide epitaxy as well, but there are additional limitations not encountered in the case of other semiconductors.

Stoichiometric high purity silicon carbide can be grown by CVD or gas source MBE at temperatures as low as 1100°C.[70,71] However, the stable form of SiC at temperatures below 1800°C is cubic 3C–SiC polytype. This implies that whenever silicon carbide is nucleated on the (0001)-type facet, the stacking sequence will be ABCABC ... or A'B'C'A'B'C' ... , corresponding to either one of two possible variants of 3C–SiC. The type of variant deposited is determined locally by the orientation of the upper-most layer in the substrate: the standard variant will grow on the surface terminated by an A, a B or C layer, while a twinned variant will grow on an A', a B', or C' terminated surface. Since the surfaces of hexagonal SiC substrates are stepped and terminated by both types of layers (i.e. normal and twinned), the 3C–SiC epilayers are a mixture of both variants. The boundaries between them are referred to as double positioning boundaries (Fig. 24), with typical densities of about 100 cm$^{-1}$. This makes the 3C–SiC/4H–SiC structure unsuitable for device applications.

One approach to elimination of double positioning boundaries and deposition of high quality 6H–SiC/6H–SiC or 4H–SiC/4H–SiC epilayers is "step controlled epitaxy".[72,73] In this method, the growth occurs on a substrate intentionally off-cut from the basal plane by 3°–8° towards either the $[11\bar{2}0]$ or the $[1\bar{1}00]$ direction. As a result, the vicinal surfaces are covered by steps separated by about 4–12 nm. During deposition, supersaturation

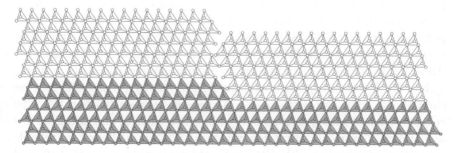

**Fig. 24.** Nucleation of 3C–SiC epilayer on 4H–SiC substrate and formation of double positioning boundary. 4H–SiC surface with a two bilayer step is shown in grey; 3C–SiC overgrowth in white.

is maintained low enough to prevent nucleation on terraces and growth proceeds by adatoms migrating to step edges. The structure of the film is determined by the stacking sequence of the exposed steps.

Commonly used sources for CVD are: silane, propane, nitrogen, and trimethyl aluminum in the flow of palladium-purified hydrogen. Since most SiC substrates are polished mechanically using diamond abrasives, the wafer surface is covered by a network of fine scratches. The subsurface layer is mechanically damaged and growth on such a surface results in high density of dislocations.[74] Because of this, the initial step in CVD growth is *in situ* etching using either hydrogen and/or HCl.[75,76] After the etching, the temperature is raised up to the growth temperature, typically 1550°– 1650°C, and growth commences by injection of hydrocarbon source. Due to the high temperatures, the reaction kinetics is fast and the growth rate is mass transport limited. Typical growth rates are between 2 and 5 $\mu$m/h.

Similar to bulk growth, nitrogen and aluminum are used as primary shallow donor and acceptor species with the carrier density controlled by the corresponding flow rates. The only problem with doping arises in the case of deposition of blocking layers for high voltage devices. For example, the 10 kV diode requires an $n$-type blocking layer at least 50 $\mu$m thick with a net donor concentration of $1 \times 10^{15}$ cm$^{-3}$. Control of such diluted flows is difficult and an alternative method called "site-competition" epitaxy was proposed by Larkin *et al.*[77,78] The approach relies on the fact that dopants in silicon carbide lattice occupy specific substitutional site. Nitrogen is known to substitute for carbon, while aluminum occupies exclusively silicon sites. During CVD, the ratio of silicon and carbon source flows determines the surface termination and number of "available" carbon and silicon sites. By increasing the Si/C ratio, the silicon sites are occupied by silicon and the incorporation of aluminum decreases. At the same growth conditions, the number of unfilled carbon sites is high and incorporation of nitrogen should be enhanced. This model is in good agreement with incorporation probabilities of numerous dopants in silicon carbide.[78] An example of the effect of Si/C ratio on nitrogen incorporation is shown in Fig. 25. During the growth run, the flow of nitrogen remained constant throughout, while the Si/C ratio was changed between 0.5 and 0.1. The concentration of nitrogen dropped by about an order of magnitude at low Si/C ratios. Control of source flow rates and Si/C ratio allowed for reliable deposition of epilayers with nitrogen concentrations of between $10^{14}$ and $10^{19}$ cm$^{-3}$.

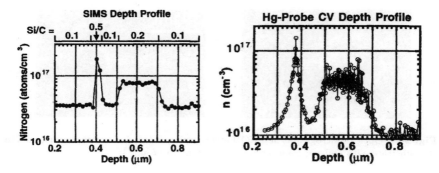

**Fig. 25.** Site competition epitaxy: direct comparison between SIMS profile analysis and mercury probe C-V scan of a structure grown by varying the carbon-source while maintaining a constant silicon-source (200 at. ppm) and constant nitrogen concentration (180 at. ppm) (after Larkin *et al.*[78]).

The densities of extended defects in SiC epilayers are determined primarily by the quality of the underlying substrate. Since dislocations present in SiC wafers cannot terminate inside the crystal, they have to replicate into the overgrowth. If there are no additional nucleation mechanisms of dislocations, the expected dislocation density in the active layers should be $10^3$ cm$^{-3}$ screw dislocations and $10^4$–$10^5$ cm$^{-2}$ threading edge dislocations. Basal plane dislocations should replicate as well. This particular defect is especially important in high voltage diodes. The basal plane dislocations in semiconductors are frequently dissociated into two partial dislocations with a stacking fault in between. Such configuration allows for minimum strain energy of the defect. Since the energy of stacking faults in silicon carbide polytypes is small (3 mJ/m$^2$ and 14.7 mJ/m$^2$ in 6H and 4H–SiC, respectively) the separation between partials is large. Such pre-existing stacking faults in diode structures are expected to serve as convenient starting sites for expansion of stacking faults leading to degradation of silicon carbide bipolar devices.[19]

As argued in Sec. 3.1, the basal plane dislocations are introduced into SiC boules as a result of plastic deformation of already solidified crystal. In the epilayers, however, they are grown-in defects. Since the intentional off-cut angle is small (3.5° for 6H–SiC and 8° for 4H–SiC), the dislocation line forms an oblique angle with the crystal surface and is strongly attracted to it by an image force. The system of a crystal and a dislocation can greatly reduce their energy if a dislocation, rather than extending in the basal plane, threads to the top of the layer. This indeed happens

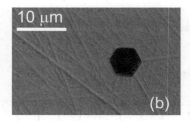

**Fig. 26.** Optical micrographs of a KOH etched epilayer (a), and corresponding area in the substrate with epilayer removed (b) (after Kamata *et al.*[80]).

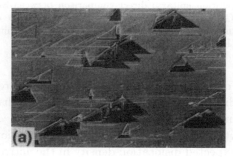

**Fig. 27.** Optical Nomarski image of triangular defects in 4H–SiC epilayer. Defects have been decorated by thermal oxidation (after Konstantinov *et al.*[81]).

frequently during epitaxy and most of the basal plane dislocations are converted into threading edges.[79] The details of this process and its dependence on growth conditions have not yet been quantified. Since micropipes correspond to open-core screw dislocations, they also must propagate into the epilayers. Most of them do, but in some growth conditions, the super-screw dislocations dissociate into several elementary dislocations (Fig. 26).[80] The mechanism of this process is under investigation.

Since there is no strain generated during homo-epitaxy, one would not expect nucleation of additional dislocations in the epilayer. This assumption has not yet been carefully verified in silicon carbide and does not have to be true for all growth conditions. In fact, preliminary data indicate that threading dislocation arrays are nucleated during epitaxy.

Yet another extended defect characteristic of SiC epitaxy is a 3C–SiC inclusion. They frequently have a shape of triangular pit with the apex at the original substrate/layer interface (Fig. 27). The arms are usually along the $\langle 11\bar{2}0 \rangle$ directions, forming either a 60° or a 120° angle. The accepted model of their formation involves a defect on the original substrate surface that impedes the step flow. Such a defect could be a deeper scratch due to polishing, a micropipe, or a contamination. Interference with step flow produces a large terrace down-step from the obstacle. Once the size of the terrace exceeds the adatom diffusion length, a 2D cubic nucleus can form. The density of 3C inclusions strongly depends on the substrate surface preparation.

Residual impurities and native point defects in epilayers have received only limited attention. The Deep Level Transient Spectroscopy data on Schottky diodes on $n$-type material revealed only one trap in as-grown 4H–SiC epilayers located at 0.65 eV below conduction band. The concentration was exceedingly small and in all reports was below $10^{13}$ cm$^{-3}$.[82,83] In 6H–SiC, two DLTS peaks have been reported corresponding to activation energies 0.39–0.43 eV and 1.17–1.2 eV, respectively.[82] The origin of these traps remains unknown. The only residual impurity in epilayers present in significant concentrations is hydrogen. It tends to form complexes with acceptors and in particular with boron. The hydrogen concentration follows that of boron and can exceed $10^{18}$ cm$^{-3}$.[78] Boron–hydrogen complex dissociates at temperatures above 1700°C, with hydrogen diffusing out of the layer.

Very low concentrations of deep traps in SiC epilayers detected by DLTS are somewhat at odds with experimental results on minority carrier lifetimes. Both room temperature, time resolved photoluminescence[84] and reverse recovery switching analysis[85] estimated carrier lifetime values in the highest quality CVD epilayers at 0.7–2.0 $\mu$s. These results indicate that carrier lifetimes are limited by a non-radiative recombination process most likely involving deep recombination centers.

# References

1. P. Pirouz and J. W. Yang, *Ultramicroscopy* **51**, 189 (1993).
2. J. A. Lely, *Ber. Deut. Keram. Ges.* **32**, 229 (1955).
3. Y. M. Tairov and V. F. Tsvetkov, *J. Cryst. Growth* **43**, 209 (1978).
4. M. Skowronski, *Mater. Sci. Forum*, in print.
5. J. Drowart, G. D. Maria and M. G. Inghram, *J. Chem. Phys.* **29**, 1015 (1958).
6. S. K. Lilov, *Diamond and Related Materials* **4**, 1331 (1995).
7. R. C. Glass, D. Henshall, V. F. Tsevetskov and C. H. Carter, *Phys. Status Solidi* **B202**, 149 (1997).
8. D. Hofmann, M. Heinze, A. Winnacker, F. Durst, L. Kadinski, P. Makarov and M. Schafer, *J. Cryst. Growth* **146**, 214 (1995).

9.  S. G. Mueller, R. C. Glass, H. M. Hobgood, V. F. Tsvetskov, M. Brady, D. Henshall, D. Malta, R. Singh, J. Palmour and C. H. Carter, *Mater. Sci. Eng.* **B80**, 327 (2001).
10. K. Koga, Y. Fujikawa, Y. Ueda and T. Yamaguchi, *Springer Proc. Phys.* **71**, 96 (1992).
11. P. G. Neudeck and J. A. Powell, *IEEE Electron. Device Lett.* **15**, 63 (1994).
12. J. A. Powell, P. G. Neudeck, D. J. Larkin, J. W. Yang and P. Pirouz, *Inst. Phys. Conf. Ser.* **137**, 161 (1994).
13. P. G. Neudeck, W. Huang and M. Dudley, *IEEE Trans. Electron. Device* **46**, 478 (1999).
14. J. P. Bergman, H. Lendenmann, P. A. Nilsson, U. Lindfelt and P. Skytt, *Mater. Sci. Forum* **353-356**, 299 (2000).
15. H. Lendenmann, F. Dahlquist, N. Johansson, R. Soderholm, P. A. Nilsson, J. P. Bergman and P. Skytt, *Mater. Sci. Forum* **353-356**, 727 (2001).
16. S. G. Sridhara, F. H. C. Carlsson, J. P. Bergman and E. Janzen, *Appl. Phys. Lett.* **79**, 3944 (2001).
17. H. Iwata, U. Lindefelt, S. Oberg and P. R. Briddon, *Mater. Sci. Forum*, in print.
18. M. S. Miao, S. Limpijumnong and W. R. L. Lambrecht, *Appl. Phys. Lett.*, in print.
19. H. Lendenmann, F. Dahlquist, J. P. Bergman, H. Bleichner and C. Hallin, *Mater. Sci. Forum*, in print.
20. J. Takahashi, M. Kanaya and Y. Fujiwara, *J. Cryst. Growth* **135**, 61 (1994).
21. A. S. Jordan, R. Caruso, A. R. VonNeida and J. W. Nielsen, *J. Appl. Phys.* **52**, 3331 (1981).
22. A. S. Jordan, A. R. V. Neida and R. Caruso, *J. Cryst. Growth* **70**, 555 (1984).
23. S. Fujita, K. Maeda and S. Hyodo, *Phil. Mag.* **A55**, 203 (1987).
24. K. Maeda, K. Suzuki, S. Fujita, M. Ichihara and S. Hyodo, *Philos. Mag.* **A57**, 573 (1988).
25. A. V. Samant, W. L. Zhou and P. Pirouz, *Phys. Status Solidi* **A166**, 155 (1998).
26. S. Ha, G. S. Rohrer, M. Skowronski, V. D. Heydemann and D. W. Snyder, *Mater. Sci. Forum* **338-342**, 67 (2000).
27. D. Hobgood, M. Brady, W. Brixius, G. Fetchko, R. Glass, D. Henshall, J. Jenny, R. Leonard, S. G. Muller, V. Tsetskov and J. C. Cartyer, *Mater. Sci. Forum* **338-342**, 3 (2000).
28. M. Pons, C. Moulin, J.-M. Dedulle, A. Pisch, B. Pelissier, E. Blanquet, M. Anikin, E. Pernot, R. Madar, C. Bernard, C. Faure, T. Billon and G. Feuillet, *Mater. Res. Soc. Symp. Proc.* **640**, H1.4.1 (2001).
29. M. Selder, L. Kadinski, F. Durst and D. Hofmann, *J. Cryst. Growth* **236**, 501 (2001).
30. S. Ha, M. Skowronski, W. M. Vetter and M. Dudley, *J. Appl. Phys.*, in print.
31. S. Ha, N. T. Nufher, G. S. Rohrer, M. D. Graef and M. Skowronski, *J. Electron. Mater.* **29**, L5 (2000).
32. R. C. Glass, L. O. Kjellberg, V. F. Tsvetskov, J. E. Sundgren and E. Janzen, *J. Cryst. Growth* **132**, 504 (1993).

33. S. Ha, N. T. Nuhfer, G. S. Rohrer, M. D. Graef and M. Skowronski, *J. Cryst. Growth* **220**, 308 (2000).
34. J. Takahashi, N. Ohtani and M. Kanaya, *J. Cryst. Growth* **167**, 596 (1996).
35. S. Ha, M. DeGraef, G. S. Rohrer and M. Skowronski, *Mater. Sci. Forum* **338–342**, 477 (2000).
36. M. Katsuno, N. Ohtani, T. Aigo, T. Fujimoto, H. Tsuge, H. Yashiro and M. Kanaya, *J. Cryst. Growth* **216**, 256 (2000).
37. X. R. Huang, M. Dudley, W. M. Vetter, W. Huang, S. Wang and J. C. H. Carter, *Appl. Phys. Lett.* **74**, 353 (1999).
38. M. Dudley, S. Wang, W. Huang, C. H. Carter, V. F. Tsvetskov and C. Fazi, *J. Phys. D: Appl. Phys.* **28**, A63 (1995).
39. J. Takahashi, N. Ohtani, M. Katsuno and S. Shinoyama, *J. Cryst. Growth* **181**, 229 (1997).
40. E. K. Sanchez, J. Q. Liu, M. D. Graef, M. Skowronski, W. M. Vetter and M. Dudley, *J. Appl. Phys.* **91**, 1143 (2002).
41. M. H. Hong, A. V. Samant and P. Pirouz, *Mater. Sci. Forum* **338–342**, 513 (2000).
42. J. Q. Liu, E. K. Sanchez and M. Skowronski, *Mater. Sci. Forum*, in print.
43. F. C. Frank, *Acta Cryst.* **4**, 497 (1951).
44. W. Si, M. Dudley, R. Glass, V. Tsvetskov and J. C. Carter, *J. Electron. Mater.* **26**, 128 (1997).
45. S. Wang, M. Dudley, J. C. H. Carter, V. F. Tsvetskov and C. Fazi, *Mat. Res. Soc. Symp. Proc.* **375**, 281 (1995).
46. W. Si, M. Dudley, R. Glass, V. Tsetskov and J. C. H. Carter, *Mater. Sci. Forum* **264–268**, 429 (1998).
47. J. Giocondi, G. S. Rohrer, M. Skoewronski, V. Balakrishna, G. Augustine, H. M. Hobggod and R. H. Hopkins, *J. Cryst. Growth* **182**, 351 (1997).
48. J. Giocondi, G. S. Rohrer, M. Skowronski, V. Balakrishna, G. Augustine, H. M. Hobgood and R. H. Hopkins, *Mater. Res. Soc. Symp. Proc.* **423**, 539 (1996).
49. J. Heindl, H. P. Strunk, V. D. Heidemann and G. Pensl, *Phys. Status Solidi* **A162**, 251 (1997).
50. P. Pirouz, *Phil. Mag.* **A78**, 727 (1998).
51. M. Dudley, X. R. Huang, W. Huang, A. Powell, S. Wang, P. Neudeck and M. Skowronski, *Appl. Phys. Lett.* **75**, 784 (1999).
52. T. A. Kuhr, E. K. Sanchez, M. Skowronski, W. M. Vetter and M. Dudley, *J. Appl. Phys.* **89**, 4625 (2001).
53. V. Balakrishna, R. H. Hopkins, G. Augustine, G. T. Dunne and R. N. Thomas, *Inst. Phys. Conf. Ser.* **160**, 321 (1997).
54. H. H. Woodbury and G. W. Ludwig, *Phys. Rev.* **124**, 1083 (1961).
55. W. Gotz, A. Schoner, G. Pensl, W. Suttrop, W. J. Choyke, R. Stein and S. Leibenzeder, *J. Appl. Phys.* **73**, 3332 (1993).
56. W. Suttrop, G. Pensl, W. J. Choyke, A. Dornen, S. Leibenzeder and R. Stein, *Amorphous and Crystalline SiC* **IV**, 1992.
57. B. J. Skromme, K. Palle, C. D. Poweleit, L. R. Bryant, W. M. Vetter, M. Dudley, K. Moore and T. Gehoski, *Mater. Sci. Forum*, in print.

58. J. Q. Liu, H. Chung, T. Kuhr, Q. Li and M. Skowronski, *Appl. Phys. Lett.*, in print.
59. L. S. Dang, K. M. Lee, G. D. Watkins and W. J. Choyke, *Phys. Rev. Lett.* **45**, 390 (1980).
60. M. Ikeda, H. Matsunami and T. Tanaka, *Phys. Rev.* **B6**, 2842 (1980).
61. O. Kordina, C. Hallin, A. Ellison, A. S. Bakin, I. G. Ivanov, A. Henry, R. Yakimova, M. Touminen, A. Vehanen and E. Janzen, *Appl. Phys. Lett.* **69**, 1456 (1996).
62. S. G. Mueller, M. Brady, B. Brixius, G. Fechko, R. C. Glass, F. D. Henshall, H. M. Hobgood, J. R. Jenny, R. Leonard, D. Malta, A. Powell, V. F. Tsvetskov and J. C. H. Carter, *Mater. Sci. Forum*, in print.
63. A. Ellison, B. Magnusson, C. Hemmingson, W. Magnusson, T. Iakimov, L. Storasta, A. Henry, N. Henelius and E. Janzen, *Mater. Res. Soc. Symp. Proc.* **640**, H1.2.1 (2001).
64. H. M. Hobgood, R. C. Glass, G. Augustine, R. H. Hopkins, J. R. Jenny, M. Skowronski, W. C. Mitchel and M. Roth, *Appl. Phys. Lett.* **66**, 1364 (1995).
65. J. R. Jenny, M. Skowronski, W. C. Mitchel, H. M. Hobgood, R. C. Glass, G. Augustine and R. H. Hopkins, *J. Appl. Phys.* **79**, 2326 (1996).
66. J. Schneider, H. D. Muller, K. Meier, W. Wilkening, F. Fuchs, A. Dornen, S. Leibenzeder and R. Stein, *Appl. Phys. Lett.* **56**, 1184 (1990).
67. J. R. Jenny, M. Skowronski, W. C. Mitchel, H. M. Hobgood, R. C. Glass, G. Augustine and R. H. Hopkins, *Appl. Phys. Lett.* **68**, 1963 (1996).
68. O. Noblanc, C. Arnodo, C. Dua, E. Chartier and C. Brylinski, *Mater. Sci. Forum* **338–342**, 1247 (2000).
69. M. E. Zvanut and V. V. Konovalov, *Appl. Phys. Lett.* **80**, 410 (2002).
70. T. Hatayama, Y. Tarui, T. Fuyuki and H. Matsunami, *J. Cryst. Growth* **150**, 934 (1995).
71. O. Kordina, L. O. Bjorketun, A. Henry, C. Hallin, R. C. Glass, L. Hultman, J. E. Sundgren and E. Janzen, *J. Cryst. Growth* **154**, 303 (1995).
72. H. S. Kong, J. T. Glass and R. F. Davis, *J. Appl. Phys.* **64**, 2672 (1988).
73. T. Ueda, H. Nishino and H. Matsunami, *J. Cryst. Growth* **104**, 695 (1990).
74. E. K. Sanchez, S. Ha, J. Grim, M. Skowronski, W. M. Vetter, M. Dudley, R. Bertke and W. C. Mitchel, *J. Electrochem. Soc.* **149**, G131 (2002).
75. A. Burk and L. B. Rowland, *J. Cryst. Growth* **167**, 586 (1996).
76. J. A. Powell, D. J. Larkin and P. B. Abel, *J. Electron. Mater.* **24**, 295 (1995).
77. D. J. Larkin, P. G. Neudeck, J. A. Powell and L. G. Matus, *Appl. Phys. Lett.* **65**, 1659 (1994).
78. D. J. Larkin, *Phys. Status Solidi* **B202**, 305 (1997).
79. S. Ha, P. Mieszkowski, L. B. Rowland and M. Skowronski, *Mater. Sci. Forum*, in print.
80. I. Kamata, H. Tsuchida, T. Jikimoto and K. Izumi, *Mater. Sci. Forum* **353–356**, 311 (2001). •
81. A. O. Konstantinov, C. Hallin, P. Pecz, O. Kordina and E. Janzen, *J. Cryst. Growth* **178**, 495 (1997).
82. T. Kimoto, A. Itoh and H. Matsunami, *Phys. Status Solidi* **B202**, 247 (1997).

83. T. Dalibor, G. Pensl, H. Matsunami, T. Kimoto, W. J. Choyke, A. Schoner and N. Nordell, *Phys. Status Solidi* **A162**, 199 (1997).

84. J. P. Bergman, O. Kordina and E. Janzen, *Phys. Status Solidi* **A162**, 65 (1997).

85. P. Neudeck, *J. Electron. Mater.* **27**, 317 (1998).

# CHAPTER 6

# HIGH-VOLTAGE SiC POWER RECTIFIERS

T. Paul Chow

*Center for Integrated Electronics, Rensselaer Polytechnic Institute,*
*Troy, NY 12180-3590, USA*

## Contents

## 1. Introduction

Silicon has long been the dominant semiconductor of choice for high-voltage power electronics applications.[1,2] However, recently, wide bandgap semiconductors, particularly SiC and GaN, have attracted much attention because they are projected to have much better performance than silicon.[3-7] Compared with silicon, these wide bandgap semiconductors, SiC, GaN and InN can be categorized into one group while diamond, BN and AlN into another because the former have bandgaps of 2–3.5 eV and the latter 5.5–6.5 eV. On the other hand, Group IV or Groups IV–IV semiconductors have indirect bandgaps, whereas most of the Group III-nitrides have direct bandgaps (except for BN). The superior physical properties of

**Table 1.**  Physical properties of important semiconductors for high-voltage power devices.

| Material | $E_g$ (eV) | $n_i$ (cm⁻³) | $\varepsilon_r$ | $\mu_n$ (cm²/Vs) | $E_c$ (MV/cm) | $v_{sat}$ (10⁷ cm/s) | $\lambda$ (W/cm K) | Direct/ Indirect |
|---|---|---|---|---|---|---|---|---|
| Si | 1.1 | $1.5 \times 10^{10}$ | 11.8 | 1350 | 0.3 | 1.0 | 1.5 | I |
| Ge | 0.66 | $2.4 \times 10^{13}$ | 16.0 | 3900 | 0.1 | 0.5 | 0.6 | I |
| GaAs | 1.4 | $1.8 \times 10^{6}$ | 12.8 | 8500 | 0.4 | 2.0 | 0.5 | D |
| GaP | 2.3 | $7.7 \times 10^{-1}$ | 11.1 | 350 | 1.3 | 1.4 | 0.8 | I |
| InN | 1.86 | $\sim 10^{3}$ | 9.6 | 3000 | 1.0 | 2.5 | - | D |
| GaN | 3.39 | $1.9 \times 10^{-10}$ | 9.0 | 900 | 3.3 | 2.5 | 1.3 | D |
| 3C–SiC | 2.2 | 6.9 | 9.6 | 900 | 1.2 | 2.0 | 4.5 | I |
| 4H–SiC | 3.26 | $8.2 \times 10^{-9}$ | 10 | $720^a$ $650^c$ | 2.0 | 2.0 | 4.5 | I |
| 6H–SiC | 3.0 | $2.3 \times 10^{-6}$ | 9.7 | $370^a$ $50^c$ | 2.4 | 2.0 | 4.5 | I |
| Diamond | 5.45 | $1.6 \times 10^{-27}$ | 5.5 | 1900 | 5.6 | 2.7 | 20 | I |
| BN | 6.0 | $1.5 \times 10^{-31}$ | 7.1 | 5 | 10 | $1.0^*$ | 13 | I |
| AlN | 6.1 | $\sim 10^{-31}$ | 8.7 | 1100 | 11.7 | 1.8 | 2.5 | D |

*Note:* $a$–mobility along $a$-axis, $c$–mobility along $c$ axis, *–estimate.

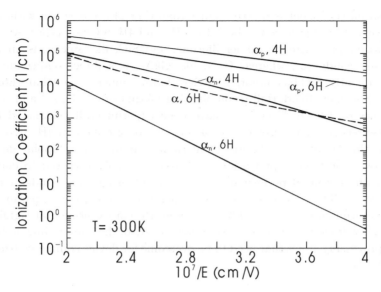

**Fig. 1.** Experimental impact ionization coefficients of electron and hole in 6H– and 4H–SiC.[8,9,12]

these semiconductors offer a lower intrinsic carrier concentration (10–35 orders of magnitude), a higher electric breakdown field (4–20 times), a higher thermal conductivity (3–13 times), and a larger saturated electron drift velocity (2–2.5 times), when compared with silicon (See Table 1). SiC has over 150 polytypes (which are different crystal structures with the same stoichiometry of a compound semiconductor). Also, only the 6H– and 4H–SiC polytypes are available commercially in both bulk wafers and custom epitaxial layers. Between the two polytypes, 4H–SiC has become the semiconductor of choice due to the more isotropic nature of many of its electrical properties. Besides these properties, the impact ionization coefficients of electron and hole are also very important for power device considerations. The experimental coefficients, usually extracted from breakdown characteristics of reverse-biased $p$-$n$ or Schottky junctions, are shown in Fig. 1 for both 6H– and 4H–SiC.[8–10] Also included in the figure is the average ionization coefficient for 6H–SiC. Such an average ionization coefficient, usually modeled by a power law dependence on the electric field ($\alpha \propto \xi^n$, where $n$ is 5 or 7), allows one to estimate analytically the breakdown voltage and depletion width at breakdown. (See Ref. 11 for the silicon case and Ref. 12 for 6H–SiC.) It should be emphasized that, unlike in silicon, the hole ionization coefficient is mostly higher than that of the electron within the electric field range shown in both the 6H– and 4H–SiC polytypes. Another important parameter is carrier lifetime, and SiC, like silicon, is an indirect semiconductor, whereas GaN, like GaAs, is a direct semiconductor. Consequently, SiC has minority carrier lifetimes much longer than those in GaN. Shockley–Read–Hall recombination lifetimes of $> 1$ $\mu s$[13] and Auger lifetimes of $4 \times 10^{-30}$ cm$^6$/s[14] have been experimentally extracted from reverse recovery measurements of *p-i-n* junction rectifiers in 4H–SiC).

In this chapter, the figures of merit for unipolar and bipolar devices for power electronic applications will first be briefly reviewed to demonstrate the potential performance improvement using wide bandgap semiconductors. Then, the basic physics of operation and key device parameters of two-terminal rectifiers will be presented. Recent experimental highlights on high-voltage SiC devices are summarized. Also, the outstanding material and processing issues that need to be overcome for device commercialization will be pointed out.

## 2. Performance Projection

Categorically, there are two families of power rectifiers — the Schottky rectifier belonging to the unipolar family and the junction rectifier belonging to the bipolar family. It is worth pointing out that, despite their simplicity, these power rectifiers are often just as important as the three-terminal switching devices in determining the total power loss and hence the power efficiency in many power electronic circuits, such as the half-bridge circuit.[2] To quantify the device performance enhancement possible with SiC and GaN, several unipolar and bipolar figures of merit have been proposed.[3-7] In Table 2, the various unipolar figures of merit for the semiconductors are shown and SiC and GaN offer more than an order of magnitude improvement over silicon. These figures assume different heat sinks and operating frequency.[3] The unipolar figures of merit apply to the Schottky rectifiers. For junction rectifiers, a bipolar figure of merit is required. Since SiC has a large diode turn-on voltage due to its larger bandgap, its forward drop is lower than that of equivalent devices only at very high current densities ($> 5000$ A/cm$^2$). Consequently, the conduction loss often cannot be less than the silicon device at low to medium current density and only yields a lower total power loss when the switching frequency exceeds a certain frequency, $f_{min}$.[5,6] This $f_{min}$, at which the conduction loss is equal to the switching loss (at 50% duty cycle), has been

**Table 2.** Normalized unipolar figures of merit of important semiconductors for high-voltage power devices.[3,7]

| Material | $\lambda$ | JM $(E_c v_{sat}/\pi)^2$ | MJM $\lambda^* JM$ | KM $\lambda(v_{sat}/\varepsilon_r)^{1/2}$ | $Q_{F1}$ $\lambda\sigma_A$ | $Q_{F2}$ $\lambda\sigma_A E_c$ | BM $(Q_{F3})$ $\varepsilon_r\mu E_c^3$ | BHFM $\mu E_c^2$ |
|---|---|---|---|---|---|---|---|---|
| Si | 1 | 1 | 1 | 1 | 1 | 1 | 1 | 1 |
| Ge | 0.4 | 0.03 | 0.012 | 0.20 | 0.06 | 0.02 | 0.2 | 0.3 |
| GaAs | 0.33 | 7.1 | 2.4 | 0.45 | 5.2 | 6.9 | 15.6 | 10.8 |
| GaP | 0.53 | 37 | 20 | 0.7 | 10 | 40 | 16 | 5 |
| InN | – | 58 | – | – | – | – | 46 | 19 |
| GaN | 0.87 | 760 | 655 | 1.6 | 560 | 6220 | 650 | 77.8 |
| 3C–SiC | 3 | 65 | 195 | 1.6 | 100 | 400 | 33.4 | 10.3 |
| 4H–SiC | 3 | 180 | 540 | 4.61 | 390 | 2580 | 130 | 22.9 |
| 6H–SiC | 3 | 260 | 780 | 4.68 | 330 | 2670 | 110 | 16.9 |
| Diamond | 13.3 | 2540 | 33800 | 32.1 | 54860 | 1024000 | 4110 | 470 |
| BN | 8.7 | 1100 | 9700 | 11 | 715 | 23800 | 83 | 4 |
| AlN | 1.7 | 5120 | 8700 | 21 | 52890 | 2059000 | 31700 | 1100 |

**Table 3.** A bipolar figure of merit applied to the power $p$-$i$-$n$ junction rectifier (calculated at $J_F = 100$ A/cm², BV $= 1000$ V).[5,6]

| Name | $N_d$ (cm⁻³) | $W_{N-}$ (μm) | $V_F$ (V) | $J_{end}$ (A/cm²) | $J_{off}$ (A/cm²) | $t_s$ (μs) | $t_{f1}$ (μs) | $t_{f2}$ (μs) | $E_{off}$ (mJ) | $f_{min}$ (kHz) |
|---|---|---|---|---|---|---|---|---|---|---|
| Si | $1.3 \times 10^{14}$ | 100 | 0.88 | 9.7 | $2 \times 10^{-5}$ | $7 \times 10^{-3}$ | 0.81 | 0 | $5.4 \times 10^{-3}$ | – |
| Ge | $4.4 \times 10^{13}$ | 285 | 0.60 | 8.3 | $8 \times 10^{-2}$ | $7.6 \times 10^{-3}$ | 1.8 | 0 | $6.2 \times 10^{-3}$ | 0 |
| 3C–SiC | $3.8 \times 10^{15}$ | 33.1 | 2.00 | 42.8 | $2 \times 10^{-15}$ | $4.4 \times 10^{-3}$ | 0.2 | $2.2 \times 10^{-3}$ | $2 \times 10^{-2}$ | 1.62 |
| 6H–SiC | $1.6 \times 10^{16}$ | 23.9 | 2.65 | 46.2 | $3 \times 10^{-21}$ | $1.0 \times 10^{-2}$ | 0.14 | $4.6 \times 10^{-3}$ | $5.5 \times 10^{-3}$ | 1.82 |
| Diamond | $1.2 \times 10^{17}$ | 72.0 | 5.21 | 93.4 | $5 \times 10^{-44}$ | $2.6 \times 10^{-5}$ | 0.02 | $1 \times 10^{-2}$ | $1.1 \times 10^{-3}$ | 4.12 |

considered as the bipolar figure of merit, and, in the case of a 1000 V *p-i-n* junction rectifier, is about 20 kHz for SiC when compared with silicon, as illustrated in Table 3. This $f_{min}$ decreases with increasing blocking voltage rating due to faster rate of increase in drift layer thickness and hence minority carrier storage for silicon devices.

The unipolar and bipolar figures of merit mentioned above compare silicon devices with the wide bandgap ones of the same type. However, for each semiconductor, depending on its bandgap, there is a crossover in voltage rating above which bipolar devices are preferred over unipolar ones due to the reduced drift-layer resistance from conductivity modulation of the bipolar carrier injection. For silicon, this voltage is about 300 V, whereas for SiC, it is about 3000 V. Also, with increasing operating temperature, this crossover voltage is expected to decrease since the ON-resistance of unipolar devices varies inversely with the second power of temperature, while the turn-on voltage decreases and carrier lifetimes increase. We will discuss this crossover in more detail below.

## 3. Termination

To design a power rectifier of a particular reverse blocking voltage, we have approximated the electron and hole ionization coefficients using a power law, as mentioned earlier. Then, we obtain the ideal breakdown voltage ($BV_{pp}$) and depletion layer width at breakdown ($W_{pp}$) as a function of background doping, as shown in Fig. 2 for 6H– and 4H–SiC. These analytical calculations have been corroborated well with numerical simulations and experimental results. It can be seen that, for the same background doping and along the *c*-axis, 6H–SiC has a 10–15% higher BV than 4H–SiC, despite the larger bandgap of the latter. Also, the effective avalanche field estimated for doping concentration of $10^{15}$ to $10^{17}$ cm$^{-3}$ is in the range of 2.5 to $5 \times 10^6$ V/cm, close to the experimental value of $2$–$3 \times 10^6$ V/cm. Also, it is now generally accepted that the avalanche voltage in both 6H– and 4H–SiC has a positive temperature coefficient (first reported in Ref. 15) like that observed in silicon. Early experimental results to the contrary are attributed to defects.

Besides the one-dimensional BV design, proper termination[1,2] must also be employed to mitigate the effect of field crowding due to junction curvatures in planar devices. While one can use the same techniques as for silicon devices, some are substantially easier to design and process than others because of the $10\times$ higher surface field. Termination techniques that

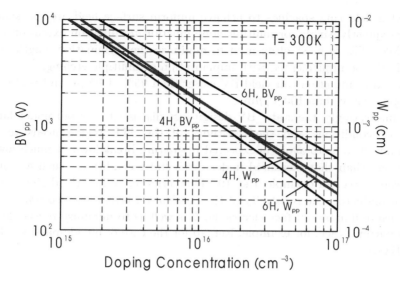

**Fig. 2.** Breakdown voltage of parallel-plane, one-sided abrupt junction ($BV_{pp}$) and its depletion layer width at breakdown ($W_{pp}$) for 6H– and 4H–SiC at 300 K.

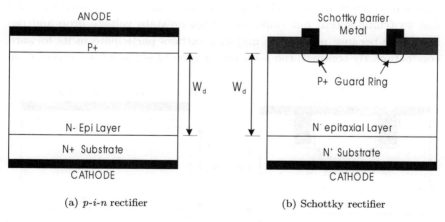

(a) *p-i-n* rectifier        (b) Schottky rectifier

**Fig. 3.** Schematic device cross-sections of planar (a) pin junction rectifier and (b) Schottky barrier rectifier.

have been used in SiC devices include field rings, field plates and junction termination extension (JTE).[1,2] While field plates and field rings are extensively utilized as terminations in silicon, they are not preferred since, in the former, the necessary oxide field is too large, whereas, in the latter, the needed field ring spacings to achieve uniform surface field are too small. Consequently, single- or multiple-zone junction termination extension (SZ or MZ JTE) is most often used to minimize the termination length. MZ JTE can be implemented with implantation or multiple etchings.[16–18] An optimal termination can yield BV closer than 95% of $BV_{PP}$ and is able to support more than 100 V/$\mu$m.

Interestingly, the static BV design just discussed may fail under large $dV/dt$ conditions and the resistivity of the JTE region needs to be optimized so as to avoid unexpected premature breakdown at much lower voltages than that measured under quasi-static conditions (such as curve tracers). (See, for example, Ref. 19.) Furthermore, besides the design in the semiconductor, the resistivity of the passivation layers are also important to ensure uniform surface electric field as well as to minimize reverse leakage current. (See, for example, Ref. 20 for the case of nitride-passivated Si IGBTs.)

## 4. Rectifier Structures

The schematic cross-sections of the basic structures of planar Schottky and junction rectifiers are shown in Fig. 3. High-voltage Schottky rectifiers offer fast switching speed, but suffer from high on-state voltage drop and on-resistance because mostly the majority carriers participate in its forward conduction. By contrast, the *p-i-n* junction rectifier has a low forward drop

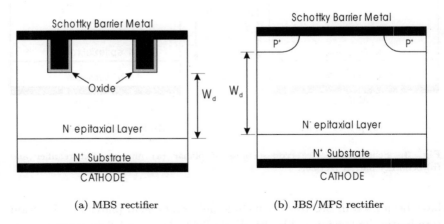

(a) MBS rectifier                 (b) JBS/MPS rectifier

**Fig. 4.** Schematic device cross-sections of (a) Trench MOS-Barrier Schottky (TMBS) rectifier, and (b) planar Junction Barrier Schottky (JBS) or Merged *p-i-n*/Schottky (MPS) rectifier.

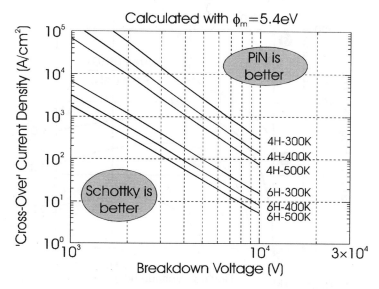

**Fig. 5.** Crossover voltage between Schottky and junction rectifiers at different operating temperature for 6H– and 4H–SiC.[25]

and high current capability due to conductivity modulation, but has slow reverse recovery characteristics due to minority carrier storage. To combine the best features of these two rectifiers, hybrid rectifier structures, such as the Junction Barrier Schottky (JBS), Merged *p-i-n*/Schottky (MPS) and MOS Barrier Schottky (MBS) rectifiers, have been proposed and have been demonstrated in silicon.[21–24] The trench version of the MBS rectifier is schematically shown in Fig. 4(a), while the planar version of the JBS/MPS rectifiers are shown in Fig. 4.(b).

Whether a unipolar or a bipolar rectifier is preferred depends on many device parameters, such as reverse blocking voltage, forward current density, maximum allowable reverse current density, operating temperature and switching frequency. The particular device type is often chosen to either minimize the total power dissipation or maximize the Safe-Operating-Area (SOA) during device turn-on or turn-off. If we only focus on forward conduction, we note that the turn-on voltage of silicon rectifier is only 0.7 V while that of the 4H–SiC rectifier is about 2.5 V (the turn-on voltage is approximately 75% of the bandgap). Consequently, as shown in Fig. 5, the crossover voltage for the 6H–SiC rectifier operating at 100 A/cm$^2$ exceeds 3000 V at room temperature but drops to about 2000 V at 300°C. (That

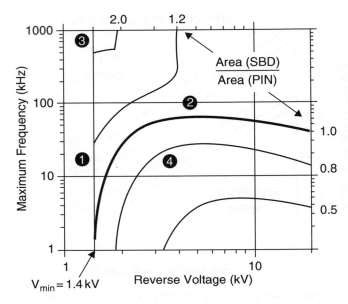

**Fig. 6.** Contour plot of the area ratio for a 4H–SiC *p-i-n* diode relative to a 4H–SiC Schottky barrier diode as a function of reverse blocking voltage and switching frequency.[26]

for 4H–SiC devices is slightly higher due to the 0.2 eV larger bandgap.) Practically, the crossover voltage is below this estimate because other factors, such as reverse leakage current density and forward over-current density tend to lower this crossover voltage. For example, while the crossover voltage based on forward conduction considerations alone for silicon would be 300 V, the maximum reverse voltage for commercial silicon Schottky rectifiers is only 100 V. We estimate that the practical upper reverse blocking voltage limit for 4H–SiC rectifiers is about 2000 V. Similar considerations also apply to the crossover voltage between unipolar and bipolar transistors as well.[25] To further refine the power loss estimate, switching loss has been included in a recent calculation, in which it is proposed that Schottky rectifiers are preferred over junction rectifiers for all reverse voltage ranges, if a switching frequency that exceeds 70 kHz is needed.[26] A contour plot of the area ratio for a 4H–SiC *p-i-n* diode relative to a 4H–SiC Schottky barrier diode as a function of reverse blocking voltage and switching frequency is shown in Fig. 6.[26] However, to include the most general power electronics circuits and systems, other factors, such as turn-on d$I$/d$t$ and reverse leakage current at elevated temperatures, must be considered. Inclusion of these factors makes the junction rectifiers more attractive with increasing reverse voltage rating. Consequently, SiC junction rectifiers are dominant for reverse voltages at or above 5 kV.

As far as the process technology is concerned, the Schottky rectifier is constructed on the metal-semiconductor junctions whereas the junction rectifier is based on the *p-n* junctions. The device process technology of the former depends on the choice of the Schottky metal as well as on the surface cleaning techniques and post-metal deposition annealing procedure. By contrast, the junction rectifier characteristics are controlled by the *p-n* junction formation technology, the most popular of which are *in-situ* doped epitaxy and compensation doping by ion implantation. Due to its bipolar nature, the junction rectifier is very sensitive to the carrier lifetimes of the material and bulk crystalline defects.

We will examine the theoretical electrical characteristics of these rectifiers and their SiC experimental realizations in the following sections. A list of recent experimental results on various high-voltage rectifiers in 6H– and 4H–SiC is shown in Table 4. At present, the highest experimental BV is 5 kV for Schottky rectifiers, 19 kV for *p-i-n* junction rectifiers, and 3.7 kV for JBS rectifiers in 4H–SiC.

**Table 4.** A list of SiC power rectifiers that have been experimentally demonstrated.

| Device Type | Polytype | Power Ratings | Features | Developer |
|---|---|---|---|---|
| Schottky | 6H–SiC | 1100 V | Ni, Field plate term. | U. of Cincinnati, 1995 |
| | 4H–SiC | 1750 V | Ti, C-Face, $p^+$ edge term. $n = 1.02$, $V_{F100} = 1.12$ V, 5 m$\Omega$ cm$^2$ | Kyoto U., 1995 |
| | 4H–SiC | 1720 V | Ni, B-implanted edge term. 5.6 m$\Omega$ cm$^2$ | Purdue U., 1997 |
| | 4H–SiC | 3 kV | Ni, Field plate term., $V_{F100} = 7.1$ V, 34 m$\Omega$ cm$^2$ | Linköping U./ABB, 1997 |
| | 4H–SiC | 4.9 kV | Ni, 17 m$\Omega$ cm$^2$ | Purdue U., 1999 |
| | 4H–SiC | 1.6 kV | Ni, $V_{F100} = 1.16$ V, 7.7 m$\Omega$ cm$^2$ | N. Carolina State U., 1999 |
| | 4H–SiC | 2.2 kV 1.4 kV | Ni, 34 m$\Omega$ cm$^2$ Ni, 13 m$\Omega$ cm$^2$ | Kansai Electric/Hitachi, 1999 |
| | 4H–SiC | 1.25 kV, 6 A | Ni, B-implanted edge term. $V_{F300} = 2.2$ V, 2.2 m$\Omega$ cm$^2$ | Purdue U., 2001 |

**Table 4**   (*Continued*)

| Device Type | Polytype | Power Ratings | Features | Developer |
|---|---|---|---|---|
| Junction | 6H–SiC | 2000 V, 1mA | | NASA, 1993 |
| | 6H–SiC | 4.5 KV, 20 mA | $V_{F100} = 6$ V | Linköping U./ABB, 1995 |
| | 6H–SiC | 1325 V | Floating field rings term. | ABB, 1995 |
| | 4H–SiC | 3.4 kV | JTE, $V_{F100} = 6$ V | ABB, 1997 |
| | 4H–SiC | > 5.5 kV | | Cree, 1998 |
| | 4H–SiC | 2 kV, 5 A | JTE, $V_{F500} = 4.0$ V, 2.2 m$\Omega$ cm$^2$, Al-implanted | Siemens, 1998 |
| | 4H–SiC | 3 kV, 7 A | JTE, $V_{F500} = 4.8$ V, 3.0 m$\Omega$ cm$^2$, Al-implanted anode | Siemens, 1998 |
| | 4H–SiC | 600 V | JTE, $V_{F100} = 6.0$ V, $n^+pp^+$ Phosphorus-implanted anode | RPI, 1998 |
| | 4H–SiC | 1.1 kV | JTE, $V_{F100} = 3.4$ V, Al/C- and B-implanted anode | RPI, 1999 |
| | 4H–SiC | 3.5 kV | JTE, $V_{F100} < 4$ V Al- and B-implanted anode | ABB, 1999 |
| | 4H–SiC | 5.5 kV | JTE, $V_{F100} = 4.2$ V Al/C- and B-implanted anode | RPI/GE, 1999 |
| | 4H–SiC | 4.9 kV | JTE, $V_{F100} = 4.2$ V Al-implanted anode | Siemens, 1999 |
| | 4H–SiC | 4.8 kV, 2.2 A | 2-zone JTE, $V_{F100} = 4.0$ V Al-implanted anode | Siemens, 2000 |
| | 4H–SiC | 4.5 kV, 10 A | JTE, $V_{F100} = 3.08$ V Epi anode JTE, $V_{F100} = 3.5$–$3.8$ V Al- and B-implanted anode | ABB, 2001 |
| | 4H–SiC | 5 kV, 20 A | JTE, $V_{F100} = 3.5$ V Epi anode | Cree, 2001 |
| | 4H–SiC | 19.5 kV | JTE, $V_{F100} = 6.5$ V Epi anode | Kansai Electric/Cree, 2001 |
| | | 14.9 kV | JTE, $V_{F100} = 4.4$ V Epi anode | |

**Table 4** (*Continued*)

| Device Type | Polytype | Power Ratings | Features | Developer |
|---|---|---|---|---|
| MPS/JBS | 4H–SiC | $\sim$ 750 V | Ti, $V_{F100} \sim 1.5$ V | Daimler-Benz, 1997 |
| | 4H–SiC | 1 kV, 13.5 A | Ni, 3-step JTE $V_{F431} = 3$ V, 6.4 m$\Omega$ cm$^2$ | Rutgers U./ United SiC, 1999 |
| JBS | 6H–SiC | 540/850 V | Ti, $V_{F100} = 5.3$ V, 43 m$\Omega$ cm$^2$ | Royal Inst./ABB, 1997 |
| | 4H–SiC | 870/1 kV | Ti, $V_{F100} = 3.1$ V, 19 m$\Omega$ cm$^2$ | Royal Inst./ABB, 1997 |
| | 4H–SiC | 2.8 kV | Ti, $V_{F100} = 2.0$ V, 7.5 m$\Omega$ cm$^2$ | ABB, 1999 |
| | 4H–SiC | 3.7 kV | Ni, $V_{F100} = 6.0$ V, 43 m$\Omega$ cm$^2$ | Kansai Electric/ Hitachi, 1999 |
| | 4H–SiC | 1.5 kV, 4 A | Ni, $V_{F200} = 3.1$ V, 10.5 m$\Omega$ cm$^2$ | Cree, 2000 |

## 4.1. *Schottky Rectifiers*

Schottky barrier rectifiers are attractive due to their fast switching speed. This device exhibits fast reverse recovery with no reverse current and no forward over-voltage transient during switching because the forward current transport is mainly carried out by the majority carriers and minority carriers are mostly absent.

The electrical parameters of the Schottky rectifier is primarily determined by the drift layer thickness ($W_d$) and doping ($N_d$), and the metal-to-semiconductor Schottky barrier height ($\phi_B$). The drift layer design can be performed using Fig. 2. For example, for a 1000 V 4H–SiC Schottky rectifier, a drift layer thickness of 12 $\mu$m and doping of $1.3 \times 10^{16}$ cm$^{-3}$ are needed. In addition to the one-dimensional breakdown design, termination is necessary to prevent edge breakdown at the device peripheries due to electric field crowding. As can be seen from Fig. 3(b), a *p-n* junction guard ring is almost always placed at the edges. In addition, termination techniques, such as JTE, are then placed beyond the guard ring to further optimize the BV by suppressing the surface and bulk fields.

In the on-state, unlike the low-voltage counterpart, the on-state voltage of the high-voltage power rectifier is determined by the voltage drops across the metal-semiconductor junction, the lightly doped drift region and the heavily doped substrate.[2] To estimate the voltage drop across the metal-semiconductor junction, a thermionic emission model[27] is often employed for Si and apparently can be extended to SiC. The Schottky barrier height, $\phi_B$, which controls the asymmetric current flow across the Schottky junctions, is ideally described as

$$\phi_B = \phi_M - \chi \tag{1}$$

where $\phi_M$ is the metal work function and $\chi$ is the electron affinity of the semiconductor. For silicon and GaAs, due to surface states, such an ideally relationship has been found not to hold. For 6H–SiC, the Schottky barrier of several metals has been experimentally measured and found to be closer to ideal with a slope of about 0.7 in the $\phi_B$ vs. $\phi_M$ plot (vs. unity for the ideal case).[28]

The specific on-resistance, $R_{on,sp}$, of the Schottky rectifier can be expressed as a sum of the drift layer and substrate specific resistances, $R_d$ and $R_{sub}$, respectively.

$$R_{on,sp} = R_d + R_{sub} = R_d + \rho_{sub} W_{sub} \tag{2}$$

where $\rho_{sub}$ and $W_{sub}$ are the resistivities and thickness of the substrate respectively. The specific drift layer resistance is

$$R_d = 4(BV)2/(\mu \, \varepsilon_S \, \xi_C^3)$$

$$= 2.78 \times 10^{-12}(BV)^{2.5} \quad \text{in } \Omega \text{ cm}^2 \text{ for 4H–SiC} \tag{3}$$

where $\mu$ is the carrier mobility, $\xi_C$ is the critical field at breakdown, and $\varepsilon_S$ is the semiconductor permittivity. Figure 7 illustrates the $R_{on,sp}$ vs. BV relationship calculated for $n$-type Schottky rectifiers on silicon, 6H–SiC and 4H–SiC. The substrate thickness and resistivity is to be 300 $\mu$m and 0.01, 0.03 and 0.015 $\Omega$ cm for Si, 6H–SiC and 4H–SiC, respectively. Also, the electron mobility in the drift layer is taken to be 450 and 1000 cm$^2$/Vs for 6H–SiC and 4H–SiC, respectively. At low values of BV ($< 500$ V for SiC), the specific on-resistance is dominated by the substrate resistance. When the BV exceeds 1000 V, the drift resistance starts to be the main limiting factor and the increase in $R_d$ is a direct consequence of the increase in drift layer thickness and reduction in drift layer doping.

We have also included recent published experimental results[29–41] in Fig. 7. As seen in the figure, experimental SiC Schottky rectifiers have achieved not only significant improvement over Si counterparts, but also

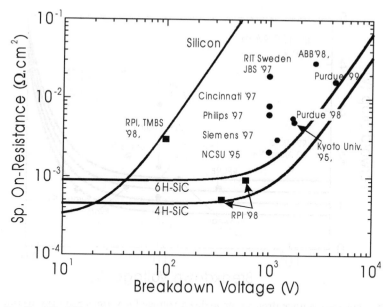

**Fig. 7.** Specific ON-resistance vs. reverse blocking voltage for Schottky rectifiers on Si, 6H– and 4H–SiC.

are getting close to the performance predicted from the theoretical predictions, particularly for high reverse voltage devices, in which the parasitic substrate resistance is not important. The highest reverse blocking voltage reported so far for a 4H–SiC Schottky rectifier is about 5000 V. This blocking voltage probably represents the highest voltage rating for a 4H–SiC Schottky rectifier. Several termination techniques[31,32,39,42,43] have been tried to maximize the BV and minimize the reverse leakage current. Also, promisingly, commercial 4H–SiC Ti Schottky rectifiers are now available with voltage ratings of 300–600 V and up to 10 A.[44] Further increase to higher blocking voltages (e.g., 1.2 kV) and current levels should be expected in the near future.

Based on the modified thermionic emission model, the total voltage drop across the Schottky rectifier can be written as[2]

$$V_{\mathrm{F}} = (nkT/q)\ln(J_{\mathrm{F}}/A^{**}T^2) + n\,\phi_{\mathrm{B}} + R_{\mathrm{on,sp}}J_{\mathrm{F}} \qquad (4)$$

where $n$ is the ideality factor, $k$ is the Boltzmann's constant, $J_{\mathrm{F}}$ is the forward conduction current density and $T$ is the temperature. $A^{**}$ is the Richardson's constant whose value has been experimentally determined to

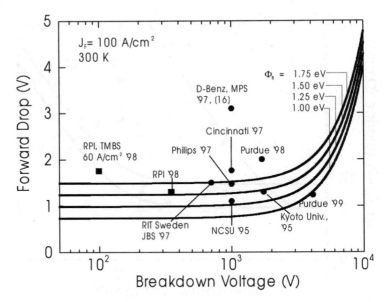

**Fig. 8.** Forward voltage drop vs. breakdown voltage for various Schottky rectifiers on 4H–SiC. Shown are the experimental results reported and analytical calculations. The forward drop has been calculated for four different barrier heights at a current density of 100 A/cm² and room temperature (300 K).

be 140 A/cm² K² on 4H–SiC using log J-V-T measurements and is in good agreement with the theoretically estimated value of 146 A/cm² K².[45] When the BV of the Schottky rectifier increases, the forward drop increases with $R_{on,sp}$ as shown in Eq. (4). Changing the barrier height also influences $V_F$ directly. Figure 8 shows the calculated forward drop of 4H–SiC Schottky rectifier at 100 A/cm² as a function of BV for barrier heights ranging from 1 to 1.75 V. $V_F$ does not increase up to a breakdown voltage rating of 2000 V, beyond which a rapid increase is observed, mainly due to the increase in $R_{on,sp}$. Also included in Fig. 8 are the recently reported values of $V_F$ of many Schottky rectifiers at 100 A/cm². While several metals have been employed to form Schottky rectifiers, the most popular metals are titanium (Ti) and nickel (Ni) and near-ideal junction characteristics have been reported for these. Barrier heights of 1.25 and 1.69 V have been measured for Ti and Ni, respectively. However, the experimental characteristics of many of the high-voltage rectifiers still fall short of the theoretically predicted values, usually attributed to parasitic contact resistance, resistive shunt defects across the metal-SiC junction, and bulk defects.

The forward drop of a Schottky rectifier is also a function of temperature (Eq. (4)). With increasing temperature, the drift layer resistance increases because of a decrease in mobility. Consequently, the forward drop increases with temperature. Although it is also possible to decrease the forward drop with a Schottky metal having a low $\phi_B$, such a choice is not utilized because it leads to a drastic increase in reverse leakage current density, and, in turn, an increase in off-state power dissipation.

Besides bulk effects, surface inhomogeneities have been proposed as possible shunt paths across the metal-semiconductor junctions and thus can degrade both forward and reverse $I$-$V$ characteristics.[46,47] From forward $I$-$V$ measurements, it has been found that, in addition to the standard barrier height mentioned above, a second and lower barrier height, is needed to model the anomalously large forward current.[47,48] However, these excessive forward current only becomes noticeable when the temperature is below 250 K, but it can vary between 100–200 K. It is not clear at present which physical mechanism is responsible for this lower Schottky barrier. Some possible ones are the difference in the crystal symmetry of the metal with respect to the semiconductor and variation in the orientation at the metal-semiconductor interface, as well as doping inhomogeneity and clustering, and interfacial contaminations.

The off-state reverse leakage current in SiC Schottky rectifiers can have bulk contributions from thermionic emission, field emission, space-charge generation, and tunneling, as well as edge and surface leakage and defect-related leakage. In addition, the thermionic emission component is affected by barrier lowering due to image force.[27] In the case, the reverse current component can be expressed as

$$J_{R,TE} = -A^{**}T^2 \exp[-q(\phi_B - \Delta\phi_B)/kT] \exp(-qV_R/kT) \qquad (5)$$

where $\Delta\phi_B$ is the decrease in Schottky barrier height due to image force and it can be written in terms of the maximum electric field at the junction

$$\Delta\phi_B = (q\xi_m/4\pi\varepsilon_S)^{1/2} \qquad (6)$$

where $\xi_m$ is the peak electric field at the Schottky junction. Experimental reverse leakage currents in SiC Schottky diodes have been reported to have orders of magnitude higher than that predicted by the thermionic emission model and a stronger voltage dependence than that given by the image-force induced barrier lowering.[44] To provide additional bulk leakage current mechanisms, electron tunneling via deep levels has been found to be important.[48,49] The tunneling reverse current can be written as

$$J_{R,TUN} = -A\exp(-BV_R).\tag{7}$$

One way to determine whether the space charge generation or tunneling current component under reverse bias is dominant in the total reverse current is to measure its temperature dependence. The tunneling process has very weak temperature dependence while the generation current from the Shockley–Read–Hall process increases exponentially with temperature. Experimental results from 300–600 V commercial 4H–SiC Schottky rectifiers with Ti metal have demonstrated that the tunneling current component is dominant for reverse voltage > 200 V. We have used a Fowler-Nordheim tunneling model to analyze the reverse characteristics of 300–600 V commercial 4H–SiC Schottky rectifiers.[48]

Figure 9 summarizes the measured reverse leakage current of 4H–SiC Schottky diodes at room temperature reported in the literature. While some of the data points have been measured at reverse biases near the breakdown voltage, where pre-avalanche multiplication has a significant contribution

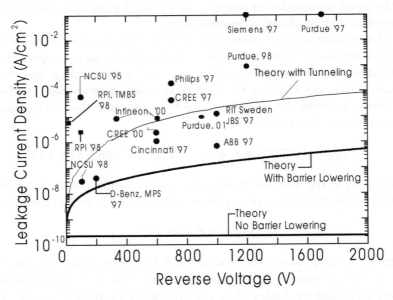

**Fig. 9.** Theoretical reverse leakage current, as predicted by the thermionic emission model and including Schottky barrier lowering, as well as Fowler-Nordheim tunneling current, as a function of reverse blocking voltage. Experimental measured reverse current is also shown for comparison.

to the leakage current, the leakage current of all of the Schottky devices are orders of magnitude higher than the theoretical prediction, even on devices with edge terminations. Recent studies on correlating the epitaxial growth conditions[37] and specific defects as measured by SWBXT and EBIC techniques[50] on reverse current level of 4H–SiC Schottky rectiifers, have indicated that bulk defects, rather than surface passivation, are probably responsible for a major part of the leakage current.

It should be pointed out that both forward and reverse currents can be strongly influenced by processing conditions, such as surface cleaning techniques[31,51] and plasma etching damage.[45] While comparisons of early experimental data from different research groups are difficult due to the absence of detailed processing conditions, state-of-art commercial devices have demonstrated that the ideal thermionic model is adequate in forward *I–V* characteristic and tunneling current can be used to describe the reverse current at room temperature.

The power dissipation ($P_D$) of a Schottky rectifier consists of only the on-state conduction loss and the off-state leakage loss and it can be optimized with the choice of Schottky barrier height.

$$P_D = J_F V_F D + J_R V_R (1 - D) \tag{8}$$

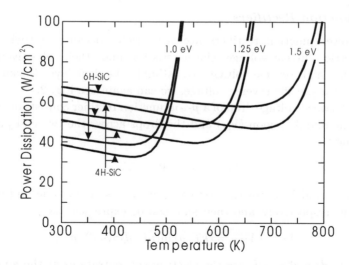

**Fig. 10.** Estimated power dissipation as a function of temperature for 1000 V Schottky rectifiers on 6H–SiC and 4H–SiC. $J_F = 100$ A/cm$^2$, $V_R = 500$ V and 50% duty cycle are assumed. The effect of the barrier height can also be seen.[52]

where $D$ is the duty cycle and no switching loss is assumed. Figure 10 shows the calculated power dissipation of 1000 V Schottky rectifiers on 6H–SiC and 4H–SiC with three different barrier heights.[52] $J_F = 100$ A/cm$^2$, $V_R = 500$ V and $D = 0.5$ (or 50%) are assumed. It can be seen that the power dissipation can be lowered by decreasing the barrier height employed but is accompanied with a lower maximum operating temperature from increased leakage current. For the silicon Schottky with the same voltage rating, the power dissipation has been estimated to be at least an order of magnitude higher. For a silicon junction rectifier with the same rating, the power dissipation increases rapidly with increasing switching frequency due to minority carrier storage. Consequently, the advantages of high-voltage SiC Schottky rectifiers are clearly established.

The reliability assessment of 600 V, packaged 4H–SiC Schottky diodes has recently been performed under overcurrent stressing and high-temperature reverse blocking and multiple temperature cycling.[53] While the long-term stability of these diodes under reverse biasing testing conditions was found to be excellent, these devices showed a limited overcurrent capability compared to the typical ultrafast Si $p$-$n$ junction diodes. The failure mechanism has been attributed to the current crowding in the edge area after the complete depletion of the $p$-type termination.[53]

### 4.2. *Junction Rectifiers*

$p$-$i$-$n$ power junction rectifiers are very useful because of their special asymmetric current-voltage characteristics and their effectiveness in supporting high reverse voltage. To estimate the drift layer thickness and doping for a specific reverse voltage, we can again use Fig. 2.

The forward voltage drop of a $p$-$i$-$n$ junction rectifier consists of the drop across the middle region ($V_m$) and the drops across the two end junctions according to

$$V_F = V_{P+} + V_m + V_{N+} \tag{9}$$

where $V_{P+}$ and $V_{N+}$ are the voltage drops across the anode and cathode junctions, respectively, and their sum can be expressed[1] as

$$V_{P+} + V_{N+} = (kT/q)\ln(n(-d)n(+d)/n_i{}^2) \tag{10}$$

where $n(-d)$ and $n(+d)$ are the electron concentrations at the anode and cathode junctions respectively. The mid-region drop, $V_m$, depends strongly on carrier recombination lifetimes and can be expressed as transcendental functions of $d/L_a$ (see Ref. 1, Eq. (3.81)). Approximately,

$$V_{\rm m} = (3kT/q)(d/L_a)^2 \quad \text{for } d < L_a \tag{11a}$$

$$V_{\rm m} = (3\pi kT/8q)\exp(d/L_a) \quad \text{for } d \le L_a \tag{11b}$$

where $2d$ is the middle drift layer width and $L_a$ is the ambipolar diffusion length. Combining Eqs. (9) and (10), we can get[1]

$$J = (2qD_a n_i/d)F(d/L_a)\exp(qV/2kT) \tag{12}$$

where $F(d/L_a)$ is a function of $d/L_a$ and $V_{\rm m}$ but not a function of the current density (see Ref. 1, Eq. (3.91)) and no end recombination is assumed. We can observe that Eq. (11) has a strikingly resemblance to the *I–V* relation of a conventional *p-n* junction under high-level injection.

The bulk off-state leakage current of a silicon junction rectifier is dominated by the space-charge generation current, which is

$$J_{\rm gen} = qn_i W/\tau_{\rm eff} \propto n_i\sqrt{V/\tau_{\rm eff}} \tag{13}$$

where the effective generation lifetime $\tau_{\rm eff}$ contains both bulk space-charge generation lifetime, $\tau_{\rm SC}$, and surface generation velocity, $s_g$. Also, $W$ becomes constant once the mid-region is completely depleted. Usually, the junction rectifiers are designed to have a mesa-isolation for epi-grown anode and planar structure for ion-implanted anode, as illustrated in Fig. 11.

The reverse recovery characteristics of *p-i-n* junction rectifiers are limited by the removal rate of the excess minority carriers. For power rectifiers, unlike the low-voltage diodes, the switching from the forward to reverse conditions usually goes through a constant $\mathrm{d}I/\mathrm{d}t$ ramp,[2,54] and

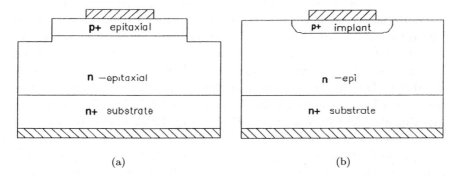

          (a)                        (b)

**Fig. 11.** Schematic cross-sections of *p-i-n* junction rectifiers with (a) mesa-isolated, epitaxially grown anode and (b) planar, ion-implanted anode.

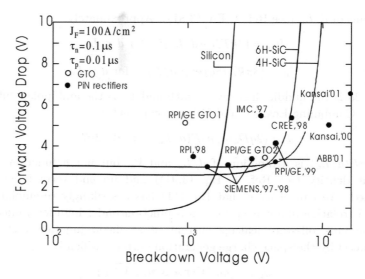

**Fig. 12.** Forward voltage drop ($V_F$) at 100 A/cm$^2$ vs. breakdown voltage for *p-i-n* junction rectifiers on Si, 6H–SiC and 4H–SiC. The solid lines are analytical calculations assuming $\tau_{n0} = 10\,\tau_{p0}$ and $\tau_{n0} = 0.1\,\mu$sec.

hence the reverse recovery times ($t_{rr}, t_A, t_B$) and charge ($Q_{rr}$) are dependent on the high-level recombination lifetime as well as the circuit-imposed $dI/dt$ ramp rate. For simple analysis, the switching of the power junction rectifier has been modeled with a charge-control model and the details of the dependence of switching times as well as the peak reverse current on the $dI/dt$ ramp rate have been obtained analytically.[54] Due to its bipolar nature, the on-state voltage of the junction rectifier is traded off its switching performance, as illustrated in Fig. 12. Figure 12 shows the estimated forward voltage at 100 A/cm$^2$ as a function of reverse blocking voltage for Si, 6H– and 4H–SiC, with $\tau_{n0} = 10$ and $\tau_{p0} = 100$ ns. Even with such a low lifetime value, 4H–SiC junction rectifiers have low forward drops up to 6000 V. Due to the reduced drift region width, a less demanding lifetime is needed to achieve conductivity modulation in SiC devices. Recent experimental data have been included in Fig. 12 to assess the degree of optimization of these devices.

Typical forward current conduction according to the widely accepted Sah–Noyce–Shockley (SNS) model consists of diffusion current and recombination current components.[1,2,27] Diffusion current ($J_d$) originates due to the recombination of electrons and holes in the quasi-neutral regions outside the

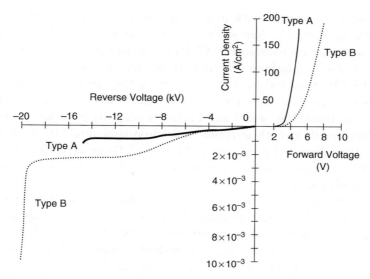

**Fig. 13.** *I–V* characteristics of high voltage 4H–SiC *p-i-n* rectifiers made on 2 × $10^{14}$ cm$^{-3}$, 120 µm (Type A) and 8 × $10^{13}$ cm$^{-3}$, 200 µm (Type B) epi layers.[67]

space-charge region, whereas the recombination current component ($J_{sc}$) is due to recombination via single-level, uniformly distributed centers located at or near the intrinsic level. The current-voltage characteristics according to the SNS model can be written as

$$J_F = J_{SC} + J_d = J_{S1} \exp(qV/2kT) + J_{S2} \exp(qV/kT) \qquad (14)$$

where $J_{s1}$, $J_{s2}$ are the saturation current densities for recombination and diffusion current components, respectively. The activation energy, obtained by an Arrehenius plot of $J_{s1}$, is approximately equal to the half of the energy bandgap of the semiconductor.

The *I–V* characteristics of 4H–SiC junction rectifiers with the highest BV today are shown in Fig. 13.[68] To model the actual forward *I–V* characteristics of SiC *p-n* junction rectifiers,[13–16,55–70] particularly those fabricated with planar implanted anode or cathode junctions, it has been found initially that the theory established for silicon presented above is often inadequate. Early experimental results yielded forward *I–V* characteristics far from those of the ideal theory, particularly the implanted diodes, regardless of whether they are $p^+/n$ or $n^+/p$ diodes. The reasons are either residual lattice damages from the implantation process or defects originated from the epi growth process. Often, instead of an exponential increase in the conduction current with increasing voltage, the power law predicted from the space charge limited current has been obtained. Also, a current-controlled negative resistance (CCNR) in the forward $I$–$V$ characteristics has been observed in both 6H– and 4H–SiC junction rectifiers.[71] This forward CCNR characteristic will appear when the carrier diffusion lengths are shorter than the drift region width.[1] Also, the triggering voltage in the CCNR region has been found to show a simple dependence on lifetime and drift length in both nitrogen- and boron-implanted 6H–SiC junction diodes,[71] as discussed previously in Si diodes.[1] Other anomalous behavior includes a charge trapping phenomenon, attributed to deep levels from nitrogen implantation.[72]

With the improvement in process technology, particularly in dopant implant activation and annealing, the CCNR and charge trapping phenomena have largely been suppressed. However, implanted SiC $p$-$i$-$n$ junction rectifiers experimentally exhibited a forward J–V relationship following $J_F \sim J_0 \exp(qV/nkT)$ and $n$, the ideality factor, lies between 1 and 2. Also, the activation energy is equal to $E_g/n$, instead of $E_g/2$ for the SNS model. The origin of this unusual ideality factor is attributed to the existence of multiple deep and shallow impurity levels in the bandgap of SiC, which act as effective recombination sites. A multiple level recombination model was formulated and proposed for wide bandgap compound semiconductors.[73–76] We have applied this multiple level model to explain the forward characteristics of 4H–SiC power junction rectifiers and have found excellent agreement in both ideality factor and activation energy in both $p^+/n/n^+$ and $n^+/p/p^+$ junction rectifiers.[64,67]

The best forward $I$–$V$ characteristics of 4H–SiC $p$-$i$-$n$ junction rectifiers have been shown for epitaxially grown anode junctions,[69] not implanted junctions. Figure 14 shows the apparent reason for this is that, so far, the complete elimination the lattice damage introduced during dopant implantation. Nevertheless, both epi and the best implanted anode $p$-$i$-$n$ junction rectifiers were found to follow the power junction rectifier model developed for silicon.[69,60]

The reverse leakage current has been found to be of orders of magnitude higher than that predicted by standard silicon theory extensions. Obviously, material defects, such as screw dislocations, play a role in determining the reverse current.[77] With improvement in material quality, fairly low reverse current densities ($< 10^{-5}$ cm$^2$) have been observed, at least for small area devices.[59–61] Figure 15 shows the reverse log $J$ vs. $V$ of a 4H–SiC, 15 kV, epi anode $p$-$i$-$n$ junction rectifier.[68] The reverse current density remains constant up to at least 200°C and only starts to increase significantly at 250°C. At high reverse bias, some of the SiC junction rectifiers show an abrupt increase in reverse current, this being attributed to trap-assisted tunneling.[63]

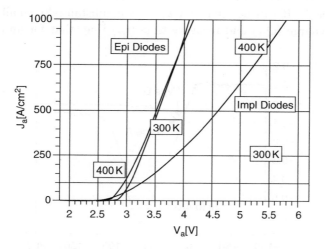

**Fig. 14.** Comparison of forward $I$–$V$ characteristics of epi- and implanted-anode 4.5 kV 4H–SiC $p$-$i$-$n$ junction rectifiers.[68]

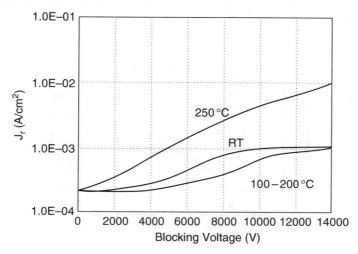

**Fig. 15.** Reverse leakage current vs. reverse voltage of 15 kV (Type A) 4H–SiC $p$-$i$-$n$ junction rectifier from room temperature to 250°C.[67]

The reverse recovery current waveforms of implanted junction rectifiers show no temperature dependence up to at least 200°C;[14,67] this trend has not been seen in Si devices. The reason for this trend is probably due to the deepness of the recombination level. In Fig. 16, typical reverse recovery current waveforms of 12 and 19 kV, $p^+n$ epi diode are shown along with Si rectifiers.[68] It can be seen that there is a substantially smaller reverse recovery charge $(Q_{rr})$ and recovery time $(t_{rr})$ than that for an equivalent

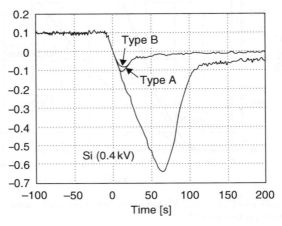

**Fig. 16.** Reverse recovery current waveforms measured on 15 kV (Type A) and 19 kV (Type B), mesa-isolated, epi anode 4H–SiC junction rectifiers and comparison to that of a 0.4 kV Si junction rectifier.[67]

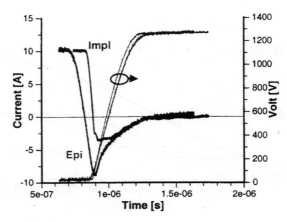

**Fig. 17.** Comparison of reverse recovery characteristics of epi- and implanted-anode 4.5 kV 4H–SiC $p$-$i$-$n$ junction rectifiers.[68]

silicon rectifier of the same voltage rating. Also, it is worth pointing that even though the implanted junction rectifiers have higher forward drop, but it also has a faster reverse recovery time (Fig. 17), so the $V_F$ vs. $t_{rr}$ tradeoffs need to be considered when deciding which type of junction rectifiers should be used for a particular application.

Recently, it was pointed out by Lendemann and co-workers of ABB for the first time that degradation in the forward voltage in junction rectifiers has been measured under forward biasing stressing conditions.[78–81] Line defects, attributed to stacking faults, can be observed with segmental metallized anode structures.[82,83] While the actual magnitude of the forward drop increase can be small (0.2 to 0.5 V),[84] it raises concern on the long-term reliability issue of SiC devices. Detailed analysis and study of this phenomenon points to the formation of bulk stacking faults with the energy from electron-hole recombination in the carrier plasma during forward conduction. These defect regions facilitate carrier recombination and quench light emission, so they can be readily monitored optically. Using real-time optical imaging, it has been found that these line defects are readily formed with fairly low (80 A/cm$^2$) current density at room temperature[82] and the density of these defects can be quantitatively correlated with the magnitude of the forward drop increase (Figs. 18 and 19). Encouragingly, Cree researchers have reported[85] that a process modification, which suppresses this degradation phenomenon, has been found but they have not released any details.

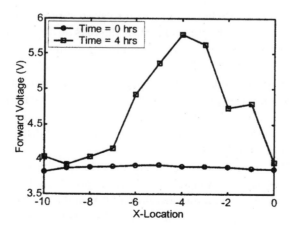

**Fig. 18.** Forward drop vs. current-stressing time at various locations at a constant current of 0.5 A (80 A/cm$^2$).[81]

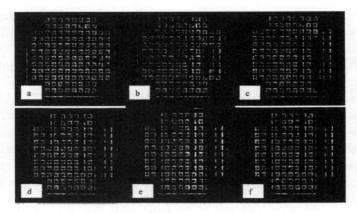

**Fig. 19.** Light emission from a pin diode with anode windows at a forward current of 0.5 A (80 A/cm$^2$) at various time intervals (a) 0 hr., (b) 0.3 hr., (c) 1 hr., (d) 2 hr., (e) 3 hr., and (f) 3.7 hr.[81]

### 4.3. *Hybrid Rectifiers*

In order to achieve good reverse blocking characteristics while maintaining a Schottky forward conduction and fast reverse recovery characteristics, several modern rectifiers, such as Junction Barrier Schottky (JBS),[21,22] Trench MOS Barrier Schottky (TMBS)[24] and Dual Metal Trench (DMT) Schottky[86] and rectifiers, have been explored. The physical principle used to reduce the leakage current is the same — to completely deplete the Schottky mesa regions under the Schottky junction so as to suppress the electric field from rising at the Schottky junction when the bias at the substrate cathode is increased. In the JBS rectifier, reverse biased $p$-$n$ junctions are used, while in the TMBS and DMT rectifiers, Schottky junctions with MOS capacitors and a higher barrier height are used respectively. An analytical model has been developed for the planar Si JBS rectifier[87] to describe the forward conduction and reverse blocking characteristics and it can be extended to SiC devices. The key design parameters are the mesa width, junction depth and the junction diffusion window width.

Several JBS rectifiers have been experimentally demonstrated in SiC.[88-92] The maximum breakdown voltage demonstrated has now been increased to 3.7 kV[91,92] and a forward drop of 6 V and a specific on-resistance of 43 m$\Omega$ cm$^2$ was reported.[91] Since the JBS rectifier has a lower reverse leakage current than the Schottky rectifier due to field shielding of the surface Schottky junction from adjacent $p$-$n$ junctions, it can operate at

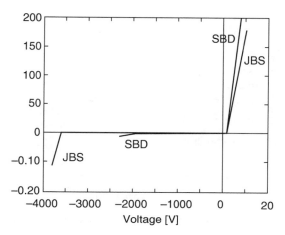

**Fig. 20.** *J–V* characteristics of a 3.7 kV 4H–SiC JBS recitifer and comparison to a Schottky rectifier made on the same 50 μm thick epi layer.[91]

higher temperatures and are useful at higher blocking voltages (up to 4 kV instead of 3 kV). Nevertheless, the lack of conductivity modulation causes the JBS rectifier to have too high an on-resistance for reverse blocking voltages higher than 4 kV. Two process considerations need to be made for the JBS rectifier implementations. First, Ti is preferred at lower blocking voltages (because of its lower barrier height) to Ni, and hence results in lower forward drops, but Ni has been employed for the 3.7 kV JBS Rectifier[91,92] to ensure low leakage current. Secondly, either boron or aluminum can be used for the $p^+$ grid of the $p$-$n$ junction regions. It is known that boron tends to diffuse faster than aluminum, particularly in the lateral direction. Hence, a larger cell pitch can be used for boron-diffused JBS rectifiers than aluminum ones for the same Schottky region width and lower implantation energy is needed. However, the percentage area available for forward conduction is also smaller.

Figure 20 shows the *J–V* characteristics of 4H–SiC JBS and Schottky rectifiers made on the same 50 μm epi.[91] It can be seen that the reverse breakdown voltage is clearly superior in the JBS rectifier due to the reduced surface field at the Schottky junction. The increase in forward drop in the 3.7 kV JBS rectifier can be observed in Fig. 21, in which the forward *J–V* characteristics are shown up to 270°C.[91] The advantages of the JBS rectifiers are further demonstrated in the reverse recovery waveforms of Fig. 22. The reverse recovery time and charge, as well as the peak reverse current

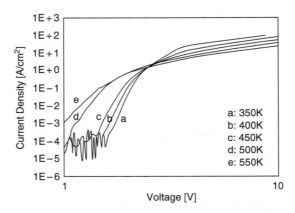

**Fig. 21.** Forward log J vs. V characteristics of 3.7 kV JBS rectifier from room temperature to 275°C.[91]

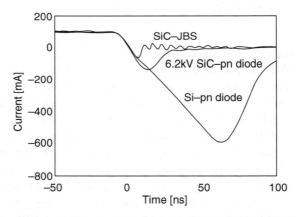

**Fig. 22.** Comparison of reverse recovery waveform of 3.7 kV 4H–SiC JBS rectifier with those of 6.2 kV 4H–SiC *p-i-n* junction rectifier and 400 V high-speed Si junction rectifier.[91]

have all been minimized in the JBS rectifier. The temperature dependence of these reverse recovery parameters is shown in Fig. 23.[91]

Very little has been reported for reliability testing of SiC JBS rectifiers but it has been observed that, similar to the Schottky rectifiers, the JBS devices do not exhibit a forward voltage drift that was the major instability issue in *p-i-n* junction rectifiers.[80] In addition, we expect that JBS rectifiers have a better overcurrent capability than Schottky rectifiers because the *p-n* junctions within the unit cell can help conduct excessive current, whereas Schottky devices have only the edge termination area for bipolar conduction.[53]

The DMT rectifier[86] utilizes a Schottky metal with a higher barrier height than the main Schottky junction to achieve field shielding. The

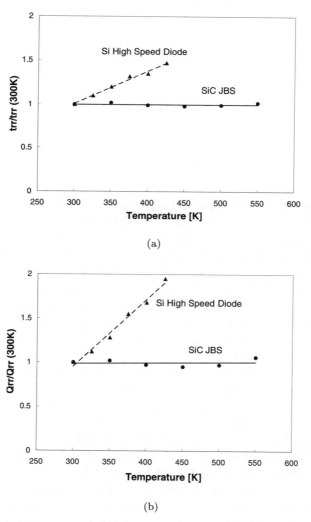

(a)

(b)

**Fig. 23.** Comparison of (a) reverse recovery time ($t_{rr}$), (b) reverse recovery charge ($Q_{rr}$) and (c) peak reverse current of 3.7 kV 4H–SiC JBS rectifier with that of a 400 V high-speed Si diode.[91]

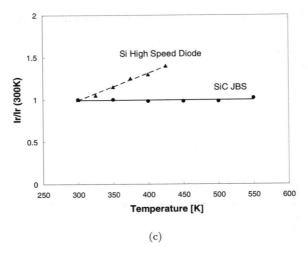

(c)

**Fig. 23.**   *Continued.*

reported device had Ti deposited to achieve a low barrier height on top
of the mesa, followed by conformal deposition of Ni over the entire device
to achieve a higher barrier height at the bottom of trenches. The forward
drop and reverse leakage current reported for this device were between the
Ti and Ni Schottky diodes. In particular, there is a forward voltage drop
close to that of the Ti Schottky junctions, but with leakage current two orders of magnitude lower than Ti SBD and about a factor of two higher than
the Ni devices. Although the leakage current measurements were shown up
to a reverse voltage of 300 V, no information on the true breakdown voltage
of the device was given.

Embedding a UMOS trench like grid instead of a *p-n* junction grid as in
JBS/MPS rectifiers, yields a structure known as the TMBS rectifier. The
original Si device reported in Ref. 24 was shown to be capable of achieving higher than parallel plane breakdown voltage with little compromise
in the forward conduction capability. We have fabricated a polysilicon planarized Ni-TMBS in 4H–SiC, the forward and reverse characteristics of
which are compared to that of simultaneously fabricated Ni SBD and *p-i-n*
rectifiers.[93] A significant improvement over the SBD leakage current is observed with little sacrifice in the forward voltage drop. The reverse leakage
current and forward voltage drop, as expected, were observed to scale with
the percentage of Schottky area. An increase in mesa width would reduce
the amount of field shielding under the Schottky contact and at the same

time it will increase the percentage of Schottky area. The combined effects would result in an increase in the leakage current and a reduction in the forward voltage drop. The depth of the trench would also alter the device performance significantly, as it will alter the aspect ratio directly thereby affecting the field shielding under the Schottky mesa. This indicates that an optimized device design would involve a careful selection of mesa width and trench depth.

By contrast, other hybrid rectifiers, such as the Merged $p$-$i$-$n$/Schottky (MPS) rectifier, are mainly $p$-$i$-$n$ junction rectifiers with an adjacent Schottky region to facilitate reverse recovery.[2] While both the JBS and MPS rectifiers have apparently similar device structures as both have deep implanted interdigitated $p^+$ grids adjacent to Schottky junctions, the main junctions in the MPS rectifier are the $p$-$i$-$n$ diode regions.[23] The $p^+$ anode injects holes into the drift region, resulting in a drastic reduction in the series resistance via conductivity modulation. Also, the auxiliary Schottky regions are placed next to the $p$-$n$ junction so as to reduce reverse peak current and to speed up reverse recovery. In addition, the Schottky regions are designed not to degrade the reverse leakage current at reverse bias. Furthermore, while the injection level in MPS is not as high as in $p$-$i$-$n$ rectifiers, the Schottky region also participate in the forward conduction by injecting a significant amount of majority carriers into the conductivity modulated drift region. One recent report on a 1000 V SiC MPS rectifier in an inductive half-bridge circuit has been published.[94] However, even though a forward voltage up to 4 V has been applied, it is not clear how much, if any, minority carrier injection has taken place.

## 5. Issues and Challenges

At present, commercial 4H–SiC substrates of up to 3 inches in diameter are available (four inch wafers have been announced) and custom $n$- and $p$-type epitaxial layers over 100 $\mu$m can be ordered so that junction rectifiers with reverse blocking voltages of up to at least 30 kV are possible. However, current levels over 25 A are difficult to achieve due to yield limitations at the moment because of structural defects, micropipes and other factors such as screw dislocations and stacking faults. The forward voltage drift observed in the $p$-$i$-$n$ junction rectifiers needs to be further studied and suppressed before these rectifiers can be incorporated into power electronics systems. The occurrence of two Schottky barriers seen in the forward $I$–$V$ characteristics of SiC Schottky rectifiers also needs to be more closely examined, controlled and correlated with process conditions. The reverse leakage current in SiC Schottky rectifiers needs to be further reduced to minimize the standby power dissipation and the JBS approach can help in this regard, but at the expense of more complex processing steps. Nevertheless, we expect that the JBS rectifiers will be the dominant rectifier structure in the near future for low- and medium-power electronics applications. These SiC rectifiers are expected to replace Si junction rectifiers and to be packaged together with Si IGBTs or MOSFETs in power modules to lower the module power loss.

## 6. Summary

We have reviewed the theory and recent progress of high-voltage SiC power rectifiers in the last few years. Besides the performance projections, we have discussed the physics of operation and the dominant current components in the forward and reverse bias conditions that are applicable to SiC. The performance potential, tradeoffs, and limitations of these SiC devices are discussed, together with their recent experimental demonstrations. The new power rectifiers, one of which has already been commercialized in the 300–600 V range, are expected to be increasingly important in power electronics systems for efficiency and performance enhancement.

## Acknowledgments

I would like to acknowledge support from the Office of Naval Research under Grant # N00014-95-1-1302, the ERC Program of the National Science Foundation under Award Number EEC-9731677 and the Army Collaborative Technology Alliance led by Honeywell.

## References

1. S. K. Ghandhi, *Semiconductor Power Devices* (Wiley, 1977, republished 1998).
2. B. J. Baliga, *Physics of Semiconductor Power Devices* (JWS Publishing, 1996).
3. K. Shenai, R. S. Scott and B. J. Baliga, *IEEE Trans. Electron Devices* **36**, 1811–1820 (1989).
4. B. J. Baliga, *IEEE Electron Device Letters* **10**, 455–457 (1989).
5. A. Bhalla and T. P. Chow, *Inst. Phys. Conf. Ser.* **137**, 621, IOP Publishing Ltd. (1994).
6. A. Bhalla and T. P. Chow, *Proc. 6th Int. Symp. Power Semiconductor Devices and ICs*, 287–292 (1994).

7. T. P. Chow and R. Tyagi, *IEEE Trans. Electron Devices* **41**, 1481–1482 (1994).
8. A. S. Kyuregyan and S. N. Yurkov, *Sov. Phys. Semicond.* **23**, 1126–1131 (1989).
9. A. O. Konstantinov, Q. Wahab, N. Nordell and U. Lindefelt, *Appl. Phys. Lett.* **71**, 90–92 (1997).
10. R. Raghunathan and B. J. Baliga, *Proc. 9th IEEE Int. Symp. Power Semiconductor Devices and ICs*, 173–176 (1997).
11. W. Fulop, *Solid-State Electron.* **10**, 39–43 (1967).
12. N. Ramungul, R. Tyagi, A. Bhalla, T. P. Chow, M. Ghezzo, J. Kretchmer and W. Hennessy, *Inst. Phys. Conf. Ser.* **142**, 773–776 (1995).
13. See, for example, R. Singh, K. G. Irvine, O. Kordina, J. W. Palmour, M. E. Levinshtein and S. L. Rumyanetsev, *56th Annual Device Res. Conf. Digest*, 86–87 (1998).
14. V. Khemka, R. Patel, N. Ramungul, T. P. Chow and R. J. Gutmann, *Proc. Int. Symp. Power Semiconductor Devices and ICs*, 137–140 (1999).
15. P. Neudeck and C. Fazi, *IEEE Electron Device Lett.* **18**, 96–98 (1997).
16. See, for example, K. H. Rottner, A. Schoner, S. M. Savage, M. Frischholz, C. Hallin, O. Kordina and E. Janzen, *Diamond and Related Mater.* **6**, 1485–1488 (1997).
17. N. Ramungul, R. Tyagi, A. Bhalla, T. P. Chow, M. Ghezzo, J. Kretchmer and W. Hennessy, *Inst. Phys. Conf. Ser.* **142**, 773–776 (1995).
18. N. Ramungul, T. P. Chow, M. Ghezzo, J. Kretchmer and W. Hennessy, *54th Annual Device Res. Conf. Digest*, 56–58 (1996).
19. D. T. Morisette and J. A. Cooper, Jr., *Tech. Dig. Int. Conf. SiC and Related Mater.* (2001).
20. R. Saitoh, *et al.*, *Proc. Int. Symp. Power Semiconductor Devices and ICs*, 206–210 (1992).
21. B. M. Wilamowski, *Solid-State Electron.* **26**, 491–493 (1983).
22. B. J. Baliga, *IEEE Electron Device Lett.* **5**, 194–196 (1984).
23. B. J. Baliga and H.-R. Chang, *IEEE IEDM, Tech. Dig.*, 658–661 (1987).
24. M. Mehrotra and B. J. Baliga, *Solid-State Electron.* **38**, 703–713 (1995).
25. T. P. Chow and M. Ghezzo, *Materials Research Society Fall Meeting* (1997).
26. D. T. Morisette and J. A. Cooper, Jr., *Tech. Dig. Int. Conf. SiC and Related Mater.*, 304–305 (2001).
27. See, for example, S. M. Sze, *Physics of Semiconductor Devices* (Wiley, 1981).
28. A. Itoh, O. Takemura, T. Kimoto and H. Matsunami, *Inst. Phys. Conf. Ser.* **142**, 685–688 (1995).
29. M. Bhatnagar, P. K. McLarty and B. J. Baliga, *IEEE Electron Device Lett.* **13**, 501–503 (1992).
30. R. Raghunathan, D. Alok and B. J. Baliga, *IEEE Electron Device Lett.* **16**, 226–228 (1995).
31. A. Itoh, T. Kimoto and H. Matsunami, *Inst. Phys. Conf. Ser.* **142**, 689–692 (1995).

32. K. Ueno, T. Urushidani, K. Hashimoto and Y. Seki, *IEEE Electron Device Lett.* **16**, 331–332 (1995).

33. C. E. Weitzel, J. W. Palmour, C. H. Carter, Jr., K. Moore, K. J. Nordquist, S. Allen, C. Thero and M. Bhatnagar, *IEEE Trans. Electron Devices* **43**, 1732–1741 (1996).

34. H. Mitlehner, W. Bartsch, M. Bruckmann, K. O. Dohnke and U. Weinert, *Proc. 9th Int. Symp. Power Semiconductor Devices and ICs*, 165–168 (1997).

35. V. Saxena and A. J. Steckl, *Mater. Science Forum* **264–268** (2), 937–940 (1997).

36. R. Singh and J. W. Palmour, *Proc. 9th Int. Symp. Power Semiconductor Devices and ICs*, 157–160 (1997).

37. Q. Wahab, T. Kimoto, A. Ellison, C. Hallin, M. Tuominen, R. Yakimova, A. Henry, J. P. Bergmann and E. Janzen, *Appl. Phys. Lett.* **72**, 445–447 (1998).

38. K. J. Schoen, J. M. Woodall, J. A. Cooper, Jr. and M. R. Melloch, *IEEE Trans. Electron Devices* **45**, 1595–1604 (1998).

39. R. K. Chilukuri and B. J. Baliga, *Proc. Int. Symp. Power Semiconductor Devices and ICs*, 161–164 (1999).

40. D. T. Morisette, J. A. Cooper, Jr., M. R. Melloch, G. M. Dolny, P. M. Shenoy, M. Zafrani and J. Gladish, *IEEE Trans. Electron Devices*, **48**, 349–351 (2001).

41. H. M. McGlothlin, D. T. Morisette, J. A. Cooper, Jr. and M. R. Melloch, *Device Res. Conf., Tech. Digest*, 42–43 (1999).

42. M. Bhatnagar, H. Nakanishi, S. Bothra, P. K. McLarty and B. J. Baliga, *Proc. 5th Int. Symp. Power Semiconductor Devices and ICs*, 89–94 (1993).

43. D. Alok and B. J. Baliga, *IEEE Trans. Electron Devices* **44**, 1013–1017 (1997).

44. Infineon SDP10S30, SDB10S30, SDP06S60, and SDB06S60.

45. V. Khemka, T. P. Chow and R. J. Gutmann, *J. Electron. Mater.* **27**, 1128–1135 (1998).

46. M. Bhatnagar, B. J. Baliga, H. R. Kirk and G. A. Rozgoni, *IEEE Trans. Electron Devices* **43**, 150–156 (1996).

47. D. Defives, O. Noblane, C. Brylinski, M. Barthula, V. Aubry-Fortuna and F. Meyer, *IEEE Trans. Electron Devices* **46**, 449–454 (1999).

48. M. Shanbhag and T. P. Chow, *Proc. Int. Symp. Power Semiconductor Devices and ICs* (2002).

49. Shinohe, *Tech. Dig. Int. Conf. SiC and Related Mater.* (2001).

50. D. T. Morisette and J. A. Cooper, Jr., *Tech. Dig. Int. Conf. SiC and Related Mater.*, 535–536 (2001).

51. D. Alok, R. Egloff and E. Arnold, *Mater. Sci. Forum* **264–268** (2), 929–932 (1998).

52. V. Khemka, R. Patel, T. P. Chow and R. J. Gutmann, *Solid-State Electron.* **43**, 1945–1962 (1999).

53. R. Rupp, M. Treu, A. Mauder, E. Griebl, W. Werner, W. Bartsch and D. Stephani, *Mater. Sci. Forum* **338–342**, 1167–1170 (2000).

54. N. Ramungul and T. P. Chow, unpublished results. The highlights of this analysis are discussed in T. P. Chow, "Silicon Carbide Power Devices,"

Chapter 7 of *Handbook of Thin Film Devices*, ed. M. H. Francombe, Vol. I: *Hetero-Structures for High Performance Devices* (Academic Press, 2000).

55. P. G. Neudeck, D. J. Larkin, C. S. Salupo, J. A. Powell and L. G. Matus, *Inst. Phys. Conf. Ser.* **137**, 475–478 (1993).

56. P. M. Shenoy and B. J. Baliga, *Inst. Phys. Conf. Ser.* **142**, 717–720 (1995).

57. O. Kordina, J. P. Bergman, A. Henry, E. Janzen, S. Savage, J. Andre, L. P. Ramberg, U. Lindefelt, W. Hermansson and K. Bergman, *Appl. Phys. Lett.* **67**, 1561–1564 (1995).

58. D. Alok and B. J. Baliga, *Proc. Int. Symp. Power Semiconductor Devices and ICs*, 107–110 (1996).

59. K. H. Rottner, A. Schoner, S. M. Savage, M. Frischholz, C. Hallin, O. Kordina and E. Janzen, *Diamond and Related Mater.*, **6**, 1485–1488 (1997).

60. D. Peters, R. Schorner, K.-H. Holzlein and P. Friedrichs, *Appl. Phys. Lett.* **71**, 2996–2997 (1997).

61. O. Takemura, T. Kimoto, H. Matsunami, T. Nakata, M. Watanabe and M. Inoue, *Mater. Sci. Forum* **264–268** (2), 701–704 (1997).

62. N. Ramungul, Y. P. Zheng, R. Patel, V. Khemka and T. P. Chow, *Mater. Sci. Forum* **264–268** (2), 1049–1052 (1998).

63. H. Mitlehner, P. Friedrichs, D. Peters, R. Schorner, U. Weinert, B. Weis and D. Stephani, *Proc. Int. Symp. Power Semiconductor Devices and ICs*, 127–130 (1998).

64. R. Patel, V. Khemka, N. Ramungul and T. P. Chow, *Proc. 10th Int. Symp. Power Semiconductor Devices and ICs*, 387–390 (1998).

65. K. Rottner, *et al.*, *European SiC Conference* (1998).

66. D. Peters, P. Friedrichs, H. Mitlehner, R. Schorner, U. Weinert, B. Weis and D. Stephani, *Proc. Int. Symp. Power Semiconductor Devices and ICs*, 241–244 (2000).

67. J. B. Fedison, N. Ramungul, T. P. Chow, M. Ghezzo and J. W. Kretchmer, *IEEE Electron Device Lett.* **22**, 130–132 (2001).

68. Y. Suguwara, D. Takayama, K. Asano, R. Singh, J. Palmour and T. Hayashi, *Proc. Int. Symp. Power Semiconductor Devices and ICs*, 27–30 (2001).

69. H. Lendenmann, A. Mukhitdinov, F. Dahlquist, H. Bleichner, M. Irvin, R. Soderholm and P. Skytt, *Proc. Int. Symp. Power Semiconductor Devices and ICs*, 31–34 (2001).

70. R. Singh, A. R. Hefner, D. Berning and J. W. Palmour, *Proc. Int. Symp. Power Semiconductor Devices and ICs*, 45–48 (2001).

71. N. Ramungul and T. P. Chow, *IEEE Trans. Electron Devices* **46**, 493–496 (1999).

72. N. Ramungul, T. P. Chow and D. M. Brown, *Proc. Int. Symp. Power Semiconductor Devices and ICs*, 161–164 (1997).

73. V. V. Evstropov, B. N. Kalinin and B. V. Tsarenkov, *Sov. Phys. Semicond.* **17**, 373–378 (1983).

74. V. V. Evstropov, K. V. Kiselev, I. L. Petrovich and B. V. Tsarenkov, *Sov. Phys. Semicond.* **18**, 1156–1159 (1984).

75. M. M. Anikin, V. V. Evstropov, I. V. Popov, V. N. Rastegaev, A. M. Strel'chuck and A. L. Syrkin, *Sov. Phys. Semicond.* **23**, 405–407 (1989).

76. M. M. Anikin, V. V. Evstropov, I. V. Popov, A. M. Strel'chuck and A. L. Syrkin, *Sov. Phys. Semicond.* **23**, 1122–1125 (1989).

77. P. Neudeck, W. Huang and M. Dudley, *Solid-State Electron.* **42**, 2157–2164 (1999).

78. H. Lendenmann, F. Dahlquist, N. Johansson, R. Soderholm, P. A. Nilsson, J. P. Bergman and P. Skytt, *3rd European Conf. Silicon Carbide and Related Mater.*, 727 (2000).

79. J. P. Bergman, H. Lendenmann, P. A. Nilsson, U. Lindefelt and P. Skytt, *Proc. 3rd European Conf. Silicon Carbide and Related Mater.*, 299 (2000).

80. H. Lendenmann, F. Dahlquist, N. Johansson, J. P. Bergman, H. Bleichner and C. Ovren, *Proc. 1st Int. Workshop Ultra-Low-Loss Power Device Technol. (UPD2000)*, 125–130, Nara, Japan (2000).

81. P. O. A. Persson, H. Jacobson, J. M. Molina-Aldareguia, J. P. Bergman, T. Tuomi, W. J. Clegg, E. Janzen and L. Hultman, *Tech. Dig. Int. Conf. SiC and Related Mater.*, 55–56 (2001).

82. R. F. Stahlbush, J. B. Fedison, S. D. Arthur, L. B. Rowland, J. W. Kretchmer and S. Wang, *Tech. Dig. Int. Conf. SiC and Related Mater.*, 57–58 (2001).

83. A. Galeckas, J. Linnros and B. Breitholtz, *Tech. Dig. Int. Conf. SiC and Related Mater.*, 59–60 (2001).

84. H. Lendenmann, F. Dahlquist, J. P. Bergman, H. Bleichner and C. Hallin, *Tech. Dig. Int. Conf. SiC and Related Mater.*, 181–182 (2001).

85. St. G. Muller, M. Brady, B. Brixius, G. Fechko, R. C. Glass, D. Henshall, H. McD. Hobgood, J. R. Jenny, R. Leonard, D. Malta, A. Powell, V. F. Tsvetkov and C. H. Carter, Jr., *Tech. Dig. Int. Conf. SiC and Related Mater.*, 191 (2001).

86. K. J. Schoen, J. P. Henning, J. M. Woodall, J. A. Cooper, Jr. and M. R. Melloch, *IEEE Electron Device Lett.* **19**, 97–99 (1998).

87. B. J. Baliga, *Solid-State Electron.* **28**, 1089–1093 (1985).

88. F. Dahlquist, C.-M. Zetterling, M. Ostling and K. Rottner, *Mater. Sci. Forum* **264–268** (2), 1061–1064 (1997).

89. R. Held, N. Kaminski and E. Neimann, *Mater. Sci. Forum* **264–268** (2), 1057–1060 (1997).

90. Y. Sugawara, K. Asano and R. Saito, *Mater. Sci. Forum*, **338–342**, 1183–1186 (2000).

91. K. Asano, T. Hayashi, R. Saito and Y. Suguwara, *Proc. Int. Symp. Power Semiconductor Devices and ICs*, 97–100 (2000).

92. R. Singh, S. H. Ryu, J. W. Palmour, A. R. Hefner and J. Lai, *Proc. Int. Symp. Power Semiconductor Devices and ICs*, 101–104 (2000).

93. V. Khemka, V. Ananthan and T. P. Chow, *Electron Device Letters*, **27**, 286–288 (2000).

94. K. Tone, J. H. Zhao, M. Weiner and M. Pan, *Mater. Sci. Forum* **338–342**, 1187–1190 (2000).

# CHAPTER 7

# SILICON CARBIDE MOSFETs

James A. Cooper, Jr.

*School of Electrical and Computer Engineering, Purdue University,*
*West Lafeyette, IN 47907, USA*

## Contents

## 1. Introduction

This chapter provides an overview of the current status of MOS device development in silicon carbide (SiC). We assume the reader is generally familiar with MOS technology in silicon, and we focus our attention on the *differences* between silicon and SiC. These differences arise because of the profound differences in crystal structure and composition between the two materials. In Sec. 2 we provide a general overview of issues peculiar to SiC MOSFETs, including the hexagonal crystal structure, the effect of surface orientation, the role of carbon at the interface, the nature of the interfacial transition layer, the effect of interface states, the role of surface morphology, and the issue of oxide reliability. In Sec. 3 we present the latest experimental data on the SiC MOS interface, including the effect of processing variations on interface quality and inversion layer mobility. Section 4 discusses important device results achieved to date, focusing mainly on high-voltage power MOSFETs, but with a brief overview of other results such as MOS-based integrated circuits and charge coupled devices. Section 5 gives a summary and conclusions.

### 1.1. *The Nature of SiC*

SiC is a IV-IV binary compound semiconductor comprised of an equal number of silicon and carbon atoms, arranged in a hexagonal lattice.[1] Each silicon atom is bonded to four nearest-neighbor carbon atoms, and each carbon atom is bonded to four nearest-neighbor silicon atoms. In the natural progression of group IV semiconductors — germanium ($E_G = 0.66$ eV), silicon ($E_G = 1.12$ eV), diamond ($E_G = 5.5$ eV) — SiC lies intermediate between silicon and diamond in terms of bandgap energy, and represents a binary combination of the two elements, silicon and carbon.

As a material, SiC is physically robust. It is one of the hardest and most thermally stable materials known, and its high thermal stability presents both advantages and challenges. On the one hand, the thermal stability increases the reliability of devices, and offers the possibility of operation at significantly higher temperatures than silicon. However, the high thermal stability presents challenges in device processing. Consider crystal growth. Since silicon melts at a relatively moderate temperature, 1412°C, large single-crystal boules can be formed by slowly pulling a seed crystal from a reservoir of molten silicon, a procedure known as the Czochralski process. Over the years, this process has been refined so that 200 mm and even 300 mm wafers can be produced at reasonable cost. By comparison, SiC does not melt, but instead it sublimes at temperatures above 2000°C, so crystal growth by the Czochralski process is not feasible. Instead, SiC boules are grown by a modified sublimation process[2] in which SiC source material at approximately 2400°C is sublimed onto a seed crystal at approximately 2200°C. This process is far more difficult to control than the Czochralski process, and consequently SiC wafers are smaller and more costly than silicon. Current technology permits commercial production of 75 mm SiC wafers, and 100 mm wafers are in the development phase.

The high thermal stability also presents challenges in device processing.[3] Thermal diffusion of dopant impurities is not practical, except under very limited circumstances, and instead, selective-area doping is accomplished by ion implantation. However, activation of the implanted species also requires very high temperatures, typically 1400–1700°C. Thermal oxidation is slower than in silicon, and ohmic contacts must be alloyed at higher temperatures (typically 850–1000°C). Although many of these fabrication challenges have been overcome, the thermal requirements of certain unit processes still place important boundary conditions on SiC process and device development.

## 1.2. *Motivation for Electronic Devices in SiC*

As a semiconductor material, SiC has several important advantages. Because of its wide bandgap, the critical field for avalanche breakdown in SiC is almost 10 times higher than in silicon. An important figure of merit for power devices is the Baliga figure of merit $\mu_N \varepsilon_S E_C^3$.[4] Because of the high critical field, the Baliga figure of merit in 4H–SiC is about 400 times higher than in silicon. The specific on-resistance of unipolar devices such as power MOSFETs is roughly proportional to the blocking voltage squared divided by the Baliga figure of merit. Since the Baliga figure of merit is 400 times higher in SiC, the specific on-resistance (i.e. the resistance-area product) in SiC can be up to 400 times lower than in silicon for the same blocking voltage.

Another advantage of SiC for power devices is its high thermal conductivity, approximately twice as high as silicon. This facilitates efficient heat removal from the device during high power operation. SiC has a saturation drift velocity about twice as high as silicon,[5,6] but the electron mobility is lower than in silicon, approximately 60% of the silicon value in the 4H polytype of SiC, and even lower in the 6H polytype.

SiC is the only compound semiconductor whose native oxide is $SiO_2$. This makes it possible to fabricate the entire range of MOS devices in the material. The unique combination of high thermal stability, high break-down field, high thermal conductivity, reasonable electron mobility, and a high quality native oxide makes SiC an attractive candidate for the next generation of MOS-based power switching devices.

## 2. Overview of Issues Particular to SiC MOSFETs

In this section we will review the unique aspects of SiC as a host material for MOS-based devices and discuss the issues that must be addressed to realize the full potential of SiC MOS devices. In this discussion, we assume the reader is familiar with silicon MOS technology, and we focus on the *differences* between SiC and silicon. In Sec. 3 we present detailed experimental results on the $SiO_2/SiC$ MOS interface.

### 2.1. *SiC Crystal Structure: Polytypism, Polarity and Anisotropy*

As already stated, SiC is a binary compound semiconductor consisting of an equal number of silicon and carbon atoms in a hexagonal lattice. SiC is drastically different from silicon both in its lattice arrangement and in the fact that half the atoms are carbon. The basic bonding arrangement of silicon and carbon atoms in SiC is illustrated in Fig. 1. The SiC lattice can be viewed as a series of alternating hexagonal planes of silicon and carbon atoms, with every silicon atom bonding to four nearest-neighbor carbon atoms and every carbon atom bonding to four nearest-neighbor

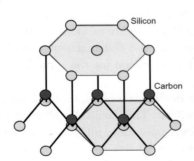

**Fig. 1.** Lattice arrangement of SiC. Each carbon atom (dark color) is tetrahedrally bonded to three silicon atoms in the plane below and one silicon atom in the plane above. This positions the carbon atoms over the centers of triangles formed by the silicon atoms in the plane below, creating an offset between successive planes.

silicon atoms. In the lattice shown in Fig. 1, each carbon atom bonds to three silicon atoms in the plane below and one silicon atom in the plane above. To equalize the bonds to the lower plane, each carbon atom is centered over the triangle formed by the underlying silicon atoms to which it is bonded. Since each carbon atom has only a single bond to the silicon atom above, the silicon atoms in the third plane are positioned directly above the carbon atoms in the second plane. Thus, the atoms in the second and third planes lie directly opposite each other, forming a silicon/carbon plane pair. The tetrahedral bonding clearly requires that successive plane-pairs be offset from each other, with the result that atoms in successive plane-pairs lie opposite open spaces in the plane-pairs above and below. Since there are only two possible positions that a second plane can assume relative to the first plane, there are a total of three positions that can be assigned to any plane-pair in the stacking sequence, as illustrated in Fig. 2. These positions are designated A, B, or C. Following the rule that each plane-pair must be offset from the plane-pair below gives rise to a set of allowed stacking sequences, known as *polytypes*. SiC therefore exhibits the phenomenon of *polytypism*, in which many similar crystal structures are possible, differentiated only by the stacking sequence of successive plane-pairs in the direction perpendicular to the basal plane, i.e. the c-axis direction.

The simplest polytype of SiC has the stacking sequence A-B-A-B- and is designated the 2H polytype. The next simplest sequence is A-B-C-A-B-C- and is designated the 3C polytype. In this designation, the number refers to the repeat sequence and the letter indicates the symmetry properties of the crystal, i.e. the 3C polytype exhibits cubic symmetry (3C is the *only* polytype of SiC that exhibits cubic symmetry). Letters H and R designate hexagonal and rhombohedral lattices, respectively. About 170 polytypes

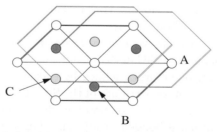

**Fig. 2.** Projection of the three possible positions of subsequent silicon planes onto the basal plane. The plane positions are designated A, B, and C. Tetrahedral bonding requires that adjacent planes lie in different positions, leading to unique stacking sequences known as polytypes.

**Table 1.** Important properties of silicon and several SiC polytypes.

| Parameter | Silicon | 3C–SiC | 4H–SiC | 6H–SiC |
|---|---|---|---|---|
| Bandgap Energy $E_G$ [eV] | 1.12 | 2.37 | 3.25 | 3.0 |
| Intrinsic Carrier Concentration $n_i$ [cm$^{-3}$] (at 300 K) | $1 \times 10^{10}$ | – | $2 \times 10^{-8}$ | $3 \times 10^{-6}$ |
| Critical Field $E_C$ [V/cm] | $3\text{–}4 \times 10^5$ | $\sim 1.8 \times 10^6$ | $1.5\text{–}4 \times 10^6$ | $1.5\text{–}4 \times 10^6$ |
| Electron Mobility $\mu_N$ [cm$^2$/Vs] | $\sim 1350$ | $\sim 900$ | – | – |
| ($\parallel$ to $c$-axis) | – | – | $\sim 1050$ | $\sim 100$ |
| ($\perp$ to $c$-axis) | – | – | $\sim 800$ | $\sim 375$ |
| Saturation Velocity [cm/s] | $1 \times 10^7$ | $\sim 2 \times 10^7$ | $2.2 \times 10^7$ | $1.9 \times 10^7$ |
| Thermal Conductivity [W/cm K] | 1.5 | 3.3 | 3.3 | 3.3 |

have been identified in nature, but only a few of these have been exploited as semiconductor materials. These include the 3C, 4H, 6H, 15R, and 21R polytypes. Only the 4H and 6H polytypes are commercially available as single crystal wafers for semiconductor device fabrication.

The various polytypes share the same atomic arrangement and lattice constant in the basal plane, but differ in the stacking sequence and repeat distance along the $c$-axis. Due to the different crystal structure, the electrical parameters are different as well. Table 1 lists important electrical parameters of the principal SiC polytypes. Primary among these is the bandgap energy, which varies from 2.37 eV in the 3C polytype to 3.25 eV in the 4H polytype. The intrinsic carrier concentration in SiC is orders of magnitude lower than in silicon, but it varies significantly among polytypes due to the exponential dependence on bandgap energy. Since the thermal generation rate is proportional to the intrinsic carrier concentration, thermally generated leakage currents are orders of magnitude lower in SiC than in silicon. For instance, in 4H–SiC the intrinsic carrier concentration is about $2 \times 10^{-8}$ cm$^{-3}$ at room temperature, so the thermal generation rate is approximately *18 orders of magnitude* lower than in silicon! This explains why SiC devices can be operated at higher junction temperatures without being overwhelmed by leakage current.

Figure 1 suggests another attribute of the SiC crystal structure, namely that the crystal is *polar* in the direction parallel to the $c$-axis. Since each hexagonal plane of atoms is bonded to adjacent planes by either three bonds per atom or one bond per atom, it follows that the surface of the crystal will be defined by the atomic plane type (silicon or carbon) that has *three bonds* to the adjacent interior plane (if only one bond per atom were present, the surface atoms would be unstable). Thus, one surface of a $c$-axis oriented wafer will be a silicon-atom plane while the surface on the opposite side of the wafer will be a carbon-atom plane. In Miller notation, the silicon face is designated the (0001) plane and the carbon face is designated the (000$\bar{1}$) plane. Many physical and electrical properties, including the oxidation rate and the electrical quality of the MOS interface, are different on the silicon and carbon faces. For reasons that will be described below, almost all device work to date has been preformed on the (0001) silicon-terminated face of 4H or 6H–SiC.

Except for the 3C polytype which exhibits cubic symmetry, all polytypes of SiC are anisotropic in both crystal structure and electrical behavior. The degree of anisotropy varies significantly among polytypes. For instance, the electron mobility in 6H–SiC is about 375 cm$^2$/Vs perpendicular to the $c$-axis, but only about 100 cm$^2$/Vs parallel to the $c$-axis. In 4H–SiC, the electron mobility is nearly isotropic at about 900 cm$^2$/Vs, being only about 10% higher parallel to the $c$-axis. Other parameters are also anisotropic, including thermal conductivity and impact ionization rates (and hence critical fields). As might be expected, certain processing operations, such as thermal oxidation and dry etching, are also anisotropic.

## 2.2. *Surface Orientation, Bonding at the Interface and the Interfacial Transition Layer*

As stated, the overwhelming majority of MOS research on SiC has been conducted on the (0001) silicon-terminated surface of 4H and 6H–SiC. Other surfaces and polytypes have been investigated to a limited extent, and still others are now coming under investigation. Early research centered on Lely crystals of the 3C polytype, whose surfaces are designated by the cubic notation (111), (110), or (100). However, work on 3C–SiC was generally discontinued after about 1991 due to the commercial availability of single crystal wafers of 6H–SiC, and later 4H–SiC. Some MOS measurements have been reported on 15R material, found as inclusions in other polytypes, and a few measurements have been performed on (000$\bar{1}$) carbon-face of 6H–SiC. However, the best electrical quality to date has been obtained on the (0001) face of 4H and 6H–SiC. Recently, several workers have reported promising results on the (11$\bar{2}$0) and (03$\bar{3}$8) surfaces of 4H–SiC, and we shall review these results in the next section.

The role of surface orientation is closely related to the nature of the chemical bonding within the transition layer at the interface. In general terms, one visualizes an interfacial transition layer a few monolayers thick lying between stoichiometric SiC and stoichiometric $SiO_2$, and providing a transition between these regions. The atomic species and bonding configurations that exist within this transition layer will be those which exhibit the *lowest energy*, while satisfying the boundary conditions imposed by the SiC below and the $SiO_2$ above. Theoretical calculations of interfacial bonding are currently underway, and only very preliminary results are available at this time. However, it is clear that the physical nature of the transition layer depends critically upon the atomic arrangement presented by the particular surface plane of the lattice, whether it be (0001), ($11\bar{2}0$), or ($03\bar{3}8$). For example, on the (100) plane of silicon it is possible to envision a stable suboxide transition layer without any direct silicon-silicon bonds. The (0001) surface of SiC is also terminated with silicon atoms, but it has a different atomic arrangement and atomic spacing than the (100) surface of silicon. It turns out that it is not possible to envision a stable transition layer on (0001) SiC without the use of silicon-silicon bonds in the interfacial sub-oxide layer.[7] Silicon-silicon bonds give rise to energy levels within the bandgap of SiC, and these energy levels may be observed as interface states in MOS devices. Clearly, the physical nature of the interfacial transition layer and the electrical quality of the resulting interface depend critically on the atomic configuration of the particular surface on which the oxide is formed.

When SiC is thermally oxidized, the oxidizing species diffuses to the interface and reacts with SiC, forming $SiO_2$ and CO. The $SiO_2$ adds to the growing oxide layer, while the CO diffuses away and escapes as a gas. During this process, carbon may become incorporated at the interface. On any surface other than (0001), a large fraction of the exposed surface atoms are also carbon. As a result, carbon-silicon and carbon-carbon bonds are almost inevitable. Theoretical studies suggest that carbon-carbon bonds can also introduce energy levels in the bandgap.[8] The density of these interface states will also depend upon surface orientation, due to the nature of the bonding configurations within the interfacial transition layer. Although most of the MOS studies to date have been performed on the (0001) silicon-terminated surface, the ($03\bar{3}8$) surface has recently attracted attention because it corresponds to the (100) surface of the cubic polytype. By analogy with silicon, one might expect a lower interface state density on this surface. However, in SiC the situation is complicated by the fact that some of the surface atoms will necessarily be carbon. More investigation is needed to determine whether the presence of carbon on these surfaces leads to electrically active defect states within the bandgap.

Transmission Electron Microscopy (TEM) and Electron Energy Loss Spectroscopy (EELS) have shown conclusively that the composition and abruptness of the transition layer is affected by post-oxidation annealing.[9,10] Thermal oxidation of SiC typically takes place at 1100–1150°C in wet $O_2$, followed by a 30 minute *in-situ* anneal in argon.[11] The argon anneal permits any residual CO to diffuse out of the oxide, resulting in a stoichiometric oxide layer with very low carbon content. The breakdown strength of the resultant oxide is about $10^7$ V/cm, similar to thermal oxides on silicon. Following the argon anneal, a re-oxidation anneal in wet $O_2$ is performed at 950°C for one hour.[12] Electron energy loss spectroscopy (EELS) has shown that the "re-ox" anneal produces a more abrupt interfacial transition layer,[9] and electrical analysis reveals a reduction in interface state density near midgap.[12] A final post-oxidation anneal is performed in nitric oxide (NO) at 1150°C for two hours to incorporate nitrogen at the interface, resulting in an order of magnitude reduction in interface state density in the upper half of the bandgap.[13] Details will be presented in the next section, but these results demonstrate that the interfacial transition layer and the electrical quality of the interface are dependent both on surface orientation and post-oxidation processing.

## 2.3. *Interface States and Fixed Oxide Charges*

Interface states (or traps) are energy levels distributed across the bandgap in energy, but located physically at the oxide/semiconductor interface.[14] They arise from defects, strained bonds, or dangling bonds at the interface. Since they can exchange charge with the semiconductor, their charge state depends upon bias. Fixed oxide charges also lie physically at the interface, but correspond to energy levels outside the bandgap, and therefore do not change their charge state with bias. Both types of defects are present to some extent on all semiconductors.

The density of interface states per unit area per unit energy is designated $D_{IT}$ and expressed in units of (cm$^{-2}$ eV$^{-1}$). The density of fixed charges is designated as $Q_F$ and expressed in units of (C/cm$^2$). In silicon, $D_{IT}$ is typically on the order of $10^{11}$ cm$^{-2}$ eV$^{-1}$ or higher immediately following oxidation, but is reduced below $10^{10}$ cm$^{-2}$ eV$^{-1}$ by a 450°C post-oxidation anneal in hydrogen.[14] Significantly, the 450°C hydrogen anneal has not proven effective in SiC. Since the bandgap of 4H and 6H–SiC is almost three times wider than silicon, the same density of interface states per unit energy corresponds to three times more charge at the SiC interface. For this reason, control of interface states is extremely important for SiC devices. As will be discussed in the next section, the interface state density in the lower half of the bandgap on 4H or 6H–SiC typically ranges from 0.5–$1.5 \times 10^{11}$ cm$^{-2}$ eV$^{-1}$. These values, while higher than in silicon, are still quite acceptable for normal device operation. However, the interface state density in the upper half of the bandgap increases exponentially towards the conduction band, reaching values in excess of $10^{13}$ cm$^{-2}$ eV$^{-1}$ within 0.2 eV of the conduction band edge on 4H–SiC.[15] This has a serious effect on $n$-channel MOSFETs in SiC. When an $n$-channel MOSFET is biased into inversion, the Fermi level lies close to the conduction band edge, and all the interface states below the Fermi level are occupied by electrons. Since these trapped electrons increase the oxide electric field, the number of electrons present in the inversion layer at a given oxide field is reduced. In addition, Coulomb scattering from the trapped electrons reduces the *mobility* of the inversion electrons. The combined effect is a significant reduction in channel current. Interface state reduction is currently a major topic of investigation in SiC, and will be discussed in more detail in the following sections.

In both silicon and SiC, the fixed oxide charge appears to be positive. In silicon, $Q_F$ is independent of oxide thickness, semiconductor doping density, and doping type ($n$ or $p$), but depends upon surface orientation and oxidation and annealing conditions.[14] In silicon, $Q_F$ can be reduced to around $10^{11}$ cm$^{-2}$ by post-oxidation annealing in nitrogen. In SiC, $Q_F$ is typically around $0.5$–$1 \times 10^{12}$ cm$^{-2}$. This positive charge shifts the flat-band voltage of both $n$-type and $p$-type MOS capacitors towards negative voltages. The net result is that the flat-band voltage on $p$-type SiC is negative, while the flat-band voltage on $n$-type SiC is close to zero.[11] Without a positive fixed charge, the flat-band voltage on $p$-type material would be less negative and the flat-band voltage on $n$-type material would be more positive. Detailed information on the dependence of $Q_F$ on SiC oxidation and anneal conditions is still lacking.

### 2.4. *Surface Morphology*

Almost all SiC MOSFETs require ion implantation, either to form the source/drain regions, to dope the channel regions, or both (a notable exception is the trench-gate UMOSFET, which can be fabricated without ion implantation). Common $n$-type dopants in SiC are nitrogen and phosphorus. Activation of nitrogen implants in 4H–SiC requires annealing at 1400°C or above, while phosphorus can be activated at 1200°C.[16] $p$-type dopants for SiC are aluminum and boron. Activation of aluminum requires annealing at 1600°C, while boron requires annealing at 1700°C.[16] At these higher temperatures, the vapor pressure of silicon is high enough to cause the loss of silicon from the surface, degrading the surface morphology.

Another problem is related to the surface steps. SiC wafers are normally cut slightly off-axis with respect to the *c*-axis (3.5° for 6H–SiC and 8° for 4H–SiC). The resulting wafer surfaces consist of broad (0001) terraces separated by steps that expose the $(11\bar{2}0)$ plane. These steps reveal the polytype stacking sequence, providing a template that preserves the polytype during subsequent epigrowth. Unfortunately, at the high temperatures of the implant activation anneal, the surface atoms are quite mobile, and the surface tends to reconfigure to reduce the surface energy. Since steps represents a higher energy state, the surface reconstructs to reduce the number of steps. This process is known as step bunching, and results in a roughening of the surface.[16,17] Step bunching is exacerbated by the lattice damage from ion implantation, and is more severe in regions that have been implanted. If such regions are used to form the channel regions for MOSFETs, as in the DMOS process to be discussed in Sec. 4, the resulting surface roughness can severely degrade MOSFET performance.

### 2.5. *Oxide Reliability and Oxide Breakdown*

The final MOS issue peculiar to SiC is the question of oxide reliability. The breakdown strength of thermally grown oxides on SiC is around $10^7$ V/cm, similar to thermal oxides on silicon, but the long-term *reliability* of oxides on SiC is a serious concern, particularly at high temperatures and high fields. Because SiC has a wider bandgap than silicon, the barrier height for charge injection from SiC into $SiO_2$ is lower than for silicon, as illustrated in Fig. 3. Since injection current depends exponentially on barrier height, even a slight reduction in barrier height produces significantly higher injection. Although the physical mechanisms for oxide degradation are not completely understood, it is clear that the rate of degradation is directly related to charge injection.

The reliability of oxides on silicon has been studied for many years, and a significant amount of statistical data is available upon which to base conclusions. In comparison, reliability data on SiC is seriously lacking, mainly because the technology for optimizing the $SiC/SiO_2$ interface is still evolving. However, the limited amount of data available at the present time suggests that adequate (i.e. multi-year) mean-time-to-failure can be insured

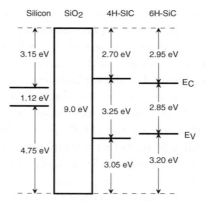

**Fig. 3.**  Conduction and valence band alignments for Si, SiO$_2$, 4H–SiC, and 6H–SiC. The barriers for electron and hole injection from SiC into SiO$_2$ are lower than for silicon.

by restricting the maximum oxide electric field to the range of 3–4 MV/cm and the junction temperature to 250°C or below.[18] This represents a severe limitation on SiC devices, since the wide bandgap of SiC would otherwise allow device operation to 500–600°C. The electric field limitation is also more severe than might first appear, for the following reason. In silicon, the critical field for avalanche breakdown is around 3–4 × 10$^5$ V/cm. If the peak field in a silicon MOS device reaches this value, the field in the adjacent oxide will be about 3 times higher due to the ratio of dielectric constants at the interface, making the oxide field around 9 × 10$^5$ V/cm. But this is still well below the oxide breakdown field of 10$^7$ V/cm, and low enough that significant charge injection will not take place. In SiC, on the other hand, the critical field for avalanche breakdown is 2–3 MV/cm, and the field in the adjacent oxide will be about 7.5 MV/cm. This is dangerously close to the breakdown strength of the oxide and high enough to cause significant charge injection into the oxide. Therefore, as a result of oxide reliability considerations, both the maximum field and the maximum temperature of SiC MOS devices must be restricted to values below those of which the semiconductor would otherwise be capable.

## 3.  Experimental Results on the MOS Interface on SiC

In this section we discuss the most recent experimental results on the SiO$_2$/SiC interface, with particular emphasis on the effect of processing conditions on the electrical quality of the interface. These issues are key to the performance of SiC MOSFETs, and they will be discussed in Sec. 4. Before

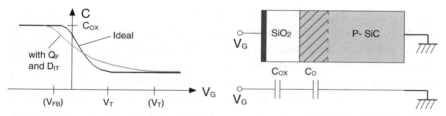

**Fig. 4.** Capacitance-voltage curve for a $p$-type MOS capacitor. The solid curve represents an ideal capacitor, while the shaded curve is a capacitor with fixed oxide charge and interface states. Voltages in parentheses correspond to the shaded curve. The components of capacitance are illustrated at right.

we present the experimental data, it is advisable to take a few moments to review the MOS fundamentals, with particular emphasis on those issues that are most relevant to the $SiO_2/SiC$ interface.[19]

### 3.1. *Review of MOS Fundamentals Pertinent to SiC*

We begin our discussion of MOS technology in SiC by considering the C–V curve of an MOS capacitor on a $p$-type semiconductor as shown in Fig. 4. The gate capacitance can be represented as a series combination of the oxide capacitance $C_{OX}$ and the depletion region capacitance $C_D$. The oxide capacitance is invariant under bias, and is simply the structural capacitance of the insulator per unit area, $C_{OX} = \varepsilon_{OX}/t_{OX}$, where $\varepsilon_{OX}$ and $t_{OX}$ are the permittivity and thickness of the insulator, respectively. The semiconductor depletion capacitance is determined by the width of the depletion region at the surface, $C_D = \varepsilon_S/x_D$, where $\varepsilon_S$ and $x_D$ are the permittivity of the semiconductor and the width of the depletion region, respectively. The width of the depletion region depends on band bending, or more precisely, on the surface potential $\phi_S$. To a reasonable approximation, $\phi_S$ can be given by

$$\phi_S = V_G^* + V_0 - \sqrt{V_G^{*2} + 2V_0 V_G^*} \qquad (1)$$

where $V_0$ is a constant determined by the oxide thickness and semiconductor doping, and $V_G^*$ is an effective gate voltage, given by

$$V_0 = \frac{q\varepsilon_S N_A}{C_{OX}^2} \qquad (2)$$

$$V_G^* = V_G - V_{FB} - \frac{\Delta Q_{IT}(\phi_S)}{C_{OX}}. \qquad (3)$$

Here $N_A$ is the doping of the semiconductor, $V_{FB}$ is the flat-band voltage, and $\Delta Q_{IT}(\phi_S)$ is the change in interface trapped charge between the current biasing point and flat band, $\phi = 0$. The flat-band voltage is determined by the gate-semiconductor work function $\Phi_{MS}$ and the total interface charge at flat band. The interface charge at flat band is comprised of the oxide fixed charge $Q_F$ and the charge in interface states at flat band, $Q_{IT}(0)$. Since it is not easy to distinguish between these latter two quantities, we represent their sum by an *effective* fixed charge $Q_{F,EFF}$. Thus,

$$V_{FB} = \Phi_{MS} - \frac{Q_{F,EFF}}{C_{OX}}. \tag{4}$$

Equation (1) strictly applies only under depletion (and deep depletion) biasing. Under accumulation biasing ($V_G < V_{FB}$) the surface potential is assumed to be zero. Under inversion biasing ($V_G > V_T$, where $V_T$ is the threshold voltage), the surface potential is assumed to be equal to $2\phi_F$, where $\phi_F = kT/q \ln(N_A/n_i)$. Here $k$ is Boltzmann's constant and $T$ is the absolute temperature. These equations constitute the well-known *delta-depletion* approximation for MOS electrostatics. We reproduce them here to draw attention to the impact of the effective fixed charge and interface states on the electrostatics of SiC MOS devices.

Consider the effect of interface charge on flat-band voltage. We see from Eq. (4) that a positive effective fixed charge shifts the flat-band voltage negative. This is equally true for both $n$-type and $p$-type MOS capacitors. The effective fixed charge $Q_{F,EFF}$ at the interface is the sum of the fixed charge $Q_F$ and the interface trapped charge at flat band $Q_{IT}(0)$. The fixed charge $Q_F$ is independent of the substrate doping polarity (at least in silicon), and therefore will be the same on both $n$-type and $p$-type MOS capacitors. However, the interface trapped charge at flat band, $Q_{IT}(0)$, is different for an $n$-type MOS capacitor and a $p$-type MOS capacitor. This is because at flat band in an $n$-type MOS capacitor, the Fermi level at the surface is close to the conduction band, while at flat band in a $p$-type MOS capacitor, the Fermi level at the surface is close to the valence band. Since all the interface states below the Fermi level are filled with electrons, the interface trapped charge $Q_{IT}(0)$ will be more negative for an $n$-type MOS capacitor than for a $p$-type MOS capacitor, *even if the interface state density is identical* for the two devices. This results in a more negative flat-band voltage for a $p$-type MOS capacitor than for an $n$-type MOS capacitor. Again, we emphasize that this conclusion applies even if the fixed oxide charge and the interface state density are *identical* on the two doping types.

The threshold voltage $V_T$ is the value of gate voltage that produces a surface potential equal to $2\phi_F$. Solving Eq. (1) for gate voltage and inserting Eq. (3) into the result yields

$$V_G^* = V_G - V_{FB} + \frac{\Delta Q_{IT}(\phi_S)}{C_{OX}} = \phi_S + \sqrt{2V_0\phi_S}\,. \tag{5}$$

Now solving Eq. (5) for the threshold voltage $V_T$ (i.e. $V_G$ when $\phi_S = 2\phi_F$), we obtain

$$V_T = V_{FB} - \frac{\Delta Q_{IT}(2\phi_F)}{C_{OX}} + 2\phi_F + \sqrt{2V_0(2\phi_F)}\,. \tag{6}$$

Equation (6) shows that the threshold voltage of an MOS capacitor or MOSFET depends on the amount of charge contained in interface traps when the bands are bent by $2\phi_F$. Thus, for a $p$-type MOS capacitor (or $n$-channel MOSFET), a positive fixed oxide charge produces a negative flat-band voltage, while a distribution of interface states across the bandgap spreads the C–V curve along the voltage axis, producing a more positive threshold voltage. This is illustrated by the shaded line in Fig. 4, and is fairly representative of the C–V curves in most $p$-type SiC MOS capacitors. The main point to be drawn from the preceding discussion is that control of both fixed oxide charge (or effective fixed charge) and interface state density is crucial to the performance of MOS devices in SiC.

We now consider the effect of interface states when the MOS device is biased into inversion, $V_G > V_T$. Our simple model for surface potential assumes that $\phi_S$ remains fixed at the value $2\phi_F$ for all $V_G > V_T$. This is not strictly correct. In actuality, $\phi_S$ continues to increase slowly as $V_G$ increases beyond $V_T$. Figure 5 shows a plot of surface potential vs. gate voltage for

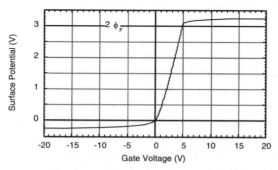

**Fig. 5.** Surface potential as a function of gate voltage for a 4H–SiC MOS capacitor at room temperature. Doping is $2 \times 10^{16}$ cm$^{-3}$ and oxide thickness is 50 nm. The surface potential increases very slightly as $V_G$ is taken above $V_T$, which is 5 V in this example.

a 4H–SiC MOS capacitor with $t_{OX} = 50$ nm and $N_A = 2 \times 10^{16}$ cm$^{-3}$. Since $\phi_S$ continues to increase in inversion, more interface states are pushed below the Fermi level, and the charge in interface states $\Delta Q_{IT}(\phi_S)$ continues to increase. The trapping of charge in interface states reduces the mobile charge in the inversion layer at a given gate voltage. This can be seen by rewriting Eq. (5) to include the inversion layer charge per unit area $Q_N$,

$$V_G^* = V_G - V_{FB} + \frac{\Delta Q_{IT}(\phi_S)}{C_{OX}} + \frac{Q_N(\phi_S)}{C_{OX}} = \phi_S + \sqrt{2V_0\phi_S}. \tag{7}$$

Equation (7) can then be solved for the inversion charge $Q_N$,

$$Q_N(\phi_S) = -C_{OX}\left[V_G - V_{FB} + \frac{\Delta Q_{IT}(\phi_S)}{C_{OX}} - \phi_S - \sqrt{2V_0\phi_S}\right]. \tag{8}$$

To use Eq. (8), we must determine the interface trapped charge $\Delta Q_{IT}$ at a given surface potential by integrating $D_{IT}$ from flat band to $\phi_S$,

$$\Delta Q_{IT}(\phi_S) = -q \int_{E_i - q\phi_F}^{E_i - q\phi_F + q\phi_S} D_{IT}(E)dE. \tag{9}$$

If we know the gate voltage and surface potential, Eqs. (8) and (9) allow us to calculate the inversion charge. However, we cannot simply set $\phi_S = 2\phi_F$, since this would not account for the increase in trapped charge as we move further into inversion. Instead, we need to solve for $\phi_S$ more rigorously. Such a solution is straightforward,[20] but solving for $\phi_S$ is not necessary at this point. The main reason for developing Eqs. (8) and (9) is to show that the negative charge trapped in interface states directly reduces the inversion charge at a given gate voltage.

Another way to view the situation is to consider the electric field in the oxide, $E_{OX}$. At threshold, $\phi_S = 2\phi_F$, and the oxide field is given by Gauss' law as

$$E_{OX} = -[Q_F + Q_D + \Delta Q_{IT}(2\phi_F)]/\varepsilon_{OX}. \tag{10}$$

The depletion charge $Q_D$ is given by

$$Q_D = -C_{OX}\sqrt{2V_0(2\phi_F)} \tag{11}$$

and the interface trapped charge $\Delta Q_{IT}(2\phi_F)$ is given by Eq. (9). Both $Q_D$ and $\Delta Q_{IT}(2\phi_F)$ are negative, and typically larger in magnitude than the positive fixed charge $Q_F$. The effect of the negative trapped charge is to increase the oxide field at threshold, thereby also increasing the threshold voltage (c.f. Eq. (6)). The maximum oxide field is limited by reliability considerations to 3–4 MV/cm, and there is now less room to increase the oxide field beyond threshold before reaching this limiting value. As the gate voltage and oxide field are increased, we would hope that all additional field lines would terminate on inversion charges, but since the trapped charge also increases beyond threshold, some of the additional field lines terminate on trapped charges. This reduces the charge in the inversion layer at a given gate voltage. An excellent discussion of these issues as they relate to 4H–SiC MOSFETs is given in Ref. 20.

Another detrimental effect of the interface trapped charge is the increased Coulomb scattering.[21] The mobility of inversion electrons is limited by several types of scattering, including phonon scattering, surface roughness scattering, and Coulomb scattering.[22] As we shall see momentarily, at the high interface state densities present in non-optimally processed SiC MOS devices, Coulomb scattering becomes the dominant scattering mechanism, reducing the inversion layer mobility by up to an order of magnitude.

### 3.2. *Thermal Oxidation of SiC*

Having reviewed the pertinent MOS fundamentals, we now turn to the growing body of experimental data on oxidation of SiC. As stated in the introduction, all polytypes of SiC except the 3C polytype are anisotropic, and the oxidation rate depends on surface orientation, as shown in Fig. 6.[23] The most rapid oxidation occurs on the (000$\bar{1}$) carbon-terminated face,

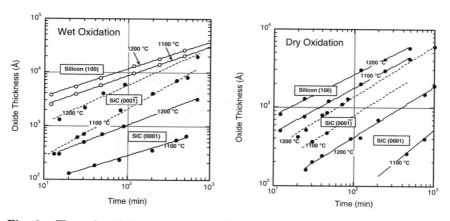

**Fig. 6.** Thermal oxidation rates in wet (left) and dry (right) oxygen for silicon (100) and two orientations of 6H–SiC. Oxidation rates for other SiC crystal orientations are intermediate between those for (0001) and (000$\bar{1}$).

while the slowest occurs on the (0001) silicon-terminated face. Other surfaces have oxidation rates intermediate between these extremes, but all SiC surfaces oxidize much slower than silicon at a given temperature.

Several techniques have been developed over the last several years to improve the quality of the MOS interface resulting from thermal oxidation. A rigorous clean of the sample is necessary before insertion into the oxidation tube,[11] and the same level of cleanliness must be applied to the tube itself and all glassware that might come into contact with the sample, including boat, push-rod, etc. The wafer is cleaned using an offshoot of the basic silicon "RCA" clean that includes a solvent rinse and repeated soaks in $H_2O_2/H_2SO_4$, buffered HF, deionized water, $H_2O_2/NH_4OH$, and $H_2O_2/HCl$. The sample is blown dry in $N_2$ and inserted immediately into the oxidation tube at 850°C in wet $O_2$. The temperature is raised slowly to the oxidation temperature, say 1150°C. The low-temperature insertion in an oxidizing ambient minimizes the loss of silicon from the surface by forming a thin oxide layer during insertion. Without such an insertion process, the loss of silicon would leave a carbon-rich surface that could include graphitic-like carbon clusters, increasing the interface state density.[24,25] After oxidation in wet $O_2$ at 1150°C, the sample receives an *in-situ* anneal in Ar for 30 minutes at 1150°C to allow for any remaining carbon to diffuse out and escape as CO. The tube is then cooled to 950°C and the sample is annealed in wet $O_2$ for an additional 120 minutes. This so-called "re-oxidation anneal"[12] produces a more abrupt interface and reduces the density of interface states in the middle part of the bandgap. Figure 7 shows the interface state density

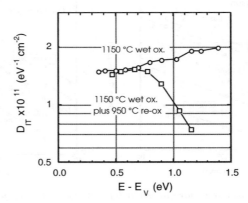

**Fig. 7.**  Interface state density in the lower half of the bandgap on 6H–SiC, with and without a 950°C re-ox anneal.[26]

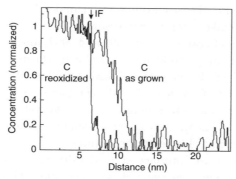

**Fig. 8.** EELS profiles of carbon at the $SiO_2$/SiC interface, with and without a 950°C re-ox anneal.[9]

and Fig. 8 shows EELS profiles on samples with and without the re-ox anneal. The re-ox anneal apparently forms a few monolayers of new oxide at the interface, and the lower oxidation rate produces a superior interface, both physically and electrically.

The thermal oxidation procedure described above is widely used, but many variations are possible. The variables include oxidation temperature, oxidation ambient (wet or dry), and post-oxidation anneal. Oxidation temperature does not have a strong effect on interface quality, at least in the range from about 1050–1150°C. As stated, a re-ox anneal at 950°C improves the interface quality, and one may argue that this represents a lower-temperature oxidation step, but it is impractical to perform the entire oxidation at such low temperature because of the long oxidation times required. Wet oxidation appears to produce interfaces with lower $Q_F$ and $D_{IT}$, but some reports suggest that dry oxidation improves oxide reliability.

Unfortunately, in many areas of SiC MOS technology it is still difficult to make definitive statements because many of the experiments reported to date do not contain sufficient control samples, i.e. experiments from one laboratory cannot be compared quantitatively to experiments from a second laboratory because the baseline processes from the two laboratories are different. Truly meaningful comparisons require control samples from which a single process deviation is made. Multiple process deviations should each be compared to a common control sample from the same laboratory. In most cases, such careful comparisons are lacking. Therefore, many of the statements in this section are based on anecdotal information, and careful experiments may yet reverse some of our conclusions. Our understanding of many issues related to SiC oxidation is still clouded by the lack of sufficient data on which to make careful comparisons.

For several years between about 1992 and 1997, most of the attention in the SiC MOS community was focused on reducing the interface state density in the lower half of the bandgap. Conventional admittance techniques, such as the high-low C–V and the AC conductance techniques, provide information on interface state density only relatively near the majority carrier band edge.[11] The range of coverage can be extended into the midgap region by measuring at elevated temperatures such as 250–350°C, but it is not possible to measure interface state density in the upper half of the bandgap on $p$-type material using these techniques. Through a variety of improvements over these years, the interface state density in the lower half of the bandgap was gradually reduced from the mid-$10^{11}$ eV$^{-1}$ cm$^{-2}$ range to the low-$10^{11}$ eV$^{-1}$ cm$^{-2}$ range, and the re-oxidation anneal further reduced interface state density near midgap to the mid-$10^{10}$ eV$^{-1}$ cm$^{-2}$ level on both 4H and 6H–SiC. However, the mobility of inversion electrons in MOSFETs did not significantly improve during this period. In 1997, Afanasev *et al.*[27] and Bassler, *et al.*[24] reported measurements of interface state density over the entire bandgap of 3C, 4H, and 6H–SiC using admittance spectroscopy and DLTS. Their data, reproduced in Fig. 9, shows an interface state density that rises rapidly towards both the

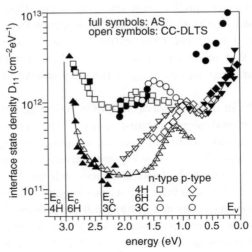

**Fig. 9.** Interface state density on 3C, 4H, and 6H–SiC samples measured at the University of Erlangen-Nürnberg.[27]

valence band and the conduction band edges. It is noteworthy that the interface state density in the lower half of the bandgap on these samples is considerably higher than on the samples of Fig. 7. This is apparently due to differences in the oxidation and anneal procedures between the different laboratories. These discrepancies notwithstanding, subsequent measurements at other laboratories confirmed that the interface state density in *all* samples increases exponentially towards the conduction band, even in samples where the interface state density in the lower half of the bandgap is low. It appears that differences in processing conditions between laboratories have an effect on the interface state density in the lower half of the bandgap, but not on interface state density in the upper half of the bandgap. Figure 10 shows $D_{IT}$ in the upper half of the bandgap[28] determined by the AC conductance[29] on $n$-type 4H–SiC MOS capacitors fabricated at several laboratories by a variety of techniques, including wet thermal oxidation, dry thermal oxidation, Low Pressure Chemical Vapor Deposition (LPCVD), and deposited Oxide/Nitride/Oxide (ONO) stacks. The behavior of all samples within about 0.4 eV of the conduction band edge is remarkably independent of the oxidation technique. From our discussion in Sec. 3.1 we know that such a high $D_{IT}$ is very detrimental to MOSFET performance, since the interface states trap electrons that would otherwise be available for conduction, and the large trapped charge reduces the mobility of free electrons due to Coulomb scattering. Since the states in the upper half of the bandgap are apparently unresponsive to oxidation techniques, the improvements in cleaning and oxidation techniques from

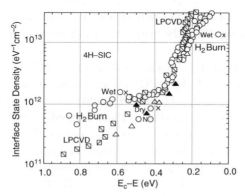

**Fig. 10.** Interface state density near the conduction band on 4H–SiC samples oxidized by a variety of techniques.[28]

1992–1997 had little effect on MOSFET mobility. Once the existence of a high $D_{IT}$ in the upper half of the bandgap was recognized, this issue quickly became the most important topic in SiC MOS research, and it remains so today. Significant progress has been made in reducing $D_{IT}$ using post-oxidation anneals, and these results will be discussed in Sec. 3.4 below. Before considering post-oxidation anneals, we will briefly discuss oxides formed by methods other than thermal oxidation.

### 3.3. *Deposited Oxides*

High quality oxides can be obtained on SiC by a variety of methods other than thermal oxidation. Sridevan *et al.*[30] and Alok *et al.*[31] report field-effect mobilities as high as 80 $cm^2/Vs$ on 4H–SiC using oxides formed by Low Pressure Chemical Vapor Deposition (LPCVD). However, in each case it was necessary to follow the LPCVD step with a wet thermal oxidation at 1100°C. Failure to perform the thermal oxidation step resulted in single-digit mobilities. It may be argued that the electrical quality of the bulk oxide is determined by the nature of the deposition process, while the electrical quality of the interface is determined by the final thermal oxidation. In this regard, the interface is essentially that of a thermal oxide, not a deposited oxide.

Wang *et al.*[32] have deposited both oxides and Oxide/Nitride/Oxide (ONO) stacked dielectrics on SiC using a Jet Vapor Deposition (JVD) process. The ONO stacked dielectrics exhibit an exponentially rising $D_{IT}$ in the upper half of the bandgap, similar to that of thermal oxides, as seen in Fig. 10. However, recent work indicates that a very thin nitride layer deposited prior to ONO deposition and subsequent post-deposition annealing yields a significantly reduced $D_{IT}$ in the upper half of the bandgap on 4H–SiC.[32]

Another technique for forming a deposited oxide is to deposit poly-crystalline silicon on SiC by LPCVD, and then to thermally oxidize the silicon layer to form $SiO_2$. The MOS quality of interfaces formed in this manner are typically comparable to those formed by thermal oxidation of SiC,[33] because the oxidation proceeds until all the silicon layer is converted to $SiO_2$ and the final interface is formed by thermal oxidation of SiC. Although confirming experiments have not yet been reported, it is expected that post-oxidation annealing, including the 950°C re-oxidation discussed above and the post-oxidation NO anneal to be discussed below, will be as effective on deposited oxides as on thermal oxides, resulting in interfaces with quality comparable to the best thermal oxides. This may not be true for stacked dielectrics such as ONO, since the nitride layer may prevent diffusion of the passivating species to the interface.

Deposited dielectrics are particularly useful in trench-gate UMOSFETs because they avoid the orientation anisotropy of thermal oxidation. UMOSFETs typically have critical dielectric layers on the $(0001)$, $(1\bar{1}00)$, and $(11\bar{2}0)$ surfaces, all in the same device. The anisotropy of thermal oxidation means that thermally grown oxides will have different thicknesses on these surfaces. This is highly undesirable, since the oxide electric field will be different on each different surface at the same gate voltage. The use of deposited dielectrics avoids this problem.

### 3.4. *Post-Oxidation Annealing*

Silicon MOS technology has long used post-oxidation annealing to improve the interface.[14] The most important example is the post-metallization anneal in $H_2$ at $450°C$. This anneal transports hydrogen atoms to the interface where they passivate dangling bonds, reducing the interface state density to the low-$10^{10}$ $eV^{-1}$ $cm^{-2}$ range. Hydrogen annealing in this temperature range has not proven effective in SiC, but Fukuda *et al.*[34] have shown that a hydrogen anneal at higher temperatures, typically around $800°C$, reduces the interface state density in the upper half of the bandgap on $n$-type 4H–SiC.

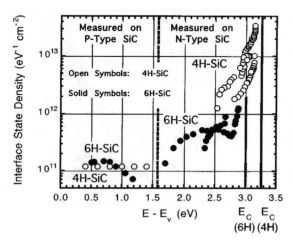

**Fig. 11.**  Interface state density across the bandgap on 4H–SiC (open symbols) and 6H–SiC (solid symbols).

Figure 11 compares the interface state density on the (0001) surface of 4H and 6H–SiC as obtained using wet thermal oxidation at 1150°C, an *in-situ* Ar anneal at 1150°C, and a re-ox anneal in wet $O_2$ at 950°C.[15] As seen, $D_{IT}$ is in the low $10^{11}$ eV$^{-1}$ cm$^{-2}$ range in the lower half of the bandgap on both polytypes, but rises exponentially in the upper half of the bandgap, reaching values above $10^{12}$ eV$^{-1}$ cm$^{-2}$ on 6H–SiC and above $10^{13}$ eV$^{-1}$ cm$^{-2}$ on 4H–SiC. As this figure shows, the higher $D_{IT}$ on 4H–SiC may be simply due to the wider bandgap of this polytype, since $D_{IT}$ appears fairly comparable on the two polytypes over most of the bandgap. Nevertheless, the higher $D_{IT}$ means that MOSFETs on 4H–SiC have higher threshold voltages and lower mobilities than comparable MOSFETs on 6H–SiC. This is unfortunate, because the 4H polytype is strongly preferred for high-voltage power devices because of its higher bulk electron mobility parallel to the *c*-axis. The high $D_{IT}$ is shown to correlate with a significantly lower inversion layer mobility on 4H–SiC.[15,35]

Post-oxidation anneals in nitrous oxide ($N_2O$) and nitric oxide (NO) have been investigated by Li *et al.*,[36] who report a 2–3 times reduction in interface state density in the upper half of the bandgap on 4H–SiC using NO annealing at temperatures up to 1130°C. Recently, Chung *et al.* have shown that a post-oxidation anneal in NO can reduce $D_{IT}$ in the upper half of the bandgap by almost an order of magnitude.[13] Figure 12 shows $D_{IT}$

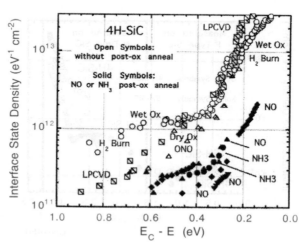

**Fig. 12.** Interface state density near the conduction band on a variety of 4H–SiC samples, with and without a post-oxidation anneal in NO or NH₃.

measured[28] on several $n$-type 4H samples that received a post-oxidation anneal in NO at 1150°C for one hour.[13] Also shown in this figure are data from Fig. 10 for similar samples without the NO anneal. The reduction in $D_{IT}$ is significant and repeatable.

Studies indicate that during the NO anneal, atomic nitrogen is incorporated at the interface at an areal density that saturates at about $1 \times 10^{14}$ cm$^{-2}$.[37] The same areal density is reached for anneals which take place at 1050°C for six hours, 1100°C for two hours, or 1150°C for one hour.[38] The ultimate saturation density is independent of anneal temperature, and longer annealing time does not increase the density of nitrogen at the interface or further reduce $D_{IT}$.

Figure 12 also contains curves for samples annealed in ammonia (NH$_3$). Although $D_{IT}$ is reduced by the NH$_3$ anneal, other studies indicate that NH$_3$ annealing results in the incorporation of nitrogen throughout the oxide.[39] This leads to increased oxide leakage current and reduced breakdown strength.

The NO anneal has also been shown to dramatically increase the electron mobility in 4H–SiC MOSFETs. Das *et al.*[40] and Chung *et al.*[41] reported field effect mobilities as high as 35 cm$^2$/Vs on samples with an NO anneal, as compared to peak mobilities around 5 cm$^2$/Vs on similar samples without the NO anneal. Lu *et al.*[42] reported a comprehensive experiment to evaluate the effect of several process parameters on MOSFET mobility, including implant anneal temperature, ohmic contact anneal, gate material, and metal deposition technique. Each process split contained identical samples processed with and without an NO anneal. Some of the results are shown in Fig. 13. While none of the other process parameters exhibited a significant effect on mobility, the NO anneal produced a dramatic ($\sim$ six times) increase in mobility in every case examined, and field effect mobilities as high as 50 cm$^2$/Vs were observed under optimum processing conditions. The NO anneal also consistently reduced the threshold voltage, indicating that the negative charge in interface states at threshold is significantly reduced.

Although $D_{IT}$ is reduced by the NO anneal, it still remains unacceptably high, reaching values in excess of $10^{12}$ eV$^{-1}$ cm$^{-2}$ within 0.2 eV of the conduction band edge, as seen in the lower curves of Fig. 12. The remaining interface states may arise from a different physical mechanism than those states that are passivated by the NO anneal, and work is currently underway to explore other passivating species that may be used in conjunction with the NO anneal to further reduce $D_{IT}$ and increase mobility.

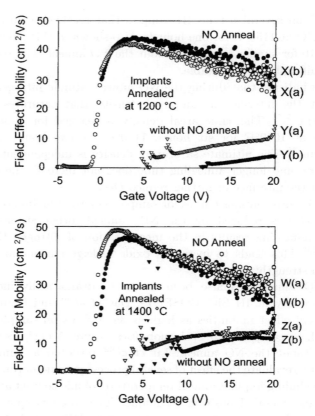

**Fig. 13.** Mobility of 4H–SiC MOSFETs, with and without a post-oxidation anneal in NO.[42] Samples X and W received the NO anneal, while samples Y and Z did not. Curves (a) are measured *after* an ohmic contact anneal at 850°C, while curves (b) are measured *before* the anneal. MOSFETs in the left figure had source/drain implants annealed at 1200°C, while those in the right figure were annealed at 1400°C.

### 3.5. *Results on Other Surface Orientations*

The discussions to this point have dealt almost exclusively with MOS interfaces on the (0001) silicon-terminated surface of 4H and 6H–SiC. However, recent results on the (11$\bar{2}$0) and (03$\bar{3}$8) surfaces are very interesting, and may provide new directions for MOS technology development in SiC. At the very least, they add to our overall understanding of the MOS interface on SiC.

The effect of crystal orientation is not a new phenomenon. In silicon, the electrical quality of the MOS interface becomes progressively better on crystal orientations that have fewer surface atoms. Both $Q_F$ and $D_{IT}$ decrease as one goes from the (111) surface to the (110) surface to the (100) surface, with the midgap $D_{IT}$ decreasing approximately in the ratio $(10 : 6 : 1)$.[43] A similar phenomenon appears to be present in SiC, although the situation is complicated by the fact that the relative number of silicon and carbon atoms on the surface also varies with orientation.

Yano *et al.* fabricated *n*-channel MOSFETs on the $(11\bar{2}0)$[44,45] surface of 4H–SiC and observed increased mobilities compared to similar MOSFETs on the (0001) orientation. They estimated mobility in three ways: the effective mobility (calculated from the drain conductance at low drain voltage), the $V_G = V_T$ intercept of the field-effect mobility (obtained from the transconductance at low drain voltage), and the saturation mobility (obtained from the transconductance in saturation). On the $(11\bar{2}0)$ orientation of 4H–SiC these three mobilities are 28, 96, and 42 cm$^2$/Vs, respectively, factors of 5.5 times, 17 times and 21 times higher than on the (0001) orientation. A companion experiment on the $(11\bar{2}0)$ surface of 6H–SiC produced mobilities of 73, 116, and 62 cm$^2$/Vs, respectively, factors of 2 times, 2.6 times and 2.4 times higher than on the (0001) orientation.[46]

The $(03\bar{3}8)$ orientation is particularly interesting because it corresponds to the (100) surface of the cubic polytype, and from analogy with silicon, might be expected to have the lowest $Q_F$ and $D_{IT}$. Hirao *et al.*[47] fabricated MOSFETs on the $(03\bar{3}8)$ surface of 4H–SiC and obtained a mobility 3–5 times higher than on the (0001) surface, although their (0001) mobility was lower than obtained by other workers.

Senzaki *et al.* have reported field-effect mobilities of 110 cm$^2$/Vs[48] and 160 cm$^2$/Vs[49] on the $(11\bar{2}0)$ face of 4H–SiC following an 800°C post-oxidation anneal in $H_2$, compared to 32 cm$^2$/Vs for similar MOSFETs without the $H_2$ anneal. Interestingly, the earlier report of electron mobilities on the $(11\bar{2}0)$ surface by Yano *et al.*[44] did not specifically quote a field-effect mobility, but their effective mobilities were about 28 cm$^2$/Vs, similar to the $\mu_{FE}$ reported by Senzaki *et al.*[48] for $(11\bar{2}0)$ samples without the $H_2$ anneal. These values are also comparable to the $\mu_{FE}$ values of 30–50 cm$^2$/Vs obtained on the (0001) surface following an NO anneal.[40,42] At this time, the lack of sufficient experimental data makes it difficult to judge whether the $(11\bar{2}0)$ or $(03\bar{3}8)$ orientations produce significantly higher mobilities than the (0001) surface. This topic is the subject of intensive investigation, and the picture may become clearer shortly. For reference, Table 2 lists the best reported mobilities on the various surfaces as of the time of this writing.

**Table 2.** Electron mobility values reported on various orientations of 4H–SiC. The mobility on the ($03\bar{3}8$) surface is low, but compares to a value of 4 cm$^2$/Vs obtained on a (0001) surface used as a control sample.

| Orientation | Mobility | Value | Processing | Ref. |
|---|---|---|---|---|
| (0001) | Field-Effect | 50 cm$^2$/Vs | wet ox 1150°C, Ar anneal 1150°C, re-ox anneal 950°C, NO post-ox anneal 1150°C | 42 |
| ($11\bar{2}0$) | Effective | 28 cm$^2$/Vs | wet ox 1150°C, Ar anneal 1150°C | 44 |
| ($11\bar{2}0$) | Field-Effect | 30 cm$^2$/Vs | wet ox 1150°C, Ar anneal 1150°C | 34 |
| ($11\bar{2}0$) | Field-Effect | 110 cm$^2$/Vs | wet ox 1150°C, Ar anneal 1150°C, H$_2$ post-ox anneal 800°C | 48 |
| ($11\bar{2}0$) | Field-Effect | 160 cm$^2$/Vs | wet ox 1050°C, Ar anneal 1050°C, H$_2$ post-ox anneal 800°C | 49 |
| ($03\bar{3}8$) | Effective | 11 cm$^2$/Vs | wet ox 1150°C, Ar anneal 1150°C | 47 |

## 3.6. *Oxide Breakdown Strength, Leakage Current and Mean Time to Failure*

The electrical quality of the bulk SiO$_2$ on SiC appears remarkably similar to that of oxides on silicon. Breakdown strength of thermal oxides is typically around 10$^7$ V/cm on both 4H and 6H polytypes.[50] However, since the barrier height to charge injection into the oxide are lower for SiC than for silicon (c.f. Fig. 3), oxide leakage tends to be higher and the mean-time-to-failure lower for oxides on SiC. Figure 14 shows the oxide current measured on an $n$-type 6H–SiC epilayer at 140°C.[18] The structure is biased into accumulation, so the current is primarily due to electrons injected from the SiC into the oxide.

Experimental data on oxide reliability on SiC is quite limited, since obtaining a statistically significant data set is time consuming, and processing conditions to optimize the SiO$_2$/SiC interface are still evolving. Mathur and Cooper[18] reported a statistical study of the mean-time-to-failure of MOS capacitors on $n$-type 6H–SiC. In this experiment, 40 identical MOS capacitors are stressed simultaneously at high field and elevated temperature, and the failure distribution is recorded at each field, where failure is defined as an oxide leakage current of 0.1 $\mu$A. Figure 15 shows the failure distributions measured at 145°C at three fields. The mean-time-to-failure at each field is determined from a fit to the failure distribution and plotted vs. field in Fig. 16. Also shown in the figure are comparable data for

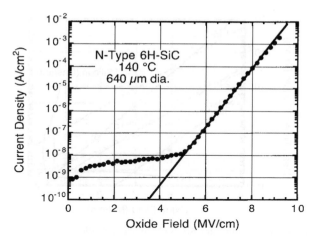

**Fig. 14.** Oxide current on *n*-type 6H–SiC at 140°C.

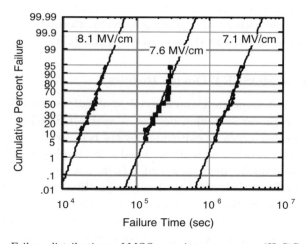

**Fig. 15.** Failure distributions of MOS capacitors on *n*-type 6H–SiC at 145°C.

silicon.[51] As seen, the mean-time-to-failure on SiC is lower than on silicon at high fields, but extrapolation indicates that the mean-times-to-failure are comparable at fields in the 3–4 MV/cm regime. These data also indicate that operation at oxide fields up to 4 MV/cm is permissible, provided the temperature is kept below 250°C. Of course, similar data needs to be taken on 4H–SiC, where the barrier height for charge injection is lower. In addition, other failure criteria need to be considered, such as, for example,

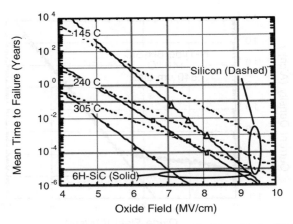

**Fig. 16.**   Mean-time-to-failure for 6H–SiC (solid) and silicon (dashed) MOS capacitors
at three temperatures.

threshold shift due to oxide charging, which might in some cases be more
severe than oxide leakage in degrading device performance. Clearly, much
work remains to be done in the area of oxide reliability on SiC.

## 4. Review of MOSFET Device Results

One of the most important commercial applications for SiC MOSFETs
is their use in power switching devices. Power MOSFETs are attractive
compared to other types of switching devices because they switch at high
frequencies and provide a high input impedance, simplifying drive circuitry.
Since they are unipolar devices, MOSFETs do not store minority carriers,
so their turn-off transient consists of only capacitive currents. This permits
high frequency operation with very low switching loss. In the following
sections, we first introduce the unipolar figure of merit for power switching
devices and then review the progress and status of the two main types of
SiC power MOSFETs: UMOSFETs and DMOSFETs. We then conclude
with a brief summary of other MOS results in SiC, including digital and
analog MOS integrated circuits and Charge Coupled Devices (CCDs).

### 4.1. *Power Device Figures of Merit*

The most important operating parameters for power switching devices
are the specific on-resistance and the blocking voltage. The specific on-
resistance $R_{\mathrm{ON,SP}}$ is the product of the resistance of the device and the

device area, expressed in units of $\Omega$ cm$^2$, or more conveniently, m$\Omega$ cm$^2$. It is obviously desirable to minimize both the resistance and the area, so this product becomes a meaningful parameter for power devices.

Let us assume that the maximum power dissipation per unit area is limited by thermal considerations to $P_{\mathrm{D}}$. With this limitation known, the maximum current density a unipolar device can control is given by $P_{\mathrm{D}} = J_{\mathrm{ON}}^2 R_{\mathrm{ON,SP}}$. The blocking voltage in the off-state determines the maximum voltage the device can control. The power rating of the device (per unit area) is then given by

$$P_{\mathrm{MAX}} = J_{\mathrm{ON}} V_{\mathrm{B}} = \sqrt{P_{\mathrm{D}} \frac{V_{\mathrm{B}}^2}{R_{\mathrm{ON,SP}}}}. \tag{12}$$

Thus, it is evident that the pertinent device figure of merit is the ratio $V_{\mathrm{B}}^2 / R_{\mathrm{ON,SP}}$, i.e. we need to maximize this quantity to optimize the power handling capability of the device.

Virtually all power switching devices utilize a reverse-biased $p$-$n$ or Schottky junction to hold off the blocking voltage in the off-state. To reach high voltages, the $p$-$n$ junction is designed as a $p^+/n^-/n^+$ one-sided step junction, similar to a $p$-$i$-$n$ diode. The $n$-drift region is thick and lightly doped, and is fully depleted in the blocking state. The peak electric field occurs at the $p^+/n^-$ junction, and under large reverse bias the field tapers very gradually through the lightly doped drift region. Since the field remains high throughout most of the drift region, and since the region is thick, most of the blocking voltage is dropped across this layer. In the on-state the depletion region only extends a short distance from the $p^+/n^-$ junction, leaving most of the $n$-drift region undepleted. Since the drift region is thick and lightly doped, its resistance is high. The doping and thickness of the drift region determine both the blocking voltage and the resistance, so increasing the blocking voltage (for example, by making the region thicker and more lightly doped) also increases the on-resistance. If the drift region doping and thickness are chosen so that it is fully depleted just at the onset of avalanche breakdown, the specific on-resistance is related to the blocking voltage by

$$R_{\mathrm{ON,SP}} = \frac{4V_{\mathrm{B}}^2}{\mu_{\mathrm{N}} \varepsilon_{\mathrm{S}} E_{\mathrm{C}}^3}. \tag{13}$$

Since the device figure of merit suggested by Eq. (12) is the ratio $V_B^2/R_{ON,SP}$, we rearrange Eq. (13) to give

$$\frac{V_B^2}{R_{ON,SP}} = \frac{\mu_N \varepsilon_S E_C^3}{4}. \tag{14}$$

The numerator on the right side is the high-frequency power device figure of merit cited by Baliga.[4] The Baliga figure of merit $\mu_N \varepsilon_S E_C^3$ is a measure of the ultimate capability of a unipolar power device, as determined by fundamental parameters of the semiconductor material. Because of the high critical field, the Baliga figure of merit in 4H–SiC is approximately 400 times higher than in silicon. This means that MOSFETs in 4H–SiC can have specific on-resistances 400 times lower than silicon MOSFETs for the same blocking voltage, as shown by Eq. (13). Since silicon devices are now approaching their theoretical limits of performance, this 400 times factor provides a powerful incentive for the development of power MOSFET technology in SiC.[4]

The design rules implied by Eqs. (13) and (14) are actually somewhat approximate, for two reasons. First, they ignore the dependence of mobility $\mu_N$ and critical field $E_C$ on doping. Second, it is possible to obtain slightly better performance if the drift region is designed to punch through before the onset of avalanche breakdown. The most accurate way to calculate the blocking voltage is to evaluate the ionization integral across the depletion region at each applied voltage. The applied voltage at which the ionization integral equals unity defines the blocking voltage. Figure 17 shows the dependence of blocking voltage on doping in 4H–SiC[52] obtained by numerically evaluating the ionization integral across the drift region using the ionization coefficients of Konstantinov *et al.*[53] As the drift region doping is reduced, the blocking voltage increases rapidly until the drift region punches through. Once the drift region is punched through, the blocking voltage increases only slightly as doping is further reduced, eventually saturating. For a given drift region thickness, punchthrough occurs at the point where the curves begin to deviate from the curve for an infinitely thick drift region.

The relationship between on-resistance and blocking voltage in 4H–SiC is shown for several drift region thicknesses in Fig. 18. As we move along each curve from left to right, the drift region doping is steadily decreasing to achieve higher blocking voltage. Of course, the reduced doping also increases the specific on-resistance. At some point the drift region becomes punched through, and beyond this point the blocking voltage begins to

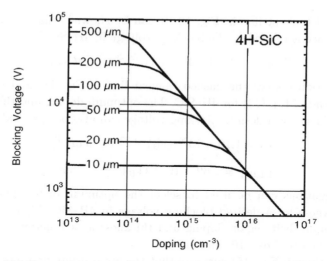

**Fig. 17.** Blocking voltage of 4H–SiC as a function of epilayer doping and thickness.

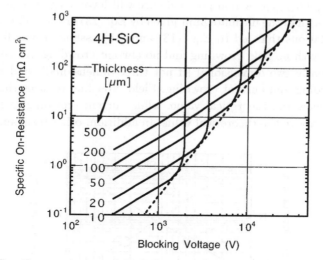

**Fig. 18.** Specific on-resistance of 4H–SiC as a function of blocking voltage for several epilayer thicknesses.

saturate. However, the resistance continues to increase as doping is reduced, making the curves tend towards the vertical on this plot. Our objective is to choose the drift region doping and thickness that provide the minimum on-resistance at a given blocking voltage. This optimum is defined by the dashed tangent line in Fig. 18. It can be shown[52] that the minimum on-resistance is related to the blocking voltage in 4H–SiC by

$$R_{\mathrm{ON,SP}} = 2.33 \times 10^{-8} \ (V_{\mathrm{B}})^{2.33} \quad [\mathrm{m\Omega cm^2}] . \tag{15}$$

This equation defines the dashed tangent line in Fig. 18, and should be used instead of Eq. (13) for 4H–SiC. The optimum doping and thickness of the drift layer for a desired blocking voltage can also be specified as[52]

$$N_{\mathrm{D}} = 1.10 \times 10^{20} \ (V_{\mathrm{B}})^{-1.27} \quad [\mathrm{cm}^{-3}] , \tag{16}$$

$$T_{\mathrm{DRIFT}} = 2.62 \times 10^{-7} \ (V_{\mathrm{B}})^{1.12} \quad [\mathrm{cm}] . \tag{17}$$

These equations can be used to select the optimum drift region doping and thickness for a desired blocking voltage in 4H–SiC. Figure 19 shows the optimum drift region doping and thickness as a function of blocking voltage given by Eqs. (16) and (17).

Even though Eq. (15) suggests that the appropriate figure of merit for 4H–SiC would be the ratio $V_{\mathrm{B}}^{2.33}/R_{\mathrm{ON,SP}}$, we will continue to use the traditional $V_{\mathrm{B}}^2/R_{\mathrm{ON,SP}}$. A real power device will have a $V_{\mathrm{B}}^2/R_{\mathrm{ON,SP}}$ figure of merit lower than the ultimate limit given by Eq. (13) or (15). This is because the calculations that led to Eqs. (13) and (15) assume a one-dimensional structure with no field crowding and no current crowding. All real devices are inherently two-dimensional (if not three-dimensional), and have both field crowding and current crowding. Field crowding reduces the blocking voltage below the theoretical value, while current crowding increases the resistance above the theoretical value. In addition, other resistances within

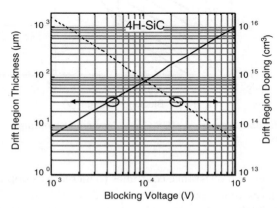

**Fig. 19.** Epilayer doping and thickness that give the maximum $V_{\mathrm{B}}^2/R_{\mathrm{ON,SP}}$ as a function of design blocking voltage for 4H–SiC.

the device add to the resistance of the drift region. These include contact resistances, and in the case of power MOSFETs, the resistance of the MOS inversion layer. For these reasons, real devices have $V_B^2/R_{ON,SP}$ values below the theoretical limit as given by Eq. (14).

SiC power MOSFETs were first reported in 1994,[54] and considerable progress has been made since that time. The earliest devices had blocking voltages around 60 V, whereas the most recent power MOSFETs have blocking voltages over 6 kV.[55] The best SiC MOSFETs now have $V_B^2/R_{ON,SP}$ figures of merit more than 25 times higher than the theoretical limit for silicon, as given by Eq. (14), but their figures of merit are still well below the theoretical limit for 4H–SiC. In the following sections we will discuss the two major types of power MOSFETs in SiC: UMOSFETS and DMOSFETs. In reviewing these devices, we will highlight the major limitations of each type and summarize the latest experimental results. We will conclude by comparing these MOSFETs with other types of power transistors, such as JFETs and BJTs, which are currently under development in SiC.

## 4.2. *SiC UMOSFETs*

The first power transistors ever reported in SiC[54] were UMOSFETs. A cross section of the basic SiC UMOSFET is shown in Fig. 20. In this structure, the $n^+$ substrate acts as the drain, and the inversion channel is formed on the sidewalls of trenches created by Reactive Ion Etching (RIE). In the

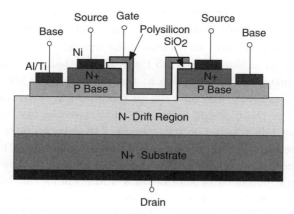

**Fig. 20.** Cross section of a basic UMOSFET in SiC. This implementation utilizes an epitaxial source that is patterned by reactive ion etching.

on-state, electrons flow from the source on the top surface, through the MOS channel formed on the trench sidewalls, then through the lightly-doped $n$-drift region to the $n^+$ substrate.

The UMOS geometry was used for the first SiC power MOSFETs because it could be formed without the use of ion implantation. In the original UMOS devices, the $n$-type drift layer, $p$-type base layer, and $n^+$ source layer were all epitaxially grown. The source was patterned by RIE, exposing the base in selected areas for ohmic contacts. The gate trench was formed by RIE; the gate oxide was grown either by thermal oxidation or chemical vapor deposition; the gate was deposited and patterned; ohmic metals were deposited on the source, drain, and base contact regions; the ohmic contacts were annealed; and a top metal layer was deposited. This completed the structure as shown. Several modifications and improvements on this basic structure have been developed over the years, and these will be discussed momentarily.

The main advantages of the UMOSFET are: (1) It can be made without ion implantation, if desired. (2) Since the MOS channel is formed on the sidewalls of trenches, the surface area is minimized. (3) Its structure avoids the resistive JFET region present in DMOSFETs (to be discussed below). However, the UMOSFET also has several disadvantages: (1) Since the channel is formed on the sidewall of an etched trench, the surface is rough, and may not be precisely aligned with a favorable crystalline plane. (2) Several crystalline planes, including the $(0001)$, $(1\bar{1}00)$, and $(11\bar{2}0)$ planes, must be covered with gate-quality oxide, and the anisotropy of thermal oxidation results in different oxide thicknesses on each of these surfaces. In particular, the horizontal $(0001)$ surfaces have thinner oxides than the vertical $(1\bar{1}00)$ and $(11\bar{2}0)$ surfaces. (3) Two-dimensional field crowding occurs at the bottom corners of the trenches in the blocking state. Because the field in the oxide is 2.5 times higher than the surface field in the semiconductor, the oxide fields at the trench corners are close to the oxide breakdown field. This limits the maximum voltage that can be applied to the device in the blocking state. (4) Field crowding also occurs in the on-state at both the top and bottom corners of the trench due to the positive voltage applied to the gate. Because the oxide is thinner over the horizontal $(0001)$ surfaces, the oxide field is higher than along the trench sidewalls. Since the oxide field must be kept below 3–4 MV/cm for reliability reasons, the field in the thicker oxide over the inversion channel will be much less than 3–4 MV/cm. This reduces the charge in the inversion layer, increasing the channel resistance.

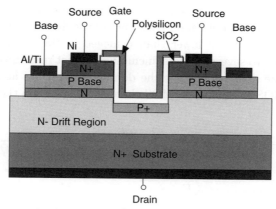

**Fig. 21.** Trench Oxide Protected (TOP) UMOSFET. The $p^+$ region at the bottom of the trench protects the oxide from high electric fields in the blocking state, while the $n$-type epilayer below the base prevents JFET pinch-off and facilitates lateral current spreading.

The problem of field crowding at the trench corners in the blocking state can be mitigated by structural modifications. Figure 21 shows a Trench-Oxide-Protected (TOP) UMOSFET[56] first introduced in 1997. Two new elements are added in this structure: a thin $n$-type epilayer between the $p$-base and the drift region, and a $p^+$ region at the bottom of the trench. The $p^+$ region is formed by self-aligned ion implantation after the trench is etched, and is grounded by a contact at the edge of the device (not shown). In the blocking state, the grounded $p^+$ region protects the trench oxide by terminating field lines that would otherwise penetrate the oxide. Although some field lines enter the oxide at the trench corners, the fields there remain below the 3–4 MV/cm safety limit. The $n$-type epilayer prevents the $p^+$ base and $p^+$ trench implant from depleting the conducting channel in the on-state. It also enhances lateral current spreading into the drift region, reducing current crowding that would otherwise occur at the trench corners. The first TOP UMOSFETs were fabricated on 10 $\mu$m drift regions and achieved blocking voltages of 1400 V with a specific on-resistance of 15.7 m$\Omega$ cm$^2$. The corresponding $V_{\mathrm{B}}^2/R_{\mathrm{ON,SP}}$ figure of merit was 125 MW/cm$^2$, about 25 times higher than the theoretical limit for silicon devices.[56]

Even though the figure of merit for the TOP UMOSFET is 25 times higher than the silicon limit, it is still at least an order of magnitude below the expected theoretical limit for 4H–SiC. The blocking voltage of 1400 V achieved in this device is about 85% of the theoretical value for the 10 $\mu$m drift region, so the main reason for the low figure of merit is the specific on-resistance. The theoretical on-resistance of the drift region in this device is 0.8 m$\Omega$ cm$^2$, 20 times lower than achieved in the actual device. The difference is due to several factors, including contact resistance and current crowding in the drift region, but the dominant source of additional resistance is the MOS channel. The MOS channel resistance in the on-state can be written as

$$
\begin{aligned}
R_{\mathrm{CH,SP}} &= \frac{L_{\mathrm{CH}}S}{\mu_{\mathrm{CH}}[\varepsilon_{\mathrm{OX}}E_{\mathrm{OX,MAX}} - Q_{\mathrm{F,EFF}} - \Delta Q_{\mathrm{IT}}(\phi_{\mathrm{S}}) - Q_{\mathrm{D}}(2\phi_{\mathrm{F}})]} \\
&= \frac{L_{\mathrm{CH}}S}{\mu_{\mathrm{CH,EFF}}\varepsilon_{\mathrm{OX}}E_{\mathrm{OX,MAX}}}
\end{aligned}
\tag{18}
$$

where $L_{\mathrm{CH}}$ is the channel length, $S$ is the half-pitch of the surface area of the cell, and $\mu_{\mathrm{CH}}$ is the channel mobility. The term in the square brackets is the charge per unit area in the inversion layer available for current flow. This is equal to the charge on the gate $\varepsilon_{\mathrm{OX}}E_{\mathrm{OX,MAX}}$ when the oxide field is at its maximum safe value, minus the other charge components within the semiconductor, including the effective fixed charge $Q_{\mathrm{F,EFF}}$, the additional charge in interface states $\Delta Q_{\mathrm{IT}}(\phi_{\mathrm{S}})$, and the charge in the semiconductor depletion region $Q_{\mathrm{D}}(2\phi_{\mathrm{F}})$. To minimize $R_{\mathrm{CH,SP}}$ we need to minimize the fixed charge and interface trapped charge, increase the channel mobility, reduce the channel length, and minimize the cell pitch. Since $Q_{\mathrm{F,EFF}}$ and $\Delta Q_{\mathrm{IT}}(\phi_{\mathrm{S}})$ are typically not known, it is customary to write Eq. (18) in terms of the maximum oxide field and an *effective* channel mobility $\mu_{\mathrm{CH,EFF}}$, as shown in the second form of the equation.

As stated earlier, in 4H–SiC MOSFETs fabricated to date the channel mobility has been far below that which would be expected based on the bulk mobility of electrons in the material. In UMOSFETs, the channel mobility may be reduced even further by surface roughness from the reactive ion etch used to form the gate trench. In cases where the mobility is less than 1 cm$^2$/Vs, the inversion layer may actually be physically discontinuous at points along the sidewall. It has been found empirically that introducing a thin $n$-type counter-doped region at the surface can dramatically improve the effective surface mobility. This thin $n$-type layer can be formed by epigrowth following trench etch. When carefully designed, the counter-doped layer is fully depleted at zero gate voltage, yielding a normally-off device. In the on-state, most of the electron conduction current flow takes place in an inversion layer at the surface, but any discontinuity in the inversion layer will be "shorted out" by parallel conduction through the

$n$-type layer. This can significantly enhance the effective mobility of the channel. The first reported use of surface counter-doping in SiC UMOSFETs was by Onda *et al.*[57] and Hara,[58] who achieved a specific on-resistance of 10.9 m$\Omega$ cm$^2$ at a blocking voltage of 450 V. The effective mobility of these devices was estimated at 108 cm$^2$/Vs. Surface counter-doping was also used in the TOP UMOSFETs described above,[56] although the effective mobility in these devices was lower than in the Onda and Hara devices.

Recent UMOSFETs have achieved blocking voltages in the 3–5 kV range. Li *et al.*[59] report counter-doped TOP UMOSFETs on 50 $\mu$m 4H–SiC drift layers with blocking voltage of 3360 V and specific on-resistance of 199 m$\Omega$ cm$^2$, and non-counter-doped TOP UMOSFETs with blocking voltage of 3055 V and specific on-resistance of 121 m$\Omega$ cm$^2$. Khan *et al.*[60] fabricated non-counter-doped TOP UMOSFETs on 100 $\mu$m 4H–SiC drift regions and achieved blocking voltages over 5 kV with specific on-resistance of 105 m$\Omega$ cm$^2$.

## 4.3. *SiC DMOSFETs*

Planar double-diffused MOSFETs (DMOSFETs) have long been used in silicon. In silicon DMOSFETs, the $p$-base region is selectively formed on the top surface by diffusion through an oxide mask, and the source is then formed by a second diffusion through the same mask opening. The spacing between the edge of the source and the edge of the $p$-base is determined by the difference in diffusion depth between the source and $p$-well impurities. This difference can be controlled with great precision, and the result is a self-aligned process utilizing a single masking operation.

The double-diffused process cannot be used in SiC, since diffusion coefficients for common dopants are extremely low in SiC at any reasonable temperature. However, in 1996 the first SiC DMOSFETs were produced by selective ion implantation.[61] This process requires two lithography steps, and therefore introduces a realignment tolerance between the source and $p$-well. A cross section of a SiC DMOSFET is shown in Fig. 22.

The main advantages of the DMOSFET process are: (1) It is a planar process, and requires no critical reactive ion etching steps. (2) The planar structure eliminates field crowding that occurs at the trench corners in the UMOS process. (3) The MOSFET is formed entirely on the (0001) face of the crystal, avoiding the oxidation anisotropy that arises in the UMOS process. However, the DMOSFET structure has its own particular

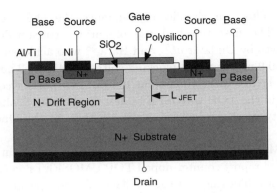

**Fig. 22.** Cross section of a double-implanted SiC DMOSFET. $L_{\text{JFET}}$ is the width of the JFET region between the ion-implanted *p*-wells.

disadvantages: (1) Since the device is planar, its area is slightly larger than an equivalent UMOSFET. (2) The *p*-well and $n^+$ source regions are formed by ion implantation. Ion implantation damages the lattice, and subsequent high-temperature activation annealing degrades the surface. These effects introduce surface roughness and reduce the inversion layer mobility of subsequently formed MOSFETs. (3) The DMOSFET structure creates a parasitic JFET in the *n*-region between *p*-wells at the surface. This parasitic JFET introduces an additional series resistance in the current path.

The design of the DMOSFET involves several conflicting considerations. The *p*-well doping-thickness product must be sufficiently high that the punchthrough voltage is above the avalanche breakdown voltage. Since the thickness of the *p*-well is limited to about 0.5 $\mu$m by the ion implantation process (or up to 1 $\mu$m if MeV energies are used), the doping must be in the range of $3$–$5 \times 10^{18}$ cm$^{-3}$. However, such a high doping would lead to an unacceptably high threshold voltage in the MOSFET, so the surface doping is usually designed to be about an order of magnitude lower, say in the mid-$10^{17}$ cm$^{-3}$ range. This is accomplished by a multiple-energy implant that produces a retrograde doping profile.[61]

One of the critical dimensions of the DMOSFET is the width of the JFET region between the *p*-wells, indicated as $L_{\text{JFET}}$ in Fig. 22. If $L_{\text{JFET}}$ is decreased, the resistance of the JFET region increases, eventually becoming the dominant resistance of the device. If $L_{\text{JFET}}$ is increased, the grounded *p*-wells are less effective in screening the oxide from high electric fields in the blocking state, and the oxide field increases. This is a serious concern, since the oxide field must be kept below at about 3–4 MV/cm, as stated

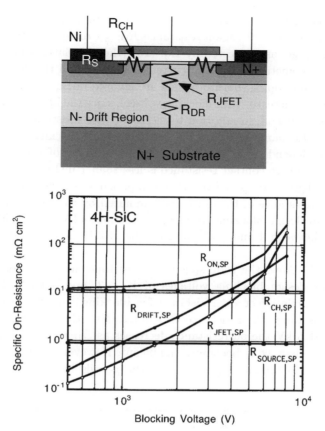

**Fig. 23.** Components of specific on-resistance for a DMOSFET. In the calculations, we assume 4H–SiC with $\mu_{CH} = 25$ cm$^2$/Vs, $L_{CH} = 3$ $\mu$m, and $L_{JFET} = 4$ $\mu$m. For these parameters, the MOS channel is the dominant resistance for $V_B < 4$ kV, the drift region is dominant between 4 and 6 kV, and the JFET region is dominant above 6 kV.

before. Thus, the maximum allowable drain voltage must be de-rated to limit the oxide field. A larger $L_{JFET}$ also increases the area of the device.

To understand these tradeoffs, we first consider the effect of $L_{JFET}$ on specific on-resistance. Figure 23 shows the different components of the on-resistance as a function of the design blocking voltage[62] for an $L_{JFET}$ of 4 $\mu$m. These calculations assume a MOSFET channel mobility of 25 cm$^2$/Vs, a channel length of 2 $\mu$m, and a layout based on 3 $\mu$m feature sizes and 2 $\mu$m alignment tolerances, leading to a cell pitch $S = (9.5$ $\mu$m $+ L_{JFET}/2)$. As the blocking voltage is increased, the drift region must be made thicker and more lightly doped in order to withstand the voltage. This increases the drift region resistance $R_{DR,SP}$ roughly as the square of blocking voltage, as suggested by Eq. (13) (the calculations in this figure include the geometrical effects of current spreading and the dependence of critical field on doping, both of which are ignored in Eq. (13)). The JFET resistance also increases with blocking voltage, since the lighter doping leads to increased pinchoff, in addition to increasing the resistivity of the material. The MOS channel resistance is independent of blocking voltage, since it does not depend on either the doping or thickness of the drift region. The specific source contact resistance is less than 1 m$\Omega$ cm$^2$, and in most cases is negligible compared to other components of resistance. The total specific on-resistance of the device is dominated by the MOSFET channel resistance at design blocking voltages below about 4 kV, by the drift region resistance between 4 and 6 kV, and by the JFET resistance above about 6 kV.

Figure 24 shows the total specific on-resistance as a function of JFET spacing for design blocking voltages of 1, 2, 3, and 5 kV.[62] Increasing $L_{JFET}$ has little effect on resistance for blocking voltages below 3 kV, since in this range the resistance is dominated by the MOSFET channel. However, at 5 kV the on-resistance decreases as $L_{JFET}$ is increased, since the JFET resistance is becoming significant. However, larger JFET spacing reduces the screening provided by the $p$-wells in the blocking state,

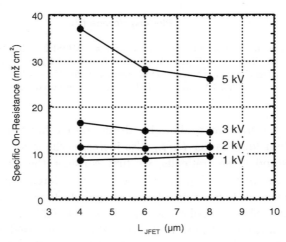

**Fig. 24.**   Specific on-resistance of a DMOSFET as a function of JFET spacing for four blocking voltages.

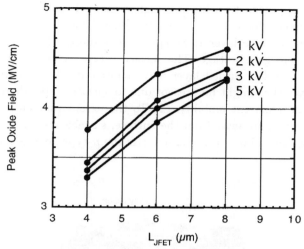

**Fig. 25.** Peak oxide field in a DMOSFET as a function of JFET spacing for four blocking voltages.

leading to higher oxide fields. Figure 25 shows peak oxide field as a function of $L_{JFET}$ for design blocking voltages of 1, 2, 3, and 5 kV, as obtained from two-dimensional computer simulations.[62] As seen, larger JFET spacings lead to higher oxide fields. Since the oxide field must be restricted to 3–4 MV/cm, it is necessary to keep the JFET spacing below a certain value. This represents a compromise between protecting the gate oxide in the blocking state and minimizing the JFET resistance in the on-state.

The major drawback of SiC DMOSFETs to date has been the low inversion layer mobility in the implanted $p$-well. In some cases, the electron mobility has been less than 1 cm$^2$/Vs. When the mobility is this low, the inversion layer is probably blocked by discontinuities at macroscopic surface steps. Such steps can be readily observed using optical microscopy, SEM, or AFM. These steps arise from step migration during the high-temperature implant activation anneal.[16] $p$-type dopants are aluminum and boron. Aluminum implants require activation temperatures of 1600°C and boron implants require 1700°C. At such temperatures, surface steps due to the 8° off-axis surface orientation can migrate and bunch, creating macro-steps that may be several tens of nm high, comparable to the oxide thickness. Several experiments have shown that step bunching can be minimized by annealing under silicon overpressure[17] to minimize loss of silicon from the surface.

Another approach to achieving higher channel mobility is the use of a counter-doped channel, as shown in generic form in Fig. 26. In this structure, the surface layer is $n$-type, and is designed to be completely depleted at zero gate voltage. The surface $n$-type layer has several beneficial effects: (1) The buried layer may "short out" discontinuities in the inversion layer that might be present due to surface steps, thereby greatly enhancing the apparent mobility. (2) To the extent that current flows in the $n$-type layer and away from the surface, surface scattering is reduced and the effective mobility is increased. (As the $n$-type layer is made thicker

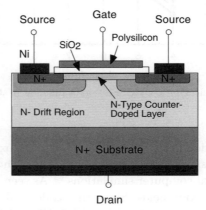

**Fig. 26.** Cross section of a counter-doped DMOSFET. The surface is converted to $n$-type by implantation or epigrowth.

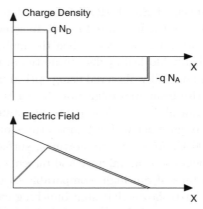

**Fig. 27.** Charge density and electric field perpendicular to the surface in a counter-doped DMOSFET (shaded lines) and conventional DMOSFET (solid lines).

and more heavily doped, the channel mobility can approach the bulk value. However, power transistors are typically designed as *normally-off* devices, so the doping-thickness product of the $n$-layer must be limited.) (3) The positively-charged donors in the $n$-layer help support the charge in the depletion region, reducing the electric field at the surface, as shown in Fig. 27. This reduces the threshold voltage and lowers the oxide field at threshold, providing more room to increase the oxide field above threshold, thereby increasing the maximum achievable inversion charge density. The lower surface field also means that inversion electrons are not as tightly bound to the surface, reducing surface scattering and increasing mobility.

The first doped-channel DMOSFETs were reported on 6H–SiC by Sridevan and Baliga[30] in 1998. In these devices, the $p$-well was implanted to lie below the surface, leaving the surface $n$-type. An on-resistance of 18 m$\Omega$ cm$^2$ was obtained at a blocking voltage of 350 V. The device was normally-off, and the low blocking voltage was thought to be caused by incomplete activation of the $p$-well implant, resulting in punchthrough of the $p$-base.

In 1999, Ueno and Oikawa[63] reported the dependence of MOSFET mobility on the dose of the counter-doped surface layer on 4H–SiC. Their test MOSFETs were lateral devices on $p$-type epilayers, so the surfaces were not degraded by a $p$-type ion implantation, as in high voltage DMOSFETs. The source, drain, and $n$-type doped channels were formed by nitrogen ion implantation. In their experiments, an $n$-type channel dose of $5 \times 10^{11}$ cm$^{-2}$ had no effect on the effective channel mobility, which was very low, around 2 cm$^2$/Vs. At a dose of $2.5 \times 10^{12}$ cm$^{-2}$, the devices were still normally off, and the field-effect mobility reached a peak value of 38 cm$^2$/Vs, falling to around 10 cm$^2$/Vs at higher gate voltages, as shown in Fig. 28. At a dose of $8 \times 10^{12}$ cm$^{-2}$, the devices were normally on, with an effective mobility close to the bulk value. In comparing the mobility curves in Fig. 28 with subsequent figures, it is important to realize that the field-effect mobility is a *differential* mobility, and the drain current that would flow at a given gate voltage is proportional to the *integral* of the field-effect mobility up to that gate voltage. Thus, the counter-doped MOSFET in Fig. 28 would carry substantially higher drain current than the conventional MOSFET, even though the mobilities are comparable at high gate voltages.

Ryu *et al.*[64,65] fabricated counter-doped vertical DMOSFETs on ion implanted $p$-wells in 4H–SiC. The drift regions in these devices were 25 $\mu$m thick, doped $3 \times 10^{15}$ cm$^{-3}$. Unlike Shrevidan and Baliga, these devices utilized an $n$-type ion implantation to form the counter-doped layer, and

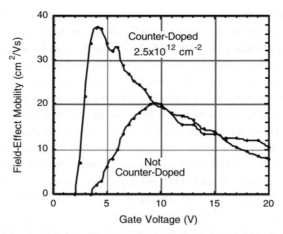

**Fig. 28.** Field-effect mobility for counter-doped and conventional 4H–SiC lateral epitaxial MOSFETs.[63]

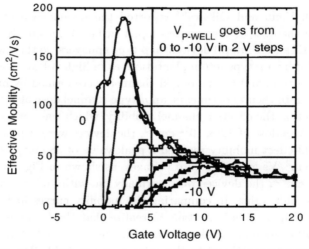

**Fig. 29.** Effective mobility for counter-doped 4H–SiC vertical DMOSFETs as a function of back-gate bias.[64,65]

unlike Ueno and Oikawa, the channels were formed on ion-implanted $p$-wells rather than on $p$-type epilayers. Figure 29 shows effective (not field-effect) mobility for a device with channel doping of $2.7 \times 10^{12}$ cm$^{-2}$ as a function of back-gate bias applied to the $p$-well. With zero back-gate bias, the device is normally on, with a peak mobility of 194 cm$^2$/Vs. Increasing the back-gate bias raises the threshold voltage, and at a back-gate bias of $-4$ V, the device becomes normally off. At this point, the peak mobility is around 65 cm$^2$/Vs, decreasing to about 30 cm$^2$/Vs at high gate voltages. One would not wish to operate a power device with a back-gate bias. However, the back-gate bias in these devices expands the depletion region of the buried $p$-$n$ junction, reducing the bulk channel in the surface $n$-layer. The same effect could be obtained by reducing the dose of the channel implant. The dependence of threshold voltage on back-gate bias in Fig. 29 suggests that a channel dose of approximately $1 \times 10^{12}$ cm$^{-2}$ would result in a normally-off device with a peak mobility around 65 cm$^2$/Vs.[64] The largest counter-doped MOSFETs had lateral dimensions of $3.3 \times 3.3$ mm$^2$ and carried a drain current of 10 A at a specific on-resistance of 43 m$\Omega$ cm$^2$.[65] The blocking voltage of the large devices was about 350 V, possibly limited by material defects.

Perhaps the most successful counter-doped DMOSFETs to date were reported by Kansai Electric Power Co. and Cree, Inc.[55,66,67] In these devices, the $n$-type surface channel is formed by epitaxial regrowth following $p$-well ion implantation. After the channel epigrowth, a $p^+$ implantation is performed to contact the buried $p$-well and a shallow $n^+$ implantation forms the source regions. In a novel twist, the devices are operated with the $p$-well under *forward bias*. This shrinks the depletion regions surrounding the buried $p$-wells, reducing the resistance of the JFET region between $p$-wells and allowing closer $p$-well spacing. At high $p$-well bias, the surface channel may also be slightly conductivity modulated due to the injection of holes from the forward biased $p$-wells. Small devices on 75 $\mu$m drift regions doped $5 \times 10^{14}$ cm$^{-3}$ achieved blocking voltages of 4580 V with a specific on-resistance of 387 m$\Omega$ cm$^2$ at a $p$-well bias of 7 V and gate bias of 40 V.[66] Larger devices on the same wafer had blocking voltages of 2030 V and specific on-resistances of 172 m$\Omega$ cm$^2$ at a $p$-well bias of 10 V and gate bias of 20 V. Subsequent devices on 75 $\mu$m drift regions doped $3 \times 10^{14}$ cm$^{-3}$ achieved blocking voltages of 6.1 kV and specific on-resistances of 732 m$\Omega$ cm$^2$ at a $p$-well bias of 7.5 V and gate bias of 20 V.[55] Recently the same group reported counter-doped DMOSFETs on 60 $\mu$m drift regions doped $7 \times 10^{14}$ cm$^{-3}$ with blocking voltages of 5 kV and specific on-resistances of 88 m$\Omega$ cm$^2$ at a $p$-well bias of 2 V and gate bias of 20 V.[67] At a $p$-well bias of only 2 V, very little charge injection takes place from the $p$-well. As a result, there is no conductivity modulation of either the channel or JFET region, and the main benefit of the forward bias is to reduce the JFET resistance by shrinking the depletion regions surrounding the $p$-wells. The $V_B^2/R_{ON,SP}$ figure of merit for the 5 kV counter-doped

MOSFETs with 2 V $p$-well bias is 284 MW/cm$^2$, the highest yet reported for a SiC MOSFET, and approximately 70 times higher than the silicon theoretical limit.[67]

## 4.4. *Comparison of Power Transistor Performance*

Figure 30 and Table 3 summarize the performance of leading power MOSFETs and other types of power transistors, including JFETs and BJTs. The figure shows the specific on-resistance vs. blocking voltage for SiC DMOSFETs (solid circles), UMOSFETs (solid squares), JFETs (open squares), and BJTs (shaded triangles). The two JFETs denoted with asterisks in the squares are normally-on devices. As stated earlier, power systems usually require normally-off devices to ensure that the system shuts down safely in the event of a failure of control power. The diagonal lines in the figure are the loci of constant figure of merit $V_B^2/R_{ON,SP}$. As seen from

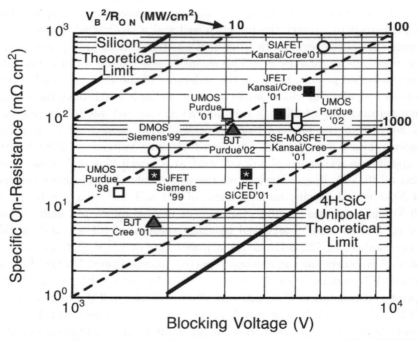

**Fig. 30.** Specific on-resistance and blocking voltage for the most recent SiC MOSFETs, JFETs, and BJTs. Open symbols are MOSFETs, solid squares are JFETs, and shaded triangles are BJTs. The diagonal lines are loci of constant figure-of-merit $V_B^2/R_{ON,SP}$. The two JFETs with asterisks within the symbols are normally-on devices.

**Table 3.** Performance of recent SiC MOSFETs, JFETs, and BJTs. Asterisks ($*$) denote devices that utilize forward-biased base regions in the on-state. Daggers ($\dagger$) denote normally-on devices.

| Device | $V_B$ (V) | $R_{ON,SP}$ (m$\bullet$ cm$^2$) | $V_B^2/R_{ON,SP}$ (MW/cm$^2$) | Reference |
|---|---|---|---|---|
| UMOSFET | 3055 | 121 | 78 | 59 |
| UMOSFET | 1400 | 15.7 | 125 | 56 |
| UMOSFET | 5050 | 105 | 243 | 60 |
| DMOSFET* | 6100 | 732 | 51 | 55 |
| DMOSFET | 1800 | 46 | 70 | 68 |
| DMOSFET* | 5000 | 88 | 284 | 67 |
| JFET$\dagger$ | 1800 | 24 | 135 | 69 |
| JFET* | 5500 | 218 | 139 | 70 |
| JFET* | 4450 | 121 | 164 | 70 |
| JFET$\dagger$ | 3500 | 26 | 471 | 71 |
| BJT | 3200 | 78 | 131 | 72 |
| BJT | 1800 | 7 | 463 | 73 |

the figure and Table 3, SiC MOSFETs are competitive in performance with both SiC JFETs and SiC BJTs.

The selection of a power device technology for a particular application depends upon a number of factors. The MOSFET has the advantage of high input impedance, which simplifies the design of the drive circuitry. On the other hand, its oxide raises potential reliability issues and limits the maximum operating temperature. Since neither the JFET or BJT require a critical oxide, they can be operated at higher temperatures, reducing the cost of heat removal. In addition, the BJT can be operated at higher current density than the MOSFET (around 250 A/cm$^2$, as compared to 100 A/cm$^2$), and the drift region resistance of the BJT can be lower than the theoretical limit for unipolar devices as given by Eq. (13) because the drift region is conductivity modulated under strong saturation biases.

In attempting to assess the ultimate winner in the MOSFET vs. JFET vs. BJT competition, it is important to realize that none of these devices have yet reached their theoretical performance limits, and the situation remains highly fluid. MOSFET performance is currently limited by the MOS channel resistance, and research now underway in several laboratories will almost certainly result in improved performance over the next several years. Similarly, it is expected that both the JFET and BJT will continue to improve in performance. In the end, the winner may be determined more by economics than by performance, and the device with the simplest fabrication and highest yield may ultimately predominate.

### 4.5. *Other SiC MOSFET Results*

Although SiC power MOSFETs have attracted the most attention in recent years, other applications for SiC MOSFETs have also been investigated, and we will briefly review them in this section.

The first monolithic integrated circuits were reported in 6H–SiC in 1994.[74] These NMOS logic circuits consisted of an $n$-channel enhancement-mode pull-down network and an $n$-channel enhancement-mode load transistor with separate $V_{DD}$ and $V_{GG}$ bias supplies. Logic inverters with beta ratios of 9, 12, and 16 were operated with $V_{DD} = 10$ V and $V_{GG} = 15$ V. Inverters with a beta ratio of 9 exhibited logic levels of 0.8 V and 10 V, with static noise margins of approximately 1.5 V for logic zero and 3.5 V for logic one. The individual logic circuits consisted of NAND and NOR gates, D-latches, RS flip flops, XNOR gates, binary counters, and half adders. All devices were functional up to 300°C.

The first monolithic analog integrated circuits were also reported in 6H–SiC in 1994,[75] and the first 6H–SiC CMOS integrated circuits were reported in 1996.[76] The initial CMOS circuits utilized an ion implanted $n$-well in a $p$-type epilayer, and required $V_{DD} > 10$ V for operation. In 1997, Ryu *et al.*[77] described a CMOS process based on implanted $p$-wells in an $n$-type epilayer. $n^+$ and $p^+$ implants formed the source/drain regions of $n$-channel and $p$-channel MOSFETs respectively, and also acted as channel-stops for devices of opposite polarity. These CMOS circuits were operational from a single power supply at voltages ranging from 10 V to as low as 2 V.

Charge Coupled Devices (CCDs)[78] are linear shift registers formed by a series of closely-spaced MOS plates on the surface of a semiconductor. Application of bias voltages to the MOS plates creates localized potential wells in the semiconductor under each plate. Charge packets can be confined in the potential wells and shifted along the surface under the influence of appropriate clocking waveforms applied to the gates. Silicon CCDs are widely used as image sensors, particularly in digital still cameras and hand-held video cameras. SiC is of interest as a solar-blind UV sensor because its wide bandgap makes it transparent to visible light. Such a sensor has applications in aerospace research, UV astronomy, and in certain military systems. Due to the wide bandgap of SiC, the thermal generation rate is extremely small at room temperature, and dark current is virtually negligible. The first SiC CCD was reported by Sheppard *et al.* in 1996.[79] This four-phase buried-channel device, shown in Fig. 31, employs

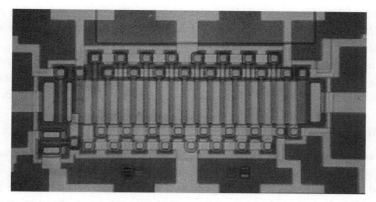

**Fig. 31.** Photograph of the first SiC CCD, a 32-bit four-phase, overlapping gate, buried channel device for use as a solar-blind UV imager.[79] Input is at the right and the output amplifier is at the left.

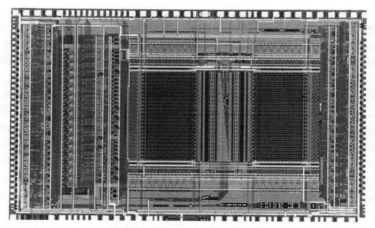

**Fig. 32.** A 1024-bit nonvolatile RAM in 6H–SiC.[80] The rectangular arrays in the center of the chip are vertically integrated bipolar NVRAM cells. Surrounding these are enhancement-load NMOS circuits for sense amplification, row and column select, and control logic. This is believed to be the most complex SiC IC developed to date.

overlapping polysilicon and aluminum gates, and is operated as an 8-bit, 32-phase shift register complete with MOS input and output amplifiers.

Perhaps the most complex SiC integrated circuit to date is a monolithic bipolar/ MOS nonvolatile random access memory (NVRAM) developed by Purdue University and Cree Research.[80] Figure 32 is a photograph of this device, fabricated in 6H–SiC. The memory array consists of 1024 vertically integrated bipolar-accessed dynamic storage cells. Surrounding this array are NMOS integrated circuits for row and column addressing, bit line sense amplifiers, and refresh circuitry.

The integrated circuits described in this section were "proof-of-concept" efforts and were not intended for commercial production, but they demonstrate that complex monolithic MOS integrated circuits are feasible in SiC.

## 5. Summary and Conclusions

In this chapter we introduced SiC as a wide bandgap semiconductor and highlighted some of the ways in which it is different from silicon. These include its hexagonal crystal structure, anisotropy, polytypism, and polar nature. We discussed the various ways that SiC can be oxidized to produce MOS devices, and considered how the unique features of SiC influence the electrical quality of the MOS interface. These are the effect of surface orientation, the nature of the sub-oxide bonding in the transition layer at the interface, the presence and role of carbon at the interface, the effect of interface states, and the high density of interface states in the upper half of the bandgap. We also discussed the low inversion layer mobility and the role of interface roughness and Coulomb scattering in reducing the mobility. We summarized recent experiments using post-oxidation annealing in NO and $H_2$ to reduce the interface state density and increase the mobility. We reviewed the basic theory of power MOSFETs, including the appropriate figures of merit and the optimum design of the drift region. We discussed power UMOSFETs and DMOSFETs, and summarized recent innovations, including the TOP UMOSFET and several versions of counter-doped DMOSFETs. We compared the performance of the latest power MOSFETs with other types of SiC power transistors, including JFETs and BJTs, and we concluded with a review of other SiC MOSFET results, including NMOS and CMOS integrated circuits and buried-channel CCDs.

In many ways the current situation with respect to SiC MOS technology is similar to that of silicon MOS technology in the early 1960s. As a semiconductor material, SiC is undergoing rapid development: material quality is improving, wafer size is increasing, and costs are declining. And like silicon in the 1960s, many fundamental materials issues still need to be resolved. Silicon technology struggled with fixed oxide charge, interface states, mobile ion contamination, oxide reliability, and control of threshold voltage. All these problems were eventually solved, but their solution required years of effort by hundreds of researchers. SiC MOS technology faces a similar situation today. Significant progress has been made over the past several years, and impressive performance has been demonstrated, but several fundamental issues remain. These include the high density of interface states in the upper half of the bandgap, the low inversion layer mobility, and the question of oxide reliability under high field and high temperature stressing.

The economic and technological impact of a viable SiC MOSFET technology would be enormous, and difficult to overstate. At this point, a viable MOS technology appears to be within our grasp, but success still hinges on continued progress in a few critical areas. It is an exciting time to be involved in MOS development in this novel and challenging material.

## References

1. W. J. Choyke and R. P. Devaty, *Naval Research Rev.* **51**, 4 (1999).
2. G. Ziegler *et al.*, *IEEE Trans. Electron Devices* **30**, 227 (1983).
3. M. R. Melloch and J. A. Cooper, Jr., *MRS Bull.* **22**, 42 (1997).
4. B. J. Baliga, *IEEE Electron Device Lett.* **10**, 455 (1989).
5. W. V. Müench and E. Pettenpaul, *J. Appl. Phys.* **48**, 4823 (1977).
6. I. A. Khan and J. A. Cooper, Jr., *IEEE Trans. Electron Devices* **47**, 269 (2000).
7. R. Buczko, S. J. Pennycook and S. T. Pantelides, *Phys. Rev. Lett.* **84**, 943 (2000).
8. S. T. Pantelides, private communication.
9. S. T. Pantelides *et al.*, *Mater. Res. Soc. Symp. Proc.* **640**, H3.3.1 (2001).
10. K. C. Chang *et al.*, *Mat. Res. Soc. Symp. Proc.* **640**, H5.45.1 (2001).
11. J. A. Cooper, Jr., *Phys. Stat. Solidi* **A162**, 305 (1997).
12. L. A. Lipkin and J. W. Palmour, *J. Electron. Mater.* **25**, 909 (1996).
13. G. Y. Chung *et al.*, *Appl. Phys. Lett.* **76**, 1713 (2000).
14. E. H. Nicollian and J. R. Brews, *MOS (Metal Oxide Semiconductor) Physics and Technology* (John Wiley & Sons, New York, 1982).
15. M. K. Das *et al.*, *Mater. Sci. Forum* **338–342**, 1069 (2000).
16. M. A. Capano *et al.*, *J. Electron. Mater.* **27**, 370 (1998).
17. M. A. Capano *et al.*, *J. Electron. Mater.* **28**, 214 (1999).
18. M. M. Maranowski and J. A. Cooper, Jr., *IEEE Trans. Electron Devices* **46**, 520 (1999).
19. MOS fundamentals are covered in most elementary semiconductor device textbooks, such as, for example, R. F. Pierret, *Semiconductor Device Fundamentals* (Addison-Wesley, Reading, MA, 1996, ISBN 0-201-54393-1) or B. G. Streetman and S. Banerjee, *Solid State Electronic Devices*, 5th Edition (Prentice Hall, Upper Saddle River, NJ, 2000, ISBN 0-13-025538-6).
20. V. R. Vathulya and M. H. White, *IEEE Trans. Electron Devices* **47**, 2018 (2000).

21. S. C. Sun and J. D. Plummer, *IEEE Trans. Electron Devices* **27**, 1497 (1980).
22. S. A. Schwarz and S. E. Russek, *IEEE Trans. Electron Devices* **30**, 1634 (1983).
23. Landolt-Börstein, *Physics of Group IV Elements and III-V Compounds* **3**, 17a (O. Madelung, 1982).
24. M. Bassler *et al.*, *Diamond and Related Mater.* **6**, 1472 (1997).
25. V. R. Vathulya *et al.*, *Appl. Phys. Lett.* **73**, 2161 (1998).
26. M. K. Das *et al.*, *J. Electron. Mater.* **27**, 353 (1998).
27. V. V. Afanas'ev *et al.*, *Phys. Stat. Sol.* **A162**, 321 (1997).
28. B. S. Um, C.-Y. Lu and J. A. Cooper, Jr., *unpublished.*
29. E. H. Nicollian and A. Goetzberger, *Bell Sys. Tech. J.* **46**, 1055 (1967).
30. S. Sridevan and B. J. Baliga, *IEEE Electron Device Lett.* **19**, 228 (1998).
31. D. Alok *et al.*, *Mater. Sci. Forum* **338–342**, 1077 (2000).
32. X. W. Wang *et al.*, *Mater. Sci. Forum* **389–393**, 993 (2002).
33. J. Tan *et al.*, *Appl. Phys. Lett.* **70**, 2280 (1997).
34. K. Fukuda *et al.*, *Appl. Phys. Lett.* **76**, 1585 (2000).
35. M. Bassler *et al.*, *Mater. Sci. Forum* **338–342**, 1065 (2000).
36. H.-F. Li *et al.*, *J. Electron. Mater.* **29**, 1027 (2000).
37. K. McDonald *et al.*, *Appl. Phys. Lett.* **76**, 568 (2000).
38. L. C. Feldman, *private communication.*
39. G. Y. Chung *et al.*, *Appl. Phys. Lett.* **77**, 3601 (2000).
40. M. K. Das *et al.*, *IEEE Device Research Conf.*, Denver, CO (2000).
41. G. Chung *et al.*, *IEEE Electron Device Lett.* **22**, 176 (2001).
42. C.-Y. Lu *et al.*, *Mater. Sci. Forum* **389–393**, 977 (2002).
43. E. Arnold *et al.*, *Appl. Phys. Lett.* **13**, 413 (1968).
44. H. Yano *et al.*, *IEEE Electron Device Lett.* **20**, 611 (1999).
45. H. Yano *et al.*, *Appl. Phys. Lett.* **78**, 374 (2001).
46. H. Yano *et al.*, *Mater. Sci. Forum* **338–342**, 1105 (2000).
47. T. Hirao *et al.*, *Mater. Sci. Forum* **389–393**, 1065 (2002).
48. J. Senzaki *et al.*, *IEEE Electron Device Lett.* **23**, 13 (2002).
49. J. Senzaki *et al.*, *Mater. Sci. Forum* **389–393**, 1061 (2002).
50. J. N. Shenoy, PhD Thesis, Purdue University, November 1996.
51. J. Pendergast *et al.*, *Proc. Int. Reliab. Phys. Symp.* **23**, 124 (1995).
52. D. T. Morisette, PhD Thesis, Purdue University, May 2001.
53. A. Konstantinov *et al.*, *Appl. Phys. Lett.* **71**, 90 (1997).
54. J. W. Palmour *et al.*, in *Silicon Carbide and Related Mater.*, Institute of Physics Conference Series (137), 499 (1994).
55. D. Takayama *et al.*, *Int. Symp. Power Semi. Devices and ICs*, Osaka, Japan (2001).
56. J. Tan *et al.*, *IEEE Electron Device Lett.* **19**, 487 (1998).
57. S. Onda *et al.*, *Phys. Stat. Solidi* **A162**, 369 (1997).
58. K. Hara, *Mater. Sci. Forum* **264–268**, 901 (1998).
59. Y. Li *et al.*, *Int. Conf. Silicon Carbide and Related Mater.*, Tsukuba, Japan (2001).
60. I. A. Khan *et al.*, *Int. Symp. Power Semi. Devices and ICs*, Sante Fe, NM, USA (2002).

61. J. N. Shenoy *et al.*, *IEEE Electron Device Lett.* **18**, 93 (1997).
62. P. M. Matin and J. A. Cooper, Jr., *unpublished.*
63. K. Ueno and T. Oikawa, *IEEE Electron Device Lett.* **20**, 624 (1999).
64. S.-H. Ryu *et al.*, *Mater. Res. Soc. Symp. Proc.* **640**, H4.5 (2001).
65. S.-H. Ryu *et al.*, *Mater. Sci. Forum*, **389–393**, 1195 (2002).
66. Y. Sugawara *et al.*, *Int. Symp. Power Semi. Devices and ICs*, Toulouse, France (2000).
67. Y. Sugawara *et al.*, *Mater. Sci. Forum* **389–393**, 1199 (2002).
68. D. Peters *et al.*, *IEEE Trans. Electron Devices* **46**, 542 (1999).
69. P. Friedrichs *et al.*, *Mater. Sci. Forum* **338–342**, 1243 (2000).
70. A. Asano *et al.*, *Int. Symp. Power Semi. Devices and ICs*, Osaka, Japan (2001).
71. P. Friedrichs, *Int. Conf. Silicon Carbide and Related Mater.*, Tsukuba, Japan (2001).
72. C.-F. Huang and J. A. Cooper, Jr., *Int. Symp. Power Semi. Devices and ICs*, Sante Fe, NM, USA (2002).
73. S.-H. Ryu *et al.*, *IEEE Electron Device Lett.* **22**, 124 (2001).
74. W. Xie *et al.*, *IEEE Electron Device Lett.* **15**, 455 (1994).
75. D. M. Brown *et al.*, *Trans. Second Int. High Temp. Electron. Conf.* **1**, XI-17 (1994).
76. D. B. Slater *et al.*, *Trans. Third Int. High Temp. Electron. Conf.* **2**, XVI-27 (1996).
77. S.-H. Ryu *et al.*, *IEEE Electron Device Lett.* **18**, 194 (1997).
78. W. S. Boyle and G. E. Smith, *Bell Syst. Tech. J.* **49**, 587 (1970).
79. S. T. Sheppard *et al.*, *IEEE Electron Device Lett.* **17**, 4 (1996).
80. G. M. Johnson, J. W. Palmour, W. Xie, Y. Wang, J. A. Cooper, Jr., M. R. Melloch, and C. H. Carter, Jr., *unpublished.*

# CHAPTER 8

# InGaAsN-BASED HBTs

Albert G. Baca

*Sandia National Laboratory, Albuquerque, NM 87123, USA*

Pablo C. Chang

*Agilent Technologies, Santa Clara, CA 95054, USA*

## Contents

## 1. Introduction

Gallium nitride (GaN) and related compounds have received a lot of attention in recent years for their wide bandgap characteristics. Rapid progress in GaN technology has led to advances in various high power devices as discussed in other chapters of this book. Indium gallium arsenide nitride (InGaAsN) is another GaN related alloy that has received a lot of attention lately. However, unlike the other GaN related materials, the interest in InGaAsN has focused on its narrow bandgap characteristics.

Incorporation of a small amount of N into indium gallium arsenide (InGaAs) results in a reduction of its lattice constant, thus reducing the strain of InGaAs layer grown on gallium arsenide (GaAs). In addition, due to a large gap bowing, the bandgap ($E_G$) decreases with incorporation of nitrogen (N). The earliest development of the InGaAsN material system was done by Kondow *et al.* in 1996.[1] The InGaAsN material system would allow a much narrower bandgap to be incorporated onto a GaAs wafer. Thus, narrow bandgap devices that have previously been suited to other substrate systems may take advantage of the mature GaAs substrate technology that generally offers larger, less costly substrates than other III-V materials.

The first early studies on this novel material system have focused on applications for optoelectronic devices. The interest on InGaAsN material was prompted by the development of a low-cost, long-wavelength laser diode at 1.3 $\mu$m for the optical fiber communication systems.[1] Another application of this narrow bandgap material is for solar cells, where greater efficiency was predicted for multiple junction cells, including an InGaAsN junction, through the incorporation of different bandgap materials that better match the solar spectrum.[2] This chapter will focus on the latest application of InGaAsN material as the base material for heterojunction bipolar transistors (HBTs).

The rapid growth of the wireless communication market has generated the need for extended battery lifetime without sacrificing device performance. To reduce the power dissipation of HBT circuits, a smaller turn-on voltage, $V_{\mathrm{ON}}$, would be desirable. The $V_{\mathrm{ON}}$ of a HBT is defined as the bias required at the input terminals of the device in order to produce a specified amount of current at the output terminals. In the common-emitter HBT circuit application, the $V_{\mathrm{ON}}$ is the base-emitter bias, $V_{\mathrm{BE}}$, required to generate 1 mA/mm$^2$ of collector current, $J_{\mathrm{C}}$. The collector current, $J_{\mathrm{C}}$, of an HBT with ideal base-emitter and base-collector junctions is given by Eq. 1,

$$J_{\mathrm{C}} = \frac{q \exp\left[\frac{qV_{\mathrm{BE}}}{kT}\right]}{\int \frac{p}{D_{\mathrm{N}} n_i{}^2} \mathrm{d}x} \tag{1}$$

where $p$ is the hole concentration in the base, $D_{\mathrm{N}}$ is the minority carrier diffusion constant, and $n_i$ is the intrinsic carrier concentration, and the integral is across the quasi-neutral base region. The effect of decreasing the $E_{\mathrm{G}}$ of the base is taken into account by the intrinsic carrier concentration defined in Eq. 2,

$$n_i{}^2 = N_{\mathrm{C}} N_{\mathrm{V}} \exp\left[\frac{-E_{\mathrm{G}}}{kT}\right] \tag{2}$$

where $N_{\mathrm{C}}$ and $N_{\mathrm{V}}$ are the density of states of the conduction band and the valence band, respectively. From Eqs. 1 and 2, it is seen that the bandgap of the base region of a HBT with ideal heterojunction gradings will define the $V_{ON}$ of the device.

GaAs is the dominant material for commercial wireless HBT applications. With a bandgap of 1.42 eV, GaAs is used as the base material in both aluminum gallium arsenide (AlGaAs)/GaAs HBTs and indium gallium phosphide (InGaP)/GaAs HBTs. One approach to lower the $V_{ON}$ of HBTs uses strained InGaAs instead of GaAs in the base layer; however, the range of indium (In) composition for growing strained InGaAs on GaAs without the formation of misfit dislocations is very limited. In addition, due to the compressive strain, the bandgap of InGaAs increases, reducing the benefits of having InGaAs in the base layer. For this reason most of the earlier works on low $V_{ON}$ HBTs has focused on the indium phosphide (InP)/InGaAs material system. But the InP material technology is expensive and its application has been limited primarily to military and space uses. Because the InGaAsN material system can take full advantage of the existing GaAs foundry technology, it may be an appropriate alternative for the low-cost, low-power electronics for the more consumer-oriented market.

The first InGaAsN base HBT was grown by metal organic chemical vapor deposition (MOCVD), demonstrated by Li *et al.* in 2000.[3] To benefit from the mature GaAs material technology, the first structure was a AlGaAs/InGaAsN/GaAs double heterojunction bipolar transistor (DHBT) using GaAs as the collector material. The potential of this novel device has led to further interest and studies by various groups.

This chapter will discuss and summarize the progress and current status of this technology. It is divided into five sections, including: Introduction (this section), Material Properties, Growth, and Processing, InGaAsN based *p-n-p* HBTs, InGaAsN based *n-p-n* HBTs, and the outlook for InGaAsN HBTs.

## 2. Material Properties, Growth and Processing

The material properties of the quaternary InGaAsN were first explored by Kondow *et al.* as early as 1997,[4] while investigation of the mixed Group-V nitride ternary alloys go back as early as 1992.[5-9] The III-V alloys have a general trend of increasing bandgap with decreasing lattice constant, as illustrated by the shaded area in Fig. 1. However, the mixed Group-V nitride alloys violate this trend due to a large bandgap bowing as seen by the gallium arsenide nitride (GaAsN) alloy in Fig. 1. Although all of the

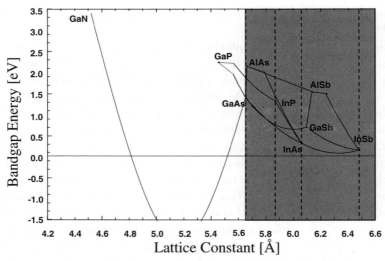

**Fig. 1.** The relationship between the lattice constant and bandgap energy in III-V alloy semiconductors.[1]

compound semiconductor alloys exhibit some degree of bandgap bowing, the larger electronegativity of N compared to the other Group V elements make the bandgap bowing of the Group-V nitride alloys much greater. The large bandgap bowing of GaAsN has also been confirmed theoretically.[10–14] Although Fig. 1 would seem to predict that GaAsN is metallic over a range of composition, it merely shows a parabolic fit to data for small amounts of N incorporation into GaAs and small amounts of arsenic (As) incorporation to GaN.[15] As GaAsN alloys with greater than a few percent N incorporation, or conversely with greater than a few percent As incorporation into GaN, do not exist, there is no assurance that the parabolic fit or bandgap theory is justified over the full alloy composition.

The band alignment of the InGaAsN material system with respect to GaAs according to this model is illustrated in Fig. 2. As N is incorporated into GaAs, the lattice constant becomes smaller and a tensile strain develops in the resulting GaAsN. Adding N to GaAs lowers both the conduction-band ($E_C$) and the valence-band ($E_V$). On the other hand, an increase in lattice constant is favored and a compressive strain builds up as In is added to GaAs to form InGaAs. However, in the case of InGaAs, the $E_C$ is lowered and the $E_V$ is raised with respect to GaAs. By incorporating the proper amount of In and N into GaAs simultaneously, InGaAsN that is lattice matched to GaAs can be obtained. A simple rule of thumb for

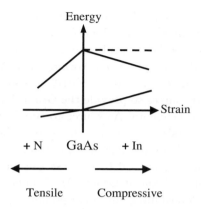

**Fig. 2.** The effect on conduction-band and valence-band edges by incorporating N and In into GaAs.[30]

strain balanced InGaAsN has about 3% of In incorporation for each 1% of N added. Higher In and N incorporation results in greater bandgap reduction. The resulting $E_C$ of the InGaAsN quaternary would be much lower because of the combined lowering effect from the incorporation of In and N. The resulting energy level of $E_V$ is nearly the same as that for GaAs for low to moderate incorporation of N and most of the $E_G$ reduction comes from the lowering of $E_C$. At even higher N levels relative to In, the InGaAsN would have a type-II alignment with GaAs because the effect of lowering $E_V$ due to N addition outweighs the effect of raising $E_V$ caused by additional In.

Incorporation of N into the InGaAs alloy, however, is not as easy as other Group V elements, because the N atom is relatively small (atomic mass = 14) compared to In (atomic mass = 115), Ga (atomic mass = 70), and As (atomic mass = 75). The relatively large difference in the sizes and crystal structure of the constituent binary materials ($a_{GaAs} = 5.65\,\text{Å}$, $a_{GaN} = 4.50\,\text{Å}$, $a_{InAs} = 6.06\,\text{Å}$, $a_{InN} = 3.54\,\text{Å}$, and $c_{InN} = 5.70\,\text{Å}$) makes it very difficult to incorporate N into GaAs. In fact, Ho and Stringfellow compared delta lattice parameter and valence force field models to calculate the interaction parameter for bulk GaAsN and found N to be miscible only up to fractions of a percent.[16] Although later work by Schlenker *et al.* found that by introducing strain, metastability for N composition up to 36% in InGaAsN is possible, actual N incorporation in the real world is still very limited.[17]

Two growth methods have been explored for making InGaAsN HBTs, MOCVD and molecular beam epitaxy (MBE). Both MBE and MOCVD have matured significantly in the 1990s, but both technologies are still at exploratory levels for the growth of InGaAsN.

While the solid source MBE has been used for optical devices, only gas source MBE (GSMBE) has produced working InGaAsN based electronic devices. Both solid source MBE and GSMBE use solid elemental gallium (Ga) and indium (In) sources to provide the Group-III atoms. Solid source MBE also uses a solid elemental As source to supply the As atoms, while GSMBE uses arsine ($AsH_3$) gas source to deliver the As needed. A RF plasma nitrogen beam is used as a source to produce atomic N; generation of the atomic N radical is critical to increase the incorporation efficiency of N atoms. The composition of Group III elements, Ga and In, can be controlled easily by determining the corresponding growth rate of Ga and In by measuring the oscillations of reflection high-energy electron diffraction (RHEED). Due to the low Group V vapor pressure, the ratio of Group V elements is more difficult to control. However, Tu has demonstrated control of the As/N ratio by a detailed procedure.[18] First, the As growth rate is determined by Group V induced RHEED oscillations. The As growth rate is then controlled to be less than the total Group III deposition rate. Finally, excess atomic N is supplied to match the total Group III flux to achieve the desired As/N composition.[18]

The MOCVD technique uses new types of reactive precursors. The most common N source for MOCVD is ammonia, which requires much higher temperatures for thermal dissociation than are allowable in GaAs-based growths. Dimethylhydrazine (DMHy) is more reactive than ammonia and was first used by Ohkouchi *et al.* for GaAsN growth by MOCVD.[19] The control of N composition can be achieved via controlling the flow rate ratio of DMHy/(DMHy+$AsH_3$). However, DMHy has some disadvantages as a N precursor. In general, MOCVD growth requires low-temperatures and a much higher DMHy/$AsH_3$ flow ratio to incorporate enough N into InGaAs. However, incomplete pyrolysis of DMHy at low growth temperatures usually introduces carbon impurities from the methyl ligands of DMHy into the InGaAsN epilayers, resulting in a higher background carrier concentration. Nevertheless, the surface morphology of MOCVD grown InGaAsN is uniform and smooth for optimized growth conditions.

A new type of precursor, tertiarybutylhydrazine (TBHy), was investigated for an N source. It was shown to be less efficient at cracking the butyl group, resulting in less carbon incorporation.[20] Another problem with DMHy is the requirement for a high DMHy/$AsH_3$ flow ratio to incorporate N into InGaAs, which may make DMHy unfeasible for low-cost mass production. Initial results with TBHy indicated that a three-fold lower flow rate and a seven-fold lower consumption rate than DMHy was sufficient to grow InGaAsN with 1% N incorporation.[20] In addition, the quality of the InGaAsN with TBHy appears to be higher as indicated by a smaller photoluminescence linewidth, 43 nm for TBHy vs. 51 nm for DMHy. Undoubtedly, the search for more efficient and practical precursors for MOCVD growth will continue.

Nitrogen incorporation leads to a high density of defects and interstitials that can deteriorate the semiconductor device performance. Both MOCVD and MBE grown InGaAsN show low photoluminescence (PL) intensity as compared to GaAs or InGaAs.[21,22] Material quality is known to improve by thermal annealing[21,23] as evidenced by a large (20 fold, typically[24]) increase in the photoluminescence intensity. However, the increase in the PL intensity is accompanied by a blueshift of the emission, which has been attributed to N outdiffusion[25] and Group III interdiffusion.[24] Depassivation of hydrogen complexes[26] doesn't appear to be a likely explanation for the PL recovery after anneal since the same low PL intensity and the same improvement after anneal was found in solid source MBE material with relatively low levels of hydrogen.[22,27] The low PL intensity of as-grown InGaAsN was attributed to other types of defects related to interstitial N incorporation.[28] These defects were studied by several techniques including nuclear reaction analysis and XPS. It was found that a Ga–N site remained after annealing while the interstitial site disappeared.[28] Recently, it has been shown by positron-annihilation and Rutherford backscattering measurements that InGaAsN grown by plasma–N MBE contains vacancy-type defects attributed to Ga vacancies and N interstitials.[24] It is now believed that these defects are responsible for the low luminescence intensity of as-grown material and the enhanced diffusion during annealing. Undoubtedly, much more will be learned about how these defects are related to the growth of InGaAsN leading to better growth technology for improving the material quality.

These types of defects are a concern for HBTs since nonradiative recombination centers in the base can lower the DC current gain, $\beta$, of the HBT. The defects can also lower $\beta$ by degrading transport and decreasing the transit time in the base. Incorporation of N into InGaAs reduces the mobility and decreases the diffusion length. However, the physics behind the degraded transport are not clear. S. R. Kurtz *et al.* reported that the carrier transport appears to be limited by large scale (much greater than the mean free path) material inhomogeneities forming potential barriers.[29]

They note that the thermal activation of the mobility is strikingly similar to inhomogeneous materials such as polycrystalline Si. More work is needed in this area.

One advantage of the InGaAsN materials is that the incorporation of N is small enough such that the chemical etch properties are indistinguishable from GaAs and low In-content InGaAs. The same wet etches that are used in AlGaAs/GaAs and InGaP/GaAs give approximately the same etch rates and selectivities as those for small amounts of N incorporation. As a result the common triple mesa process for AlGaAs/GaAs or InGaP/GaAs HBT fabrication works well for InGaAsN HBTs. A common implementation of this process uses a wet etch for etching emitter, base, and collector mesas.[30] Non selective etches can be used for AlGaAs, InGaAs, and GaAs with a $H_3PO_4 : H_2O_2 : H_2O$ etch, commonly used with a 1 : 4 : 45 ratio of etchants.[30] A selective etch of InGaP over InGaAsN can be used for an emitter etch with $H_3PO_4 : HCl$ in a 7 : 1 ratio.

Likewise, the ohmic contact formation for InGaAsN base material can use recipes already optimized for GaAs. A commonly used $p$-type ohmic contact for the base of $p$-$n$-$p$ HBTs contains thin layers of Pt/Ti/Pt/Au that do not spike through the base after alloying.[31] Ohmic metal for the base of $p$-$n$-$p$ HBTs will often utilize the common Ge/Au/Ni recipes.[32] Another option is the use of Pd/Ge/Au which is alloyed at 175°C in order to prevent spiking of the Au through the base.[33] In a $p$-$n$-$p$ HBT, the $p$-type emitter and collector contacts can be non-alloyed, reducing the thermal budget to prevent contact spiking. Other aspects of processing InGaAsN HBTs that don't impact the base layer are well known and will depend on the investigators' discretion.

Although InGaAsN HBTs benefited from the earlier development of InGaAsN optoelectronic device materials, some material requirements for InGaAsN HBTs are very different from those of optoelectronic devices. As thicker InGaAsN layers would be needed for HBT structures than in quantum well laser diode structures, lattice matching is necessary. Thus the material requirement of InGaAsN for HBT is much closer to those used for InGaAsN solar cells, where minority carrier diffusion length and mobility are more important than in laser diodes. Photoluminescence intensity and radiative recombination times are important in laser diodes, solar cells, and HBTs, while metastable strain effects are important in laser diodes and not relevant to solar cells and HBTs.

## 3. InGaAsN Based *p-n-p* HBTs

The use of *p-n-p* bipolar transistors is desirable because of the ability to make complementary circuits. Field effect transistors (FET) can be operated as complementary circuits as well but the large discrepancy between the speed of a *n*-channel FET and a *p*-channel FET has constrained the used of complementary FETs for compound semiconductors. This limitation need not be so great for HBTs. The speed of a FET is determined by the transit time from the source to the drain. *n*-channel FETs have mobilities and saturated velocities much greater than *p*-channel FETs, leading to shorter transit times and higher speeds. The same transport issues affect the HBT, but being a vertical device, the distances can be much smaller and not as much of a limiting factor, compared to the charging time for junction capacitors. As the FET is a lateral device the source-to-drain distance can be large compared to the vertical distances in a HBT without resorting to heroic and costly methods of lithography. The RF performance of *p-n-p* HBTs is nearly 10 times better than that reported for *p*-FETs.[34,35] The main limitation for the adoption of complementary HBTs is the process complexity for integrating the epitaxial structures needed for *p-n-p* and *p-n-p* structures.

The InGaAsN material system is an excellent choice for *p-n-p* HBTs due to its unique band alignment. As presented in the previous section, addition

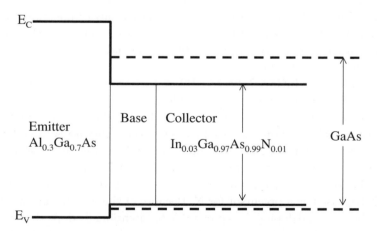

**Fig. 3.** The band alignment of an $Al_{0.3}Ga_{0.7}As/In_{0.03}Ga_{0.97}As_{0.99}N_{0.01}$ HBT compared to an AlGaAs/GaAs HBT. The InGaAsN HBT has a greater $\Delta E_C$ than does the GaAs HBT, making it more suited for *p-n-p* HBTs.[30]

**Table 1.**   The structure of a *p-n-p* AlGaAs/InGaAsN HBT. The collector doping was high to compensate the high unintentional *p*-type background doping encountered in MOCVD-grown InGaAsN.[30]

| Layer | Material | Thickness | Doping |
|---|---|---|---|
| Contact Cap | $p^+$-GaAs | 200 nm | $2.0 \times 10^{19}$ cm$^{-3}$ |
| Emitter | $p$-Al$_{0.3}$Ga$_{0.7}$As | 100 nm | $3.0 \times 10^{18}$ cm$^{-3}$ |
| Spacer | $u$-Al$_{0.3}$Ga$_{0.7}$As | 5 nm | Undoped |
| Base | $n$-In$_{0.03}$Ga$_{0.97}$As$_{0.99}$N$_{0.01}$ | 100 nm | $1.2 \times 10^{18}$ cm$^{-3}$ |
| Collector | $p$-In$_{0.03}$Ga$_{0.97}$As$_{0.99}$N$_{0.01}$ | 300 nm | $2.0 \times 10^{18}$ cm$^{-3}$ |
| Subcollector | $p^+$-GaAs | 500 nm | $2.0 \times 10^{19}$ cm$^{-3}$ |
| Substrate | | S. I. GaAs | |

of both N and In lowers the bandgap of InGaAsN while a lattice matched condition can be maintained. Almost all of the bandgap reduction occurs in the lowering of $E_C$. The band alignment for a *p-n-p* AlGaAs/InGaAsN HBT is illustrated in Fig. 3 and contrasted to that for AlGaAs/GaAs HBTs. The unique physics of predominant conduction band lowering in InGaAsN is very favorable to *p-n-p* InGaAsN HBTs. The original concept for the HBT relied on the suppression of minority carrier injection into the emitter by a valence band offset in an *n-p-n* HBT and by the conduction band offset for a *p-n-p* HBT. The increased conduction band offset of about 0.4 eV for AlGaAs/InGaAsN can further suppress minority carrier injection into the emitter. At the same time, the valence band offset remains unchanged, so that a relatively low barrier for majority carrier conduction from the emitter to the base is maintained.

The first *p-n-p* InGaAsN based HBT was reported by Chang *et al.*[30] The structure for this device is given in Table 1. The Al content in the emitter was 30%, while the In and N content in the base and collector were 3% and 1%, respectively. The only unusual aspect of the structure is the relatively high collector doping of $2.0 \times 10^{18}$ cm$^{-3}$, which increases the base-collector capacitance. This structure was grown by MOCVD which leads to a high *p*-type background doping, possibly due to carbon (C) incorporation from the DMHy used as a precursor for N. Because of the variable nature of the *p*-type background doping, it is preferable to intentionally dope the collector at a high enough level to improve the reproducibility in the collector doping.

The Gummel plot and the common emitter current–voltage (I–V) plots for large area HBTs ($100 \times 100$ $\mu$m$^2$) are shown in Figs. 4 and 5. As shown in

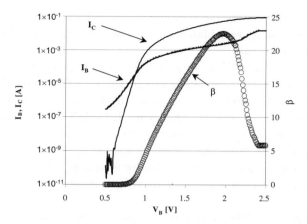

**Fig. 4.** Gummel plot of a large area $Al_{0.3}Ga_{0.7}As/In_{0.03}Ga_{0.97}As_{0.99}N_{0.01}$ $p$-$n$-$p$ HBT. The $\beta$ of this device is 23. $V_{ON}$ is only 0.77 V, but $\eta_b$ is 3.1, indicating a high level of recombination.[30]

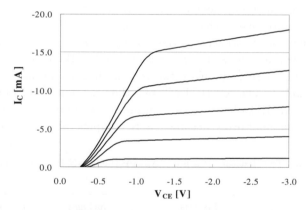

**Fig. 5.** Common-emitter I–V plot of a large area $Al_{0.3}Ga_{0.7}As/In_{0.03}Ga_{0.97}As_{0.99}$-$N_{0.01}$ $p$-$n$-$p$ HBT. The five curves correspond to $I_B$ of 0.2, 0.4 ,0.6, 0.8, and 1.0 mA.[30]

Fig. 4, the AlGaAs/InGaAsN HBT has a peak current gain, $\beta$, of 23 before the base push-out effect takes place. This current gain is sufficient to be useful for many circuit applications, but it is much lower than expected for an AlGaAs/GaAs HBT. However, the turn-on voltage, $V_{ON}$, as defined by the bias level at which the collector current exceeds 1 $\mu$A, is significantly lower than that for an AlGaAs/GaAs HBT of a similar structure, having 0.77 V as compared to 1.03 V. Both the base sheet resistance ($R_s = 3$ k$\Omega$/sq.) and the electron mobility ($\mu_n = 350$ cm$^2$V$^{-1}$s$^{-1}$) in the InGaAsN base layer are degraded compared to the GaAs values. These values contribute to a higher offset voltage, $V_{\text{offset}}$, and higher knee voltage for the AlGaAs/InGaAsN HBT as seen in Fig. 5.

The AlGaAs/GaAs HBT, with its better base transport properties, has a $\beta$ greater than 100. Considering the presence of the larger $\Delta E_C$ in the AlGaAs/InGaAsN material system and the fact that $\beta$ should increase exponentially with increasing $\Delta E_C$, $\beta$ for InGaAsN HBTs should ideally be greater than 23. A likely cause is the recombination centers in the InGaAsN base layer from the defects discussed in Sec. 2. One indication of excess recombination is the base ideality factor, $\eta_b$, which is measured to be 3.1, indicating the presence of more than just intrinsic base recombination effects.

This early $p$-$n$-$p$ HBT result is very promising because of the reduced $V_{\text{ON}}$ and because the $\beta$ is high enough for many circuit applications. However, the degraded base transport properties and excess base recombination result in tradeoffs between turn-on characteristics and other important device parameters. Clearly, InGaAsN HBTs require either improved material or aggressive device design to maximize the beneficial aspects and minimize the drawbacks of the InGaAsN material. One method for addressing the limitations of the $p$-$n$-$p$ InGaAsN HBT is to design an electric field in the base to improve the transit time in the base. Such techniques are well known for HBTs,[36] but they are not commonly adopted for GaAs-based HBTs because most applications do not require them and because the manufacturing becomes more complex after they are adopted.

**Table 2.** The structure of a $p$-$n$-$p$ AlGaAs/InGaAsN HBT with a graded base doping.[37]

| Layer | Material | Thickness | Doping |
|---|---|---|---|
| Contact Cap | $p^+$-GaAs | 300 nm | $2.0 \times 10^{19}$ cm$^{-3}$ |
| Emitter | $p$-Al$_{0.3}$Ga$_{0.7}$As | 120 nm | $3.0 \times 10^{18}$ cm$^{-3}$ |
| Spacer | $u$-GaAs | 5 nm | Undoped |
| Base | $n$-In$_{0.03}$Ga$_{0.97}$As$_{0.99}$N$_{0.01}$ | 100 nm | $4.5 \times 10^{18}$ cm$^{-3}$ graded to $7.0 \times 10^{17}$ cm$^{-3}$ |
| Collector | $p$-In$_{0.03}$Ga$_{0.97}$As$_{0.99}$N$_{0.01}$ | 300 nm | $2.0 \times 10^{17}$ cm$^{-3}$ |
| Subcollector | $p^+$-GaAs | 875 nm | $2.0 \times 10^{19}$ cm$^{-3}$ |
| Substrate | S. I. GaAs | | |

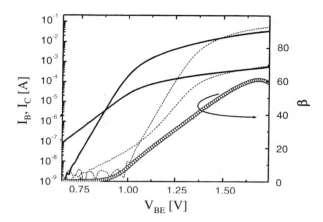

**Fig. 6.** Gummel plot of a conventional *p-n-p* Al$_{0.3}$Ga$_{0.7}$As/GaAs HBT and an Al$_{0.3}$Ga$_{0.7}$As/In$_{0.03}$Ga$_{0.97}$As$_{0.99}$N$_{0.01}$ HBT with graded base doping.[37]

Monier *et al.* reported a *p-n-p* AlGaAs/InGaAsN HBT with graded doping in the base to provide the electric field.[37] The structure for this HBT is given in Table 2. The InGaAsN was grown by MOCVD with a rotating disk reactor using DMHy as the precursor. Figure 6 shows a Gummel plot for both AlGaAs/InGaAsN and AlGaAs/GaAs HBTs with $3 \times 12$ $\mu$m$^2$ emitter areas. The current gain, $\beta$, reaches 60 as compared to 23 for the uniform base, although it is still less than that of an AlGaAs/GaAs HBT with uniform base doping. The base current ideality factor is also improved with $\eta_b = 2.2$ as compared to 3.1 for the uniform base. This value is closer to the space-charge recombination value, which suggests that much of the non-ideal recombination has been eliminated. Like the uniform base HBT, the $V_{ON}$ of the InGaAsN base HBT is reduced by about 0.29 V for 1% of N and 3% of In. The common-emitter I–V plot is shown in Fig. 7 for the $3 \times 12$ $\mu$m$^2$ InGaAsN HBT with graded base doping. The offset voltage of 0.16 V is slightly better than that of the original uniform base HBT. The RF characteristics of this device are shown in Fig. 8. With cutoff frequency $f_T$ of 15 GHz (as compared to 10 GHz for the uniform base InGaAsN HBT), the graded base demonstrates enhanced performance over the uniform base. However, the $f_{MAX}$ of 10 GHz is limited by the high base sheet resistance 2.3 k$\Omega$/sq.

Although these results are promising, further improvements are possible. The effects of annealing InGaAsN improve PL intensity and the quality of laser diodes as discussed in the last section. Rather than improving HBT

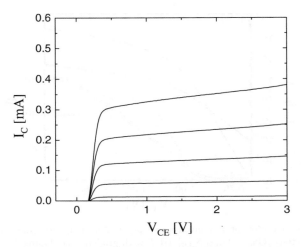

**Fig. 7.** Common-emitter I–V plot of a $3 \times 12$ $\mu$m$^2$ Al$_{0.3}$Ga$_{0.7}$As/In$_{0.03}$Ga$_{0.97}$As$_{0.99}$-N$_{0.01}$ HBT graded base doping.[37]

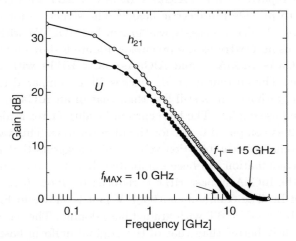

**Fig. 8.** Small signal gain, $h_{21}$, and unilateral power gain, $U$, of a $3 \times 12$ $\mu$m$^2$ Al$_{0.3}$Ga$_{0.7}$As/In$_{0.03}$Ga$_{0.97}$As$_{0.99}$N$_{0.01}$ HBT with graded base doping.[37]

performance, post-growth annealing reduces $\beta$ for AlGaAs/InGaAsN/GaAs DHBTs grown by MOCVD with a DMHy precursor.[38] It was suggested that annealing increases the $p$-type background doping from $1 \times 10^{17}$ cm$^{-3}$ to near $5 \times 10^{17}$ cm$^{-3}$, which results in a highly compensated base layer.[38] Further work is needed with higher purity material, using MBE or MOCVD

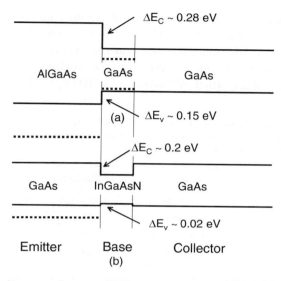

**Fig. 9.** Band alignments for *p-n-p* HBTs with GaAs collectors using (a) AlGaAs emitter and GaAs or InGaAsN bases and (b) InGaAsN base and AlGaAs or GaAs emitter. These HBT use $Al_{0.3}GaAs$ and $In_{0.03}Ga_{0.97}As_{0.99}N_{0.01}$.

with better precursors, in order to establish whether annealed, high purity InGaAsN can be used to improve $\beta$, $f_T$, and $f_{MAX}$.

InGaAsN material lends itself to some unique *p-n-p* InGaAsN HBT designs, as illustrated by the band diagrams in Fig. 9. These designs exploit the fact that the bandgap discontinuity occurs predominantly in the conduction band offset. First, it is seen that a DHBT is very advantageous with GaAs as the collector. Because GaAs and InGaAsN have very little valence band offset, there is no transport barrier for a *p-n-p* DHBT in contrast to the situation for GaAs and InP-based *p-n-p* DHBTs. It makes sense to incorporate GaAs in place of InGaAsN in the collector due to its better breakdown properties and its superior transport properties. Second, a GaAs emitter may become feasible as well. The large conduction band discontinuity between AlGaAs and InGaAsN, which provides a large barrier to electron-back-injection into the emitter, is not effective in improving current gain because defect associated recombination is so much larger. The lower $\Delta E_C$ of GaAs/InGaAsN may be sufficient to control electron-back-injection into the emitter. A GaAs emitter can improve the base-emitter junction quality by eliminating Al and improving hole transport in the emitter as well.

$p$-$n$-$p$ DHBTs using emitters with both AlGaAs and an Al-free design have been recently reported.[39,40] Due to the presence of DX centers associated with Al in AlGaAs, reliability and the overall quality of the base-emitter junction are a concern. For example, InGaP/GaAs $n$-$p$-$n$ HBTs show better reliability than AlGaAs/GaAs HBTs.[41] The structure for a

**Table 3.** The structure of the $p$-$n$-$p$ GaAs/InGaAsN/GaAs DHBT. The improved emitter-base junction leads to better carrier transport characteristics and superior device performance.[40]

| Layer | Material | Thickness | Doping |
|---|---|---|---|
| Contact Cap | $p^+$-GaAs | 300 nm | $2.0 \times 10^{19}$ cm$^{-3}$ |
| Emitter | $p$-GaAs | 70 nm | $2.0 \times 10^{18}$ cm$^{-3}$ |
| Base | $n$-In$_{0.03}$Ga$_{0.97}$As$_{0.99}$N$_{0.01}$ | 100 nm | $3.0 \times 10^{18}$ cm$^{-3}$ |
| Collector | $p$-GaAs | 500 nm | $3.0 \times 10^{16}$ cm$^{-3}$ |
| Subcollector | $p^+$-GaAs | 750 nm | $2.0 \times 10^{19}$ cm$^{-3}$ |
| Substrate | | S. I. GaAs | |

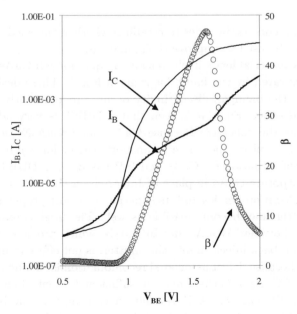

**Fig. 10.** Gummel plot of a $3 \times 12$ $\mu$m$^2$ $p$-$n$-$p$ GaAs/In$_{0.03}$Ga$_{0.97}$As$_{0.99}$N$_{0.01}$/GaAs DHBT. Th base-collector bias is set at 0 V.[40]

GaAs/InGaAsN/GaAs DHBT with a base bandgap of approximately 1.2 eV is given in Table 3. The InGaAsN was grown by MOCVD with a DMHy precursor using a rotating disk reactor. No grading was used in the base. A Gummel plot for this DHBT with a $3 \times 25 \ \mu m^2$ emitter is shown in Fig. 10. A peak $\beta$ of 45 was demonstrated, which is higher than the earlier $p$-$n$-$p$ InGaAsN HBTs with uniform base doping but comparable to an AlGaAs/InGaAsN/GaAs DHBT used as a control. This improved $\beta$ is obtained while maintaining the same reduction in turn-on voltage (approximately 0.27 V) as the earlier work for the InGaAsN. The improved performance of the DHBT compared to the single HBT may be due to the higher material quality of the GaAs collector as compared to InGaAsN.

The base current ideality factor for the GaAs/InGaAsN/GaAs DHBT is $\eta_b = 1.9$, which is smaller than that for the AlGaAs/InGaAsN/GaAs DHBT ($\eta_b = 2.38$). Though both DHBTs have a large space charge recombination component, the Al-free DHBT is free of much of the non-ideal recombination present in the AlGaAs emitter DHBT.

The common emitter I–V characteristics are shown in Fig. 11. It is seen that both the offset voltage and the knee voltage are improved as compared to the earlier InGaAsN HBT. The offset voltage is lower because of the well-known fact that a DHBT with no emitter and collector conduction band discontinuity produces the lowest offset voltage. The reduction in the knee voltage is attributed to the improved emitter transport properties since the base sheet resistance was comparable in AlGaAs and GaAs emitter devices.

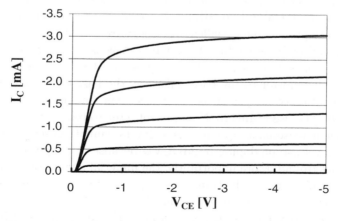

**Fig. 11.** Common-emitter I–V plot of a $3 \times 25 \ \mu m^2$ $p$-$n$-$p$ GaAs/In$_{0.03}$Ga$_{0.97}$As$_{0.99}$-N$_{0.01}$/GaAs DHBT. The base current varies from 20 $\mu$A to 100 $\mu$A at 20 $\mu$A/step.[40]

The breakdown voltage, $BV_{CEO}$, is about 12 V, which is comparable to that for an AlGaAs/GaAs HBT, as expected since both have GaAs collectors. The RF performance is comparable for GaAs/InGaAsN/GaAs DHBTs and AlGaAs/GaAs HBTs with comparable base sheet resistance. The InGaAsN DHBT shows both $f_T$ and $f_{MAX}$ of approximately 12 GHz, while the GaAs HBT shows $f_T$ of 12 GHz and $f_{MAX}$ of 10.5 GHz. Both devices can have improved $f_T$ if the base thickness is reduced.

It is clear that great strides have been made in the design and fabrication of $p$-$n$-$p$ InGaAsN based HBTs. Further improvements are possible through material quality improvements and by combining some of the most promising design approaches. For example, an Al-free DHBT with a graded base doping may considerably improve both the DC current gain and the RF performance.

## 4. InGaAsN Based $n$-$p$-$n$ HBTs

The first InGaAsN based $n$-$p$-$n$ HBT was demonstrated using MOCVD. Li *et al.* demonstrated the device with 1% N with 3% In composition.[3] The InGaP/InGaAsN/GaAs $n$-$p$-$n$ DHBT was grown using DMHy as the N source. For 1% N incorporation, the effects on $E_V$ from incorporation of N and In are compensated, and the $E_V$ is relatively unchanged as compared to the $E_V$ level of GaAs. The resulting band alignment (shown in Fig. 12), whose unique physics is favorable for $p$-$n$-$p$ HBTs, presents more challenges for $n$-$p$-$n$ HBTs. The valence band offset is little changed from InGaP/GaAs, making the reverse current injection characteristics comparable to InGaP/GaAs HBTs. However, the large conduction-band discontinuity presents new challenges for forward conduction. The conduction band offset between InGaP and InGaAsN is even larger than that between AlGaAs and GaAs, and it can present a barrier to electron injection from the emitter to the base. Likewise, the base-collector conduction band discontinuity $\Delta E_C$ is a barrier to electron conduction and must be dealt with. The use of InGaAsN as the collector to reduce the second barrier is not a reasonable option. These layers cannot presently be grown with low level $n$-type doping by MOCVD because of a high unintentional $p$-type background. More importantly, the material quality of an InGaAsN collector would degrade the transport properties too much. Also, for many applications, a lower breakdown voltage from an InGaAsN collector would not be acceptable. For all of these reasons, the early reports of InGaAsN based $n$-$p$-$n$ HBTs use the DHBT design.

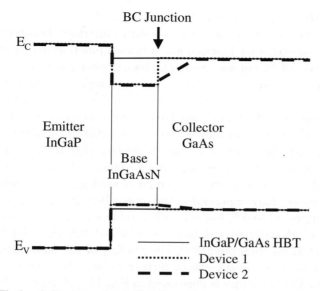

**Fig. 12.** The band alignment of an $InGaP/In_{0.03}Ga_{0.97}As_{0.99}N_{0.01}/GaAs$ DHBT compared to an InGaP/GaAs HBT. The InGaAsN DHBT uses either an abrupt or graded base-collector junction for $n$-$p$-$n$ HBT designs.[31]

**Table 4.** The structure of the first MOCVD grown $n$-$p$-$n$ InGaP/InGaAsN/GaAs DHBT.[3]

| Layer | Material | Thickness | Doping |
|---|---|---|---|
| Contact Cap | $n^+$-$In_{0.53}Ga_{0.47}As$ | 30 nm | $2.0 \times 10^{19}$ cm$^{-3}$ |
| Graded Contact | $n^+$-$In_{0-0.53}Ga_{1-0.47}As$ | 30 nm | $2.0 \times 10^{19}$ cm$^{-3}$ |
| Superemitter | $n^+$-GaAs | 100 nm | $2.0 \times 10^{18}$ cm$^{-3}$ |
| Emitter | $n$-InGaP | 50 nm | $2.0 \times 10^{17}$ cm$^{-3}$ |
| Base | $p$-$In_{0.03}Ga_{0.97}As_{0.99}N_{0.01}$ | 70 nm | $1.0 \times 10^{19}$ cm$^{-3}$ |
| Spacer | $u$-$In_{0.03}Ga_{0.97}As_{0.99}N_{0.01}$ | 5 nm | Undoped |
| Graded Collector (optional) | $n$-$In_{0-0.1}Ga_{1-0.9}As$ | 30 nm | $3.0 \times 10^{16}$ cm$^{-3}$ |
| Collector | $n$-GaAs | 500 nm | $3.0 \times 10^{16}$ cm$^{-3}$ |
| Subcollector | $n$-GaAs | 500 nm | $2.0 \times 10^{18}$ cm$^{-3}$ |
| Substrate | | S. I. GaAs | |

The DHBT structure reported by Li *et al.*[3] is given in Table 4. An abrupt junction was used for the base-emitter interface. Designs with both an abrupt base-collector junction and a graded junction were compared. A base thickness of 70 nm was used. The base doping was conservative so as not to introduce excess recombination due to high base doping. With In and N content set at 3% and 1%, respectively, the bandgap of the base was approximately 1.2 eV. Large area DHBTs were fabricated and characterized. DC current gain was only 3.5 for the structure with abrupt junctions and varied with the collector voltage, indicating that the current blocking effect was present. The structure with a graded collector, from GaAs to $In_{0.1}Ga_{0.9}As$, used a conservative amount of grading, and reduced, but did not eliminate the barrier present at the base-collector junction. The current gain was improved to 5.5, but was still varying with the collector voltage.

Chang *et al.* reported a similar *n-p-n* InGaP/InGaAsN/GaAs DHBTs with improved junction designs.[31] The grading of the base-collector junction was made more aggressive to further reduce the barrier to electron conduction. This structure is illustrated in Fig. 12 and Table 5. The 30 nm base-collector grading layer requires InGaAs with 26% In composition at the interface with the base to eliminate the barrier, but this was thought to be too close to the strain relaxation limit and therefore 20% In composition was used. Delta-doping layers were also inserted near the emitter and collector interfaces. This structure with improved interfaces succeeded

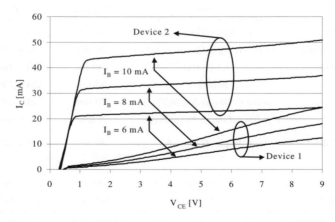

**Fig. 13.** Common-emitter I–V plots of the two large area *n-p-n* InGaP/$In_{0.03}Ga_{0.97}$-$As_{0.99}N_{0.01}$/GaAs DHBTs. Three curves of $I_C$ corresponding to $I_B$ of 6, 8, and 10 mA are plotted for each device. The I–V characteristic of device 2 is nearly ideal due to its improved BC junction.[31]

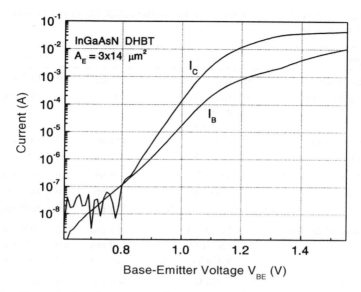

**Fig. 14.** Gummel plot of a $3 \times 14$ $\mu m^2$ $n$-$p$-$n$ InGaP/In$_{0.03}$Ga$_{0.97}$As$_{0.99}$N$_{0.01}$/GaAs DHBT.[42]

in removing the DC current blocking effects. The DC common-emitter I–V characteristics for the InGaAsN DHBT with abrupt and improved interfaces are shown in Fig. 13 for large area transistors. The DHBT with improved interfaces showed a larger current gain of 7 compared to 5.5 and much less variation of collector current with collector voltage. Small area InGaAsN DHBTs were reported with 1% N incorporation.[42,43] The structure is shown in Table 5. DC current gain of 16, shown in the Gummel plot of Fig. 14, was reported for a self-aligned DHBT with a $3 \times 14$ mm$^2$ emitter area, which is significantly better than $b = 7$ result for the large area DHBT. The difference between the small and large area results may be due to the higher current density reported for the small area devices, as b does not peak until 50 kA/cm$^2$ current density. The small area DHBT also had significantly smaller offset voltage and knee voltage as seen by the I–V curves in Fig. 15. Most significantly, the turn-on voltage is reduced by 0.25 V, similar to a control HBT whose structure only differs by having a GaAs base layer. This reduction in turn-on voltage is slightly greater than the reduction of the base bandgap (0.22 V). High frequency characterization of the InGaAsN DHBT gave fT of 40 GHz and fMAX of 72 GHz. The peak $f_T$ and $f_{MAX}$ were achieved at 55 kA/cm$^2$ current density.

**Table 5.** The structure of the $n$-$p$-$n$ InGaP/InGaAsN/GaAs DHBT reported by Chang *et al.*[31]

| Layer | Material | Thickness | Doping |
| --- | --- | --- | --- |
| Contact Cap | $n^+$-In$_{0.35}$GaAs | 30 nm | $2.0 \times 10^{19}$ cm$^{-3}$ |
| Graded Contact | $n^+$-In$_{0.35}$GaAs graded to $n^+$-GaAs | 30 nm | $2.0 \times 10^{19}$ cm$^{-3}$ |
| Superemitter | $n^+$-GaAs | 100 nm | $2.0 \times 10^{18}$ cm$^{-3}$ |
| Emitter | $n$-InGaP | 45 nm | $2.0 \times 10^{17}$ cm$^{-3}$ |
| $\delta$-Doping | InGaP | 0 nm | $5.0 \times 10^{12}$ cm$^{-2}$ |
| Spacer | $n$-InGaP | 5 nm | $2.0 \times 10^{17}$ cm$^{-3}$ |
| Base | $p$-In$_{0.03}$Ga$_{0.97}$As$_{0.99}$N$_{0.01}$ | 70 nm | $1.0 \times 10^{19}$ cm$^{-3}$ |
| Graded Collector | $n$-In$_{0.2}$GaAs graded to $n$-GaAs | 5 nm 0 nm 25 nm | $3.0 \times 10^{16}$ cm$^{-3}$ $5.0 \times 10^{12}$ cm$^{-2}$ $3.0 \times 10^{16}$ cm$^{-3}$ |
| Collector | $n$-GaAs | 500 nm | $3.0 \times 10^{16}$ cm$^{-3}$ |
| Subcollector | $n$-GaAs | 500 nm | $2.0 \times 10^{18}$ cm$^{-3}$ |
| Substrate | | S. I. GaAs | |

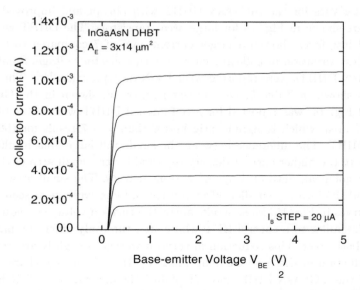

**Fig. 15.** Common-emitter I-V plot of a $3 \times 14$ $\mu$m$^2$ In$_{0.03}$Ga$_{0.97}$As$_{0.99}$N$_{0.01}$ DHBT.[42]

The RF results of these small area InGaAsN DHBTs show high enough performance for typical digital or wireless applications. They could certainly be engineered to develop products that meet specifications for these applications. However, they do not achieve the same performance as control samples with a GaAs base. InGaP/GaAs HBT control samples achieve $f_T$ of 60 GHz and $f_{MAX}$ of 110 GHz when fabricated with the same process, layout, and structure. The higher performance of the GaAs base layer comes from a better base sheet resistance (250 $\Omega$/sq. vs. 1.1 k$\Omega$/sq.) and a shorter base transit time. From the following equations,

$$f_T = \frac{1}{2\pi\tau_{EC}} \tag{3}$$

where,

$$\tau_{EC} = \tau_E + \tau_B + \tau_C + \tau_{CC}, \tag{4}$$

$\tau_{EC} = 4.4$ ps and $\tau_{EC} = 2.9$ ps are derived for the InGaAsN base and the GaAs base HBTs, respectively. The difference between the two is due to the base transit time. Because of this difference in base transit times, the InGaAsN DHBT must be engineered differently to achieve circuit specifications. Unless further improvements are made in the material quality, the base thickness should be reduced aggressively, along with aggressively increasing the base doping and introducing an electric field in the base. These measures that reduce the transit time will also increase $\beta$. Although these are drastic measures, this approach has been successful for another HBT material with lesser base transport properties, the SiGe material.

The payoff for adopting InGaAsN will be to reduce the turn-on voltage of the DHBT and enable more efficient circuits at low supply voltages. In Fig. 16, the $f_T$ is plotted as a function of supply voltage for the special case where the collector voltage is tied to the base voltage. It is seen that the InGaAsN HBT gives the same $f_T$ as InGaP/GaAs at voltages that are reduced by 150–255 mV. This low voltage performance should lead to greater efficiency. As of this writing, no amplifiers have been demonstrated to verify this expected improvement.

Other work on InGaAsN HBTs has focused on different N compositions. Welser *et al.* has also demonstrated MOCVD grown InGaAsN with significantly less than 1% nitrogen.[44] The rationale for this approach is to maintain the ratio of the current gain to base sheet resistance $R_{SB}$ below 0.3 (a common figure of merit for InGaP/GaAs HBTs) so that InGaAsN DHBTs can be adopted as a drop in replacements for

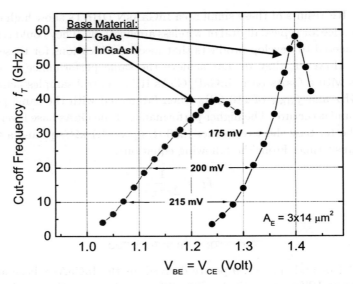

**Fig. 16.**   Cutoff frequency $f_T$ for $3 \times 14$ $\mu m^2$ $In_{0.03}Ga_{0.97}As_{0.99}N_{0.01}$ and GaAs base *n-p-n* HBTs with $V_{CE} = V_{BE}$. The InGaAsN DHBT shows better low voltage RF performance.[42]

InGaP/GaAs HBTs. This constraint resulted in about 0.3% N incorporation for their InGaP/InGaAsN/GaAs DHBT. The material shows much improved quality, resulting in device performance much closer to that of GaAs based devices. However, the minimal incorporation of N has reduced the $V_{ON}$ by only 25 mV.

Welty *et al.* have explored much greater N incorporation.[45] That group was also the first to demonstrate the InGaAsN HBT using GSMBE, with elemental Ga and In sources for Group III elements, AsH$_3$ for producing As, and a N-plasma produced by a RF generator for incorporating N. Relatively high N content of 2% along with 11% In was demonstrated. In is utilized in excess of the 3 to 1 ratio normally used to achieve a lattice match in order to be more aggressive in the base bandgap reduction. The bandgap of the base in this structure was approximately 1.0 eV.

The structure for this InGaAsN DHBT is shown in Table 6. For a number of reasons including the strain and the high recombination in the base, the base thickness was limited to 40 nm. Chirped superlattices with a 1.1 nm period and 30 nm total thickness were employed to grade the base-emitter and the base-collector junctions. Delta-doping was performed on either side of the chirped superlattice to counteract the large electric field

**Table 6.** The structure of the first GSMBE grown *n-p-n* GaAs/InGaAsN/GaAs DHBT.[45]

| Layer | Material | Thickness | Doping |
|---|---|---|---|
| Contact Cap | $n^+$-GaAs | 200 nm | $5.0 \times 10^{18}$ cm$^{-3}$ |
| Emitter | $n$-GaAs | 200 nm | $5.0 \times 10^{17}$ cm$^{-3}$ |
| $\delta$-Doping | $n^+$-GaAs | 0.5 nm | $3.0 \times 10^{19}$ cm$^{-3}$ |
| Emitter Grading | $n$-GaAs $\rightarrow$ | 30 nm | $3.0 \times 10^{17}$ cm$^{-3}$ |
| | $n$-In$_{0.11}$Ga$_{0.89}$As$_{0.98}$N$_{0.02}$ | | |
| Spacer | $u$-In$_{0.11}$Ga$_{0.89}$As$_{0.98}$N$_{0.02}$ | 5 nm | undoped |
| Base | $p$-In$_{0.11}$Ga$_{0.89}$As$_{0.98}$N$_{0.02}$ | 40 nm | $8.0 \times 10^{18}$ cm$^{-3}$ |
| Spacer | $u$-In$_{0.11}$Ga$_{0.89}$As$_{0.98}$N$_{0.02}$ | 5 nm | Undoped |
| Collector Grading | $n$-In$_{0.11}$Ga$_{0.89}$As$_{0.98}$N$_{0.02}$ | 30 nm | $3.0 \times 10^{16}$ cm$^{-3}$ |
| | $\rightarrow n$-GaAs | | |
| $\delta$-Doping | $n^+$-GaAs | 5 nm | $1.5 \times 10^{18}$ cm$^{-3}$ |
| Collector | $n$-GaAs | 400 nm | $3.0 \times 10^{16}$ cm$^{-3}$ |
| Subcollector | $n^+$-GaAs | 700 nm | $5.0 \times 10^{18}$ cm$^{-3}$ |
| Substrate | S. I. GaAs | | |

caused by the compositional grading. After crystal growth, the sample was annealed at 700°C for 1 minute in order to reduce the hydrogen passivation of the InGaAsN layer. The hydrogen can significantly reduce the free hole concentration in the base.

DC and RF characteristics were reported for DHBTs with $2 \times 3 \times 3$ $\mu$m$^2$ emitter areas. A large reduction in the turn-on voltage of 0.4 V was achieved, which is the highest value for InGaAsN HBTs and begins to approach the values for InP/InGaAs HBTs. However, peak current gain was limited to 8 and the Gummel plot showed a base and collector current crossover of only 0.75 V, as seen in Fig. 17. It is seen that a high series resistance limits the collector current as well. This series resistance is attributed to the high base sheet resistance (7 k$\Omega$/square) combined with high contact resistance and low current gain. Either better material quality or a graded base doping can improve the current gain.

The common emitter I–V curves are shown in Fig. 18. They show a low offset voltage and a reasonable knee voltage. The slope in the saturation region of the I–V curves is attributed to base thickness modulation with collector voltage due to the relatively low base doping level. The

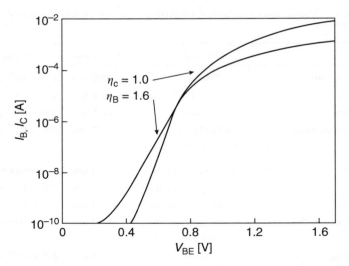

**Fig. 17.** The Gummel plot of a $2 \times 3 \times 3 \ \mu m^2$ *n-p-n* GaAs/In$_{0.11}$Ga$_{0.89}$As$_{0.98}$-N$_{0.02}$/GaAs DHBT.[45]

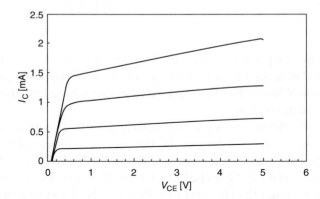

**Fig. 18.** Common-emitter I–V plot of a $2 \times 3 \times 3 \ \mu m^2$ *n-p-n* GaAs/In$_{0.11}$Ga$_{0.89}$-As$_{0.98}$N$_{0.02}$/GaAs DHBT. The curves corresponds to base current of 100 $\mu$A/step.[45]

*s*-parameters were measured and $f_T$ of 23 GHz and $f_{MAX}$ of 10 GHz were reported.[45]

Although the InGaAsN HBT work is in its early stages, a reasonable range of N incorporation has been reported with base bandgaps in the range of 1.0–1.4 eV for N composition in the range of 0.3% to 2%. Turn-on are voltages monotonically reduced as N incorporation is increased. Other figures of merit such as $\beta$, $f_{\mathrm{T}}$, and $f_{\mathrm{MAX}}$ monotonically degrade with N incorporation. At this point, optimum amounts of N tend towards the low to middle end of the range, or $< 1\%$ for reasonable tradeoff between turn-on voltage reduction and performance degradation. Whether these tradeoffs are favorable at this point is yet to be determined. Further understanding of the mechanisms of transport and their relation to point defects and other factors that also degrade photoluminescence efficiency are needed. Such an understanding may lead to improvements in material quality.

Welty *et al.* studied the minority carrier transport in *n-p-n* InGaAsN DHBTs with 2% N incorporation in the base.[45] By measuring the temperature dependence of the current gain and modeling it, they were able to gain insight into the current transport issues. They found that current gain increased with temperature in contrast to the usually observed degradation. The equation for current gain under high current density (where bulk recombination should dominate) is given as,

$$\beta = \frac{\tau_{\mathrm{rec}}}{\tau_{\mathrm{B}}} = 2D_n \frac{\tau_{\mathrm{rec}}}{w^2} \qquad (5)$$

where $\tau_{\mathrm{rec}}$ is the electron recombination time, $\tau_{\mathrm{B}}$ is the base transit time, $D_n$ is the minority carrier diffusion constant in the base, and $w$ is the base thickness. Since $\beta$ increases with temperature, either $D_n$ or $\tau_{\mathrm{rec}}$ is increasing with temperature. They were able to model a temperature dependent $D_n$ according to the following equations,

$$D_n = \frac{\mu k T}{q} \qquad (6)$$

where

$$\mu = \mu_o \exp\left[\frac{-E_{\mathrm{A}}}{kT}\right] \qquad (7)$$

is the minority carrier mobility. This dependence can be attributed to alloy fluctuations in the InGaAs that cause localized potentials with activation energy $E_{\mathrm{A}}$ in the base region. In a manner similar to hopping limited conductivity, electrons are trapped and detrapped while diffusing across the base. This process reduces $D_n$ as compared to GaAs and makes it highly temperature dependent. The current gain is plotted against the reciprocal temperature in Fig. 19 for three current densities.[45] A similar activation energy, $E_{\mathrm{A}}$, of 40 meV is extracted for each current density. Of course the model is only a rough guide unless the temperature dependence of $\tau_{\mathrm{rec}}$ is also accounted for. From the model, it is seen that $D_n$, and therefore $\beta$, is only 20% of the room temperature value of GaAs.

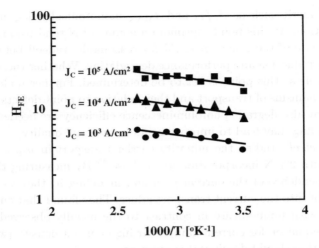

**Fig. 19.** Activation energy plot of a $2 \times 3 \times 3$ $\mu m^2$ $n$-$p$-$n$ GaAs/In$_{0.11}$Ga$_{0.89}$As$_{0.98}$-N$_{0.02}$/GaAs DHBT. The activation energy for this device is approximately 40 meV.[45]

Short minority carrier diffusion lengths were also observed in InGaAsN solar cells with 1–2% N.[29] The temperature dependence of the hall mobility data in that work were also suggestive of long range (greater than the mean free path) potential barriers and large scale material inhomogeneities were proposed as the cause.[29] However, direct observation of such inhomogeneities are lacking at this point. Also it is not clear if such potential barrier transport is limited to certain composition ranges or determined by unique growth conditions. $n$-$p$-$n$ InGaAsN based DHBTs with 1% N incorporation have not shown increasing current gain with temperature[46] as those reported by Welty *et al.*[45] Given that other types of point defects, such as those described in Sec. 2 of this chapter, can be recombination centers in InGaAsN, Eq. 4 offers other possibilities for low current gain in HBTs.

## 5. Outlook for InGaAsN HBTs

We will now make an attempt to summarize the prospects for future adoption of InGaAsN in GaAs-based HBTs. As a drop-in replacement for InGaP/GaAs HBTs, the InGaP/InGaAsN/GaAs DHBT is a promising candidate. In order to achieve the drop-in status, there must be no degradation in the current gain or the base sheet resistance. The RF performance must also be equivalent to that of InGaP/GaAs. Consequently, the N incorporation will be minimal and the reduction in turn-on voltage will only be of the order of 25–50 mV.[44] Such an improvement may not be that attractive since qualification and reliability must be established and they can lead to lengthy and costly procedures.

Another prospect is to use InGaP/InGaAsN/GaAs DHBTs with moderate N incorporation for the development of new products. A reduction of turn-on voltage of the order of 200–300 mV could be achieved. In this case there would be a tradeoff between the base sheet resistance and the $V_{ON}$ reduction. Due to the limitations of the base transport properties, the device would be aggressively engineered to thin the base or to apply an electric field in the base, or both, in order to achieve acceptable RF performance. SiGe is a good example where both of these techniques have been applied along with aggressive scaling of all device dimensions to achieve good RF performance for wireless and digital applications. It is fairly likely that these approaches would be successful for InGaAsN DHBTs if no manufacturability issues arise. However, the reduction in $V_{ON}$ is still fairly modest and InGaAsN development will compete with other alternatives for development funds.

The main alternatives are InP/InGaAs or InAlAs/InGaAs HBTs. These HBT technologies can compete with GaAs-based HBTs through cost reduction efforts. It is always possible that high volume markets will drive the cost of InP substrates and the InP/InGaAs HBT technology down to competitive levels with GaAs technology.

Another alternative is the development of metamorphic substrates for InAlAs/InGaAs HBTs. Metamorphic material has been successfully developed for InAlAs/InGaAs HEMTs. The metamorphic buffer is a quaternary (InAlGa)As alloy that is grown strain-relaxed on GaAs. The buffer is formed by several sequential "step graded" growths. The composition and lattice constant are increased in several steps and growth proceeds for each composition until the strain is relaxed. Misfit dislocations are formed when the layer is thick enough that strain can no longer be accommodated; they later converge and form vertical threading dislocations as growth is continued. However, many of the threading dislocations are annihilated at the interface between steps in composition (lattice constant) and they do not reach the extremely high levels that are often seen in the growth of other mismatched material, e.g. GaAs on Si or GaN on sapphire or SiC. Nevertheless, the HBT is a minority carrier device and threading dislocations are known to cause a failure mechanism known as the dark line defects.[47] At this writing, metamorphic HBTs are just beginning to be investigated[48] and little is known about their reliability.

Reliability is also an issue affecting the outlook for InGaAsN HBTs and no studies have yet been published. In order to gain some insight, we will review the degradation mechanisms of GaAs HBTs. Many different degradation mechanisms have been observed including midgap trap formation, hydrogen depassivation, dislocation propagation, contact degradation, contact spiking, and dopant diffusion.[47] Most of the degradation mechanisms in GaAs HBTs such as those involving hydrogen passivation, dislocation propagation, contact degradation, and contact spiking are likely to have the same outcomes for GaAs or InGaAsN HBTs, i.e. they can be controlled with proper design and manufacturing techniques. The InGaAsN HBT development will benefit from the collective knowledge gained for addressing these types of degradation in GaAs HBTs. However, these mechanisms are possibly more detrimental to InGaAsN HBTs until proven otherwise. Furthermore, the increased base transit times in InGaAsN HBTs will give incentives to aggressively reduce the base thickness beyond what is now common (about 70–100 nm) in commercial GaAs-based HBTs. Given that, contact degradation may become a renewed concern.

Two other degradation mechanisms, midgap trap formation and base dopant diffusion, can uniquely affect InGaAsN HBTs and should be considered in more detail. The dopant diffusion issue would mainly be a concern for Be doping by MBE. Problems with C diffusion have not been an issue under a wide variety of growth conditions. Because of the availability of C-doping for InGaAsN using both MOCVD and MBE, dopant diffusion should not be a fundamental concern for InGaAsN HBTs.

Midgap states are formed during bias stress of HBTs, light emitting diodes, and laser diodes. In the HBT, excess base leakage current is created due to a serial tunneling recombination process.[49] Electrons tunnel from the emitter into midgap level traps in the base and base-emitter depletion regions. Once in the traps, the electrons then recombine with the holes. The midgap states that make this leakage current possible are formed during the bias stress. The excess base leakage current results in $\beta$ degradation, leading to device failure.

The physical mechanism for the formation of midgap states can give insight to what may happen during long-term operation. The energy emitted by nonradiative recombination may be localized and transformed into lattice vibrations. These vibrations may result in low-temperature solid-state changes, including defect motion. Low temperature defect motion resulting from nonradiative recombination has been termed recombination enhanced defect reaction (REDR) and has been described by Kimmerling.[50]

The process has positive feedback such that the generation of a point defect associated with a non-radiative recombination center leads to more energy transfer to lattice vibrations, which can generate more point defects. Eventually, these point defects will join together to form larger defects and larger recombination centers which leads to the gradual degradation of the device. This process has been described for LEDs[51] and HBTs.[47] The implication for InGaAsN HBTs is that the process may start from an advanced level because the formation of many point defects already occurred during the growth process. On the other hand, the energy associated with each recombination event may be smaller due to the smaller bandgap of the InGaAsN. The relative importance of these issues will need to be sorted out by carefully designed stress tests (bias, temperature) in order to uncover the degradation mechanisms. From the information found in the stress testing, meaningful reliability tests can be made for InGaAsN HBTs.

From these thoughts, it can be seen that the InGaAsN HBT outlook depends very much on future developments in areas of active fundamental research.

## References

1. M. Kondow, K. Uomi, A. Niwa, T. Kitatani, S. Watahiki and Y. Yazawa, *Jpn. J. Appl. Phys.*, Part 1 **35** (2B), 1273 (1996).
2. H. Q. Hou, K. C. Reinhardt, S. R. Kurtz, J. M. Gee, A. A. Allerman, B. E. Hammons, P. C. Chang and E. D. Jones, *2nd World Conference and Exhibition on Photovoltaic Solar Energy Conversion* (Vienna, Austria, 1998) p. 3600.
3. N. Y. Li, P. C. Chang, A. G. Baca, X. M. Xie, P. R. Sharps and H. Q. Hou, *Electron. Lett.* **36**, 81 (2000).
4. M. Kondow, T. Kitatani, S. Nakatsuka, M. C. Larson, K. Nakahara, Y. Yazawa, M. Okai and K. Uomi, *IEEE J. Sel. Top. Quantum. Electron.* **3**, 719 (1997).
5. J. N. Baillargeon, K. Y. Cheng, G. E. Hofler, P. J. Pearah and K. C. Hsieh, *Appl. Phys. Lett.* **60**, 2540 (1992).
6. O. Igarashi, *Jpn. J. Appl. Phys.* **31**, 3791 (1992).
7. S. Miyoshi, H. Yaguchi, K. Onabe, R. Ito and Y. Shiraki, *Appl. Phys. Lett.* **63**, 3506 (1993).
8. M. Sato and M. Weyers, *Inst. Phys. Conf. Ser.* **129** (Philadelphia, Inst. of Physics, Ltd., 1993) p. 555.
9. N. Ohkouchi, S. Miyoshi, H. Taguchi, K. Onabe, Y. Shiraki and R. Ito, in *12th Alloy Semicond. Phys. Electron. Symp.* (Izu-Nagaoka, Japan, 1993) p. 337.
10. S. Sakai, Y. Ueta and Y. Terauchi, *Jpn. J. Appl. Phys.* **32**, 4413 (1993).
11. L. Bellaiche, S.-H. Wei and A. Zunger, *Phys. Rev.* **B54**, 17568 (1996).

12. J. D. Perkins, A. Mascarenhas, Y. Zhang, J. F. Geisz, D. J. Friedman, J. M. Olsen and S. R. Kurtz, *Phys. Rev. Lett.* **82**, 3312 (1999).
13. W. Shan, W. Walukiewicz, J. W. Ager III, E. E. Haller, J. F. Geisz, D. J. Friedman, J. M. Olsen and S. R. Kurtz, *Phys. Rev. Lett.* **82**, 1221 (1999).
14. E. D. Jones, N. A. Modine, A. A. Allerman, S. R. Kurtz, A. F. Wright, S. T. Tozer and X. Wei, *Phys. Rev.* **B60**, 4430 (1999).
15. M. Kondow, K. Uomi, K. Hosomi and T. Mozume, *Jpn. J. Appl. Phys.* **33**, L1056 (1994).
16. I.-H. Ho and G. B. Stringfellow, *J. Cryst. Growth* **178**, 1 (1997).
17. D. Schlenker, T. Miyamoto, Z. Pan, F. Koyama and K. Iga, *J. Cryst. Growth* **196**, 67 (1999).
18. C. W. Tu, *195th Electrochem. Soc. Meeting*, eds. C. R. Abernathy, A. Baca, D. N. Buckley, K. H. Chen, R. Kopf and R. E. Sah (Electrochemical Society PV99-4, Pennington, NJ, 1999) p. 250.
19. N. Ohkouchi, S. Miyoshi, H. Yaguchi, K. Onabe, Y. Shiraki and R. Ito, in *12th Alloy Semicond. Phys. Electron. Symp.* (Izu-Nagaoka, Japan, 1993) pp. 337–340.
20. N. Y. Li, P. R. Sharps, F. Newman, P. C. Chang, A. G. Baca, R. Kanjolia and H. Q. Hou, *200th Electrochem. Soc. Meeting*, eds. P. C. Chang, S. N. G. Chu, D. N. Buckley (Electrochemical Society PV2001-20, Pennington, NJ, 2001) p. 17.
21. E. V. K. Rao, A. Ougazzaden, Y. Le Bellego and M. Juhel, *Appl. Phys. Lett.* **72**, 1409 (1998).
22. H. P. Xin and C. W. Tu, *Appl. Phys. Lett.* **72**, 2442 (1998).
23. H. P. Xin, K. L. Kavanagh, M. Kondow and C. W. Tu, *J. Cryst. Growth* **201**, 419 (1999).
24. W. Lei, M. Pessa, T. Ahlgren and J. Decker, *Appl. Phys. Lett.* **79**, 1094 (2001).
25. S. G. Spruytte, C. W. Coldren, A. F. Marshall and J. S. Haris, *2000 MRS Spring Meeting, San Francisco* (Materials Research Society, Pittsburgh, 2000) K9-1.
26. H. Saito, T. Makimoto and N. Kobayashi, *J. Cryst. Growth* **195**, 414 (1998).
27. S. G. Spruytte, C. W. Coldren, A. F. Marshall, M. C. Larson and J. S. Harris, *197th Electrochem. Soc. Meeting*, eds R. E. Kopf, A. G. Baca and S. N. G. Chu (Electrochemical Society PV2000-18, Pennington, NJ, 2000) p. 195.
28. S. C. Spruytte, C. W. Coldren, J. S. Harris, W. Wampler, P. Krispin, K. Ploog and M. C. Larson, *J. Appl. Phys.* **89**, 4401 (2001).
29. S. R. Kurtz, A. A. Allerman, C. H. Seager, R. M. Sieg and E. D. Jones, *Appl. Phys. Lett.* **77**, 400 (2000).
30. P. C. Chang, A. G. Baca, N. Y. Li, P. R. Sharps, H. Q. Hou, J. R. Laroche and F. Ren, *Appl. Phys. Lett.* **76**, 2788 (2000).
31. P. C. Chang, A. G. Baca, N. Y. Li, X. M. Xie, H. Q. Hou and E. Armour, *Appl. Phys. Lett.* **76**, 2262 (2000).
32. R. E. Williams, *Gallium Arsenide Processing Techniques* (Artech House, Boston, 1984) p. 236.
33. L. C. Wang, P. H. Hao and B. J. Wu, *Appl. Phys. Lett.* **67**, 509 (1995).

34. A. G. Baca, J. C. Zolper, M. E. Sherwin, P. J. Robertson, R. J. Shul, A. J. Howard, D. J. Rieger and J. F. Klem, *IEEE GaAs IC Symp. Tech. Digest* (IEEE, Piscataway, NJ, 1994) p. 59.
35. D. B. Slater, P. M. Enquist, J. A. Hutchby, A. S. Morris and R. J. Trew, *IEEE Electron. Device. Lett.* **15**, 91 (1994).
36. W. Liu, *Handbook of III-V Heterojunction Bipolar Transistors* (Wiley Interscience, New York, 1998) pp. 192-193.
37. C. Monier, A. G. Baca, P. C. Chang, N. Y. Li, H. Q. Hou, F. Ren and S. J. Pearton, *Electron. Lett.* **37**, 198 (2001).
38. P. W. Li, N. Y. Li and H. Q. Hou, *Solid State Electron.* **44**, 1169 (2000).
39. A. G. Baca, P. C. Chang, N. Y. Li, H. Q. Hou, C. Monier, J. Laroche, F. Ren and S. J. Pearton, *IEEE GaAs IC Symp. Tech. Digest* (IEEE, Piscataway, NJ, 2000) p. 63.
40. P. C. Chang, N. Y. Li, A. G. Baca, H. Q. Hou, C. Monier, J. R. Laroche, F. Ren and S. J. Pearton, *IEEE Electron. Device. Lett.* **22**, 113 (2001).
41. K. T. Feng, L. Runshing, P. Canfield and W. Sun, *IEEE GaAs IC Symp. Tech. Digest* (IEEE, Piscataway, NJ, 2001) p. 273.
42. P. C. Chang, C. Monier, A. G. Baca, N. Y. Li, H. Q. Hou, E. Armour and S. Z. Sun, *Solid State Electron.* **46** (2002) to be published.
43. A. G. Baca, C. Monier, P. C. Chang, N. Y. Li, F. Newman, E. Armour and S. Z. Sun and H. Q. Hou, *IEEE GaAs IC Symp. Tech. Digest* (IEEE, Piscataway, NJ, 2001) p. 192.
44. R. E. Welser, P. M. DeLuca and N. Pan, *IEEE Electron. Device. Lett.* **21**, 554 (2000).
45. R. J. Welty, H. P. Xin, C. W. Tu and P. M. Asbek, *200th Electrochem. Soc. Meeting*, eds. P. C. Chang, S. N. G. Chu and D. N. Buckley (Electrochemical Society PV2001-20, Pennington, NJ, 2001) p. 52.
46. C. Monier and A. G. Baca, unpublished results.
47. T. Henderson, *Microelectron. Reliability* **39**, 1033 (1999).
48. H.-P. Hwang, J.-L. Shieh and J.-I. Chyi, *Solid State Electron.* **43**, 463 (1999).
49. T. Henderson *et al.*, *Int. Electron. Device. Mtg. Tech. Digest* (1994) p. 187.
50. L. C. Kimmerling, *Solid State Electron.* **22**, 1391 (1978).
51. O. Ueda, *J. Electrochem. Soc.* **135**, 11C (1988).

# CHAPTER 9

# ULTRAVIOLET PHOTODETECTORS BASED UPON III-N MATERIALS

Russell D. Dupuis and Joe C. Campbell

*Microelectronics Reserach Center*
*The University of Texas at Austin*
*Austin, TX 78758, USA*

## Contents

## 1. Introduction to Ultraviolet Photodetectors

While semiconductor-based photodetectors have been studied for many years,[1] the "state of the art" for UV detectors in the spectral range shorter than $\lambda \sim 400$ nm has not made many significant advances until recently. The analysis of emitted and absorbed photons in the ultraviolet region of the spectrum can be used to provide important information regarding many chemical, biological, and physical processes. In current applications, UV photons are often detected using specially designed Si photodetectors (typically $p$-$i$-$n$ diodes with special anti-reflection coatings and filters designed for UV wavelengths) or by photomultiplier tubes with specially fabricated (and typically expensive) UV filters. While such devices have been in use for many years, there are problems with each of these approaches and they are not suitable for all UV detection applications.

In recent years, the possibility of the analysis of light in the UV spectral region with compact, efficient photodetectors has generated a lot of interest in the application of the wide-bandgap semiconductors in the InAlGaN system, with direct bandgap energies ranging from $\sim 2.0$ eV to $\sim 6.2$ eV, for the development of high-performance UV-specific photodiodes and other types of photodetectors.[2,3] While the first such III-N detectors were photoconductors made of GaN,[4] subsequent work has concentrated on wider-bandgap $p$-$i$-$n$ diode photodetectors based upon $Al_xGa_{1-x}N$ alloys.[5] This chapter will review recent work in this area and discuss future directions for research in this field.

## 2. Applications for UV and Solar-Blind Photodiodes

In many applications for UV photodetectors on Earth, the Sun is NOT the "source" of the photons — it provides the "background" radiation that obscures the desired photon flux. Consequently, knowledge of the "local" solar spectrum is a critically important factor in the ability for the detection system to perform to the desired specifications. The "ultraviolet" spectral region is commonly referred to as containing the wavelengths shorter than $\lambda \sim 400$ nm (violet). The Earth's atmosphere strongly absorbs the UV emission from the Sun and modifies the "solar spectrum" at the Earth's surface. In general, the UV portion of the spectrum is often described in terms of three spectral bands: UVA: $\lambda = 400$–320 nm; UVB: $\lambda = 320$–280 nm; UVC: $\lambda = 280$ nm-shorter. Photons in the spectral range at wavelengths shorter than $\lambda \sim 290$ are strongly absorbed by the ozone layer. As a result of this absorption, the solar irradiance at $\lambda = 290$ nm is six orders of magnitude below that in the visible region of the spectrum. Consequently, the Sun contributes no significant energy at the Earth's surface in the spectral range shorter than $\sim 290$ nm and photodetectors with no significant spectral response for photons at wavelengths longer than 290 nm are called "solar-blind" photodetectors.[6] Such a photodetector would not produce a response at or near the Earth's surface even if it were exposed to bright sunlight. Obviously, a solar-blind photodetector located on Earth that picks up a signal is responding to an event that is not due to solar activity. Furthermore, because the UVC photons are strongly absorbed by the atmosphere, this event must have occurred relatively close to the detector's location. This makes solar-blind photodetectors useful in several important applications as discussed below. In a similar manner, photodetectors with no response for photons with wavelengths longer than $\sim 400$ nm are called "visible-blind" detectors. The "solar-blind" and "visible-blind" photodetectors are thus special classes of UV detectors, and both have important applications.

While Si-based UV-enhanced photodetectors are commercially available and are relatively cheap, the UV "solar-blind" performance of these devices is relatively poor. There are three major types of Si UV photodetectors: the *p-i-n* diode, the Schottky barrier diode and the inversion-layer diode. Si photodetectors have several inherent advantages in that they can be fabricated using a mature semiconductor material with advanced processing technology. In addition, they can be made very small, they can be made into large device arrays and they can be integrated with other Si-based VLSI devices. However, they suffer in their quantum efficiency in the UV since they have relatively large dark currents (in the nA range) and slow speed of response. Furthermore, for operation in the UVC range, the $Si$–$SiO_2$ interface degrades after long exposure times, creating a reliability problem.

## 2.1. *Commercial Applications*

Important commercial applications for UV photodetectors include flame sensors that can be used for monitoring the status of pilot lights, burner flames in large industrial furnaces and other applications for the control of high-temperature systems. In these applications, a solar-blind detector is of use because stray sunlight or artificial room lighting could cause a control system to falsely believe that the pilot light is ON when it is, in fact, OFF. Flame monitoring is a potentially important "mass market" if these systems could be made cheap enough for home use. Several companies are selling industrial versions of these types of devices, including APA Optics and Honeywell. Other applications include ozone and atmospheric pollution monitors. One particularly interesting application is the monitoring of the UV exposure for people working and playing outside. The commercially available SUN$^{UV}$ Watch[7] monitors the integrated UV photon flux to warn the wearer of potentially dangerous levels of UVA and UVB photon exposure.

## 2.2. *Military Applications*

There are many important applications for solar-blind UV photodetectors for the detection of missiles and other threats, and current systems are deployed using PMT-based systems.[2,8,9] The primary military applications are for missile and high-speed aircraft detection and for secure UV optical communications. In addition, hypervelocity projectiles exhibit a shock wave with a UV signature, and can therefore be identified with high-performance solar-blind detectors. In these applications, high sensitivity (and detectivity), low dark current, relatively high speed of response, robustness and thermal sensitivity can all be extremely important characteristics of solar-blind photodetectors. Furthermore, the "semiconductor solution" for UV SBD systems enables the development of UV imaging systems that can provide more information rather than to merely "track" a UV source. The III-N SBDs that are most useful for imaging systems are "back-illuminated" through the transparent sapphire substrate and are mounted in "flip-chip" form onto a specially designed Si integrated circuit with power and signal detection elements designed to mate with the SBD array. Such Si readout circuits (ROICs) have been used for many years in infrared (IR) imaging systems based upon HgCdTe and other detector materials. However, the large bandgap and corresponding ultra-low leakage current and low-noise characteristics of nitride *p-i-n* photodiodes have necessitated the development of a new generation of Si ROICs having even lower noise characteristics. Further discussions of device characteristics are given below in the device section.

The missile detection applications are extremely demanding, requiring ultra-low-noise solar-blind detectors. In many applications, due to the hypervelocity of the incoming projectile or missile, the response time is extremely short, making detection and tracking at low photon fluxes extremely critical. Since the standard tactic for engagements using missiles is to come at your opponent from the direction of the sun, a solar-blind detector is an important device to be used in the protection of the war-fighter and valuable platforms. For secure UV optical communications, the "point-to-point" solar-blind UV optical link is inherently secure due to the fact that atmospheric absorption will limit the ability for optical "access" to the low-level signal due to scattering or reflections.

High-sensitivity high-energy UV detectors are also important in bio-warfare agent detection. Many biological molecules of interest fluoresce in the UV spectral region and can be detected by using a laser-induced fluorescence (LIF) technique. LIF typically employs illumination of the bio-molecules of interest with a high-intensity UV photon pulse in the 260–290 nm spectral range that causes the bioagent to fluoresce in the 300–400 nm wavelength regime (i.e. an $\sim$ 50–100 nm Stokes shift). Since fluorescence processes are generally relatively slow, the speed of detection is not a critical issue. However, sensitivity (and detectivity) is extremely important. Small, efficient UV sources and high-performance UV photodetectors for compact, low-power optoelectronic detection systems are of extreme importance for many military field-deployed systems, as well as for "hand-held" biological warfare agent detection in domestic settings, e.g. in police, emergency medical, and fire department units.

## 2.3. *Medical Applications*

UV photodetectors are being developed for use in the important application of the detection of precancer. The American Cancer Society estimates that 1.2 million people were diagnosed with cancer in 1999 and in this population, there will be a resulting 563 000 deaths. Despite significant advances in the treatment and detection of cancer and its curable precursors, early detection remains the best way to ensure patient survival and improved quality of life. Thus, highly sensitive and cost-effective screening and diagnostic techniques for identification of curable pre-cancerous lesions are desperately needed. Early detection of curable pre-cancers has the potential to significantly lower cancer mortality and morbidity. Many visual exam procedures, such as colonoscopy and bronchoscopy, are routinely used to identify pre-malignant changes and early cancers. It has been discovered that by the use of cancer-specific "chemical tags" that the optical detection of precancer is possible. These tags are excited by light from a UV source and the fluorescence spectra of the tags are analyzed to identify the specific type of precancer. This approach would benefit greatly by the availability of compact, reliable, photodetectors in the UV and short-wavelength spectral range. Hence the development of III-N-based UV photodetectors could be of great significance in this application.

## 3. High-Performance UV Photodetectors

### 3.1. *UV Photodetectors*

Ultraviolet photodetectors fabricated from wide-bandgap semiconductors, e.g. GaN, are attractive alternatives to the well-developed vacuum-tube-based technology of photomultiplier tubes (PMTs). However, before III-N-based photodiodes can be considered viable replacements for PMTs in many applications, they must be able to achieve high sensitivity combined with low noise and reasonably high speed of response. PMTs are able to attain high responsivity owing to very high internal gain (typically $> 10^6$), but they are bulky, fragile, and require high bias voltages for optimum operation (typically $> 1000$ V). For certain applications, the limited reliability and excessive power requirements of PMTs make them undesirable; however, the low-noise front-end gain typically provided by the PMTs is very desirable. Consequently, it is of interest to study the performance limitations of semiconductor UV photodetectors. Early reports of GaN-based photodetectors focused on photoconductors and metal-semiconductor-metal (MSM) devices owing to their simplicity and potentially high internal gain.[4,10] Unfortunately, the photoconductive gain mechanism that is typically responsible for the high responsivity in such photodetectors can be a relatively slow and noisy process. Furthermore, these simple device structures are often characterized by relatively large dark current densities. Although some very acceptable performance from nitride-based MSMs has been reported, in general for MSM photodiodes there is a significant tradeoff among device gain, bandwidth, and noise.

An attractive alternative for MSM photodetectors operating at high gain and high speed is the avalanche photodiode (APD). However, the APD structure places severe requirements upon the epitaxial materials quality and it is a more complicated device to be reproduced. Because of the severe materials requirements for high-performance APDs, the development of III-N photodiodes having avalanche gain is only in its infancy. Nevertheless, APDs with useful gains have been demonstrated.[11,12]

### 3.2. *Solar-Blind Photodetectors*

The Group III-nitride compound semiconductors in the AlGaN system have recently been extensively studied largely as the result of the rapid progress in the use of metalorganic chemical vapor deposition (MOCVD) for the growth of high-performance visible light-emitting diodes and lasers. However, as discussed above, another important application of the III-N materials includes the AlGaN-based ultraviolet and solar-blind detectors (SBDs).[13-15] $Al_xGa_{1-x}N$ compounds are the most suitable candidates for truly solar-blind UV photodetectors, especially in airborne and space applications, because they do not require heavy and expensive filtering systems for long wavelengths and are simple to fabricate. Many detailed studies of the growth and fabrication of low Al-content ternary compound AlGaN photodetectors have been reported.[16-18] The challenge with the fabrication of true SBDs comes from the fact that AlGaN alloys having a high Al mole fraction are required. Indeed, it has been determined by theoretical work that back-illuminated SBDs should contain $Al_{0.60}Ga_{0.40}N$ material as a window layer and $Al_{0.45}Ga_{0.55}N$ material as an absorbing layer. The main difficulties for such devices are cracking problems due to the substantial lattice mismatch between $Al_{0.6}Ga_{0.4}N$, $Al_{0.4}Ga_{0.6}N$, and GaN,[19] and also $p$-type doping difficulties due to a large activation energy for the ionization of Mg acceptors with high Al-content AlGaN alloys.[20]

## 4. III-V Nitride Materials Growth by Chemical Metalorganic Vapor Deposition

One of the most successful epitaxial growth techniques for devices based on semiconductors, composed of elements from Groups III and V, has been the metalorganic chemical vapor deposition (MOCVD) technique.[21,22] This vapor-phase epitaxial growth process was first experimentally developed by Manasevit[21] in 1968 and first extended to the III-N materials by Manasevit *et al.*[23] in 1971. The first high-performance III-V devices grown by MOCVD were demonstrated by Dupuis *et al.* in 1977.[24,25] The MOCVD process was subsequently extended and improved for AlGaN growth by Amano *et al.* in 1986,[26] and by Nakamura.[27] In general, the MOCVD process involves sequences of complex reactions between the particular precursors that are employed. Suitable precursors in MOCVD growth must exhibit sufficient volatility and stability to be conveniently transported to the surface of the growing crystal. Moreover, these precursors should also have appropriate reactivity to decompose thermally into the desired solid and to generate readily removable gaseous by-products. Thus, ideally the precursors should be stable and have a reasonably high vapor pressure near room temperature (e.g. 1–100 Torr).They should also be non-pyrophoric, insensitive to water and oxygen, non-corrosive, and non-toxic. While there are many options to deliver the desired atoms, trimethylgallium (TMGa), trimethylaluminium (TMAl) and trimethylindium (TMIn) are the most commonly used Column

III precursors for the MOCVD growth of III-V nitrides. For the Column V precursor, the hydride $NH_3$ is most commonly used; however, a few groups have used hydrazine[28] or microwave-activated nitrogen[29] in their experimental growth of nitride films. Purified $H_2$ is generally used as a carrier gas during MOCVD epitaxial growth, except for alloys containing In. When In precursors are incorporated in the growth process, $N_2$ carrier gas is used rather than $H_2$, because it increases the incorporation of In, which tends to desorb easily from the substrate. The overall chemical reaction used to form Group III-nitride compounds in MOCVD growth, can be generally described by Eq. (1).[30]

$$R_3M + NH_3 \Rightarrow MN + 3RH \qquad (1)$$

In this equation, R is an alkyl group, e.g. a methyl ($CH_3$) or ethyl ($C_2H_5$) radical, and M is the Group III metal such as gallium (Ga), indium (In), or aluminium (Al). Of course, in reality, many competing and intermediate reactions such as various adduct formation and carrier gas interactions reduce the overall growth efficiency for the MN epitaxial film.[31]

### 4.1. *Sources for MOCVD Growth of III-N Materials*

The common Group III metalorganic sources that are used for MOCVD growth exist in either a liquid or a solid form at or near room temperature. These precursor sources are typically highly purified before use. Typical sources are trimethylgallium (TMGa), trimethylaluminium (TMAl) and trimethylindium (TMIn). Some workers report the use of triethylgallium (TEGa) and triethylaluminium (TEAl). Dopant sources can be in hydride form or metalorganic form, e.g. silane gas ($SiH_4$) for *n*-type Si doping and (bis)cyclopentadienylmagnesium ($Cp_2Mg$) for Mg for *p*-type doping. Since oxygen is known to be a donor in III-N films, low-oxygen containing metal alkyls are preferred for III-N growth. Oxygen is also known to foster the incorporation of unwanted C impurities. The metalorganic precursors are commercially available in stainless-steel vessels (or bubblers) that are mounted on the MOCVD reactors. The newer large-scale production MOCVD reactors can accommodate metal alkyl bubblers with capacities of $\sim 1$ kg. Recently, an automatic filling system has been developed for the bubblers containing liquid-phase metal alkyls, permitting production MOCVD reactors to be fitted with over 20 kg of source materials.[32] This is especially important for high-volume production of III-N LEDs, which typically consume large quantities of TMGa and, to a lesser extent, TMAl. Electronically controlled flow rates (volume per unit time) of a carrier gas, e.g. purified $H_2$ or $N_2$, flow though the source cylinders and transport the metalorganic precursor vapors into the reaction chamber. The Group V precursor, on the other hand, comes from a high-purity liquid $NH_3$ source stored remotely in a high-pressure gas cylinder. Since high thermal energies are required to crack $NH_3$, a heater system is incorporated inside the chamber to heat the substrates to temperatures above 1000°C.

## 4.2. *Choice of Substrates for UV Photodetectors*

As a result of a current lack of commercially available "bulk" lattice-matched substrates for the growth of nitride-related materials, basal plane (*c*-plane) $\alpha$-$Al_2O_3$ has become the substrate of choice for many III-N device applications because of wide availability, hexagonal symmetry, elevated temperature stability and low cost. The choice of sapphire over SiC substrates for the solar-blind application is also made due to the transparency of the substrate to UV radiation, permitting a backside-illuminated device geometry. Nevertheless, substantial lattice and thermal mismatch exist between sapphire and wurtzite GaN, creating difficult problems for the direct nucleation of high-quality heteroepitaxial materials on sapphire. Consequently, a low-temperature AlN or GaN buffer layer has to be grown directly on the sapphire in order to promote the formation of a relatively good quality nitride-based crystal. The insertion of the quasi-crystalline buffer layer between the mismatched sapphire substrate and the epitaxial films reduces the interfacial free energy that would exist for this combination of materials. This nucleation technique of using a low-temperature AlN (LT) buffer layer was first investigated and developed by Amano and Akasaki,[26,33,34] and later a GaN LT buffer was developed by Nakamura and others.[27,35]

## 4.3. *MOCVD Growth of UV Photodetectors*

Since most nitride-based materials are grown on substrates with a fairly large lattice and thermal expansion coefficient mismatch relative to the III-N semiconductor, many threading dislocations are formed in the heteroepitaxial layers during the growth as well as during the cool down of the wafer from the growth temperature. This arises from the fact that the epitaxial film relaxes when its thickness becomes greater than the critical layer thickness of the material under strain. To accommodate the lattice mismatch between the growing GaN film and the substrate, three-dimensional island growth is first initiated on the top of the LT buffer layer. As the growth proceeds, the islands become bigger and eventually coalesce. The intersection between the islands generally becomes the origin of threading dislocations. The threading dislocations are mainly responsible for the formation of an $n$-type background carrier concentration typically observed for unintentionally doped GaN (GaN:ud). This $n$-type background carrier concentration has also been correlated with impurities in the precursors,[36,37] as well as with a variety of native defects caused by the non-ideal growth conditions.[38-40] The typical net electron background concentration at 300 K has been measured to be $n \sim 5 \times 10^{16}$ cm$^{-3}$ for our thick ($\sim 3$ $\mu$m) unintentionally doped GaN heteroepitaxial material. Of course, this electron background concentration can be controlled to a certain extent by altering the buffer layer growth conditions. Thus, the change of the growth rate, the temperature, the pressure, the total thickness, and the V/III ratio for the buffer layer and for the epitaxial film enable us to reduce the free-electron background concentration to $< 10^{15}$ cm$^{-3}$. However, a net electron background concentration as low as $\sim 10^{15}$ cm$^{-3}$ is not always desirable because this low concentration is achieved by exploiting a carrier-compensation mechanism related to deep trap states. Thus, the (102) X-ray full width at half maximum (FWHM) increases for materials with low electron background concentration because of the formation of acceptor-like defects. Therefore, a free electron background concentration of $n \sim 5 \times 10^{16}$ cm$^{-3}$ is typical for good quality nitride-based material that has a reduced number of defects, while "insulating" GaN generally has a much higher dislocation density.

As a result of the fact that the initial buffer layer (BL) plays a key role in the growth of high-quality Group III-nitride material for heteroepitaxial growth on both sapphire and SiC substrates, optimization of this layer is critically important. Several parameters during the growth of the BL have a drastic influence on the mobility. This includes the X-ray rocking curve full width at half maximum (FWHM), the defect density, the mobility, the free-carrier lifetime, and the luminescence efficiency of the subsequent films grown on top of it.[41,42] Among these parameters are the surface preparation of the wafer, the nitridation of the wafer before growth, the growth temperature, the growth pressure, the growth rate, the III to V ratio, and the BL thickness.

The surface preparation for sapphire substrates typically consists of surface cleaning with wet-chemical etches, followed by a thermal clean in the reactor chamber prior to the growth of nitride layers. A schematic drawing of the temperature-time cycle for the growth of III-N material on sapphire is displayed in Fig. 1. As shown here, prior to growth, the substrates are

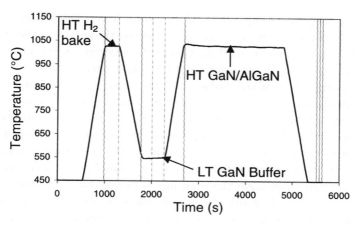

**Fig. 1.** Typical temperature profile for the growth of nitride-based material on Al$_2$O$_3$.[43]

heated in H$_2$ and brought to high temperature. Then, the substrates are baked for $\sim$ 5–15 minutes in H$_2$ at 1050°C. The purpose of the bake step is to slightly etch the sapphire wafer in such a way that the subsequent growth occurs on a clean substrate. After the bake, the wafers are cooled down in H$_2$ to the BL growth temperature of about 530°C. In some reports, improved growth has been obtained by employing an *in-situ* surface preparation or "nitridation" of the sapphire surface immediately prior to the BL growth when the surface of the sapphire is exposed to NH$_3$ and H$_2$ at high temperatures.

Once the nitridation step is completed (if it is used) and the temperature is stabilized at the optimal BL growth temperature and pressure, TMGa (or TMAl) is injected into the chamber and the growth of the GaN (AlN) BL starts. To provide the required transparency for backside-illuminated SBDs, a LT AlN buffer layer is used. The parameter space for the growth of BLs includes temperature, pressure, growth rate, III-V ratio and thickness. The optimization of the BL is obtained by varying one of these parameters around a reasonable value, keeping all the other parameters constant. Then, the quality of the BL is evaluated by analyzing the quality of a thick high-temperature (HT) AlGaN:Si layer grown on top of it. Characterization tools such as atomic force microscopy (AFM), X-ray diffraction (XRD), Hall effect (HE), photoluminescence (PL), time-resolved PL (TRPL), and transmission electron microscopy (TEM) give useful information that help to determine the choice of the right parameters for the BL growth.

For a low-growth temperature, the adatoms have a small kinetic energy and have a corresponding low surface mobility on the surface of the epitaxial layer. As a result, only a small fraction of the adatoms on the surface have enough mobility to locate appropriate bonding sites. Therefore, most of the atoms desorb from the surface and the growth rate remains low and highly dependant on the temperature. In addition, at low temperatures, the growth is limited by the cracking of $NH_3$. As the growth temperature is increased, the adatoms gain kinetic energy and therefore gain mobility on the surface of the crystal. Consequently, more appropriate bonding sites are found and the growth rate increases. At about 530°C, the temperature is high enough to provide the adatoms sufficient kinetic energy to find an appropriate bonding site. Of course, there is always a probability of desorption, but it is no longer the limiting mechanism for the reaction. Consequently, for temperatures greater than $\sim$ 530°C, the growth rate is only limited by the mass transport rate of the precursors on the surface of the wafer. Therefore, the growth rate no longer depends on the temperature, but depends linearly on the TMGa or TMAl flow, as shown Fig. 2. In this figure, the transition from the kinetic energy-limited regime to the mass transport-limited regime can be observed by plotting the BL growth rate against the growth temperature.

The thickness of the BL is a very important parameter that needs to be controlled as accurately as possible. Growing the BL at 530°C near the transition between the kinetic-energy-limited regime and the mass-transport-limited regime allows a better control of the uniformity of the

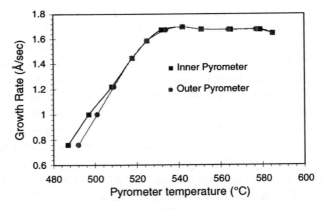

**Fig. 2.** Growth rate dependence of the buffer layer upon growth temperature measured by pyrometers located at the inner and outer regions of the wafer.[43]

layer. In this case, slight variations of the temperature across the wafer will not affect the growth rate. Therefore, the BL thickness is relatively uniform across the whole wafer. It is crucial to control the buffer layer thickness with a high accuracy, because it has been shown previously that the best material quality for GaN is obtained with a BL between 21 and 25 nm thick.[41,42] In our case, 530°C is the lowest temperature that allows for the growth of a uniform BL. Varying the total thickness of the BL results in a dramatic change in the quality of the subsequent high-temperature GaN layers.[44] As noted above, it is sometimes desirable to use an AlN BL to benefit from the optical transparency of AlN for increased light transmission through the backside of the wafer. For that reason, an optimum LT AlN BL has been developed for the growth of SBDs on sapphire substrates in the same way as for the GaN BL.

Once the buffer layer growth is completed, the temperature is ramped up quickly to about 1030–1070°C in a $H_2 + NH_3$ ambient. During this growth phase, small GaN or AlN crystallites are formed on the surface of the BL in a "recrystallization" process. As soon as the temperature reaches 1030–1070°C, the growth of high-quality high-temperature (HT) nitride-based material can begin. At that point, the crystallites act as nucleation sites and three-dimensional islands begin to form. The islands eventually coalesce to form a smooth and continuous layer.

The two-step growth technique can be used to optimize the optical and electrical characteristics of the materials used to form the SBDs. The first set of growth conditions involves the low-temperature ($\sim$ 550°C) growth of a pseudomorphic $\sim$ 20 nm thick AlN buffer layer (BL). Then the temperature is ramped to $\sim$ 1080°C for the growth of the undoped and silicon-doped AlGaN layers. The first high-temperature layer grown is a $\sim$ 700 nm-thick $Al_{0.6}Ga_{0.4}N$:ud epitaxial layer to improve the material quality for the subsequent device layers through the reduction of the defect density. In addition, the $Al_{0.6}Ga_{0.4}N$ layer, as well as the AlN BL, plays the role of an optical window for radiation in the $\lambda = 240$ to 280 nm solar-blind wavelength band. After the thick $Al_{0.6}Ga_{0.4}N$ "window" layer is grown, the Al composition, $x$, is graded from $x = 0.6$ to 0.4 over $\sim$ 36 nm to allow for the growth of the AlGaN $p$-$i$-$n$ photodiode structure, which typically consists of a 100–200 nm-thick $n$-type $Al_xGa_{1-x}N$ cathode, a 150–200 nm-thick $Al_{0.4}Ga_{0.6}N$ intrinsic region ($i$-layer) and a 100–200 nm-thick $p$-type $Al_{0.4}Ga_{0.6}N$ anode. The device structure is then completed with the growth of an AlGaN layer in which the composition is graded from $x = 0.4$ to 0.0 over 10 nm to allow for the growth of a 10 nm thin, heavily-doped $p^+$ GaN:Mg

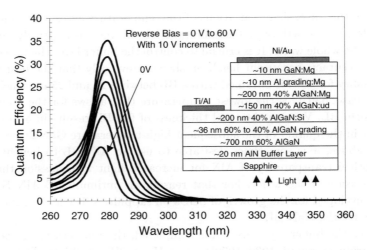

**Fig. 3.** Device structure and spectral response for back-illuminated solar-blind photodetector composed of an AlGaN $x = 0.45$ $p$-$i$-$n$ diode grown on an AlGaN $x = 0.60$ window layer.[5]

contact layer. Figure 3 illustrates the complete structure of an early SBD after processing.[5] The UV photoresponse of such a "first-generation" SBD is also shown; these early SBDs exhibit zero-bias quantum efficiencies of $\eta_{\text{ext}} \sim 12\%$ at 278 nm.

*In-situ* monitoring of the reflectivity of the wafer vs. time with an Epimetric® system is used to maintain a growth-rate of $\sim 74$ Å/min and $\sim 100$ Å/min for $Al_{0.6}Ga_{0.4}N$ and $Al_{0.4}Ga_{0.6}N$, respectively. After the growth, the surface morphology is examined using a Digital Instruments Dimension 3000 AFM. The surfaces are relatively smooth with $\sim 0.640$ nm RMS roughness on a $1 \times 1 \mu m^2$ area. This indicates the relatively high quality of the surfaces of the epitaxial layers considering the high concentrations of Al and Mg in the films, and the lack of a thick GaN "template" layer.

The high Mg doping concentrations used for the $p$-type regions of the device are indicated in the SIMS profile shown in Fig. 4. In this figure, the vertical lines indicate the expected location of the $p$-$i$ and $i$-$n$ junctions. The large concentration of Mg atoms and the relatively high density of defects in the AlGaN films are probably responsible for the shape of the impurity profile observed for the back-diffusion of Mg atoms into the unintentionally doped $i$-type material. This effect is not desirable because it hampers the formation of an abrupt $p$-$i$ junction. For optimum performance, the photodiode structure is intended to absorb the desired radiation into

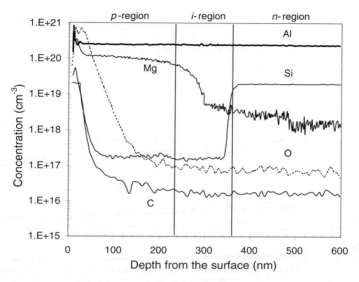

**Fig. 4.** SIMS data taken on an AlGaN SBD wafer. Note the relatively low [O] for the $Al_xGa_{1-x}N$ : Si ($x = 0.45$) region (at approximately at the detection limit for this scan).[5]

the $i$-layer. Consequently, one issue that arises in the growth of SBDs is the location and thickness of the $i$-layer of the $p$-$i$-$n$ junction. The presence of a large concentration of dislocations can contribute to the rapid diffusion of impurities, as shown by the vertical markers in Fig. 4. Obviously, Mg atoms can diffuse through a relatively long distance in the crystal during the growth of the $p$-layer. The resulting diffusion profile does not fit the conventional "error-function" type of diffusion because the diffusion kinetics are much different from that of normal diffusion of impurities through vacancies or interstitials. While the diffusion effects are probably enhanced by the large solid-phase concentration of Mg impurities, this high level of Mg doping is required for nitride-based solar-blind $p$-$i$-$n$ photodetectors, due to the problem of low activation of Mg impurity atoms for AlGaN with a large Al content. It has been observed that the activation energy for the ionization of Mg acceptors in AlGaN increases by $\sim 3.2$ meV for each 1% increase of the Al content in the alloy.[20] Thus, for $Al_{0.4}Ga_{0.6}N$, the Mg acceptor activation energy is estimated to be $E_A \sim 36$ meV. Consequently, only $\sim 0.007\%$ Mg acceptors are expected to be ionized for $Al_{0.4}Ga_{0.6}N$ at 300 K. Therefore inserting up to $10^{20}$ Mg atoms/cm$^3$ in $Al_{0.4}Ga_{0.6}N$ should lead to the formation of lightly doped $p$-type material with a free hole concentration of $p \sim 7 \times 10^{15}$ cm$^{-3}$ at 300 K. If the donor-like behavior of native defects, impurities, and N vacancies in the material does not fully compensate this low hole concentration, there is a chance in obtaining $p$-type Al$_{0.4}$Ga$_{0.6}$N at the cost of the requirement of a high concentration of Mg atoms.

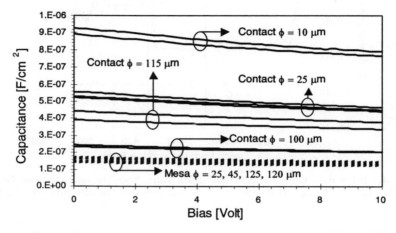

**Fig. 5.** Comparison of capacitance per unit area scaled with the AlGaN SBD mesa diameter and Ohmic contact diameter. The good agreement when scaled with mesa diameter indicates that these devices contain a $p$-$i$-$n$ photodetector structure.[5]

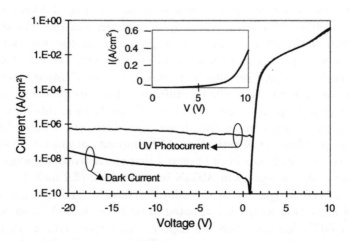

**Fig. 6.** Forward- and reverse-bias $I$–$V$ characteristics in the dark and under UV illumination of an "early" AlGaN back-illuminated SBD.[5]

As discussed above, since the Mg acceptor states are so deep, the actual hole concentration in the AlGaN films is expected to be small at room temperature. Furthermore, due to the fact that the Ohmic contacts to this material are quite resistive, and the AlGaN:Mg film is itself very resistive, conventional Hall-effect analysis of the carrier type and concentrations cannot be made. Lambert *et al.* studied the question of Mg doping in SBDs having $Al_xGa_{1-x}N$ $x = 0.45$ *p-i-n* structures.[5] To study the effectiveness of the *p*-type doping, these authors studied the dependence of the capacitance per unit area if circular-mesa-geometry SBDs were scaled with both the mesa area and the *p*-type Ohmic contact area (see Fig. 5).[5] It was determined that the capacitance/area scaled with the mesa area, not the contact area, which supported the claim that the device was a true *p-i-n* diode and not a form of Schottky-barrier photodetector. The "dark" and "light" *I–V* curves and UV spectral response for such an AlGaN SBD are shown in Fig. 6. The combination of the *I–V* curves and capacitance vs. area data indicates with high probability that the detector is performing like a *p-i-n*, as expected.

The (004) $\omega$–$2\theta$ X-ray scan of the as-grown SBD device structure of Fig. 3 is shown in Fig. 7.[5] The relatively small linewidths of the diffraction peaks indicate that the material quality is relatively good for both the $Al_{0.6}Ga_{0.4}N$ and $Al_{0.4}Ga_{0.6}N$ epitaxial layers. The (004) full width at half maximum (FWHM) is $\sim$ 449 arcseconds and $\sim$ 384 arcseconds for

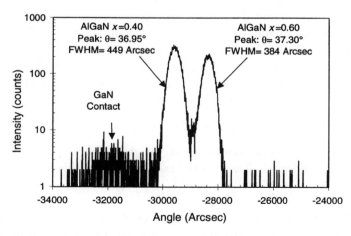

**Fig. 7.** (004) $\omega$–$2\theta$ X-ray diffraction scan for $Al_xGa_{1-x}N$ ($x = 0.45$)–$Al_yGa_{1-y}N$ ($y = 0.60$) SBD heterostructure showing peaks due to the "*c*-axis" lattice constant for both layers.[5]

$Al_{0.4}Ga_{0.6}N$ and $Al_{0.6}Ga_{0.4}N$, respectively. The alloy composition for the AlGaN films are deduced from the (004) peak position located at $\theta = 36.95°$ for $x = 0.40$ AlGaN and at $\theta = 37.30°$ for $x = 0.60$ AlGaN. The alloy compositions are calculated using the (0012) $Al_2O_3$ with the reference peak located at $\theta = 45.18°$.

## 5. UV Photodetector Device Design and Performance

### 5.1. *Device Processing and Testing*

The performance of photodetectors can be evaluated in terms of three primary characteristics: responsivity (or quantum efficiency), noise, and bandwidth. The responsivity is, perhaps, the most fundamental parameter in that it is a measure of how well the photodiode converts the incident optical signal into current. In an ideal semiconductor photodetector, each incident photon would result in the charge of one electron flowing in the external circuit, which would yield an external quantum efficiency of 100%. In practice, there are several physical effects, such as incomplete absorption, recombination through defect sites, reflection from the semiconductor surface, and contact shadowing that tend to reduce the responsivity. For solar-blind applications, the spectral shape of the responsivity curve is particularly important in that it is desirable to have high quantum efficiency in the wavelength range of 240 nm $< \lambda <$ 280 nm with a sharp cutoff toward the visible (long-wavelength) side. Typically, these measurements are made with a broadband UV source (e.g. a Xenon lamp), a monochrometer, which serves as a tunable filter, and a lock-in amplifier with a chopper. The input signal can be calibrated with a commercial reference photodetector. It is also helpful to incorporate UV optics to minimize signal loss.

The primary components of the noise are the shot noise and the thermal noise. The shot noise is given by the expression

$$i_{shot}^2 = 2q(i_{ph} + i_d)\Delta f$$

where $i_{ph}$ is the photocurrent, $i_d$ is the dark current, and $\Delta f$ is the observation bandwidth. The thermal noise current can be written as

$$i_{thermal}^2 = 4kT\Delta f/R$$

where $R$ is the resistance. For the AlGaN/GaN photodiodes, the dark current is usually sufficiently low such that the thermal noise dominates. Figures of merit for the signal to noise performance for a photodetector are the noise equivalent power, NEP, and the detectivity, $D^*$. The noise equivalent power is the input power that yields a signal-to-noise ratio of unity and $D^*$ is defined as $D^* = \sqrt{A\Delta f}/\text{NEP}$, where $A$ is the device area. When the thermal noise is larger than the background radiation, $D^*$ is given by

$$D^* = \text{Responsivity} \times \sqrt{R_0 A/kT}$$

where $R_0$ is the dynamic resistance of the photodetector. This is the detectivity that will be referred to in later sections. $R_0$ can be determined by computing the slope of the dark $I$–$V$ characteristic at the operating voltage (frequently, at 0 V bias) and the noise current can be measured with a low-noise preamplifier and a Fourier-transform spectrum analyzer.

For a photodetector in which carrier diffusion has been minimized, the bandwidth is determined by the RC time constant, capture and emission times associated with deep-level traps, and the time it takes for the carriers to transit the depletion region. For most imaging applications, high-speed performance is not essential, however, temporal response measurements can provide important information about the material quality, specifically, the density and nature of trap centers.

## 5.2. *Photoconductors*

Photoconductors were among the first III-N photodetectors studied because of their simple structures and their very high responsivities ($>$ $10^3$ A/W).[45–53] Typically, an MSM structure consists of interdigitated Ohmic contacts. However, the difficulty of achieving pure Ohmic contacts in the III-N materials, particularly AlGaN, has frequently resulted in devices that are a hybrid between a true photoconductor and back-to-back Schottky barriers. Figure 8 shows the responsivity of three $Al_xGa_{1-x}N$ photoconductors with different Al alloy compositions.[54] It is clear that the absorption edge can be shifted by changing the Al concentration in the photoconducting layer. However, all three samples exhibited strong responses to below-bandgap radiation. The UV/visible rejection was $< 100$. While high optical gains have been achieved, a sublinear dependence of the photocurrent on the incident optical power has also been reported[55]; the DC responsivity decreased with optical power as $P^{-0.9}$. Essentially, all of the III-N photoconductors that have been reported to date have exhibited a persistent photocurrent, i.e. the conductivity continues for minutes to hours after the optical excitation has been removed. The conventional explanation for persistent photocurrent invokes long hole trapping time constants, which results in a minority recombination time much longer than

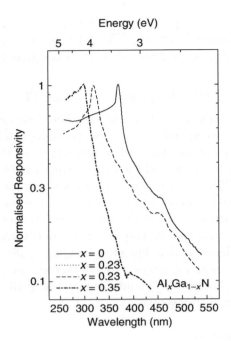

**Fig. 8.** Responsivity of $Al_xGa_{1-x}N$ photoconductors with $x = 0$, 0.23, and 0.35.[54]

the electron transit time.[48,51,52] For the III-N materials, the persistent photocurrent and the sublinear dependence of the response on optical power have been successfully modeled by taking into account the modulation of the device effective conduction area by band-bending in the vicinity of dislocations. This introduces a spatial variation of carriers.[56] It has been shown that sharper spectral cutoffs and a significant reduction of the persistent photocurrent can be achieved by low-frequency ($< 700$ Hz) modulation of the input signal. However, this also results in decreased responsivity.[56] The combination of slow photoresponse (i.e. persistent photoconductivity), sublinear power response, and below-bandgap responsivity has eliminated III-N photoconductors from consideration in many applications, but they may still be viable candidates for low-cost monitoring such as flame and fire detection.

### 5.3. *Metal-Semiconductor-Metal Photodiodes*

The metal-semiconductor-metal (MSM) photodiode is a planar device consisting of two interdigitated electrodes on a semiconductor surface. It achieved prominence for fiber optic applications because it is well suited for integration with FET preamplifier circuits.[57] MSMs have numerous positive attributes, including the potential for high quantum efficiency, low internal gain, and high speed that make them attractive for many other applications including those in solar-blind photodetectors. A principle advantage is the simplicity of the planar contact configuration, which facilitates fabrication and does not add significantly to the circuit complexity. Typically, the MSM contact structure is characterized by a very low capacitance per unit area. A particular advantage for solar-blind photodetectors is that MSMs require only a single-dopant (typically, $n$-type) active layer. The two potential limiting factors in the performance of MSMs are the dark current and, for top-illumination, shadowing of the optical active region by the electrodes.

Carrano *et al.*[58] reported on the material, electrical, and optical properties of MSM ultraviolet photodetectors that were fabricated on single-crystal GaN, with active layers of 1.5 $\mu$m and 4.0 $\mu$m thickness. It was found that trap-related processes influenced both the electrical and optical characteristics of the devices. The traps in the GaN are related to a combination of surface defects (possibly threading dislocations), and deep-level bulk states that are within a tunneling distance of the interface. The measured dark current at 300 K was as low as $\sim 800$ fA at $-10$ V reverse bias. Using thermionic field emission theory for the 1.5 $\mu$m devices and thermionic emission theory for the 4.0 $\mu$m devices, good fits to the experimental data were achieved. The external responsivity (quantum efficiency) of the 4.0 $\mu$m-thick MSMs with 2 $\mu$m finger width and 10 $\mu$m finger spacing was $\sim 0.15$ A/W ($\eta_{ext} \sim 50\%$). These were reported to be the first GaN MSMs that exhibited high $\eta_{ext}$ without internal gain due to trapping effects. For photon energies greater than the bandgap, the responsivity was flat as shown in Fig. 8.[53] There was a sharp visible-blind cutoff of more than three orders of magnitude at the bandedge energy. As a function of voltage, the ultraviolet photoresponse $I$–$V$ curves of the 4.0 $\mu$m-thick MSMs were flat out to $V_R > -25$ V (see Fig. 9), which indicated an absence of photoconductive gain. These photodiodes also exhibited excellent temporal response.[59] For the 2 $\mu$m finger spacing the 10%–90% rise time was $\sim 23$ ps at 10 V. Devices with 10 $\mu$m finger spacing had rise times approximately double that of the 2 $\mu$m devices.

Walker *et al.*[60] reported similar MSMs with different layer thickness and contact geometry fabricated from unintentionally doped GaN:ud and GaN:Mg. While both the GaN:ud and GaN:Mg devices exhibited visible-blind response, the cutoff in the wavelength range of 350 nm <

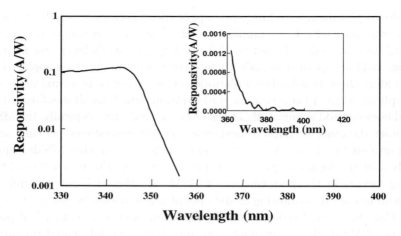

**Fig. 9.** Responsivity of a GaN MSM photodetector versus wavelength.[58]

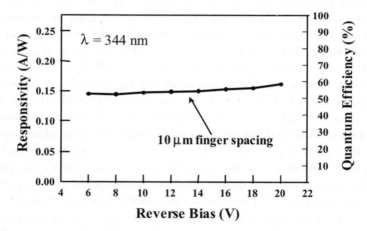

**Fig. 10.** Responsivity of GaN MSM versus bias voltage. The "flat" response is consistent with the absence of photoconductive gain.[58]

$\lambda < 450$ nm was much more abrupt for the GaN:ud devices. It was asserted that the more gradual fall off for the GaN:Mg devices was due to a high density of Mg trap states below the bandgap. The GaN:ud devices also exhibited lower dark currents, $\sim 2$ nA at 10 V, as compared to $\sim 12$ nA at 10 V for the GaN:Mg MSMs. The response times were estimated to be 10 ns and 200 ns for GaN:ud and GaN:Mg, respectively. The noise power spectral density was $< 10^{-24}$ A$^2$/Hz for bias levels up to 5 V.

In order to shift the cutoff wavelength to achieve a solar-blind response, MSMs have been fabricated from unintentionally doped $Al_xGa_{1-x}N$ layers grown on sapphire substrates. Monroy *et al.*[61] reported a cutoff at $\lambda = 310$ nm for $Al_{0.25}Ga_{0.75}N$ MSMs with Ni/Au Schottky contacts. While the responsivity was lower than that of the GaN MSMs ($\sim 0.3$ mA/W), it was flat for wavelengths of $\lambda < 310$ nm and the cutoff was abrupt. The dark current was $< 300$ nA at 40 V with a noise equivalent power of $2.4 \times 10^{-13}$ W/Hz$^{1/2}$. Yang *et al.* reported truly solar-blind AlGaN MSMs with back illumination through the sapphire substrate.[62] The epitaxial structure consisted of an AlN buffer layer ($\sim 25$ nm thick), an 0.8 $\mu$m-thick $Al_{0.6}Ga_{0.4}N$ "window" layer, and the 0.2 $\mu$m-thick $Al_{0.45}Ga_{0.55}N$ active layer, as shown schematically in Fig. 11.[62] The dark current of 40 $\mu$m $\times$ 40 $\mu$m devices was $\leq 20$ fA (instrument limited) for bias of $< 100$ V. The dark current and photoresponse vs. bias for one of these MSMs is shown in Fig. 12. The photocurrent increased sharply from 0 V and was flat above 12 V, the point at which the active region was completely depleted. The flat photocurrent is indicative of negligible photoconductive gain. Figure 13 shows the device spectral responsivity for a solar-blind MSM photodetector. At 12 V bias, the peak responsivity (quantum efficiency) was 0.105 A/W ($\eta_{ext} = 48\%$) at $\lambda = 262$ nm. There was negligible change in response up to 100 V, which is consistent with the photoresponse curve in Fig. 12. At 12 V bias, the device showed a solar-blind rejection $> 300$ times and $> 2000$ times from the peak (262 nm) to 280 nm and 300 nm, respectively. The noise characteristics on similar devices were measured and reported by

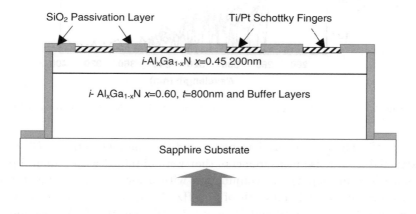

SiO$_2$ Passivation Layer     Ti/Pt Schottky Fingers

*i*-Al$_x$Ga$_{1-x}$N $x$=0.45 200nm

*i*- Al$_x$Ga$_{1-x}$N $x$=0.60, $t$=800nm and Buffer Layers

Sapphire Substrate

**Fig. 11.** Schematic cross section of a solar-blind $Al_xGa_{1-x}N$ MSM photodiode.[62]

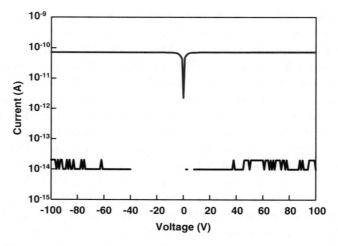

**Fig. 12.** Photocurrent and dark current of solar-blind $Al_xGa_{1-x}N$ MSM photodiode versus bias.[62]

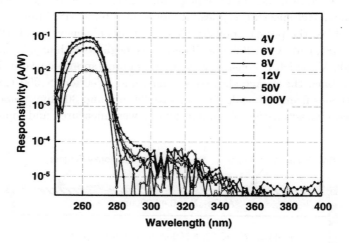

**Fig. 13.** Spectral responsivity of solar-blind $Al_xGa_{1-x}N$ MSM photodiode.[62]

Li, *et al.*[63] The noise equivalent power and the Jones detectivity, $D^*$, were extracted from noise measurements that were obtained with a digital processing lock-in amplifier in conjunction with a low-current probe station and a measurement bandwidth of 500 Hz. At 37 V, the noise equivalent power was $3.8 \times 10^{-12}$ W and $D^*$ was $3.3 \times 10^{10}$ cm $Hz^{1/2}$/W, assuming a $1/f$ noise spectrum.

Solar-blind MSM photodiodes have also been fabricated on 1 $\mu$m-thick $Al_{0.39}Ga_{0.61}N$ that was grown by MBE on Si(111) substrates, a promising approach for low-cost UV detection.[64] For "back-side" illumination, however, the substrate would have to be removed. The contacts employed were Ti/Al and the device area was 250 $\mu$m $\times$ 250 $\mu$m. The dark current density at 300 K was approximately $1.6 \times 10^{-6}$ A/cm$^2$ at 5 V. The spectral response of these devices exhibited a sharp cutoff at $\sim$ 285 nm with peak responsivities of $R = 5$ mA/W and 15 mA/W at 2 V and 4 V bias, respectively. Time response measurements showed an exponential decay with a minimum decay time of 150 ns.

### 5.4. *p-i-n Photodiodes*

A *p-i-n* photodiode, which consists of a single *p-n* junction with the depletion region serving as the "*i*" region, is designed to respond, without gain, to photon absorption. It has become the most ubiquitous photodetector for a wide range of applications because of its relative ease of fabrication and excellent performance characteristics. Typically, the *p-i-n* photodiode is operated with a reverse bias of < 20 V. Once the photons have entered the semiconductor, the photodetection process consists of (1) optical absorption and (2) collection of the resulting photogenerated electron-hole pairs. The primary absorption processes are free carrier absorption, band-to-band absorption, and band-to-impurity absorption.

As shown in the previous section, the MSM, which is one of the simplest photodetector structures, can achieve good responsivity, high speed, and relatively low noise. However, the requirement of a relatively large bias to achieve flat-band operation and full depletion of the lateral regions between the electrodes limits the suitability of MSMs for photovoltaic and other photodetector applications. Consequently, since a primary motivation for research on solar-blind photodetectors is the development of focal-plane arrays, GaN and AlGaN *p-i-n* photodiodes, designed for operation at zero bias, have been the subject of numerous recent studies. Initially, owing primarily to the superior material quality of GaN as compared to that of AlGaN, most of the work in this area concentrated on front-illuminated GaN photodiodes. While these devices did not provide a true solar-blind response, they did serve as useful vehicles for the development of III-N device processing and characterization technologies. For front illumination, a thick (> 1 $\mu$m) GaN "buffer" layer can be first grown on the sapphire substrate. This layer helps to reduce the number of defects that reach the

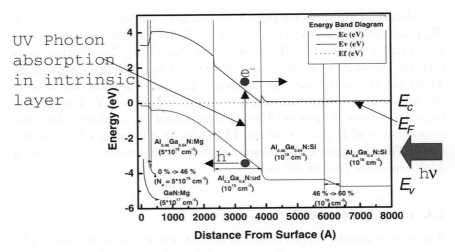

**Fig. 14.**    Band diagram for a back-illuminated solar-blind AlGaN $p$-$i$-$n$ photodetector.

subsequently grown device active layers. Back illumination through the sapphire substrate, on the other hand, facilitates the integration of focal-plane photodiode arrays with readout and image processing circuitry that are essential elements in UV imaging and tracking systems. For this configuration, heterojunction "window layers" with a larger bandgap energy than the intended operating photon energies of the device are essential in order to minimize photon absorption below the $p$-$i$-$n$ layers and the maximization of the device photoresponse. This is shown schematically in the simplified band diagram of Fig. 14.

### 5.4.1. $Al_x Ga_{1-x} N GaN$ visible-blind photodiodes

The first GaN visible-blind $p$-$i$-$n$ photodiode to achieve useful responsivity was reported by Xu *et al.*[65] These devices were "top-illuminated". The MBE-grown structure consisted of an AlN buffer layer, a 3 $\mu$m-thick Si-doped $n$-type GaN layer, a 1 $\mu$m-thick unintentionally-doped "$i$" layer, and a 0.2 $\mu$m Mg-doped GaN $p$-layer with a 10 nm AlGaN "passivation layer" (or a 0.2 $\mu$m $Al_x Ga_{1-x}$ N : Mg $x = 0.12$ $p$-type layer). The $n$- and $p$-type contacts were Ti/Al/Ni/Au and Ni/Au, respectively. The peak responsivity for the "all GaN" devices at $V = 0$ V was $R = 0.07$ A/W at a wavelength of $\lambda = 360$ nm. The responsivity showed a relatively sharp peak near the bandedge, with a strong drop at shorter wavelengths. This type of spectral response was typical of early reports on these devices.[65,66] Xu also described a *p-i-n* with an AlGaN $x = 0.12$ *p*-type "window" layer having improved zero-bias performance of $R = 0.12$ A/W at $\lambda = 364$ nm and a much less pronounced "bandedge peak".[65]

Li *et al.* explained the relatively low efficiency and sharp bandedge peak response of the "all GaN" *p-i-n* diodes in terms of an "optical dead space" at the top surface of the photodiode.[66,67] It was suggested that the dead space is a consequence of the internal electrical field, which originates from band bending at the surface of the *p*-GaN layer. Effects that could contribute to the field include a non-Ohmic *p*-type contact,[68] surface contamination,[68] and the Ga-rich, N-deficient surface that results from Mg doping.[69] Since the absorption coefficient in GaN is $> 10^5$ cm$^{-1}$ for $\lambda < 360$ nm, the thickness and material composition of the *p*-layer become important design parameters in a front-surface illumination configuration in order to mitigate the deleterious impact of the optical dead space on the quantum efficiency. One way to reduce the optical absorption and subsequent carrier recombination in the *p*-type layer is to reduce its thickness. Li *et al.* accomplished this with a recessed window structure as shown in Fig. 15.[70] The as-grown *p*-GaN layer was 0.2 $\mu$m thick. It was then thinned in the center of the mesa to 0.06 $\mu$m by RIE. The quantum efficiency of a device without the recessed window is shown in Fig. 16(a).[67] This device exhibited a sharp peak at the band edge but the efficiency dropped abruptly to $\eta_{\text{ext}} < 25\%$ for $\lambda < 355$ nm. In contrast, a recessed-window device Fig. 16(b) exhibited a high peak efficiency of $\eta_{\text{ext}} = 63\%$ at $\lambda = 357$ nm ($R = 0.18$ A/W) and it

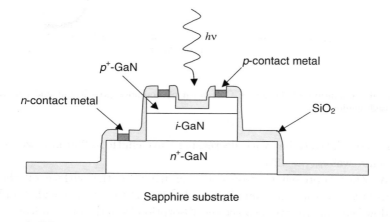

**Fig. 15.** Schematic cross section of GaN *p-i-n* photodiode with recessed window.[70]

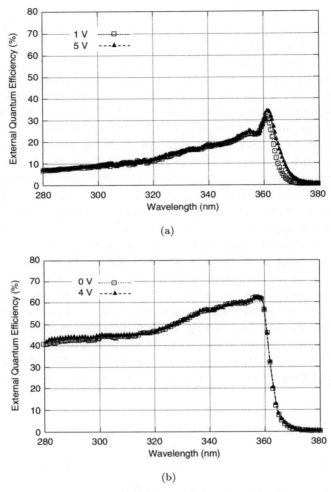

**Fig. 16.** Quantum efficiencies of GaN *p-i-n* photodiodes (a) without and (b) with a recessed window.[67]

remained relatively flat with decreasing wavelength to 280 nm, where the efficiency was $\eta_{ext} > 42\%$.

Another approach to reduce the influence of the surface on the responsivity of a top-illuminated photodetector is to utilize a wide-bandgap window layer in order to displace the absorption from the air/semiconductor interface. Figure 17 shows a cross section of an AlGaN/GaN heterojunction *p-i-n* photodiode described by Li *et al.* that utilized a recessed $Al_{0.13}Ga_{0.87}N$

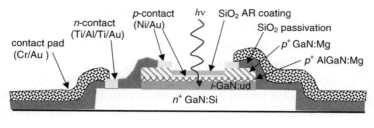

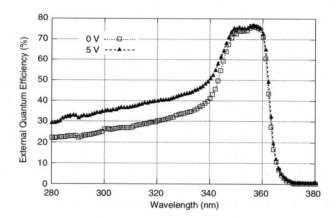

**Fig. 17.** Schematic cross section of GaN $p$-$i$-$n$ photodiode with $Al_{0.13}Ga_{0.87}N$ window layer.[67]

**Fig. 18.** Quantum efficiency of top-illuminated GaN $p$-$i$-$n$ photodiode with an $Al_{0.13}Ga_{0.87}N$ window layer. The decrease in efficiency for $\lambda < 350$ nm is due to absorption in the $Al_{0.13}Ga_{0.87}N$ region.[67]

window layer.[67] The structure consisted of four epitaxial layers grown on basal plane sapphire using a low-temperature GaN nucleation layer. The first layer grown was a 3.6 $\mu$m-thick, Si-doped ($N_d \sim 10^{19}$ cm$^{-3}$) GaN layer, followed by a 0.8 $\mu$m-thick, unintentionally doped ($N_d \sim 10^{19}$ cm$^{-3}$) absorption region, and a semi-transparent window layer of 0.25 $\mu$m-thick Mg-doped $Al_{0.13}Ga_{0.87}N$. The wafers were capped with a $p^+$-GaN contact layer. The quantum efficiency of a typical top-illuminated photodiode is shown in Fig. 18.[67] Even at zero bias, the quantum efficiency was as high as $\eta_{ext} \sim 75\%$ for the region of 350 nm $< \lambda <$ 360 nm. The decrease in efficiency at shorter wavelengths is due to the absorption in the

$Al_{0.13}Ga_{0.87}N$ "window" layer. These results show that the surface recombination velocity for the AlGaN/GaN interface is much lower than that of the air/semiconductor interface and it can be used to achieve very high efficiencies in III-N-based photodetectors.

While the use of an AlGaN window layer improved the responsivity, it also led to detrimental lateral spatial variations in the photoresponse and the temporal characteristics. This was due to the very high resistivity of the $p$-type AlGaN:Mg layers, which resulted in the "crowding" of the electric field near the $p$-contacts.[70,71] Collins *et al.*[71] showed that these effects could be eliminated by forming a thin semi-transparent 50 Å Ni/100 Å Au contact over the entire surface of the mesa. These devices exhibited excellent forward $I$–$V$ characteristics (forward current $>$ 10 mA at 5 V) and no lateral variation in the quantum efficiency or the FWHM of the temporal response. An alternative approach was reported by Yang *et al.*[72] In order to eliminate the diffusion-limited carrier transport associated with illumination through a top $p$-type GaN layer that has limited the responsivity of top-illuminated GaN homojunction devices and the spatial non-uniformities that can result from the use of a top $p$-type AlGaN window layer, they developed a back-illuminated structure in which the UV signal is transmitted through the sapphire substrate and an $n$-type $Al_{0.28}Ga_{0.72}N$ "window" layer to a 0.2 $\mu$m undoped "$i$" GaN absorption region layer. The top layer was a 0.7 $\mu$m-thick $p$-type GaN:Mg. This structure has the advantage that $n$-type AlGaN has much lower resistivity than $p$-type AlGaN and it also avoids surface recombination. Responsivities of 0.2 A/W ($\eta_{ext} \sim 70\%$) and 0.1 A/W ($\eta_{ext} \sim 50\%$) were achieved at $\lambda = 355$ nm and 300 nm, respectively.

The measurement of low-frequency noise in a photodetector is an essential characterization procedure that provides valuable information on detection limits, key physical mechanisms, and reliability. The detectivity, a widely used figure of merit for photodetectors, is frequently determined by the level of internal noise, which is expressed as the noise equivalent power (NEP). Kuksenkov *et al.* characterized the low-frequency (1 Hz to 100 kHz) noise of GaN $p$-$\pi$-$n$ photodiodes.[73] Under reverse bias ($-5$ V to $-30$ V), they have found that the dark current satisfied the relationship $S_n = s_0 \frac{I_d^2}{f^\gamma}$ where $S_n$ is the spectral density of the noise current, $I_d$ is the total current, $f$ is the frequency, and $s_0$ and $\gamma$ are fitting parameters. The value of $\gamma$ was found to be between 1 and 1.1. The noise curves obeyed the Hooge relation, $S_n = \alpha \frac{I_d^2}{f \bar{N}}$ where $\bar{N}$ is the average number of electrons that take part in the conduction process.[74] The best fit to the measured noise was obtained with the Hooge parameter, $\alpha = 3$. For the forward bias, $1/f$ noise was not observed; the primary noise mechanism was generation-recombination associated with a trap level with an activation energy of 0.49 eV. Under UV illumination, shot noise dominated the noise spectrum. The dark current of a 200 $\mu$m × 200 $\mu$m device was ~ 2.7 pA at −3 V which resulted in $S_n$ (1 Hz) = $7.3 \times 10^{-29}$ A$^2$/Hz. For frequencies > 100 Hz, the noise equivalent power was $6.6 \times 10^{-15}$ W/Hz$^{1/2}$.

The reported temporal responses of GaN and AlGaN photodiodes have varied from a few microseconds to tens of picoseconds[65,71,75,76]; the differences probably reflect improvements in material quality more than device design or structural issues. Xu et al.[65] measured the response of GaN $p$-$i$-$n$'s with 8 ns pulses from a frequency-tripled Nd:YAG laser. The $1/e$ decay time was 29 ns. Carrano et al. reported a rise-time of ~ 43 ps at 15 V reverse bias for a 60 $\mu$m-diameter mesa with 1 $\mu$m-thick intrinsic region.[75,77] The fast Fourier transform (FFT) of the $p$-$i$-$n$ pulse response gave a bandwidth of 1.4 GHz. However, these devices exhibited a spatial dependence of the time-response: the pulse width increased as the illumination was scanned across the mesa. Later, Collins et al. demonstrated that this spatial variation could be eliminated by introducing a semitransparent metal contact that covered the top of the mesa.[71] At a wavelength of $\lambda = 310$ nm they achieved a narrow FWHM of ~ 80 ps at −20 V as shown in Fig. 19. Note the symmetric pulse response and the almost negligible slow-component tail which disappears completely after ~ 300 ps.

For imaging applications, high detectivity photodetector arrays are desirable. Since the dislocation density of GaN and AlGaN films is typically about $10^7$ to $10^9$ cm$^{-2}$, device yield and spatial uniformity are concerns for focal plane arrays. Yang et al. measured the $I$–$V$ characteristics in the dark and under illumination of a 32 × 32 GaN $p$-$i$-$n$ photodetector array. The device structure was similar to the Al$_{0.13}$Ga$_{0.87}$N-window-layer devices as reported by Li et al.[67] The mean dark current density was ~ 4 nA/cm$^2$ at 5 V reverse bias, and the dark current distribution for over 100 devices was very uniform (~ 98% of the devices exhibited dark current density < 90 nA/cm$^2$). The external quantum efficiency was as high as $\eta_{\text{ext}} = 72\%$ at 357 nm. The photocurrent distribution was also characterized and the detectivity was estimated to be as high as $D^* = 8 \times 10^{14}$ cm Hz$^{1/2}$/W, assuming that the detectivity was limited by the thermal noise. UV "visible-blind" digital cameras based on 32 × 32 and 128 × 128 arrays of AlGaN $p$-$i$-$n$ photodiodes have been demonstrated by Brown et al.[78] These back-illuminated visible-blind photodiodes had a 0.2 $\mu$m-thick GaN absorption

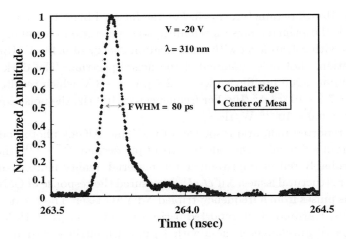

**Fig. 19.**  Temporal response of GaN *p-i-n* photodiode with semitransparent metal contact.[71]

region and an *n*-type $Al_{0.2}Ga_{0.8}N$ window layer. Visible-blind peak responsivities of $R \sim 0.21$ A/W at $\lambda \sim 360$ nm with corresponding detectivities of $D^* = 6.1 \times 10^{13}$ cm $Hz^{1/2}$/W were reported. The 32 × 32 and 128 × 128 photodiode arrays were hybridized to Si readout integrated circuits using flip-chip bonding techniques in which In bump bonds were employed. Ultraviolet movies of pulsed xenon lamps, as well as wandering welding torch flame imagery have been demonstrated. However for real-world imaging applications, the detectivity of these arrays will need to be greatly improved. Advanced high-resolution UV imagers have also been reported by Lamarre *et al.* who have fabricated 256 × 256 back-illuminated focal plane arrays of AlGaN *p-i-n*'s bump bonded to matching 256 × 256 silicon CMOS readout integrated circuits.[79,80] Some of these arrays, operating at a peak response wavelength of $\lambda \sim 265$–285 nm achieved $> 98\%$ operability. Detailed high-resolution UV reflection images have been made of various objects, including a US half-dollar coin.

### 5.4.2. $Al_x Ga_{1-x}N$ solar-blind photodiodes

True solar-blind operation can be achieved with $Al_x Ga_{1-x}N$ provided that the Al content is sufficient to increase the bandgap to transparency for wavelengths $\lambda > 280$ nm, i.e. for alloy compositions $x \geq 0.4$. As discussed above, visible-blind photodetectors were studied and developed initially as a consequence of difficulties associated with the growth of $Al_x Ga_{1-x}N$ with high Al alloy compositions. The primary problems are the lattice and thermal expansion coefficient mismatch between AlN and GaN, which often leads to the cracking of the $Al_xGa_{1-x}N$ films grown on GaN templates. In addition, as discussed above, as a result of the high activation energy of Mg acceptors in high Al-content $Al_xGa_{1-x}N$, it has been difficult to achieve high $p$-type free hole concentrations and high quality $p$-type $Al_xGa_{1-x}N$ films. Tarza *et al.* have addressed these issues in a top-illuminated structure with an "inverted-heterostructure" design.[81] This structure (shown in Fig. 20) utilizes an $Al_xGa_{1-x}N$ ($x > 0.3$) intrinsic or lightly doped active layer surrounded by $p$- and/or $n$-type contact layers having a narrower bandgap than the active layer. By utilizing GaN contact layers, the difficulties associated with achieving high doping efficiencies and good Ohmic contacts in $Al_xGa_{1-x}N$ were circumvented. Solar-blind photodiodes without an antireflection coating were demonstrated with a peak responsivity of $R = 0.08$ A/W at $\lambda = 285$ nm with a sharp spectral cutoff greater than three orders of magnitude by 325 nm. The inverted heterojunction structure has also been demonstrated on lateral epitaxy overgrowth (LEO) GaN templates.[82] Compared to similar photodiodes grown by the same workers directly on sapphire substrates, the top-illuminated LEO devices showed slightly lower responsivity (as high as $R = 0.050$ A/W at $\lambda = 285$ nm for LEO as compared to $R = 0.057$ A/W at $\lambda = 287$ nm for "regular"

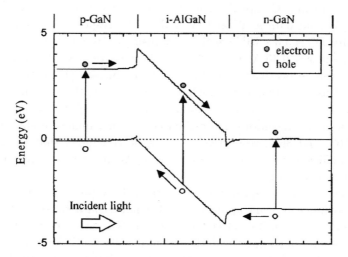

**Fig. 20.** Band structure of top-illuminated inverted heterostructure solar-blind $p$-$i$-$n$ photodiode.[81]

devices) but better temporal response, orders of magnitude lower dark current density ($\sim 10$ nA/cm$^2$, as compared to $\sim 0.3$ A/cm$^2$ at $-5$ V), and a sharper spectral cutoff.[83] In response to excitation by a frequency-tripled Ti/sapphire laser, the 90% to 10% fall time was 4.5 ns. The responsivity peak occurred at 285 nm and dropped by three orders of magnitude over 26 nm. MEDICI simulations by Pulfrey *et al.* have shown that the visible-light rejection of these inverted heterostructure photodiodes can be further improved by the insertion of In$_x$Ga$_{1-x}$N quantum wells in the GaN layers or a $p$-type Al$_{0.33}$Ga$_{0.67}$N $\delta$-doped layer in the $p$-type GaN window layer.[84] This may be particularly applicable to a back-illuminated inverted heterostructure device since the valence-band discontinuity does not provide as large a barrier to holes as the conduction-band discontinuity does for electrons.

Solar-blind response has also been achieved with conventional wide-bandgap window layer structures.[85–89] J. D. Brown *et al.* have grown AlGaN SBDs using MOCVD and have achieved a peak responsivity of $R = 0.051$ A/W (corresponding to an estimated *internal* quantum efficiency of $\eta_{\text{int}} = 27\%$) and FWHM spectral width of $\sim 21$ nm at $\lambda = 273$ nm with a back illuminated $p$-$i$-$n$ consisting of a 1 $\mu$m-thick $n$-type Al$_{0.64}$Ga$_{0.36}$N layer underneath a 0.2 $\mu$m-thick Al$_{0.47}$Ga$_{0.53N}$ "$i$" region and a 0.2 $\mu$m $p$-type Al$_{0.47}$Ga$_{0.53}$N top layer.[86] $R_0A$ values up to $8 \times 10^7$ W cm$^2$ and a detectivity of $D^* = 3.5 \times 10^{12}$ cm Hz$^{1/2}$/W were obtained. At a wavelength of 279 nm, Campbell *et al.* have achieved higher responsivities of 0.058 A/W and 0.07 A/W at 0 V and $-5$ V, respectively, and a higher detectivity $D^* = 5.3 \times 10^{12}$ cm Hz$^{1/2}$/W using a similar structure that had $x = 0.45$ and $x = 0.35$ Al$_x$Ga$_{1-x}$N alloys in the $n$-type window layer and in the "$i$" layer, respectively.[90] In this work, the measured spectral response was fitted using a model that included the wavelength dependence of the absorption coefficient and it was found that active regions having 20% and 35% Al alloy compositions had estimated *internal* quantum efficiencies of $\eta_{\text{int}} = 70\%$ and 36%, respectively. It can be concluded that the internal quantum efficiency degrades with increasing Al concentration as a result of higher defect densities.

As mentioned above, the noise properties of UV photodetectors is of critical importance for some applications. The noise characteristics of the photodiodes described in Ref. 63 were reported by Kuryatkov *et al.*[91] For diodes with a diameter of 50 $\mu$m, the dark current density was below $10^{-10}$ A/cm$^2$, and they exhibited a noise spectral density as low as $3.6 \times 10^{-32}$ A$^2$/Hz (limited by the sensitivity of the equipment). Based on current–voltage and noise measurements, $R_0$ was greater than $4 \times 10^{15}$ $\Omega$ and the room-temperature thermally-limited detectivity was estimated to be greater than $D^* = 2.4 \times 10^{14}$ cm Hz$^{1/2}$/W, and the background-limited detectivity was in excess of $3.5 \times 10^{13}$ cm Hz$^{1/2}$/W. These values are the highest detectivities reported to date for nitride-based UV photodetectors.[92]

Recently, using improved MOCVD growth techniques, Chowdhury *et al.* have demonstrated improved electrical conductivity in the *n*-type $Al_xGa_{1-x}N$ $x = 0.55$ alloys which they have used in the growth of higher-performance solar-blind detectors.[93] This study included growth of *n*-type $Al_xGa_{1-x}N$ alloys having various Al compositions in the range of $x = 0.45$–0.65. The back-illuminated device structure consisted of 700 nm-thick lightly Si-doped $Al_{0.60}Ga_{0.40}N$ grown as the window layer followed by a 100 nm $Al_{0.55}Ga_{0.45}N$:Si *n*-contact layer. Then, the Al composition was graded down to $\sim 45\%$ while maintaining a high Si doping over a thickness of $\sim 15$ nm. Subsequently a 150/200 nm $Al_{0.45}Ga_{0.55}N$ unintentionally doped *i*-layer was deposited followed by the *p*-layer of 10 nm $Al_{0.45}Ga_{0.55}N$:Mg. A schematic diagram of the SBD structure is shown in Fig. 21. The structure was terminated with a 25 nm GaN:Mg *p*-contact layer. The back-illuminated SBD photoresponse and the transmission vs. wavelength for various AlGaN alloy layers are shown in Fig. 22. Uncoated solar-blind *p-i-n* photodetectors fabricated from these materials exhibited a peak external quantum efficiency at a wavelength of 270 nm of $\eta_{ext} = 45.5\%$ ($R \sim 0.10$ A/W). Further improvement of these SBD device structures resulted in even higher performance of UV solar-blind

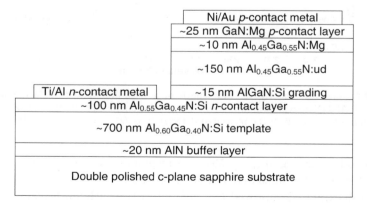

**Fig. 21.** Schematic diagram of the structure of an improved AlGaN solar-blind photodetector.[93]

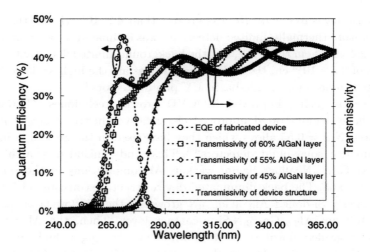

**Fig. 22.** Transmission vs. wavelength of AlGaN calibration samples and external quantum efficiency vs. wavelength of the as-grown AlGaN solar-blind *p-i-n* photodiode.[93]

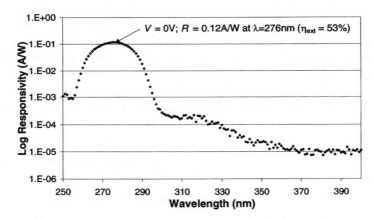

**Fig. 23.** Log responsivity vs. wavelength of a back-illuminated AlGaN SBD with improved AlGaN *n*-type window layer. The peak responsivity is $R = 0.12$ A/W at $\lambda = 276$ nm ($\eta_{ext} = 53\%$); the measured detectivity is $D^* = 3.2 \times 10^{14}$ cm Hz$^{1/2}$/W.[94]

photodiodes, having $\eta_{ext} = 53\%$, corresponding to $R \sim 0.12$ A/W at $\lambda = 276$ nm (for uncoated devices), as shown in Fig. 23.[94] These devices also exhibited improved $I$–$V$ characteristics with low leakage currents near zero bias and a larger detectivity, $D^* = 3.2 \times 10^{14}$ cm Hz$^{1/2}$/W. These values represent the current State-of-the-Art for semiconductor SBD performance, and actually exceed that achieved for the photocathodes in UV-enhanced "reflection-mode" photomultiplier tubes, which exhibit peak values of $\eta_{ext} \sim 30\%$ and $R \sim 0.047$ A/W at $\lambda = 260$ nm.[95]

## 5.5. *Avalanche Photodiodes*

As discussed above, ultraviolet photodetectors fabricated in III-N materials are potentially attractive alternatives to the aging vacuum-tube-based technology of photomultiplier tubes (PMTs). However, in order for GaN-based photodiodes to be considered as viable replacements for PMTs in many systems, they must be able to achieve high sensitivity comparable to that of PMTs. PMTs are able to attain high overall responsivity owing to very high internal gain (typicall $y > 10^6$). However they are bulky, fragile, and require high bias voltages (typically $> 1000$ V). For certain applications, the reliability and power requirements of PMTs make them undesirable; however, the low-noise front-end gain typically provided by PMTs is very desirable. Consequently, it is of interest to study the performance limitations of semiconductor UV photodetectors. Early reports of GaN-based photodetectors focused on metal-semiconductor-metal (MSM) devices owing to potentially high internal gain.[96] Unfortunately, as discussed above, the photoconductive gain mechanism that is responsible for the high responsivity in MSM photodetectors is a slow and noisy process. Furthermore, these simple device structures are typically characterized by relatively large dark current densities. Thus, for MSM photodiodes there is a significant tradeoff between gain, bandwidth, and noise. Another approach for photodetectors providing high internal gain is the avalanche photodiode (APD).

Microplasmas were a dominant factor in early observations of avalanche gain in GaN-based photodiodes.[97] The existence of microplasmas in GaN-based photodiodes biased at high reverse voltages was not surprising, since is well known that microplasmas are associated with defects such as threading dislocations that perpendicularly penetrate a *p-n* junction.[98,99] Since GaN typically has a high density of dislocations ($\sim 10^9$ cm$^{-2}$), and the most common device structure used is a vertical *p-i-n*, it is reasonable to expect spatially non-uniform, pre-mature breakdown in these devices. The first report of nitride APDs with uniform gain, free of microplasmas, utilized the photodiode structure shown in Fig. 24.[100] The $\pi$-*i*-*n* GaN epitaxial structures were grown by hydride vapor-phase epitaxy. First, a 10–15 $\mu$m-thick unintentionally doped ($\sim 10^{17}$ cm$^{-3}$ *n*-type) GaN layer was grown. This was followed by the growth of a thin ($\sim 0.25$ $\mu$m) Zn-doped layer; the

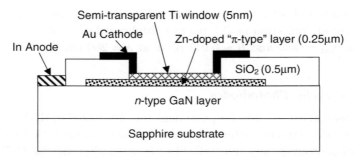

**Fig. 24.** Schematic diagram of a GaN avalanche photodiode grown by hydride vapor-phase epitaxy.[100]

Zn-doped layer (called "π-type" because it is not *p*-type) was produced by the introduction of diethylzinc into the reactor for the last 15 seconds of growth. Secondary ion mass spectrometry analysis revealed a Zn concentration in the low $10^{19}$ cm$^{-3}$ range. For GaN films grown by hydride vapor-phase epitaxy, Zn doping is believed to introduce a deep acceptor level at approximately 340 meV above the valence-band edge, which results in highly compensated material. The unity-gain external quantum efficiency of these devices was 35%. An avalanche gain of 10 was achieved as the bias was increased from $-60$ V to $-220$ V. In this voltage range, the gain was relatively independent of wavelength between 320 nm and 360 nm. Photon counting has also been achieved with these APDs operating in the Geiger-mode.[101] At room temperature, a maximum photon detection efficiency of 13% was measured at 325 nm with a dark count rate of 400 kHz.

Avalanche gain in a true GaN *p-i-n* structure has been reported by Yang *et al.*[102] These devices were grown by MOCVD. The first deposited layer was a 3.6 μm-thick, Si-doped ($N_d = 10^{19}$ cm$^{-3}$) GaN layer. This was followed by a 0.2 μm-thick, unintentionally-doped ($N_d = 10^{16}$ cm$^{-3}$) high-field region. The last layer was a 0.2 μm-thick Mg-doped $p^+$ GaN top layer. The estimated net free hole concentration at 300 K was $\sim 3 \times 10^{17}$ cm$^{-3}$, as determined by Hall-effect measurements. After growth, 30 μm-diameter mesas were defined by RIE to the underlying $n^+$-GaN contact layer. SiO$_2$ deposited by plasma-enhanced CVD was used as a passivation layer. Figure 25 shows the current–voltage characteristics of one of these APDs. For low to intermediate reverse bias ($< 80$ V), the dark current was $< 1$ pA. Above 80 V reverse bias, the dark current increased exponentially with voltage, which is consistent with tunneling processes. All the devices exhibited uniform abrupt breakdown at $\sim 105$ V. The photoresponse was independent

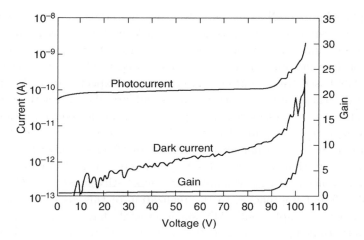

**Fig. 25.** Photocurrent, dark current, and gain of a GaN avalanche photodiode.[100]

of bias voltage between 10 V and 80 V. For this voltage range, the gain was unity and the external quantum efficiency was $\eta_{ext} = 33\%$. Avalanche gain appeared for bias voltage greater than 90 V. The highest gain achieved was 23. Most of devices degraded rapidly when biased above the abrupt dark-current-breakdown voltage. Capacitance–voltage measurements and numerical simulations indicated that the depletion region thickness was approximately 0.22 $\mu$m, from which it followed that the onset avalanche gain field was $\sim 4$ MV/cm, and that the abrupt, dark-current-breakdown field was $\sim 5$ MV/cm. The breakdown voltage showed a clear positive shift of 0.03 V/K with temperature, clearly indicating avalanche multiplication processes dominated the photocurrent response.

## 6. Future Developments

The primary emphasis in the future development of III-N-based SBDs will be on the realization of improved materials and, as a direct consequence, improved device performance. It should be noted that the application of specialized growth techniques, e.g. selective-area epitaxy and lateral epitaxial overgrowth (SAELEO, or LEO, or ELOG),[103] Pendeoepitaxy,[104] or cantilever epitaxy,[105] will probably not be viable solutions to the improvement of materials for *back-illuminated* solar-blind detectors, due to the fact that a transparent window layer is required between the substrate and the photodetector structure. To date, these specialized epitaxial processes have all employed pure GaN as the "first-to-grow" materials; unless this material is somehow removed, the absorption of solar-blind UV energy in this region will severely limit the performance of the SBDs. Owing to the greatly reduced surface mobility of Al on N as compared to Ga on N surfaces, the use of AlN-based "first-to-grow" materials is not expected to lead to planar coalesced surfaces in these selective-areas and lateral overgrowth processes. However, these processes can be expected to provide reduced dislocation density materials for photodetectors that rely on GaN-based epitaxy on sapphire or SiC, e.g. avalanche photodiodes or "front-illuminated" devices.

We expect that III-N-based UV photodetector performance will be greatly advanced by the availability of "bulk" AlN substrates. Currently, several different approaches to the realization of this goal have been developed. For example, bulk AlN single crystals have been produced by Crystal IS using a process originally developed at General Electric Research Labs.[106] We expect that the use of such substrates, after they are more fully developed, will provide improved performance over the SBDs demonstrated to date using sapphire substrates.

## 7. Summary and Conclusions

The performance of nitride-based UV visible-blind and solar-blind photodetectors has advanced rapidly in the past few years. While early work was focused on the use of GaN metal-semiconductor-metal and *p-i-n* "visible-blind" devices, more recent advances in the growth of AlGaN on sapphire by MOCVD have resulted in rapid advances in the development of true solar-blind photodetectors, including MSMs, photoconductors, and *p-i-n* diodes. For example, solar-blind MSMs with $R = 0.105$ A/W ($\eta_{ext} = 48\%$) at $\lambda = 262$ nm and $D^* = 3.3 \times 10^{10}$ cm Hz$^{1/2}$/W have been reported.

The performance of photodiodes employing *p-n* junctions has also advanced recently. For example, the first GaN photovoltaic visible-blind *p-i-n* photodetectors exhibited zero-bias responsivities of 0.07 A/W at $\lambda = 360$ nm,[65] while the latest solar-blind AlGaN *p-i-n* photodetectors have $R = 0.12$ A/W ($\eta_{ext} = 53\%$) at $\lambda = 276$ nm at $V = 0$ V and detectivities as high as $D^* = 3.2 \times 10^{14}$ cm Hz$^{1/2}$/W[94] have been reported for uncoated devices. These results actually exceed the performance of the photocathodes employed in some UV-enhanced photomultiplier tubes.[98] The development of high-density $256 \times 256$ element solar-blind photodiode imaging arrays has also recently been demonstrated. Avalanche photodiodes operating in the "visible-blind" region and fabricated from III-N materials have also been

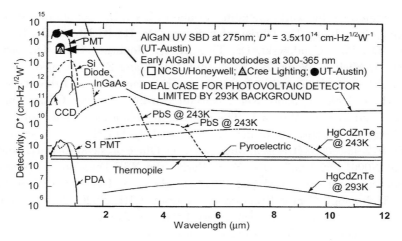

**Fig. 26.** Detectivity of several types of photodetectors vs. wavelength. Note that the best AlGaN SBD described here has a detectivity equal to the photocathode of the photomultiplier at $\lambda \sim 260$ nm.[107]

demonstrated with external quantum efficiencies as high as $\sim 35\%$ and gains as high as 23.

Shown in Fig. 26 is the detectivity vs. wavelength for various types of photodetectors, including the $D^* = 3.2 \times 10^{14}$ cm Hz$^{1/2}$/W data point for the recently developed improved AlGaN SBD discussed above.[107] Note that these AlGaN SBD devices have by far the highest $D^*$ measured for any semiconductor photodetector and, in fact, higher than the photocathodes of most PMTs in the solar-blind spectral region. Further research will produce increasing values of solar-blind performance and improved noise and detectivity characteristics for III-nitride SBDs, including solar-blind APDs.

The principal contributions to these device advances have been the further development of III-nitride MOCVD and the resulting improvement of the materials quality for AlGaN films grown on sapphire substrates. The development of high-quality "bulk" III-N substrates, especially AlN, can be expected to provide an improved "template" for the growth of solar-blind photodetectors and to lead to devices with near theoretical performance.

## Acknowledgments

We thank the graduate students and postdoctoral fellows, past and present, in our research groups at The University of Texas at Austin who have contributed to this work, including Ariane L. Beck, Lt. Col. John Carrano, Uttiya Chowdhury, Dr. Christopher Eiting, Dr. Ho Ki Kwon, Dr. Ki Soo Kim, Dr. Paul Grudowski, Dr. Ting Li, Dr. Joongseo Park, Dr. Bryan Shelton, Michael Wong, Bo Yang, and Ting Gang Zhu. We acknowledge helpful technical discussions regarding nitride materials and devices with Drs. Lou Cook, R. Scott Kern, Nathan Gardner, Werner Götz, Marion Reine, and Michael Wraback. The research in this area at The University of Texas at Austin was primarily supported by DARPA under contract MDA972-95-3-0008 (Dr. E. J. Martinez), the Office of Naval Research under grant N00014-99-1-0304 (Dr. Y. S. Park), the NSF Division of Materials Research under grant DMR-93-12947 (L. Hess), and by the NSF Science and Technology Center Program under grant CHE-89-20120, and by the State of Texas Advanced Technology Program. Additional support from BAE Systems and AVYD Devices is also gratefully acknowledged. RDD acknowledges the support of the Judson S. Swearingen Regents Chair in Engineering; JCC acknowledges the support of the Cockrell Family Regents Chair in Engineering.

## References

1. For a recent treatment, see S. Donati, *Photodetectors: Devices, Circuits, and Applications* (Prentice Hall, Upper Saddle River, NJ) 2000.
2. J. C. Carrano, T. Li, P. A. Grudowski, R. D. Dupuis and J. C. Campbell, *IEEE Circuits and Devices* **15**(5), 15–24 (1999).
3. M. Razeghi and A. Rogalski, *J. Appl. Phys.* **79**, 7433 (1996).
4. M. A. Khan, J. N. Kuznia, D. T. Olson, J. M. Van Hove, M. Blasingame and L. F. Reitz, *Appl. Phys. Lett.* **60**, 2917 (1992).
5. D. J. H. Lambert, M. M. Wong, U. Chowdhury, C. Collins, T. Li, H. K. Kwon, B. S. Shelton, T. G. Zhu, J. C. Campbell and R. D. Dupuis, *Appl. Phys. Lett.* **77**, 1900 (2000).
6. According to the NOAA website, www.noaa.gov, the solar flux at sea level on the Earth's surface at noon with a clear sky on June 22 at $\lambda = 290$ nm is over four orders of magnitude below that in the visible region of the spectrum.
7. Manufactured and marketed by APA Optics, Inc.
8. P. Schreiber, T. Dang, T. Pickenpaugh, G. Smith, P. Gehred and C. Litton, *Proc. SPIE* **3629**, 230 (1999).
9. C. W. Litton, H. Morkoc, G. A. Smith, T. Dang and P. J. Schreiber, to be published in *Proc. 46th SPIE Symp.* **4454**, (2001).
10. J. C. Carrano, P. A. Grudowski, C. J. Eiting, R. D. Dupuis and J. C. Campbell, *Appl. Phys. Lett.* **70**, 1992 (1997).
11. J. C. Carrano, D. J. H. Lambert, C. J. Eiting, C. J. Collins, T. Li, S. Wang, B. Yang, A. L. Beck, R. D. Dupuis and J. C. Campbell, *Appl. Phys. Lett.* **76**, 924 (2000).

12. K. A. McIntosh, R. J. Molnar, L. J. Mahoney, A. Lightfoot, M. W. Geis, K. M. Molvar, I. Melngailis, R. L. Aggarwal, W. D. Goodhue, S. S. Choi, D. L. Spears and S. Verghese, *Appl. Phys. Lett.* **75**, 3489 (1999).
13. D. Walker, V. Kumar, K. Mi, P. Sandvik, P. Kung, X. H. Zhang and M. Razeghi, *Appl. Phys. Lett.* **76**, 403 (2000).
14. D. Walker, A. Saxler, P. Kung, X. Zhang, M. Hamilton, J. Diaz and M. Razeghi, *Appl. Phys. Lett.* **72**, 3303 (1998).
15. E. Monroy, F. Calle, E. Muñoz, F. Omnès, P. Gibart and J. A. Muñoz, *Appl. Phys. Lett.* **73**, 2146 (1998).
16. K. A. McIntosh, R. J. Molnar, L. J. Mahoney, A. Lightfoot, M. W. Geis, K. M. Molvar, I. Melngailis, R. L. Aggarwal, W. D. Goodhue, S. S. Choi, D. L. Spears and S. Verghese, *Appl. Phys. Lett.* **75**, 3485 (1999).
17. A. M. Streltsov, K. D. Moll, A. L. Gaeta, P. Kung, D. Walker and M. Razeghi, *Appl. Phys. Lett.* **75**, 3778 (1999).
18. E. Monroy, F. Calle, E. Muñoz and F. Omnès, *Appl. Phys. Lett.* **74**, 3401 (1999).
19. J. Han, M. H. Crawford, R. J. Shul, S. J. Hearne, E. Chason, J. J. Figiel and M. Banas, *MRS Internet J. Nitride Semicond. Res.* **4S1**, G7.7 (1999).
20. W. Götz, S. Kern and J. Rosner, unpublished data.
21. H. M. Manasevit, *Appl. Phys. Lett.* **12**, 156 (1968).
22. For a recent review of the first application of MOCVD to the growth of advanced III-V semiconductor devices, see R. D. Dupuis, *IEEE J. Sel. Topics Quantum Electron.*, Millennium Issue **6**, 1040 (2000).
23. H. M. Manasevit, F. M. Erdman and W. I. Simpson, *J. Electrochem. Soc.* **118**, 1864 (1971).
24. R. D. Dupuis, P. D. Dapkus, R. D. Yingling and L. A. Moudy, *Appl. Phys. Lett.* **31**, 201 (1977).
25. R. D. Dupuis and P. D. Dapkus, *Appl. Phys. Lett.* **31**, 466 (1977).
26. H. Amano, N. Sawaki, I. Akasaki and Y. Toyoda, *Appl. Phys. Lett.* **48**, 353 (1986).
27. S. Nakamura, *Jap. J. Appl. Phys.* **30**, L1705 (1991).
28. D. K. Gaskill, N. Bottka and M. C. Lin, *J. Cryst. Growth* **77**, 418 (1986).
29. S. Meikle, H. Nomura, Y. Nakanishi and Y. Hatanak, *J. Appl. Phys.* **67**, 483 (1990).
30. R. D. Dupuis, *J. Cryst. Growth* **55**, 213 (1981).
31. G. B. Stringfellow, *Organometallic Vapor Phase Epitaxy: Theory and Practice, Second Edition* (San Diego, Academic Press, Inc., 1999).
32. EpiFill® System designed and marketed by Epichem Ltd. and Epichem Inc.
33. H. Amano, T. Asahi, and I. Akasaki, *Jap. J. Appl. Phys.* **29**, L205 (1990).
34. Y. Koide, N. Itoh, X. Itoh, N. Sawaki and I. Akasaki, *Jap. J. Appl. Phys.* **27**, 1156 (1988).
35. S. Nakamura, *Jap. J. Appl. Phys.* **30**, L1705 (1991).
36. J. F. Chen, N. C. Chen, W. Y. Huang, W. I. Lee and M. S. Feng, *Jap. J. Appl. Phys.* **35**, L810 (1996).
37. A. Ishibashi, H. Takeishi, M. Mannoh, Y. Yabuuchi and Y. Ban, *J. Electron. Mater.* **25**, 799 (1996).

38. M. Ramsteiner, J. Menniger, O. Brandt, H. Yang and K. H. Ploog, *Appl. Phys. Lett.* **69**, 1276 (1996).
39. J. Neugebauer and C. Van de Walle, *Solid-State Phys.* **35**, 25 (1996).
40. S. Fischer, D. Volm, D. Kovalev, B. Averboukh, A. Graber, H. C. Alt and B. K. Meyer, *Mater. Sci. & Eng.* **B43**, 195 (1992).
41. P. A. Grudowski, PhD Dissertation (The University of Texas at Austin, 1998).
42. C. J. Eiting, PhD Dissertation (The University of Texas at Austin, 1999).
43. D. J. H. Lambert, PhD Dissertation (The University of Texas at Austin, 2000).
44. I. Akasaki, H. Amano, Y. Koide, K. Hiramatsu and N. Sawaki, *J. Cryst. Growth* **98**, 209 (1989).
45. E. Monroy, F. Calle, E. Muñoz and F. Omnès, *III-V Nitride Semiconductors: Applications and Devices*, eds. E. T. Yu and M. O. Manasreh (New York, Gordon and Breach, 2000).
46. P. Kung, X. Zhang, D. Walker, A. Saxler, J. Piotrowski, A. Rogalski and M. Razeghi, *Appl. Phys. Lett.* **67**, 3792 (1996).
47. M. A. Kahn, J. N. Kuznia, D. T. Olson, J. M. Van Hove, M. Blaingame and L. F. Reitz, *Appl. Phys. Lett.* **60**, 2917 (1992).
48. C. V. Reddy, K. Balakrishnan, H. Okumura and S. Yoshida, *Appl. Phys. Lett.* **73**, 244 (1998).
49. C. H. Qiu, W. Melton, M. W. Leksono, J. I. Pankove, B. P. Keller and S. P. DenBaars, *Appl. Phys. Lett.* **69**, 1282 (1996).
50. J. Z. Li, J. Y. Lin, H. X. Jiang, A. Salvador, A. Botchkarev and H. Morkoc, *Appl. Phys. Lett.* **69**, 1474 (1996).
51. C. H. Qiu and J. I. Pankove, *Appl. Phys. Lett.* **70**, 1983 (1997).
52. H. M. Chen, Y. F. Chen, M. C. Lee and M. S. Feng, *J. Appl. Phys.* **82**, 899 (1997).
53. B. Beardie, W. S. Barinovich, A. E. Wickenden, D. D. Koleske, S. C. Binari and J. A. Freitas, Jr., *Appl. Phys. Lett.* **71**, 1092 (1997).
54. E. Monroy, F. Calle, J. A. Garrido, P. Youinou, E. Muñoz, F. Omnès, B. Beaumont and P. Gibart, *Semicond. Sci. Tech.* **14**, 685 (1999).
55. E. Monroy, F. Calle, C. Angulo, P. Vila, A. Sanz, J. A. Garrido, E. Muñoz, E. Calleja, F. Omnès, B. Beaumont and P. Gibart, *Appl. Opt.* **37**, 5058 (1998).
56. E. Monroy, F. Calle, E. Muñoz, F. Omnès, B. Beaumont and P. Gibart, *J. Electron. Mater.* **28**, 240 (1999).
57. J. B. D. Soole and J. Schumacher, *IEEE J. Quantum Electron.* **27**, 737 (1991).
58. J. C. Carrano, T. Li, P. A. Grudowski, C. J. Eiting, R. D. Dupuis and J. C. Campbell, *J. Appl. Phys.* **83**, 6148 (1998).
59. J. C. Carrano, T. Li, D. L. Brown, P. A. Grudowski, C. J. Eiting, R. D. Dupuis and J. C. Campbell, *Appl. Phys. Lett.* **73**, 2405 (1998).
60. D. Walker, E. Monroy, P. Kung, J. Wu, M. Hamilton, F. J. Sanchez, J. Diaz and M. Razeghi, *Appl. Phys. Lett.* **74**, 762 (1999).

61. E. Monroy, F. Calle, E. Munoz, F. Omnes and P. Gibart, *Electron. Lett.* **35**, 240 (1999).
62. B. Yang, D. J. H. Lambert. T. Li, C. J. Collins, M. M. Wong, U. Chowdhury, R. D. Dupuis and J. C. Campbell, *Electron. Lett.* **36**, 1866 (2000).
63. T. Li, D. J. H. Lambert, A. L. Beck, C. J. Collins, B. Yang, M. M. Wong, U. Chowdhury, R. D. Dupuis and J. C. Campbell, *Electron. Lett.* **36**, 1581 (2000).
64. J. L. Pau, E. Monroy, E. Munoz, F. Calle, M. A. Sanchez-Garcia and E. Calleja, *Electron. Lett.* **37**, 239 (2001).
65. G. Y. Xu, A. Salvador, W. Kim, Z. Fan, C. Lu, H. Tang, H. Morkoc, G. Smith, M. Estes, G. Goldenberg, W. Yang and S. Krishnankutty, *Appl. Phys. Lett.* **71**, 2154 (1997).
66. T. Li, J. C. Carrano, M. Schurman, I. Ferguson and J. C. Campbell, *IEEE J. Quantum Electron.* **35**, 1203 (1999).
67. T. Li, S. Wang, A. L. Beck, C. J. Collins, B. Yang, R. D. Dupuis, J. C. Carrano, M. J. Schurman, I. T. Ferguson and J. C. Campbell, *Proc. SPIE: Optoelectronics 2000* (San Jose, CA, 2000).
68. C. I. Wu, A. Kahn, N. Taskar, D. Dorman and D. Gallagher, *J. Appl. Phys.* **83**, 4249 (1998).
69. T. Mori, T. Ohwaki, Y. Taga, N. Shibata, M. Koike and K. Manabe, *Thin Solid Films* **287**, 184 (1996).
70. T. Li, A. L. Beck, C. J. Collins, R. D. Dupuis, J. C. Campbell, J. C. Carrano, M. J. Schurman and I. A. Ferguson, *Appl. Phys. Lett.* **75**, 2421 (1999).
71. C. J. Collins, T. Li, A. L. Beck, R. D. Dupuis, J. C. Campbell, J. C. Carrano, M. J. Schurman and I. A. Ferguson, *Appl. Phys. Lett.* **75**, 2138 (1999).
72. W. Yang, T. Nohava, S. Krishnankutty, R. Torreano, S. McPherson and H. Marsh, *Appl. Phys. Lett.* **73**, 1086 (1998).
73. D. V. Kuksenkov, H. Temkin, A. Osinsky, R. Gaska and M. A. Khan, *J. Appl. Phys.* **83**, 924 (1998).
74. F. N. Hooge, T. M. Kleinpenning and L. J. K. Vaandamme, *Rep. Progr. Phys.* **44**, 479 (1981).
75. J. C. Carrano, T. Li, C. J. Eiting, R. D. Dupuis and J. C. Campbell, *J. Electron. Mater.* **28**, 325 (1999).
76. J. M. Van Hove, R. Hickman, J. J. Klaassen, P. P. Chow and P. P. Ruden, *Appl. Phys. Lett.* **70**, 2282 (1997).
77. J. C. Carrano, PhD Dissertation (The University of Texas at Austin, 1999).
78. J. D. Brown, J. Matthews, C. Boney, P. Srinivasan, J. F. Schetzina, J. D. Benson, K. V. Dang, T. Nohava, W. Yang and S. Krishnankutty, *Program of the 2000 Electronic Materials Conference, Denver, CO* (TMS, Warrenville, PA, 2000), p. 64
79. P. Lamarre, A. Hairston, S. Tobin, K. K. Wong, M. F. Taylor, A. K. Sood, M. B. Reine, M. J. Schurman, I. T. Ferguson, R. Singh and C. R. Eddy, Jr., *Mater. Res. Soc. Proc.* **369**, G10.9.1 (2001).
80. P. Lamarre, A. Hairston, S. P. Tobin, A. K. Sood, M. B. Reine, M. Pophristic, R. Birkham, I. T. Ferguson, R. Singh, C. R. Eddy, Jr., U. Chowdhury, M. M. Wong, R. D. Dupuis, P. Kozodoy and E. Tarsa, *Proc.*

*4th Int. Conf. Nitride Semicond.* (ICNS-4), Denver, CO, July 2001, to be published in *Phys. Stat. Sol.*

81. E. J. Tarsa, P. Kozodoy, J. Ibbetson, B. P. Keller, G. Parish and U. Mishra, *Appl. Phys. Lett.* **77**, 316 (2000).

82. G. Parish, S. Keller, P. Kozodoy, J. P. Ibbetson, H. Marchand, P. T. Fini, S. B. Fleischer, S. P. DenBaars, U. K. Mishra and E. J. Tarsa, *Appl. Phys. Lett.* **75**, 247 (1999).

83. It should be noted that these values of leakage current density for "regular" GaN photodiodes were not nearly "state-of-the-art" at that time (1999) and other workers using conventionally grown AlGaN/GaN/sapphire photodiodes reported leakage currents about a factor of $10^6$ lower, i.e. below $\sim 0.1$ $\mu A/cm^2$. See for example, Ref. 70.

84. D. L. Pulfrey, J. J. Kuek, M P. Leslie, B. D. Nener, G. Parish, U. K. Mishra, P. Kozodoy and E. J. Tarza, *IEEE Trans. Electron. Device* **48**, 486 (2001).

85. C. Pernot, A. Hirano, M. Iwaya, T. Detchprohm, H. Amano and I. Akasaki, *Jpn. J. Appl. Phys.* **39**, L387 (2000).

86. J. D. Brown, J. Li, P. Srinivasan, J. Matthews and J. F. Schetzina, *MRS Internet J. Nitride Semicond. Res.* **5**, 9 (2000).

87. D. J. H. Lambert, M. M. Wong, U. Chowdhury, C. Collins, T. Li, H. K. Kwon, B. S. Shelton, T. G. Zhu, J. C. Campbell and R. D. Dupuis, *Appl. Phys. Lett.* **77**, 1900 (2000).

88. T. Li, D. J. H. Lambert, M. M. Wong, C. J. Collins, B. Yang, A. L. Beck, U. Chowdhury, R. D. Dupuis and J. C. Campbell, *IEEE J. Quantum Electron.* **37**, 538 (2001).

89. D. Walker, V. Kumar, K. Mi, P. Sandvik, P. Kung, X. H. Zhang and M. Razeghi, *Appl. Phys. Lett.* **76**, 403 (2000).

90. J. C. Campbell, C. J. Collins, M. M. Wong, U. Chowdhury, A. L. Beck and R. D. Dupuis, *Phys. Stat. Solid*, to be published.

91. V. V. Kuryatkov, H. Temkin, J. C. Campbell and R. D. Dupuis, *Appl. Phys. Lett.* **78**, 3340 (2001).

92. As of January 1, 2002.

93. U. Chowdhury, M. M. Wong, C. J. Collins, B. Yang, T. G. Zhu, A. L. Beck, J. C. Campbell and R. D. Dupuis, *Proc. 2001 Fall MRS Meeting Symp. I*, to be published.

94. U. Chowdhury, C. J. Collins, M. M. Wong, T. G. Zhu, J. C. Campbell and R. D. Dupuis, unpublished data.

95. "Photomultiplier Tubes" Hamamatsu Photonics KK Corporation, Catalog TMPO0003E03 (2000) p. 89. Responsivity and $\eta_{ext}$ for UV-enhanced PMT Model 250S.

96. M. A. Khan, J. N. Kuznia, D. T. Olson, J. M. Van Hove, M. Blasingame and L. F. Reitz, *Appl. Phys. Lett.* **60**, 2917 (1992).

97. A. Osinsky, M. S. Shur, R. Gaska and Q. Chen, *Electron. Lett.* **34**, 691 (1998).

98. R. J. McIntyre, *J. Appl. Phys.* **32**, 983 (1961).

99. A. G. Chynoweth and G. L. Pearson, *J. Appl. Phys.* **29**, 1103 (1956).

100. K. A. McIntosh, R. J. Molnar, L. J. Mahoney, A. Lightfoot, M. W. Geis, K. M. Molvar, I. Melngailis, R. L. Aggarwal, W. D. Goodhue, S. S. Choi, D. L. Spears and S. Verghese, *Appl. Phys. Lett.* **75**, 3485 (1999).

101. K. A. McIntosh, R. J. Molnar, L. J. Mahoney, K. M. Molvar, N. Efremow, Jr. and S. Verghese, *Appl. Phys. Lett.* **76**, 3938 (2000).

102. B. Yang, T. Li, K. Heng, C. J. Collins, S. Wang, J. C. Carrano, R. D. Dupuis, J. C. Campbell, M. J. Schurman and I. T. Ferguson, *IEEE J. Quantum Electron.* **36**, 1389 (2000).

103. T. S. Zheleva, O. H. Nam, M. D. Bremser and R. F. Davis, *Appl. Phys. Lett.* **71**, 2472 (1997).

104. T. S. Zheleva, S. A. Smith, D. B. Thomson, K. J. Linthicum, P. Rajagopal and R. F. Davis, *J. Electron. Mater.* **28**, L5 (1999).

105. C. I. H. Ashby, C. C. Mitchell, J. Han, N. A. Missert, P. P. Provencio, D. M. Follstaedt, G. M. Peake and L. Griego, *Appl. Phys. Lett.* **77**, 2333 (2000).

106. Crystal IS, 25 Cord Dr., Latham, NY 12110; www.crystal-is.com.

107. General photodetector detectivity data from Oriel Instruments "The Book of Photon Tools", p. 6-3.

# CHAPTER 10

## DILUTE MAGNETIC GaN, SiC AND RELATED SEMICONDUCTORS

Jihyun Kim and Fan Ren

*Department of Chemical Engineering, University of Florida,*
*Gainesville, FL 32611, USA*


Stephen J. Pearton, Cammy R. Abernathy, Mark E. Overberg
and Gerald T. Thaler

*Department of Materials Science and Engineering, University of Florida,*
*Gainesville, FL 32611, USA*


Yun Daniel Park

*Center for Strongly Correlated Materials Research, Seoul National University,*
*Seoul, 151–747, Korea*

## Contents

## 1. Introduction

Two of the most successful technologies in existence today have created the Si integrated circuit (ICs) industry and the data storage industry. Both continue to advance at a rapid pace. In the case of ICs, the number of transistors on a chip doubles about every 18 months according to Moore's Law. For magnetic hard disk drive technology, a typical desk-top computer drive today has a 40 Gbyte/disk capacity, whereas in 1995 this capacity was $\sim$ 1 Gbyte/disk. Since 1991, the overall bit density on a magnetic head has increased at an annual rate of 60–100% and is currently $\sim$ 10.7 Gbits/in$^2$.[1] The integrated circuits operate by controlling the flow of carriers through the semiconductor by applied electric fields. The key parameter therefore is the charge on the electrons or holes. For the case of magnetic data storage, the key parameter is the spin of the electron, as spin can be thought of as the fundamental origin of magnetic moment. The characteristics of ICs include high speed signal processing and excellent reliability, but the memory elements are volatile (the stored information is lost when the power is switched-off, as data is stored as charge in capacitors (i.e. DRAMs). A key advantage of magnetic memory technologies is that they are non-volatile since they employ ferromagnetic materials which by nature have remanence.

The emerging field of semiconductor spin transfer electronics (spintronics) seeks to exploit the spin of charge carriers in semiconductors. It is widely expected that new functionalities for electronics and photonics can be derived if the injection, transfer and detection of carrier spin can be controlled above room temperature. Among this new class of devices are spin transistors operating at very low powers for mobile applications that rely on batteries, optical emitters with encoded information through their polarized light output, fast non-volatile semiconductor memory and integrated magnetic/electronic/photonic devices ("electromagnetism-on-a-chip"). Since the magnetic properties of ferromagnetic semiconductors are a function of carrier concentration in the material in many cases, it will be possible to have electrically or optically-controlled magnetism through field-gating of transistor structures or optical excitation to alter the carrier density. This novel control of magnetism has already been achieved electronically and optically in an InMnAs metal-insulator semiconductor structure at low temperatures[2,3] and electronically in Mn : Ge.[4] A number of recent reviews have covered the topics of spin injection, coherence length and magnetic properties of materials systems such as (Ga, Mn)As,[5–7] (In, Mn)As[5–7] and $(Co,Ti)O_2$,[8] as well as the general areas of spin injection from metals into semiconductors and applications of the spintronic phenomena.[9-12] The current interest in magnetic semiconductors can be traced to difficulties in injecting spins from a ferromagnetic metal into a semiconductor,[13,14] which the idea can be traced to fruitful research in epitaxial preparation of ferromagnetic transitional metals on semiconductor substrates.[15] A theory first proposed by Schmidt *et al.*[16] points out that due to the dissimilar materials properties of a metal and semiconductor, an efficient spin injection in the diffusive transport regime is difficult unless the magnetic material is nearly 100% spin polarized i.e. half-metallic.[17] Although there have been recent reports of successful and efficient spin injection from a metal to a semiconductor even at room temperature by ballistic transport (i.e. Schottky barriers and tunneling),[18] the realization of functional spintronic devices requires materials with ferromagnetic ordering at operational temperatures compatible with existing semiconductor materials.

## 2. Materials Selection

There are two major criteria for selecting the most promising materials for semiconductor spintronics. First, the ferromagnetism should be retained to practical temperatures (i.e. $> 300$ K). Second, it would be a major advantage if there were already an existing technology base for the material in other applications. Most of the work in the past has focused on (Ga, Mn)As and (In, Mn)As. There are indeed major markets for their host materials in infra-red light-emitting diodes and lasers and high speed digital electronics (GaAs) and magnetic sensors (InAs). Most of the past attention on ferromagnetic semiconductors focussed on the (Ga, Mn)As[19-42] and (In, Mn)As[43-50] systems. In samples carefully grown single-phase by molecular beam epitaxy (MBE), the highest Curie temperatures reported are $\sim$ 110 K for (Ga, Mn)As and $\sim$ 35 K for (In, Mn)As. For ternary alloys such as $(In_{0.5}Ga_{0.5})_{0.93}Mn_{0.07}As$, the Curie temperature is also low at $\sim$ 110 K.[51] A tremendous amount of research on these materials systems has led to some surprising results, such as the very long spin lifetimes and coherence times in GaAs[4] and the ability to achieve spin transfer through a heterointerface,[52-69] either of semiconductor/semiconductor or metal-semiconductor. One of the most effective methods for investigating spin-polarized transport is by monitoring the polarized electroluminescence output from a quantum-well light-emitting diode into which the spin current is injected. Quantum selection rules relating the initial carrier spin polarization and the subsequent polarized optical output can provide a quantitative measure of the injection efficiency.[67,69,70]

There are a number of essential requirements for achieving practical spintronic devices in addition to the efficient electrical injection of spin-polarized carriers. These include the ability to transport the carriers with high transmission efficiency within the host semiconductor or conducting oxide, the ability to detect or collect the spin-polarized carriers and the ability to control the transport through external means such as the biasing of a gate contact on a transistor structure. The observation of spin current-induced switching in magnetic heterostructures is an important step in realizing practical devices.[71] Similarly, Nitta *et al.*[72] demonstrated that a spin-orbit interaction in a semiconductor quantum well could be controlled by applying a gate voltage. These key aspects of spin injection, spin-dependent transport, manipulation and detection form the basis of current research and future technology. The use of read sensors based on metallic spin valves in disk drives for magnetic recording is already a \$US100 B per year industry. It should also be pointed out that spintronic effects are inherently tied to nanotechnology, because of the short ($\sim 1$ nm) characteristic length of some of the magnetic interactions. Combined with the expected low power capability of spintronic devices, this should lead to extremely high packing densities for memory elements. A recent review of electronic spin injection, spin transport and spin detection technologies has recently been given by Van Molner *et al.*[7] as part of a very detailed and comprehensive study of the status and trends of research into spin electronics in Japan, Europe and the US. The technical issues covered fabrication and characterization of magnetic nanostructures, magnetism and spin control in these structures, magneto-optical properties of semiconductors and magneto-electronics and devices. The non-technical issues covered included industry and academic cooperation and long-term research challenges. The panel findings are posted on the web site.[7]

In this review we focus on a particular and emerging aspect of spintronics, namely recent developments in achieving practical magnetic ordering temperatures in technologically useful semiconductors.[73-79] While the progress in synthesizing and controlling the magnetic properties of III-arsenide semiconductors has been astounding, the reported Curie temperatures are too low to have significant practical impact. A key development that focused attention on wide bandgap semiconductors as being the most promising for achieving high Curie temperatures was the work of Dietl *et al.*[80] They employed the original Zener model of ferromagnetism[81] to predict $T_C$ values exceeding room temperature for materials such as GaN and ZnO containing 5% of Mn and a high hole concentration $(3.5 \times 10^{20}$ cm$^{-3})$.

Other materials for which room temperature ferromagnetism has been reported include $(Cd,Mn)GeP_2$,[74] $(Zn,Mn)GeP_2$,[75] $ZnSnAs_2$,[76] $(Zn,Co)O$[77] and $(Co,Ti)O_2$,[8,78] as well as Eu chalcogenides and others that have been studied in the past.[79] Some of these chalcopyrites and wide bandgap oxides have interesting optical properties, but they lack a technology and experience base as large as that of most semiconductors.

The key breakthrough that focused attention on wide bandgap semiconductors as being the most promising for achieving practical ordering temperatures was the theoretical work of Dietl *et al.*[80] They predicted that cubic GaN doped with $\sim 5$ at.% of Mn and containing a high concentration of holes $(3.5 \times 10^{20}$ cm$^{-3})$ should exhibit a Curie temperature exceeding room temperature. In the period following the appearance of this work, there has been tremendous progress on both the realization of high-quality (Ga, Mn)N epitaxial layers and on the theory of ferromagnetism in these so-called dilute magnetic semiconductors (DMS). The term DMS refers to the fact that some fraction of the atoms in a non-magnetic semiconductor like GaN are replaced by magnetic ions. A key, unanswered question is whether the resulting material is indeed an alloy of (Ga, Mn)N or whether it remains as GaN with clusters, precipitates or second phases that are responsible for the observed magnetic properties.[82]

## 3. Mechanisms of Ferromagnetism

Figure 1 shows some of the operative mechanisms for magnetic ordering in DMS materials. Two basic approaches to understanding the magnetic properties of dilute magnetic semiconductors have emerged. The first class of approaches is based on mean-field theory which originates in the original model of Zener.[81] The theories that fall into this general model implicitly assume that the dilute magnetic semiconductor is a more-or-less random alloy, e.g. (Ga, Mn)N, in which Mn substitutes for one of the lattice constituents. The second class of approaches suggests that the magnetic atoms form small (a few atoms) clusters that produce the observed ferromagnetism.[82] A difficulty in experimentally verifying the mechanism responsible for the observed magnetic properties is that depending on the growth conditions employed for growing the DMS material, it is likely that one could readily produce samples that span the entire spectrum of possibilities from

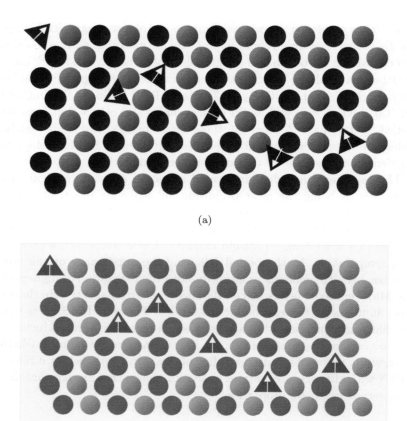

(a)

(b)

**Fig. 1.** Semiconductor matrix with high concentrations of magnetic impurities (i.e. Mn), randomly distributed (defects), can be insulators (a) for Group II-VI materials where divalent Mn ions occupy Group II sites. At high concentrations, Mn ions are antiferromagnetically coupled, but at dilute limits, atomic distances between magnetic ions are large, and antiferromagnetic coupling is weak. For the cases where there is high concentrations of carriers (b) (i.e. (Ga, Mn)As where Mn ions behave as acceptors and provide magnetic moment as they occupy trivalent Ga sites), the carriers are thought to mediate ferromagnetic coupling between magnetic ions. Between near insulating and metallic cases, at low carrier regimes, hole carrier concentrations are localized near the magnetic impurity. Below certain temperatures, a percolation network (c) is formed in which clusters the holes are delocalized and hop from site to site, which energetically favors the maintaining of the carriers' spin orientation during the process, an effective mechanism for aligning Mn moments within the cluster network. Alternatively, at percolation limits, localized hole near the magnetic impurity is polarized, and the energy of the system is lowered when the polarization of the localized holes are parallel (d).

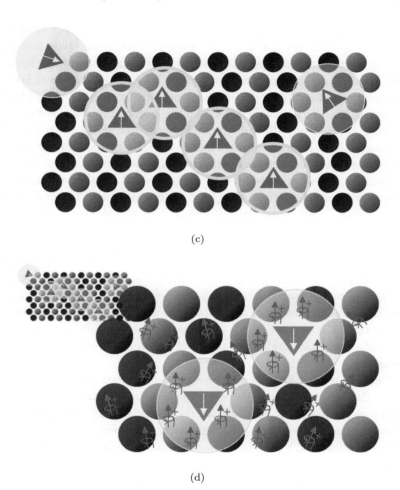

(c)

(d)

**Fig. 1** (*Continued*)

single-phase random alloys to nanoclusters of the magnetic atoms to precipitates and second phase formation. Therefore, it is necessary to decide on a
case-by-case basis which mechanism is applicable. This can only be achieved
by a careful correlation of the measured magnetic properties with materials
analysis methods that are capable of detecting other phases or precipitates.
If, for example, the magnetic behavior of the DMS is characteristic of that
of a known ferromagnetic second phase (such as MnGa or $Mn_4N$ in (Ga,
Mn)N), then clearly the mean field models are not applicable. To date,

most experimental reports concerning room temperature ferromagnetism in DMS employ X-ray diffraction, selected-area diffraction patterns, transmission electron microscopy, photoemission or X-ray absorption (including extended X-ray absorption fine structure, EXAFS, as discussed later) to determine whether the magnetic atoms are substituting for one of the lattice constituents to form an alloy. Given the level of dilution of the magnetic atoms, it is often very difficult to categorically determine the origin of the ferromagnetism. Indirect means such as SQUID magnetometer measurements, to exclude any ferromagnetic inter-metallic compounds as the source of magnetic signals and even the presence of what is called the anomalous or extraordinary Hall effect, that have been widely used to verify a single phase system, may be by itself insufficient to characterize a DMS material. It could also certainly be the case that magnetically-active clusters or second phases could be present in a pseudo-random alloy and therefore that several different mechanisms could contribute to the observed magnetic behavior. There is a major opportunity for the application of new, element- and lattice position-specific analysis techniques, such as the various scanning tunneling microscopies and Z-contrast scanning transmission electron microscopy (Z-contrast STEM) amongst others for revealing a deeper microscopic understanding of this origin of ferromagnetism in the new DMS materials.

The mean field approach basically assumes that the ferromagnetism occurs through interactions between the local moments of the Mn atoms, which are mediated by free holes in the material. The spin-spin coupling is also assumed to be a long-range interaction, allowing the use of a mean-field approximation.[80,83,84] In its basic form, this model employs a virtual-crystal approximation to calculate the effective spin-density due to the Mn ion distribution. The direct Mn-Mn interactions are antiferromagnetic so that the Curie temperature, $T_C$, for a given material with a specific Mn concentration and hole density (derived from Mn acceptors and/or intentional shallow level acceptor doping), is determined by a competition between the ferromagnetic and anti-ferromagnetic interactions. In the presence of carriers, $T_C$ is given by the expression[80,85]

$$T_C = \left\lfloor N_0 X eff \cdot S(S+1)\beta^2 A_F P_S \frac{(T_C)}{12k_B} \right\rfloor - T_{AF}$$

where $N_0 X eff$ is the effective spin concentration, $S$ is the localized spin state, $\beta$ is the p-d exchange integral, $A_F$ the Fermi liquid parameter, $P_S$ the total density of states, $k_B$ the Boltzmann's constant and $T_{AF}$ describes

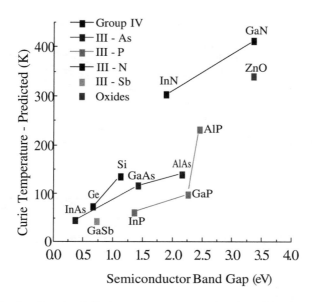

**Fig. 2.**    Predicted Curie temperatures as a function of bandgap.[80]

the contribution of antiferromagnetic interactions. Numerous refinements of this approach have appeared recently, taking into account the effects of positional disorder,[86,87] indirect exchange interactions,[88] spatial inhomogeneities and free-carrier spin polarization.[89,90] Figure 2 shows a compilation of the predicted $T_C$ values, together with the classification of the materials (e.g. Group IV semiconductor, etc.). In the subsequent period after the appearance of the Dietl *et al.*[80] paper, remarkable progress has been made on the realization of materials with $T_C$ values at or above room temperature.

The mean-field model and its variants produces reliable estimates of $T_C$ for materials such as (Ga, Mn)As and (In, Mn)As and it predicts that (Ga, Mn)N will have a value above room temperature.[80] Examples of the predicted ferromagnetic transition temperatures for both (Ga, Mn)As and (Ga, Mn)N are shown in Fig. 3 for four different variants of the mean-field approach.[91] These are the standard mean-field theory ($T_C^{MF}$), a version that accounts for the role of Coulomb interactions with holes in the valence band (exchange-enhanced, $T_C^X$). Another version that accounts for correlations in the Mn ion orientations (collective, $T_C^{coll}$) or an estimate based on where excited spin waves cancels out the total spin of the ground state ($T_C^{est}$).[91] Note that the dependence of any of the calculated $T_C$ values on hole density

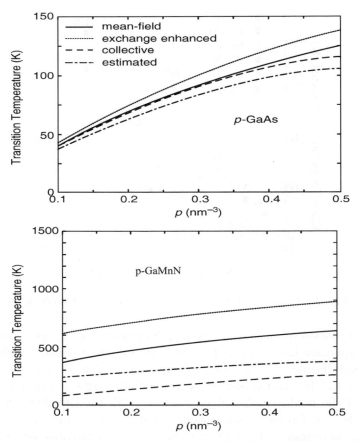

**Fig. 3.** Predicted ferromagnetic transition temperatures in (Ga, Mn)As (left) or (Ga, Mn)N containing 5 at.% Mn, as a function of hole density. The four different curves in each graph represent results obtained from different variants of mean-field theory.[91]

in the material is much steeper for (Ga, Mn)As than for (Ga, Mn)N. The range of predicted values for GaAs has a much higher distribution than for GaN. This data emphasizes the point that the mean field theories produce fairly reliable predictions for (Ga, Mn)As, but at this stage are not very accurate for (Ga, Mn)N.

A second point largely overlooked in the theoretical work to date is the fact that the assumed hole densities may not be realistic. While GaAs can be readily doped with shallow acceptors such as C to produce hole densities of around $10^{21}$ cm$^{-3}$ (Ref. 92) and the Mn acceptors also contribute holes, the *p*-doping levels in GaN are limited to much lower values under normal conditions. For example, the ionization level ($E_a$) of the most common acceptor dopant in GaN, namely Mg, is relatively deep in the gap ($E_V + 0.17$ eV). Since the number of holes ($P$) is determined by the fraction of acceptors that are actually ionized at a given temperature $T$ through a Boltzmann factor

$$P \propto \exp\left(\frac{-E_a}{kT}\right)$$

then for Mg at room temperature only a few percent of acceptors are ionized. While the Mg acceptor concentration in GaN can exceed $10^{19}$ cm$^{-3}$, a typical hole concentration at 25°C is $P \sim 3 \times 10^{17}$ cm$^{-3}$. Initial reports of the energy level of Mn in GaN show that it is very deep in the gap, $E_V + 1.4$ eV,[93] and thus would be an ineffective dopant under most conditions. Some strategies for enhancing the hole concentration do exist, such as the co-doping of both acceptors and donors to reduce self-compensation effects[94] or the use of selectively-doped AlGaN/GaN superlattices in which there is a transfer of free holes from Mg acceptors in the AlGaN barriers to the GaN quantum wells.[95] These methods appear capable under optimum conditions of increasing the hole density in GaN to $> 10^{18}$ cm$^{-3}$ at 25°C.

A further issue that needs additional exploration in the theories is the role of electrons, rather than holes, in stabilizing the ferromagnetism in DMS materials. All of the reports of ferromagnetism in (Ga, Mn)N, for example, occur for material that is actually *n*-type. Since the material has to be grown at relatively low temperatures to avoid Mn precipitation and therefore only molecular beam epitaxy (MBE) can be used, there is always the possibility of unintentional *n*-type doping from nitrogen vacancies, residual lattice defects or impurities such as oxygen. Therefore stoichiometry effects, crystal defects or unintentional impurities may control the final conductivity, rather than Mn or the intentionally-introduced acceptor dopants. Once again, this is much less of an issue in materials such as GaAs, whose low temperature growth is relatively well-understood and controlled.

While most of the theoretical work for DMS materials has focused on the use of Mn as the magnetic dopant, there has been some progress on identifying other transition metal atoms that may be effective. Figure 4 shows the predicted stability of ferromagnetic states in GaN doped with different 3*d* transition metal atoms.[96] The results are based on a local spin-density approximation which assumed that Ga atoms were randomly substituted with the magnetic atoms and did not take into account any additional

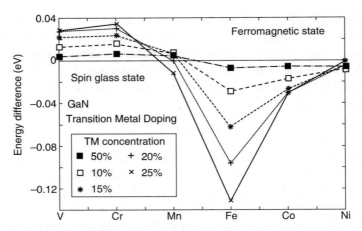

**Fig. 4.** Predicted stability of the ferromagnetic states of different transition metal (TM) atoms in GaN as a function of transition metal concentration. The vertical axis represents the energy difference between the ferromagnetic and spin-glass states for each metal atom.[96]

carrier doping effects. In this study it was found that (Ga, V)N and (Ga, Cr)N showed stable ferromagnetism for all transition metal concentrations whereas Fe, Co or Ni doping produced spin-glass ground states.[96] For the case of Mn, the ferromagnetic state was the lowest energy state for concentrations up to ∼ 20%, whereas the spin-glass state became the most stable at higher Mn concentrations.

### 3.1. *(Ga, Mn)P*

Ferromagnetism above room temperature in (Ga, Mn)P has been reported for two different methods of Mn incorporation, namely ion implantation[97] and doping during MBE growth.[97,98] The implantation process is an efficient one for rapidly screening whether particular combinations of magnetic dopants and host semiconductors are promising in terms of ferromagnetic properties. We have used implantation to introduce ions such as Mn, Fe and Ni into a variety of substrates, including GaN, SiC and GaP.

The temperature-dependent magnetization of a strongly $p$-type ($p \sim 10^{20}$), carbon-doped GaP sample implanted with $\sim 6$ at.% of Mn and then annealed at 700°C, is shown in Fig. 5. The diamagnetic contribution was subtracted from the background. A Curie temperature ($T_C$) of $\sim 270$ K is indicated by the dashed vertical line, while the inset shows a ferromagnetic Curie temperature, of 236 K.

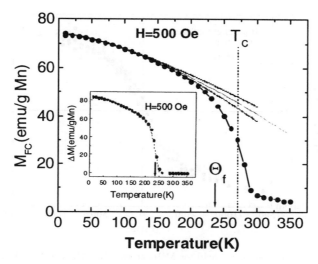

**Fig. 5.** Field-cooled magnetization of (Ga, Mn)P as a function of temperature. The solid line shows a Bloch law dependence, while the dashed lines are 95% confidence bands. The vertical dashed line at $T_C = 270$ K is the field-independent inflection point and the vertical arrows in the main panel and inset mask to ferromagnetic Curie temperature $\theta_f$. The inset shows the temperature dependence of difference in magnetization between field-cooled and zero field-cooled conditions.

Examples of hysteresis loops from MBE grown samples are shown in Fig. 6. The hysteresis could be detected to 330 K. No secondary phases (such as MnGa or MnP) or clusters were determined by transmission electron microscopy, X-ray diffraction or selected-area diffraction pattern analysis.

While mean-field theories predict relatively low Curie temperatures ($< 110$ K) for (Ga, Mn)P,[97,98] recent experiments show ferromagnetism above 300 K.[97,98] In other respects, the magnetic behavior of the (Ga, Mn)P was consistent with mean-field predictions. For example, the magnetization versus temperature plots showed a more classical concave shape than observed with many DMS materials. In addition, the Curie temperature was strongly influenced by the carrier density and type in the material, with highly $p$-type samples showing much higher values than the $n$-type or undoped samples. Finally, the Curie temperature increased with Mn concentration up to $\sim 6$ at.% and decreased at higher concentrations.[98] No secondary phases or clusters could be detected by transmission electron microscopy, X-ray diffraction or selected area diffraction patterns. Similar results were achieved in samples in which the Mn was incorporated during MBE growth or directly implanted with Mn.

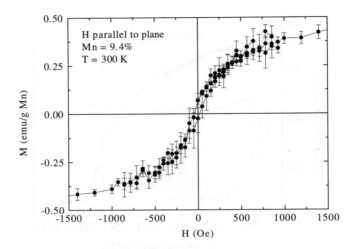

**Fig. 6.** Magnetization vs. field for MBE grown (Ga, Mn)P with 9.4 at.% Mn.

GaP is a particularly attractive host material for spintronic applications because it is almost lattice-matched to Si. One can therefore envision integration of (Ga, Mn)P spintronic magnetic sensors or data storage elements to form fast non-volatile magnetic random access memories (MRAM). Although it has an indirect bandgap, it can be made to luminescence through the addition of isoelectric dopants such as nitrogen, or else one could also employ the direct bandgap ternary InGaP, which is lattice matched to GaAs. The quaternary InGaAlP materials system is used for visible light-emitting diodes, laser diodes, heterojunction bipolar transistors and high electron mobility transistors. An immediate application of the DMS counterparts to the component binary and ternary materials in this system would be to add spin functionality to all of these devices. A further advantage to the wide bandgap phosphides is that they exhibit room temperature ferromagnetism even for relatively high growth temperatures during MBE.

Obviously, the Mn can also be incorporated during MBE growth of the (Ga, Mn)P. The $p$-type doping level can be separately controlled by incorporating carbon from a $CBr_4$ source while P is obtained from thermal cracking of $PH_3$. A phase diagram for the epi growth of this materials system has been developed and this can be used to tailor the magnetic properties of the (Ga, Mn)P. For samples grown at 600°C with 9.4 at.% Mn, hysteresis is still detectable at 300 K, with a coercive field of $\sim$ 39 Oe.

## 3.2. *(Ga, Mn)N*

The initial work on this material involved either microcrystals synthesized by nitridization of pure metallic Ga in supercritical ammonia or bulk crystals grown in reactions of Ga/Mn alloys on GaN/Mn mixtures with ammonia at $\sim$ 1200°C.[99] These samples exhibit paramagnetic properties over a broad range of Mn concentrations, as did some of its early MBE-grown films. By contrast, Fig. 7 shows room temperature ferromagnetism from more recent *n*-type (Ga, Mn)N samples.

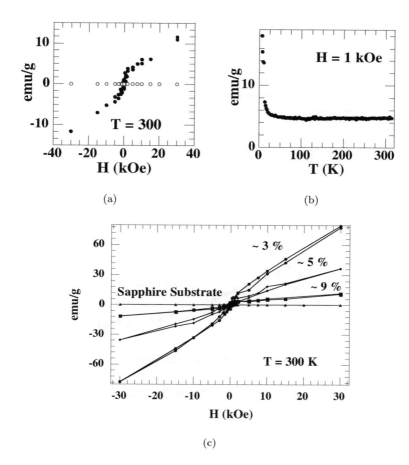

**Fig. 7.** (a) B–H graph from MBE grown (Ga, Mn)N with 9.4 at.% Mn (closed circles) and from sapphire substrate (upper circles), (b) M–T graph of (Ga, Mn)N, (c) B–H graph from (Ga, Mn)N as a function of Mn concentration.

The first reports of the magnetic properties of (Ga, Mn)N involved bulk microcrystallites grown at high temperatures ($\sim 1200°$C), but while percent levels of Mn were incorporated, the samples exhibited paramagnetic behavior.[99] By sharp contrast, in epitaxial GaN layers grown on sapphire substrates and then subjected to solid state diffusion of Mn at temperatures from 250–800°C for various periods, clear signatures of room temperature ferromagnetism were observed.[100,101] Figure 8 shows anomalous Hall effect

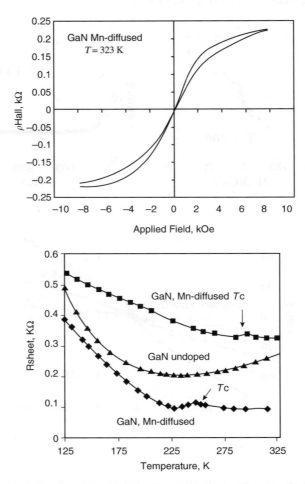

**Fig. 8.** Temperature dependence of sheet resistance at zero magnetic field for Mn-diffused GaN and as-grown GaN (right) and room temperature anomalous Hall effect hysteresis curves for Mn-diffused GaN (left).[101]

data (left) at 323 K and the temperature dependence of sheet resistance at zero applied field for 2 different Mn-diffused samples and an undoped GaN control sample (right).[101] The Curie temperature was found to be in the range of 220–370 K, depending on the diffusion conditions. The use of ion implantation to introduce the Mn produced lowers magnetic ordering temperatures.[102]

In (Ga, Mn)N films grown by MBE at temperatures between 580–720°C with Mn contents of 6–9 at.%, magnetization (M) versus magnetic field (H) curves showed clear hysteresis at 300 K, with coercivities of 52–85 Oe and residual magnetizations of 0.08–0.77 emu/g at this temperature.[103] Figure 9 shows the temperature dependence of the magnetization for a sample with 9 at.% Mn, yielding an estimated $T_C$ of 940 K using a mean field approximation. Note that while the electrical properties of the samples were not measured, they were almost certainly $n$-type. As we discussed above, it is difficult to obtain high Curie temperatures in $n$-type DMS materials according to the mean-field theories and this is something that needs to be addressed in future refinements of these theories. Room temperature ferromagnetism in $n$-type (Ga, Mn)N grown by MBE has also been reported by Thaler *et al.*[104] In that case, strenuous efforts were made to exclude any possible contribution from the sample holder in the super-conducting quantum interference device (SQUID) magnetometer or other

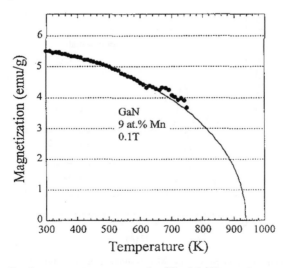

**Fig. 9.** Magnetization versus temperature for (Ga, Mn)N sample grown by MBE with ~ 9 at.% Mn. The extrapolation of the curve is based on a mean-field approximation.[103]

spurious effects. It is also worthwhile to point out that for the studies of (Ga, Mn)N showing ferromagnetic ordering by magnetization measurements, a number of materials characterization techniques did not show the presence of any second ferromagnetic phases within detectable limits. In addition, the values of the measured coercivities are relatively small. If indeed there were undetectable amounts of nano-sized clusters, due to geometrical effects, the expected fields at which these clusters would switch magnetically would be expected to be much larger than what has been observed. Extended X-ray absorption fine structure (EXAFS) measurements performed on (Ga, Mn)N samples grown by MBE on sapphire at temperatures of 400–650°C with Mn concentrations of $\sim 7 \times 10^{20}$ cm$^{-3}$ (i.e. slightly over 2 at.%) are shown in Fig. 10.[105] The similarity of the experimental data with simulated curves for a sample containing this concentration of Mn substituted for Ga on substitutional lattice positions indicates that Mn is in fact soluble at these densities. In the samples grown at 650°C, $\leq 1$ at.% of the total amount of Mn was found to be present as Mn clusters. However at lower

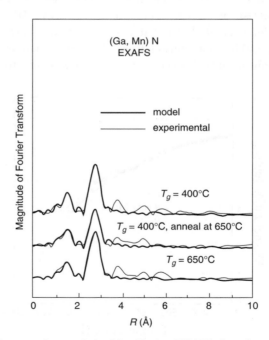

**Fig. 10.** Fourier transforms of the Mn $K$-edge EXAFS data from Mn-doped (Ga, Mn)N. The fine lines are the experimental data in these curves, while the coarse lines are the calculated curves assuming 2 at.% of Mn in the GaN.[105]

growth temperatures (400°C), the amount of Mn that could be present as clusters increased up to $\sim$ 36 at.% of the total Mn incorporated. The ionic state of the substitutional Mn was found to be primarily Mn(2), so that these impurities act as acceptors when substituting for the Ga with valence three. However, when the electrical properties of these samples were measured, they were found to be resistive.[105] This result emphasizes how much more needs to be understood concerning the effects of compensation and unintentional doping of (Ga, Mn)N, since the EXAFS data indicated that the samples should have shown very high $p$-type conductivity due to the incorporation of Mn acceptors.

Other reports have also recently appeared on the magnetic properties of GaN doped with other transition metal impurities. For initially $p$-type samples directly implanted with either Fe or Ni, ferromagnetism was observed at temperatures of $\sim$ 200 K[106] and 50 K,[107] respectively. (Ga, Fe)N films grown by MBE showed Curie temperatures of $\leq$ 100 K, with EXAFS data showing that the majority of the Fe was substitutional on Ga sites.[108] (Ga, Cr)N layers grown in a similar fashion at 700°C on sapphire substrates showed single-phase behavior, clear hysteresis and saturation of magnetization at 300 K and a Curie temperature exceeding 400 K.[109]

Epi growth of (Ga, Mn)N has produced a range of growth conditions producing single-phase material and the resulting magnetic properties.[100–107] In general, no second phases are found for Mn levels below $\sim$ 10% for growth temperatures of $\sim$ 750°C. The (Ga, Mn)N retains $n$-type conductivity under these conditions.

In accordance with most of the theoretical predictions, magneto-transport data showed the anomalous Hall effect, negative magnetoresistance and magnetic resistance at temperatures that were dependent on the Mn concentration. For example, in films with very low ($< 1\%$) or very high ($\sim 9\%$) Mn concentrations, the Curie temperatures were between 10–25 K. An example is shown in Fig. 7 for an $n$-type (Ga, Mn)N sample with Mn $\sim$ 7%. The sheet resistance shows negative magnetoresistance below 150 K, with the anomalous Hall coefficient disappearing below 25 K. When the Mn concentration was decreased to 3 at.%, the (Ga, Mn)N showed the highest degree of ordering per Mn atom.[104] Figure 7(a) shows the hysteresis present at 300 K, while the magnetization as a function of temperature is shown in Fig. 7(b). Data from samples with different Mn concentrations is shown in Fig. 7(c). The data indicates ferromagnetic coupling, leading to a lower moment per Mn. Data from field-cooled and zero field-cooled conditions was further suggestive of room temperature magnetization.[104]

The significance of these results is that there are many advantages from a device viewpoint to having $n$-type ferromagnetic semiconductors.

The local structure and effective chemical valency of Mn in MBE-grown (Ga, Mn)N samples has been investigated by extended X-ray absorption fine structure.[105] It was concluded that most of the Mn was incorporated substitutionally on the Ga sub-lattice with effective valency close to $+2$ for samples with $\sim 2$ at.% Mn.[105] There was also evidence that a fraction (from 1–36%, depending on growth condition) of the total Mn concentration could be present as small Mn clusters.[105]

### 3.3. *Chalcopyrite Materials*

The chalcopyrite semiconductors are of interest for a number of applications. For example, $ZnGeP_2$ exhibits unusual non-linear optical properties and can be used in optical oscillators and frequency converters. $ZnSnAs_2$ shows promise for far-IR generation and frequency converters. The wide bandgap chalcopyrites $ZnGeN_2$ and $ZnSiN_2$ have lattice parameters close to GaN and SiC, respectively and the achievement of ferromagnetism in these materials would make it possible for direct integration of magnetic sensors and switches with blue/green/UV lasers and light-emitting diodes, UV solar-blind detectors and microwave power electronic devices fabricated in the GaN and SiC. The bandgap of $ZnGe_xSi_{1-x}N_2$ varies linearly with composition from 3.2 eV ($x = 1$) to 4.46 eV ($x = 0$).

Numerous reports of room temperature ferromagnetism in Mn-doped chalcopyrites have appeared. A compilation of these results and those from transition metal doped GaN and GaP are shown in Table 1. The $ZnSnAs_2$ is somewhat of an anomaly due to its small bandgap, but little theory is available at this point on the chalcopyrites and their expected magnetic properties as a function of bandgap, doping or Mn concentration. In the only case in which electrical properties were reported, the $ZnGeSiN_2$ : Mn was $n$-type.[110] There is also no information available on the energy level of Mn in the bandgap.

### 3.4. *SiC*

Very little attention has been paid to potential dilute magnetic semiconductor behavior in SiC, which is at a relatively mature state of development for high power, high temperature electronics. With its wide bandgap (3.0 eV for the 6H polytype), excellent transport properties and dopability, it would

**Table 1.** Compilation of semiconductors exhibiting room temperature ferromagnetism.

| Material | Bandgap (eV) | Synthesis | $T_C$ (K) | Ref. |
|---|---|---|---|---|
| $Cd_{1-x}Mn_xGeP_2$ | 1.72 | Solid-phase reaction of evap. Mn | $> 300$ | 74 |
| (Ga, Mn)N | 3.4 | Mn incorporated by diff[n] | 228–370 | 100,101 |
| (Ga, Mn)N | 3.4 | Mn incorporated during MBE; $n$-type | $> 300$ | 104 |
| (Ga, Mn)N | 3.4 | Mn incorporated during MBE | 940 | 103 |
| (Ga, Cr)N | 3.4 | Cr incorporated during MBE | $> 400$ | 109 |
| (Ga, Mn)P : C | 2.2 | Mn incorporated by implant or MBE; $p \sim 10^{20}$ cm$^{-3}$ | $> 330$ | 97,98 |
| $(Zn_{1-x}Mn_x)GeP_2$ | 1.83–2.8 | Sealed ampule growth; insulating; 5.6% Mn | 312 | 74 |
| $(ZnMn)GeP_2$ | $< 2.8$ | Mn incorporated by diff. | 350 K | 75 |
| $ZnSnAs_2$ | 0.65 | Bridgman bulk growth | 329 K | 76 |
| $ZnSiGeN_2$ | 3.52 | Mn-implanted epi | $\sim 300$ | 106 |

be a good candidate for spintronic applications. In this section, we report on the structural and magnetic properties of 6H–SiC implanted with Ni, Fe or Mn at doses designed to produce peak concentrations of these elements up to $\sim 5$ at.%. At these concentrations, ferromagnetic ordering temperatures between 50–270 K were observed.

Bulk, Al-doped ($p = 10^{17}$ cm$^{-3}$ at 25°C) 6H–SiC substrates were used for all these experiments. The samples were directly implanted with Fe, Ni or Mn ions into the Si face at doses from $3$–$5 \times 10^{16}$ cm$^{-2}$ at 250 keV energy in all cases. To avoid amorphization the sample temperature was held at $\sim 350$°C during the implant step. Calculated ion profiles showed a projected range of $\sim 1300$ Å (an example is shown in Fig. 11 for Ni at a dose of $3 \times 10^{16}$ cm$^{-2}$). The peak concentration corresponded to roughly 3 or 5 at.% for the respective doses of 3 and $5 \times 10^{16}$ cm$^{-2}$. Following implantation, the samples were annealed for 5 minutes at 700–1000°C under flowing $N_2$ in a Heatpulse 410T system. The structural quality of the material was examined by X-ray diffraction (XRD), selected area diffraction pattern (SADP) analysis and cross-sectional transmission electron microscopy (TEM). The magnetic properties were measured on a Quantum Design SQUID magnetometer.

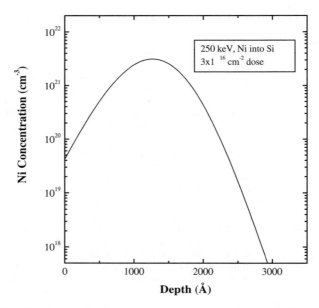

**Fig. 11.** Calculated Ni profile (from the commercially available PROFILE-CODE program)for implantation into 6H–SiC at a dose of $3 \times 10^{16}$ cm$^{-2}$ and an energy of 250 keV.

### 3.4.1. *Ni implantation*

Figure 12 (top) shows a cross-sectional TEM micrograph from a SiC sample implanted with $3 \times 10^{16}$ cm$^{-2}$ Ni$^{+}$ and annealed at 700°C. There is a buried band of defects formed at $\sim$ 2500 Å deep, commonly referred to as end-of-range damage. This consists of a variety of defect types, including dislocation loops and other extended clusters of point defects. The same basic features were present for the higher dose ($5 \times 10^{16}$ cm$^{-2}$) samples and for those annealed at 1000°C, indicating that the damage is stable to at least that temperature. The bottom of Fig. 12 shows the temperature dependence of the difference in magnetization signal between field-cooled and zero field-cooled conditions at a field of 500 Oe. The large paramagnetic background signal has been subtracted in this data. The transition temperature is $\sim$ 50 K based on an extrapolation of where the magnetization is equal to zero within experimental error and similar results were obtained for the higher Ni dose and higher annealing temperature.

A variety of secondary phases could potentially be present in Ni-implanted SiC, including $Ni_1Ni_xSi_y$ and $Ni_xC_y$ components with cubic, hexagonal, orthorhombic, tetragonal, monoclinic or rhombohedral

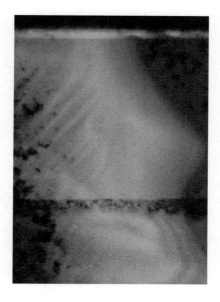

SiC Ni 3%

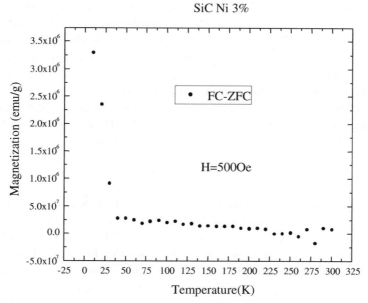

**Fig. 12.** TEM cross-section from SiC implanted with $3 \times 10^{16}$ cm$^{-2}$ Ni$^+$ ions at 250 keV and annealed at 700°C (top) and temperature dependence of difference between field-cooled and zero field-cooled magnetization from the same material (bottom). As a length scale marker, the defect band is 170 angstroms wide.

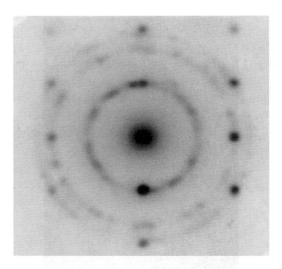

**Fig. 13.** SADP from Ni implanted (250 keV, $5 \times 10^{16}$ cm$^{-2}$) region in SiC.

symmetries. We did not observe any additional peaks in the XRD spectrum of the SiC after Ni implantation. We saw no precipitates visible to the 20 Å resolution of the TEM and no extra spots in the SADP. As an example, Fig. 13 shows a SADP from the implanted region of a sample implanted with $5 \times 10^{16}$ cm$^{-2}$ Ni$^+$ and annealed at 700°C. The rings originate from some polycrystalline regions in the implanted region.

From these results, Ni does not appear to be a promising candidate for producing high transition temperatures in SiC. There is as yet no theory to guide the choice of magnetic dopants for this material.

### 3.4.2. *Fe implantation*

Figure 14 (top) shows a magnetization curve at 10 K for SiC implanted with $5 \times 10^{16}$ cm$^{-2}$ Fe$^+$ and annealed at 700°C. Once again the diamagnetic background has been subtracted. The coercive field is $\sim 50$ Oe at 10 K, and the magnetization went to zero at $\sim 270$ K for this Fe$^+$ dose. The samples implanted with the lower dose of $3 \times 10^{16}$ cm$^{-2}$ were paramagnetic. The existing theories suggest that the presence of ferromagnetism and its associated ordering temperature are strongly dependent on the concentration of the magnetic ions. In past work on Mn-implanted GaN we have observed that doses up to 1 at.% lead to paramagnetic behavior, then the ferromagnetic ordering temperature increases with dose up to 3 at.% and then

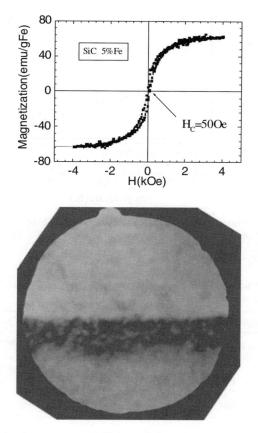

**Fig. 14.** Magnetization curve at 10 K from SiC implanted with $5 \times 10^{16}$ cm$^{-2}$ Fe$^+$ at 250 keV and annealed at 700°C (top) and TEM of damage region from same sample (bottom).The defect band is approximately 190 angstroms wide.

decreases at higher concentrations. The bottom part of Fig. 14 shows a close-up view of the buried damage layer noted at the end-of-range. The width of this layer is $\sim 260$ Å and once again no other phases were detected in the implanted region.

### 3.4.3. *Mn implantation*

Figure 15 shows the temperature dependence of the difference between the field cooled (FC) and zero field-cooled (ZFC) magnetization for a SiC sample implanted with $5 \times 10^{16}$ cm$^{-2}$ Mn$^+$ and annealed at 700°C. The apparent

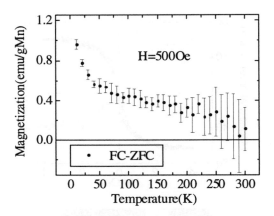

**Fig. 15.** Temperature dependence of difference between field-cooled and zero field-cooled magnetization from SiC implanted with $5 \times 10^{16}$ cm$^{-2}$ Mn$^+$ at 250 keV and annealed at 700°C.

**Table 2.** Characteristics of $p$-SiC samples used in these experiments.

| Implanted Ion | Concentration (at.%) | $T_C(K)$ |
|:---:|:---:|:---:|
| Mn | 3 | Paramagnetic |
| Mn | 5 | 250 |
| Fe | 3 | Paramagnetic |
| Fe | 5 | 270 |
| Ni | 3 | 50 |
| Ni | 5 | 50 |

ordering temperature is $\sim$ 250 K at this dose (again based on where the magnetization goes to zero within experimental error), while the samples implanted at the lower dose remained paramagnetic. The hysteresis loop for the $5 \times 10^{16}$ cm$^{-2}$ sample showed a coercive field of $\sim$ 150 Oe at 10 K.

Table 2 shows a comparison of the characteristics of the samples implanted with the three different elements. It is clear that the results for Ni are much less promising than those for Fe and Mn. In summary, the direct implantation of Fe, Ni and Mn into SiC produced significantly different magnetic characteristics. While fairly high apparent ordering temperatures were observed for Mn and Fe (between 250 and 270 K), the Ni led to low values of the ordering temperature (about 50 K). The origin of the ferromagnetic contributions in implanted SiC is still to be determined, as it is in other materials systems such as (Ga, Mn)N and (Ga, Fe)N that show similar behavior. Future work should focus also on the effects of carrier density and type on the magnetic properties. While the Dietl theory requires a high hole concentration for achievement of high $T_C$ values, recent papers indicate that ferromagnetism can be observed in $n$-type or insulating semiconductors. The use of ion implantation to introduce the magnetic dopants is attractive because of its versatility in controlling the element implanted and its concentration.

## 4. Potential Device Applications

Previous articles have discussed some spintronic device concepts such as spin junction diodes and solar cells,[9] optical isolators and electrically-controlled ferromagnets.[10] The realization of light-emitting diodes with a degree of polarized output has been used to measure spin injection efficiency in heterostructures.[111–113] Such structures can reveal much about spin transport through heterointerfaces after realistic device processing schemes involving etching, annealing and metallization. The spin transfer in such situations has proven surprisingly robust.[114] It is obviously desirable that spintronic devices are operable at or above room temperature. As an initial demonstration that (Ga, Mn)N layers can be used as the $n$-type injection layer in GaN/InGaN blue light-emitting diodes, Fig. 16 shows the LED during operation (top) and the spectral output (left). It is necessary to next establish the extent of any degree of polarization of the light emission, which might be difficult to observe in GaN/InGaN LEDs, since it has been shown that the free exciton components in the EL spectrum contribute mostly to the observed circular polarization of the emitted light.[115] While

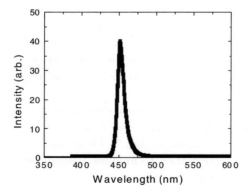

**Fig. 16.** Spectral output from GaMnN/InGaN light-emitting diode.

the expected advantages of spin-based devices include non-volatility, higher integration densities, lower power operation and higher switching speeds, there are many factors still to consider in whether any of these can be realized. These factors include whether the signal sizes due to spin effects are large enough at room temperature to justify the extra development work needed to make spintronic devices and whether the expected added functionality possible will materialize.

In addition to active and/or optical devices, wide bandgap DMS materials may also be used as passive devices. LeClair *et al.*[116] have recently shown an artificial half-metallic structure by using a polycrystalline sputtered ferromagnetic semiconductor (EuS) as a tunneling barrier. This barrier can function as an effective spin filter, since a tunneling electron encounters a differing barrier height depending on its spin below $T_c$ of the barrier material (for EuS, $T_C \sim 16.8$ K). At low temperatures, the spin-filtering efficiencies were found to be $\sim 90\%$. Room temperature DMS materials for these spin- filtering effects could be used to increase magnetoresistance changes in current magnetic tunnel junctions and metallic spin-valve structures.

## 5. Issues to be Resolved

As described earlier, there are a number of existing models for the observed ferromagnetism in semiconductors. The near-field models consider the ferromagnetism to be mediated by delocalized or weakly localized holes in the $p$-type materials. The magnetic Mn ion provides a localized spin and acts as an acceptor in most III-V semiconductors so that it can also provide holes. In these models, the $T_C$ is proportional to the density of Mn ions and the hole density. Many aspects of the experimental data can be explained by the basic mean-field model. However, ferromagnetism has been observed in samples that have very low hole concentrations, in insulating material and more recently in $n$-type material. Models in these regimes are starting to appear.[117]

An alternative approach using local density functional calculations suggests that the magnetic impurities may form small nano-size clusters that produce the observed ferromagnetism.[82] These clusters would be difficult to detect by most characterization techniques. Clearly there is a need to more fully characterize the materials showing room temperature ferromagnetism and correlate these results to establish on a case-by-case basis which is the operative mechanism and also to refine the theories based on experimental input. More work is also needed to establish the energy levels of the Mn, whether there are more effective magnetic dopant atoms and how the magnetic properties are influenced by carrier density and type. Even basic measurements such as how the bandgap changes with Mn concentration in GaN and GaP have not been performed. The control of spin injection and manipulation of spin transport by external means such as voltage from a gate contact or magnetic fields from adjacent current lines or ferromagnetic contacts is at the heart of whether spintronics can be exploited in device structures and these areas are still in their infancy. A concerted effort on the physics and materials science of the new dilute magnetic semiconductors is underway in many groups around the world, but fresh insights, theories and characterization methods would greatly accelerate the process.

## Acknowledgments

The work at UF was partially supported by NSF-DMR 0101438 and by the US Army Research Office under grants nos. ARO DAAD 19-01-1-0710 and DAAD 19-02-1-0420, while the work at SNU was partially supported by KOSEF and Samsung Electronics Endowment through CSCMR and by the Seoul National University Research Foundation. The authors are very grateful to their collaborators A. F. Hebard, N. A. Theodoropoulou, R. G. Wilson, J. M. Zavada, D. P. Norton, S. N. G. Chu, J. S. Lee and Z. G. Khim.

## References

1. See for example http://www.almaden.ibm.com/sst/.
2. H. Ohno, D. Chiba, F. Matsukura, T. Omiya, E. Abe, T. Dietl, Y. Ohno and K. Ohtani, *Nature* **408**, 944 (2000).
3. A. Oiwa, Y. Mitsumori, R. Moriya, T. Slupinski and H. Munekata, *Phys. Rev. Lett.* **88**, 137202 (2002).
4. Y. D. Park, A. T. Hanbicki, S. C. Erwin, C. S. Hellberg, J. M. Sullivan, J. E. Mattson, A. Wilson, G. Spanos and B. T. Jonker, *Science* **295**, 651 (2002).
5. H. Ohno, *J. Vac. Sci. Technol.* **B18**, 2039 (2000).
6. S. A. Wolf, D. D. Awschalom, R. A. Buhrman, J. M. Daughton, S. von Molnar, M. L. Roukes, A. Y. Chtchelkanova and D. M. Treger, *Science* **294**, 1488 (2001).
7. S. Von Molnar *et al.*, *World Technology (WTEC) Study on Spin Electronics: Highlights of Recent US Research and Development Activities*, http://www.wtec.org/spin_US_summary.pdf (2001).
8. S. A. Chambers, *Materials Today* pp. 34–39 April 2002.

9. S. D. Sarma, *American Scientist* **89**, 516 (2001).

10. H. Ohno, F. Matsukura and Y. Ohno, *JSAP Int.* **5**, 4 (2002).

11. D. D. Awschalom and J. M. Kikkawa, *Science* **287**, 473 (2000).

12. C. Gould, G. Schmidt, G. Richler, R. Fiederling, P. Grabs and L. W. Molenkamp, *Appl. Surf. Sci.* **75**, 302 (2002).

13. P. R. Hammar, B. R. Bennett, M. J. Yang and M. Johnson, *Phys. Rev. Lett.* **83**, 203 (1999).

14. F. G. Monzon, H. X. Tang and M. L. Roukes, *Phys. Rev. Lett.* **84**, 5022 (2000).

15. G. A. Prinz, *Ultrathin Magnetic Structures II*, Eds. B. Heinrich and J. A. C. Bland, (Springer-Verlag, New York, 1994).

16. G. Schmidt, D. Ferrand, L. W. Molenkamp, A. T. Filip and B. J. van Wees, *Phys. Rev.* **B62**, R4793 (2000).

17. E. I. Rashba, *Phys. Rev.* **B62**, R16267 (2000).

18. H. J. Zhu, M. Ramsteiner, H. Kostial, M. Wassermeier, H.-P. Schönherr and K. H. Ploog, *Phys. Rev. Lett.* **87**, 016601 (2001).

19. F. Matsukura, H. Ohno, A. Shen and Y. Sugawara, *Phys. Rev.* **B57**, R2037 (1998)

20. H. Ohno, A. Shen, F. Matsukura, A. Oiwa, A. Endo, S. Katsumoto and Y. Iye, *Appl. Phys. Lett.* **69**, 363 (1996).

21. R. Shioda, K. Ando, T. Hayashi and M. Tanaka, *Phys. Rev.* **B58**, 1100 (1998).

22. Y. Satoh, N. Inoue, Y. Nishikawa and J. Yoshino, *Proc. 3rd Symp. Phys. Appl. Spin-Related Phenomena Semicond.*, Eds. H. Ohno, J. Yoshino and Y. Oka, November 1997, Sendai, Japan, p. 23.

23. T. Hayashi, M. Tanaka, T. Nishinaga, H. Shimoda, H. Tsuchiya and Y. Otsuka, *J. Cryst. Growth* **175**, 1063 (1997).

24. A. Van Esch, L. Van Bockstal, J. de Boeck, G. Verbanck, A. S. vas Steenbergen, R. J. Wellman, G. Grietens, R. Bogaerts, F. Herlach and G. Borghs, *Phys. Rev.* **B56**, 13103 (1997).

25. B. Beschoten, P. A. Crowell, I. Malajovich, D. D. Awschalom, F. Matsukura, A. Shen and H. Ohno, *Phys. Rev. Lett.* **83**, 3073 (1999).

26. M. Tanaka, *J. Vac. Sci. Technol.* **B16**, 2267 (1998).

27. Y. Nagai, T. Kurimoto, K. Nagasaka, H. Nojiri, M. Motokawa, F. Matsukura, T. Dietl and H. Ohno, *Jpn. J. Appl. Phys.* **40**, 6231 (2001).

28. J. Sadowski, R. Mathieu, P. Svedlindh, J. Z. Domagala, J. Bak-Misiuk, J. Swiatek, M. Karlsteen, J. Kanski, L. Ilver, H. Asklund and V. Sodervall, *Appl. Phys. Lett.* **78**, 3271 (2001).

29. A. Shen, F. Matsukura, S. P. Guo, Y. Sugawara, H. Ohno, M. Tani, A. Abe and H. C. Liu, *J. Cryst. Growth* **201/202**, 379 (1999).

30. H. Shimizu, T. Hayashi, T. Nishinaga and M. Tanaka, *Appl. Phys. Lett.* **74**, 398 (1999).

31. B. Grandidier, J. P. Hys, C. Delerue, D. Stievenard, Y. Higo and M. Tanaka, *Appl. Phys. Lett.* **77**, 4001 (2000).

32. R. K. Kawakami, E. Johnson-Halperin, L. F. Chen, M. Hanson, N Guebels, J. S. Speck, A. C. Gossard and D. D. Awschalom, *Appl. Phys. Lett.* **77**, 2379 (2000).

33. K. Ando, T. Hayashi, M. Tanaka and A. Twardowski, *J. Appl. Phys.* **83**, 65481 (1998).
34. D. Chiba, N. Akiba, F. Matsukura, Y. Ohno and H. Ohno, *Appl. Phys. Lett.* **77**, 1873 (2000).
35. H. Ohno, F. Matsukura, T. Owiya and N. Akiba, *J. Appl. Phys.* **85**, 4277 (1999).
36. T. Hayashi, M. Tanaka, T. Nishinaga and H. Shimada, *J. Appl. Phys.* **81**, 4865 (1997).
37. T. Hayashi, M. Tanaka, K. Seto, T. Nishinaga and K. Ando, *Appl. Phys. Lett.* **71**, 1825 (1997).
38. A. Twardowski, *Mater. Sci. Eng.* **B63**, 96 (1999).
39. T. Hayashi, M. Tanaka and A. Asamitsu, *J. Appl. Phys.* **87**, 4673 (2000).
40. N. Akiba, D. Chiba, K. Natata, F. Matsukura, Y. Ohno and H. Ohno, *J. App. Phys.* **87**, 6436 (2000).
41. S. J. Potashnik, K. C. Ku, S. H. Chun, J. J. Berry, N. Samarth and P. Schiffer, *Appl. Phys. Lett.* **79**, 1495 (2001).
42. G. M. Schott, W. Faschinger and L. W. Molenkamp, *Appl. Phys. Lett.* **79**, 1807 (2001).
43. H. Munekata, H. Ohno, S. von Molnar, A. Segmuller, L. L. Chang and L. Esaki, *Phys. Rev. Lett.* **63**, 1849 (1989).
44. Akai, *Phys. Rev. Lett.* **81**, 3002 (1998).
45. Ohno, H. Munekata, T. Penney, S. von Molnar and L. L Chang, *Phys. Rev. Lett.* **68**, 2864 (1992).
46. H. Munekata, A. Zaslevsky, P. Fumagalli and R. J. Gambino, *Appl. Phys. Lett.* **63**, 2929 (1993).
47. S. Koshihara, A. Oiwa, M. Hirasawa, S. Katsumoto, Y. Iye, C. Urano, H. Takagi and H. Munekata, *Phys. Rev. Lett.* **78**, 4617 (1997).
48. Y. L. Soo, S. W. Huang, Z. H. Ming, Y. H. Kao and H. Munekata, *Phys. Rev.* **B53**, 4905 (1996).
49. A. Oiwa, T. Slupinski and H. Munekata, *Appl. Phys. Lett.* **78**, 518 (2001).
50. Y. Nishikawa, A. Tackeuchi, M. Yamaguchi, S. Muto and O. Wada, *IEEE J. Sel. Topics Quantum Electron.* **2**, 661 (1996).
51. H. Munekata, presented at ICCG-13, August 2001.
52. I. Malajovich, J. M. Kikkawa, D. D. Awschalom, J. J. Berry and N. Samarth, *Phys. Rev. Lett.* **84**, 1015 (2000).
53. P. R. Hammar, B. R. Bennet, M. Y. Yang and M. Johnson, *J. Appl. Phys.* **87**, 4665 (2000).
54. A. Hirohata, Y. B. Xu, C. M. Guetler and J. A. C. Bland, *J. Appl. Phys.* **87**, 4670 (2000).
55. M. Johnson, *J. Vac. Sci. Technol.* **A16**, 1806 (1998).
56. S. Cardelis, C. G. Smith, C. H. W. Barnes, E. H. Linfield and J. Ritchie, *Phys. Rev.* **B60** 7764(1999).
57. R. Fiederling, M. Kein, G. Rerescher, W. Ossan, G. Schmidt, A. Wang and L. W. Molenkamp, *Nature* **402**, 787 (1999).
58. G. Borghs and J. De Boeck, *Mater. Sci. Eng.* **B84**, 75 (2001).
59. Y. Ohno, D. K. Young, B. Bescholen, F. Matsukura, H. Ohno and D. D. Awschalom, *Nature* **402**, 790 (1999).

60. B. T. Jonker, Y. D. Park, B. R. Bennett, H. D. Cheong, G. Kioseoglou and A. Petrou, *Phys. Rev.* **B62**, 8180 (2000).
61. Y. D. Park, B. T. Jonker, B. R. Bennett, G. Itskos, M. Furis, G. Kioseoglou and A. Petrou, *Appl. Phys. Lett.* **77**, 3989 (2000).
62. G. Schmidt, D. Ferrand, L. W. Molenkamp, A. T. Filip and B. J. van Wees, *Phys. Rev.* **B62**, R4790 (2000).
63. Y. Q. Jin, R. C. Shi and S. J. Chou, *IEEE Trans. Magn.* **32**, 4707 (1996).
64. C. M. Hu, J. Nitta, A. Jensen, J. B. Hansen and H. Takayanagai, *Phys. Rev.* **B63**, 125333 (2001).
65. F. G. Monzon, H. X. Tang and M. L. Roukes, *Phys. Rev. Lett.* **84**, 5022 (2000).
66. S. Gardelis, C. G. Smith, C. H. W., F. Matsukura and H. Ohno, *Jpn. J. Appl. Phys.* **40**, L1274 (2001).
67. H. Breve, S. Nemeth, Z. Liu, J. De Boeck and G. Borghs, *J. Magn. Mater.* **226–230**, 933 (2001).
68. H. Munekata, H. Ohno, S. von Molnar, A. Segmuller, L. L. Chang and L. Esaki, *Phys. Rev. Lett.* **63**, 1849 (1989).
69. H. J. Zhu, M. Ramsteiner, H. Kostial, M. Wassermeier, H. P. Schononherr and K. H. Ploog, *Phys. Rev. Lett.* **87**, 016601 (2001).
70. M. Kohda, Y. Ohno, K. Takamura, F. Matsukura and H. Ohno, *Jpn. J. Appl. Phys.* **40**, L1274 (2001).
71. J. A. Katine, F. J. Albert, R. A. Buhrman, E. D. Myers and D. C. Ralph, *Phys. Rev. Lett.* **84**, 319 (2000).
72. J. Nitta, T. Ahazaki, H. Takayanngi and T. Enoki, *Phys. Rev. Lett.* **78**, 1335 (1997).
73. S. Cho, S. Choi, G. B. Cha, S. C. Hong, Y. Kim, Y.-J. Zhao, A. J. Freeman, J. B. Ketterson, B. J. Kim, Y. C. Kim and B. C. Choi, *Phys. Rev. Lett.* **88**, 257203-1 (2002).
74. G. A. Medvedkin, T. Ishibashi, T. Nishi and K. Hiyata, *Jpn. J. Appl. Phys.* **39**, L949 (2000).
75. G. A. Medvedkin, K. Hirose, T. Ishibashi, T. Nishi, V. G. Voevodin and K. Sato, *J. Cryst. Growth* **236**, 609 (2002).
76. S. Choi, G. B. Cha, S. C. Hong, S. Cho, Y. Kim, J. B. Ketterson, S.-Y. Jeong and G. C. Yi, *Solid-State Commun.* **122**, 165 (2002).
77. K. Ueda, H. Tahata and T. Kawai, *Appl. Phys. Lett.* **79**, 988 (2001).
78. Y. Matsumoto, M. Murakami, T. Shono, H. Hasegawa, T. Fukumura, M. Kawasaki, P. Ahmet, T. Chikyow, S. Koshikara and H. Koinuma, *Science* **291**, 854 (2001).
79. F. Holtzberg, S. von Molnar and J. M. D. Coey, *Handbook on Semiconductors*, Ed. T. Moss (North-Holland, Amsterdam, 1980).
80. T. Dietl, H. Ohno, F. Matsukura, J. Cibert and D. Ferrand, *Science* **287**, 1019 (2000).
81. C. Zener, *Phys. Rev.* **B81**, 440 (1951).
82. M. Van Schilfgaarde and O. N. Myrasov, *Phys. Rev.* **B63**, 233205 (2001).
83. T. Dietl, H. Ohno and F. Matsukura, *Phys. Rev.* **B63**, 195205 (2001).
84. T. Dietl, *J. Appl. Phys.* **89**, 7437 (2001).

85. T. Jungwirth, W. A. Atkinson, B. Lee and A. H. MacDonald, *Phys. Rev.* **B59**, 9818 (1999).
86. M. Berciu and R. N. Bhatt, *Phys. Rev. Lett.* **87**, 108203 (2001).
87. R. N. Bhatt, M. Berciu, M. D. Kennett and X. Wan, *J. Superconductivity: Incorporating Novel Magn.* **15**, 71 (2002).
88. V. I. Litvinov and V. A. Dugaev, *Phys. Rev. Lett.* **86**, 5593 (2001).
89. J. Konig, H. H. Lin and A. H. MacDonald, *Phys. Rev. Lett.* **84**, 5628 (2001).
90. J. Schliemann, J. Konig and A. H. MacDonald, *Phys. Rev.* **B64**, 165201 (2001).
91. T. Jungwirth, J. Konig, J. Sinova, J. Kucera and A. H. MacDonald (in press).
92. C. R. Abernathy, *Mater. Sci. Rep.* **R16**, 203 (1995).
93. R. Y. Korortiev, J. M. Gregie and B. W. Wessels, *Appl. Phys. Lett.* **80**, 1731 (2002).
94. H. Katayama-Yoshida, R. Kato and T. Yamamoto, *J. Cryst. Growth* **231**, 438 (2001).
95. I. D. Goepfert, E. F. Schubert, A. Osinsky, P. E. Norris and N. N. Faleev, *J. Appl. Phys.* **88**, 2030 (2000).
96. K. Sato and H. Katayama-Yoshida, *Jpn. J. Appl. Phys.* **40**, L485 (2001).
97. N. Theodoropoulou, A. F. Hebard, M. E. Overberg, C. R. Abernathy, S. J. Pearton, S. N. G. Chu and R. G. Wilson, *Phys. Rev. Lett.* **89**, 107203 (2002).
98. M. E. Overberg, B. P. Gila, G. T. Thaler, C. R. Abernathy, S. J. Pearton, N. Theodoropoulou, K. T. McCarthy, S. B. Arnason, A. F. Hebard, S. N. G. Chu, R. G. Wilson, J. M. Zavada and Y. D. Park, *J. Vac. Sci. Technol.* **B20**, 969 (2002).
99. M. Zajac, J. Gosk, M. Kaminska, A. Twardowski, T. Szyszko and S. Podliasko, *Appl. Phys. Lett.* **79**, 2432 (2001).
100. M. L. Reed, M. K. Ritums, H. H. Stadelmaier, M. J. Reed, C. A. Parker, S. M. Bedair and N. A. El-Masry, *Mater. Lett.* **51**, 500 (2001).
101. M. L. Reed, N. A. El-Masry, H. Stadelmaier, M. E. Ritums, N. J. Reed, C. A. Parker, J. C. Roberts and S. M. Bedair, *Appl. Phys. Lett.* **79**, 3473 (2001).
102. N. Theodoropoulou, A. F. Hebard, M. E. Overberg, C. R. Abernathy, S. J. Pearton, S. N. G. Chu and R. G. Wilson, *Appl. Phys. Lett.* **78**, 3475 (2001).
103. S. Sonoda, S. Shimizu, T. Sasaki, Y. Yamamoto and H. Hori, *J. Cryst. Growth* **237–239**, 1358 (2002).
104. G. T. Thaler, M. E. Overberg, B. Gila, R. Frazier, C. R. Abernathy, S. J. Pearton, J. S. Lee, S. Y. Lee, Y. D. Park, Z. G. Khim, J. Kim and F. Ren, *Appl. Phys. Lett.* **80**, 3964 (2002).
105. Y. L. Soo, G. Kioseouglou, S. Kim, S. Huang, Y. H. Kaa, S. Kubarawa, S. Owa, T. Kondo and H. Munekata, *Appl. Phys. Lett.* **79**, 3926 (2001).
106. N. A. Theodoropoulou, A. F. Hebard, S. N. G. Chu, M. E. Overberg, C. R. Abernathy, S. J. Pearton, R. G. Wilson and J. M. Zavada, *Appl. Phys. Lett.* **79**, 3452 (2001).

107. S. J. Pearton, M. E. Overberg, G. Thaler, C. R. Abernathy, N. Theodoro-poulou, A. F. Hebard, S. N. G. Chu, R. G. Wilson, J. M. Zavada, A. Y. Polyakov, A. Osinsky and Y. D. Park, *J. Vac. Sci. Technol.* **A20**, 583 (2002).

108. H. Akinaga, S. Nemeth, J. De Boeck, L. Nistor, H. Bender, G. Borghs, H. Ofuchi and M. Oshima, *Appl. Phys. Lett.* **77**, 4377 (2000).

109. M. Hashimoto, Y. Z. Zhou, M. Kanamura and H. Asahi, *Solid-State Commun.* **122**, 37 (2002)

110. S. J. Pearton, Overberg, C. R. Abernathy, N. A. Theodoropoulou, A. F. Hebard, S. N. G. Chu, A. Osinsky, V. Zuflyigin, L. D. Zhu, A. Y. Polyakov and R. G. Wilson, *J. Appl. Phys.* **92**, 2047 (2002).

111. R. Fiederling, M. Kein, G. Resescher, W. Ossau, G. Schmidt, W. Wang and L. W. Molenkamp, *Nature* **402**, 787 (1999).

112. Y. Ohno, D. K. Young, B. Beschoten, F. Matsukura, H. Ohno and D. D. Awschalom, *Nature* **402**, 790 (1999).

113. B. T. Jonker, Y. D. Park, B. R. Bennet, H. D. Cheong, G. Kioseoglou and A. Petrou, *Phys. Rev.* **B62**, 8180 (2000).

114. Y. D. Park, B. T. Jonker, B. R. Bennet, G. Itzkos, M. Furis, G. Kioseoglou and A. Petrou, *Appl. Phys. Lett.* **77**, 3989 (2000).

115. B. T. Jonker, A. T. Hanbicki, Y. D. Park, G. Itskos, M. Furis G. Kioseoglou and A. Petrou, *Appl. Phys. Lett.* **79**, 3098 (2001).

116. P. LeClair, J. K. Ha, H. J. M. Swagten, J. T. Kohlhepp, C. H. van de Vin and W. J. M. de Jonge, *Appl. Phys. Lett.* **80**, 625 (2002).

117. A. Kaminski and S. Das Sarma, *Phys. Rev. Lett.* **88**, 247202-1 (2002).

# INDEX